HONEY

A comprehensive survey

Sic nos non nobis mellificamus apes. | Omnia in libris

All plants yeild honey as you see
To the Industrious Chymick Bee

The woodcut reproduced dates from the late 1600s. The heading is a Latin motto: 'So we the bees make honey, but not for ourselves'. The cleric on the left wears Geneva bands, like those of Swiss Calvinists; he is eating a piece of honeycomb as a token of his preaching: 'Pleasant words are as an honeycomb, sweet to the soul, and health to the bones' (Proverbs xvi:24). The cultivated rose and the wild thistle symbolize the extremes of plant life, all yielding honey to the bee, who is a 'chymick bee', being something of an alchemist in mysteriously producing gold-like honey.

On the right, 'Everything is to be found in books' refers to the scholar in his study below. He, like the bee, gathers knowledge from all available sources and stores it for use by others.

"A pound of honey on the breakfast table necessitates a total flight path equivalent to three orbits round the earth, each orbit using up an ounce of honey as fuel."

Bees were producing honey long before man appeared on earth. Honey was one of man's earliest foods, and has been valued throughout successive civilizations. Today its popularity is increasing as never before.

Here is a lively but authoritative book written by eminent world experts covering the whole subject of honey, its plant sources, its production, marketing and economics. The contributors deal with, for example, the flowers that honey comes from, and the role of the beekeeper; the chemical, physical and biological properties of honey; modern methods of processing, storing, and quality control; legislations; world trade; the manifold uses of honey, including its fermentation; the language of honey. In fact, each chapter is almost a book in itself: that on history, for instance, describes the part played by honey - before the existence of man, among primitive peoples, in ancient civilisations, in medieval Europe, and up to the present day-with beautiful illustrations for every period.

Beekeepers, honey producers and traders, food chemists and biochemists, entomologists and botanists are not the only ones catered for: mead makers, those who like natural foods, social historians, and language students will find plenty to interest them.

HONEY
A comprehensive survey

Edited by

Eva Crane
M.Sc., Ph.D.

Apart from the cover and the four introductory pages,
this book is an unedited reproduction of the original content
of the 1975 publication of the same title (SBN 434 90270 5).

I B R A

INTERNATIONAL BEE
RESEARCH ASSOCIATION

First published in 1975
by William Heinemann Ltd, London

Reprinted in 2020
by
International Bee Research Association
and
Northern Bee Books

ISBN Softback Version 978-0-86098-294-4

Front cover top image © Richard Rickitt
Front cover bottom image and baking © by Ruth Burbidge

Cover design and artwork by DM Design & Print

This book is available from:
IBRA Bookshop at www.ibra.org.uk
tel: (0044) (01)1769 572401
or at
NBB www.northernbeebooks.co.uk
tel: (0044) (01)1422882751

EDITOR'S PREFACE

Honey is produced in almost every country of the world. It was one of man's earliest foods, and it has been valued throughout successive civilizations. Within the last half-century intensive scientific investigations have been devoted to problems concerning honey: its plant sources and origin; how the beekeeper can 'manage' his bees to get the maximum honey harvest from them; and how the honey itself can best be extracted from the comb and made ready for market. These problems are of concern to all beekeepers, whether they keep a few hives in their garden, or operate five thousand or ten thousand hives as a means of livelihood. Research on them has yielded much knowledge that is useful to the people who deal with bees and honey, and this knowledge has been sifted in preparing the present book.

Honey is, however, of interest to a wider circle than the honey producers, handlers, and traders. In its infinite variety it is a connoisseur's food, and the only food that can bring with it the aroma of the world's flowers. The story of its production is a fascinating one: a flower's secretion of nectar, the bee's collection of this nectar—a flight path of three orbits round the earth for a jar of honey to put on the table—and the work in the hive that changes the nectar into honey, a food that will keep almost indefinitely. Environmental biologists find rich material in the interactions between plant, soil, weather, and insect, that are involved in the bees' production of honey.

Many of the chapters in this book are written by the world specialist on the subject concerned. These subjects vary greatly in their complexity. Some chapters—for instance 19, on the history of honey—can be read by anyone who is interested, whereas some other chapters are necessarily in more technical language; for it is intended that the book should be used for reference by those concerned professionally with honey, including producers, traders and packers, food chemists and public analysts. Research workers, teachers, and students should find much that is of interest throughout the book.

Many readers may be astonished that so much is now known about honey. But in the course of writing my own chapters for the book I have often felt frustrated by the *lack* of knowledge, especially where evidence is needed from many different parts of the world. The book attempts a synthesis of the 'world of honey', past and present, for the first time, and the result of such an attempt will not be free from faults. I hope that readers who find errors and omissions will correct them and let me know, and that the very incompleteness of certain tentative approaches to some aspects of the subject will ultimately lead to an enrichment of our knowledge about them.

A volume published at the end of 1974, to mark the 25th anniversary of the Bee Research Association, is prefaced by a quotation from the writings of Abu'l-Rayhan Muhamad al-Biruni who lived in the first century AD. His words have been much in my mind in editing the present book:

> It is our duty to proceed from what is near to what is distant, from what is known to that which is less known, to gather the traditions from those who have reported them, to correct them as much as possible and to leave the rest as it is, in order to make our work help anyone who seeks truth and loves wisdom.

Bee Research Association, EVA CRANE
Hill House,
Chalfont St. Peter,
Gerrards Cross, Bucks., England SL9 0NR

ACKNOWLEDGMENTS

Acknowledgment must first be given to the resources of the Bee Research Association. The text, the illustrations, and especially the bibliography, show that this debt is a large one. Without these resources it would have been difficult to produce the book, and impossible to indicate where readers can obtain access to publications cited and to English summaries of them published in *Apicultural Abstracts*.

The Chairman and several other Members of Bee Research Association Council are among the authors (Chapters 9, 2, 3, 7, 16), and other authors are also long-standing members of the Association. Most of the staff of the Association have been involved in building up the 'information bank' used by all authors, and several of them worked with me on the Indexes and Bibliography; Susan Henriques has dealt with the records and correspondence since 1969.

Mildred Bindley translated Chapter 2, originally written in German. Dr. Gertrude Kolisko translated Chapter 7, and made the English translation from mediaeval Latin in Plate 26. She has also translated source material from a number of languages, and helped in many other ways.

The original impetus to produce the book came from Mr. R. B. Willson in New York, who contributes Chapters 14 and 15. Two other authors, Dr. Anna Maurizio and Dr. Jonathan W. White, have contributed much more than the chapters appearing under their names; their extensive knowledge about nectar and honey has been freely shared and is much valued.

Many libraries have provided material and information. In particular, the Chalfont St. Peter Branch of the Buckinghamshire County Library gave efficient and willing help in obtaining books from different sources. Staff of the Royal Botanic Gardens, Kew, helped with questions of botanical nomenclature.

I am indebted to authors and publishers whose works are quoted in the text of the book, and in particular to the following: the executors of the estate of C. Day Lewis, and Jonathan Cape Ltd, who permitted use of a passage from *The Georgics of Virgil* on page 463; Agricultural Press Ltd, who allowed the entertaining account of honey-getting (pages 470–471) to be quoted from *A diary of a farmer's wife 1796–1797*; Dr. Ludwig Bieler, who gave permission to quote the lines on page 468 from the poem *King and hermit*, translated from Irish by Professor James Carney.

Permission is also acknowledged for the use of quotations at the head of chapters, especially from Martin Armstrong (6, 10), *Australian Bee Journal* (3), Mrs. Arthur Guiterman and E. P. Dutton (4), Mr. Nigel Nicolson and William Heinemann Ltd (9).

The Subject Index is based on terms in *EASI 2* (*English Alphabetical Subject Index to Universal Decimal Classification Numbers used by the Bee Research Association in* Apicultural Abstracts *and in Subject Indexes*), published in 1971 and 1974.

The book is enriched by the many drawings and photographs, and I thank all who have made them, and agreed to their publication here. It is a pleasure to be able to publish for the first time the five drawings made by Dorothy Hodges to show bees' foraging behaviour on different types of flowers (Figures 1.12/1–1.12/5). I am glad of the opportunity to reproduce Harald Pager's drawing of a rock painting in Figure 19.21/3, and Celia Hart's preparation of Figure 17.41/1.

Portions of two illuminated manuscripts in the Bodleian Library, Oxford, are reproduced in Chapter 19: the thirteenth century Bodleian 764 in Figure 19.41/1, and Rawl. 937 (1370) in Figure 19.41/3. Colour transparencies of the portions shown are now available as filmstrip from the Bodleian Library. The Vatican Library gave permission to publish Figure 19.41/2, reproduced from a manuscript in the Library. The copy of a wall painting in Ørslev church in Denmark (Figure 19.41/4) is reproduced from a publication by Ole Højrup.

I am pleased that it has been possible to use one of Frank Vernon's photographs on the dust jacket. For the identification and interpretation of the woodcut reproduced on the endpapers I am indebted to Dr. L. M. Goldstein; the original is in the B.R.A. Collection.

Publishers have allowed reproduction of illustrations from various books. I thank Cambridge University Press for use of the map on page 95 of *World vegetation* by D. Riley and A. Young, 1966 (Figure 1.4/1); Dadant and Sons for Figures 60 and 240 from *The hive and the honey bee* edited by R. A. Grout, 1949 (Figures 2.222/1, 2.222/2, and Figure 2.211/1); VEB Hermann Haack for two maps from *Agrarwirtschaftsatlas der Erde* by B. Skibbe, 1958 (Figure 4.28/1, 4.28/2); H. R. Sauerländer AG for Figure 52 from *Die Biene* by F. Leuenberger and O. Morgenthaler, 1954 (Figure 2.211/3); R. Voigtländer Verlag for Karte III from *Biene und Honig im Volksleben der Afrikaner* by C. Seyffert, 1930 (Figure 19.23/4).

I much appreciate the patience shown by William Heinemann Ltd, during the long period of preparation of the book, and the attention and personal interest devoted to it by Sheila Ralph, Assistant Technical Editor.

On a less formal note, I should like to offer my thanks to the unnamed beekeepers in different parts of the world who have let me taste their honeys—each one different from every other. Language has never proved a barrier to opening a hive and taking honey from the comb. I have vivid memories of very many such occasions: in Europe, from Norway to Italy and from Ireland to the Caucasus mountains; in Africa, from Saharan oases and Ethiopian highlands to the coast of Zululand; in America, from the northern edge of human habitation to the Mexican jungle and the plains of Argentina; from the east to the west coast of Australia; and, in Asia, from the Black Sea to islands offshore from China.

Finally, I thank my husband, who enabled me to do the work, and who has tolerated living with bees and honey for more than thirty years.

Eva Crane

CONTENTS

xv

SECTION 5. OTHER ASPECTS OF HONEY

CHAPTER 17.   HONEY FROM OTHER BEES
by Dr. Eva Crane

CHAPTER 18.   THE LANGUAGE OF HONEY
by D. E. Le Sage

CHAPTER 19.   HISTORY OF HONEY
by Dr. Eva Crane

SECTION 1

HONEY PRODUCTION

CHAPTER I

THE FLOWERS HONEY COMES FROM

by Dr. Eva Crane

DIRECTOR, BEE RESEARCH ASSOCIATION

All the breath and the bloom of the year in the bag of one bee!
ROBERT BROWNING (1812–1889),
SUMMUM BONUM

1.1 A FEW FACTS ABOUT BEES AND HONEY

Honey is made by bees, and their raw material for nearly all the world's supply of honey is nectar produced in the nectaries of flowers. A much smaller amount comes from plants which have nectaries elsewhere (extra-floral nectaries). Bees frequently also make honey from honeydew; this is derived from plant fluids, as nectar is, but flowers are not involved. In many plants, nectar and pollen are produced which may be regarded as 'baits' offered by flowers that can be pollinated only if they are visited by an insect—or more rarely by a bird, bat, or other animal. Nectar is, however, also produced by some plants (like dandelion or buttercup) which need no such pollinating agent, and many insect-pollinated plants (poppy, for instance) produce pollen, but no nectar.

Bees are by far the most important pollinating insects, and the production of honey is thus a by-product of the survival of plant-species by insect pollination. The annual value of world crops produced with the aid of insect pollination is £1 000m (Crane, 1972a) or much more; Chapter 4 shows that the annual value of the honey produced is perhaps £20m.

In general only the social bees that make permanent colonies store honey in sufficient quantity to be of economic importance. These are the honeybees (*Apis*), and the tropical stingless bees, meliponins. There are four species of *Apis*, of which *Apis mellifera*, commonly known as the honeybee or the hive bee, makes the bulk of the world's honey. It has a close relative in south and east Asia, *Apis cerana*, which is also a hive bee, and is 'domesticated'. Most statements made in this book about honey-bees are true also for this Asian species.

The world's honey crop is thus produced mainly from flowers of plants that require insect pollination, by the honeybee *Apis mellifera*, main-

3

tained in hives by beekeepers. Throughout the book 'bee' normally refers
to this honeybee; all chapters are concerned with the honey it produces,
except Chapter 17 which deals with honey from the three other *Apis*
species, the meliponins, bumble bees, and other insects.

The performance of bees is truly astonishing. The fuel consumption of
a flying bee is about $\frac{1}{2}$ mg honey per kilometre, or 3 million km to the
litre. In providing one kilogram of surplus honey for market, the colony
has had to consume something like a further 8 kg to keep itself going,
and the foraging has probably covered a total flight path equal to six
orbits round the earth—at a fuel consumption of about 25 g of honey
for each orbit. In English units, this means 7 million miles to the gallon;
a pound of honey on the breakfast table necessitates a total flight path
equivalent to three orbits round the earth, each orbit using up an ounce
of honey as fuel.

The major part of the honey made by bees is also used by them, and
the beekeeper's harvest can only be the surplus they do not require. The
partition of honey between bees and beekeeper is dealt with in 4.4. In this
chapter we are concerned with the *total* amount of honey that can be
produced from a given area of land, for both bees and beekeeper, and this
depends on four main factors:

1. the nectar, honeydew, and pollen yield of the plants in the area
   (the pollen provides the protein necessary for rearing young bees);
2. the foraging range and foraging ability of the bees in the neigh-
   bourhood;
3. whether there are enough bees to exploit all the nectar resources;
4. the weather, which determines how much of the plant potential and
   the bee potential can be realized.

1.11   The honey potential of plants

The amount of honey that can be produced from the nectar of a single
flower depends on the total amount of nectar secreted and on the sugar
concentration of the nectar. In most flowers only a minute amount of
nectar is available at any one time. The amount in a white clover floret
is no more than a twentieth the size of a pin's head. One of the most
prolific flowers is that of the tulip tree (*Liriodendron tulipifera*), which
may contain nearly half a teaspoonful. A flower that is open and secreting
nectar for a long period is likely to produce more nectar than one which
opens only briefly. Some plants, for instance heather and willowherb
(*Calluna vulgaris*, *Chamaenerion angustifolium*), have a long succession
of flowers on the same stem, so a single plant may be highly productive.
A large tree may have so many flowers that it yields enough nectar to
make a pound or even a kilogram of honey.

The significance of the sugar concentration of the nectar is explained fully in Chapter 2. It varies widely in flowers of different species. Pear and plum nectars contain only about 15% sugar, and some nectars have less than 10%. Other nectars contain 50% sugar or more: as much as 76% has been recorded for marjoram (*Origanum vulgare*), and 79% for *Grevillea robusta*. Bees choose the richer nectars, which they identify by experience, and they do not usually continue to collect nectar containing less than 15% sugar. Table 2.123/1 gives the *sugar value* of a number of plant species; this is an objective measure of the rate of sugar secretion: the number of milligrams of sugar secreted by one flower in 24 hours. It does not, however, take into account the number of days during which the flower produces nectar.

In most flowering plants new flowers or florets come into bloom on each successive day. In a field carrying a crop such as rape or clover, individual plants will not all flower simultaneously, but in succession; the same is true of a prairie or mountain-side in flower. It is useful to measure the *honey potential* of a plant species by reference to a standard unit area of land occupied wholly with the plant. Table 1.3/1 gives the honey potential of 200 species, measured as the number of kilograms of honey that might be obtained from one hectare under optimal conditions. Such estimates are inevitably imprecise, because so many variable factors are involved, but they are nevertheless useful. Since 1 pound = 0·454 kg, and 1 acre = 0·405 hectare, the honey potential measured in pounds per acre is near enough to the honey potential measured in kilograms per hectare.

Figures 1.11/1 and 1.11/2 exemplify how the state of affairs in the hive reflects the honey potential of the land within flight range of the bees. They show the average weekly gain or loss in weight per hive (over 5 years) in two apiaries in Arkansas, U.S.A. The apiaries were about 10 km apart, but Emmet (Figure 1.11/1) has a much greater plant potential for honey production than Hope (Figure 1.11/2), because the strong spring nectar flow from holly (*Ilex opaca*) is followed by a long summer flow from sweet clover (*Melilotus*), before the autumn flow from Spanish needle (*Bidens*). The total gain in weight from April to September is 75 kg. At Hope there is almost no summer flow; in many weeks the colonies actually lose weight. The autumn flow brings the net gain up to only 11 kg, too little to see the colonies through the winter, and if they are not to starve the beekeeper must feed them.

Only the quantities of nectar, sugar, and honey have been considered here. There are differences in quality, too, and most of the variations between honeys originate in differences between the nectars from which the honeys are derived. These are discussed in Chapter 2.

The forage available to bees in any area changes when the land is put to a different use: when virgin land is cleared and crops are planted (or

not planted), and when existing crops are replaced by others, or by urban
or industrial development. Sometimes new nectar-producing crops are
introduced that produce a bonus for bees—like safflower (*Carthamus
tinctorius*), grown for its oil which has a high percentage of unsaturated

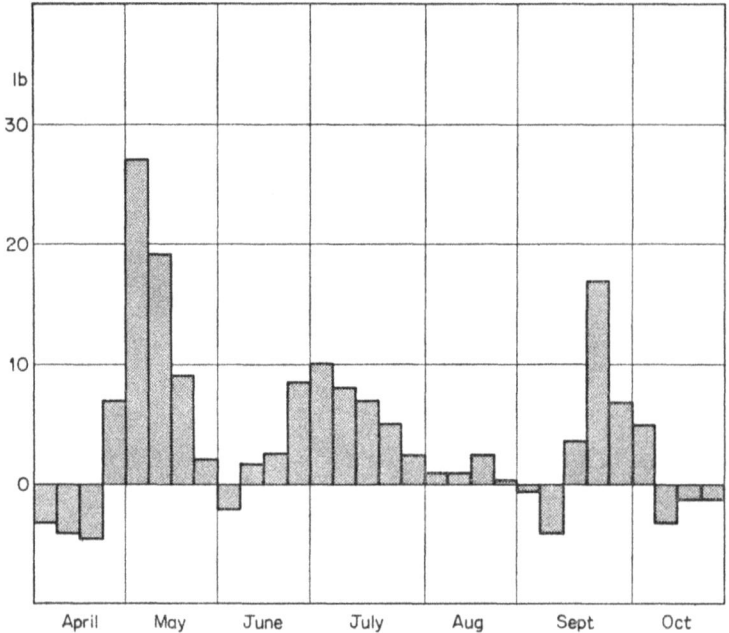

Figure 1.11/1   Average weekly gain or loss in weight (in pounds) of
five hives at Emmet, Arkansas, U.S.A., for the four years 1948–1951
[1 lb=0·454 kg].   ('Thompson, 1960)

fatty acids. Bees are the camp followers of agriculture; they harvest
what they can, but are unable to control what is available to them,
except by migrating, which can happen in the tropics; elsewhere many
beekeepers augment their honey crops by moving their hives from place
to place. Figure 1.13/1 shows the flows each year from 1917 to 1936 in
the same apiary in the south of England. As well as the year-to-year
variations there is also a general decline, which has continued in the area
since 1936. The orchard trees that flowered in May have been grubbed
out to make way for houses; effective weed killing has cleared up the
charlock that gave a June flow; the July flow from the hedgerow brambles
has largely disappeared with the hedges themselves.

Ontario in Canada provides a more dramatic example. The beekeeping
industry there was built up in the 1920s and early 1930s, when sweet
clover (*Melilotus*) and alsike clover (*Trifolium hybridum*) were grown for

hay and pasture: the first reached a peak in 1928 when more than 100 000 ha were sown, and the second in 1929 with almost 50 000 ha. Damage by the sweet clover weevil and root rot, and other factors, led to the replacement of these crops with others providing less honey. After

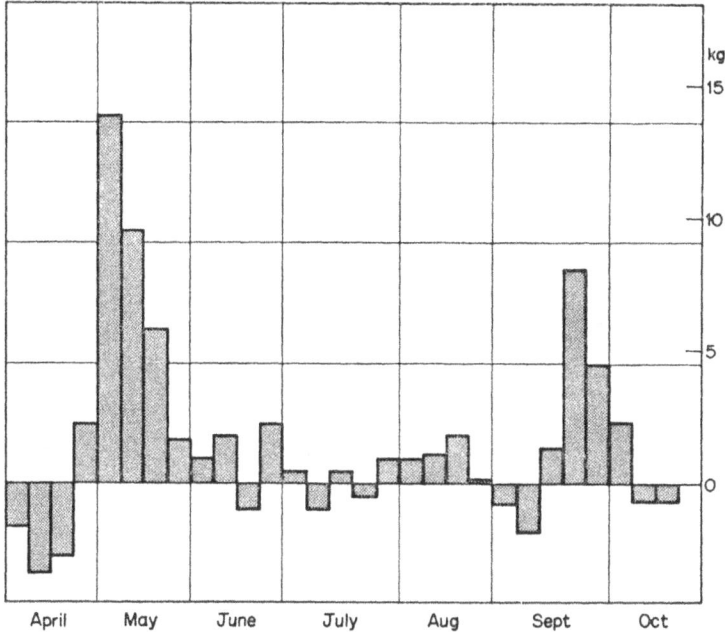

Figure 1.11/2 Similar to Figure 1.11/1, but for hives at Hope, 10 km from Emmet, where there was no summer flow. This type of record is discussed further in 1.14 [1 lb = 0.454 kg].

1947 virtually no sweet clover or alsike was grown; buckwheat (*Fagopyrum*), a good source of darker honey, occupied some 70 000 ha in 1929, but this showed a similar decline, dropping to 15 000 ha by 1952. By then, the number of beekeepers had dropped to 40% of its 1930 level (Townsend, 1956).

1.12 The bees' foraging potential

Most foraging is done within 1 or 2 km of the hive ($\frac{1}{2}$–1 mile), although bees have been recorded flying 13 km across barren desert to irrigated crops beyond. The foraging range of a colony is not a fixed circle round the hive, but a variable and irregular area depending on the lie of the land, the vegetation, and the weather. Forests and mountains can limit the range in any direction, and bees tend to ignore a stretch of land lying in the 'shadow' of a hill as they fly from their hive. The Ancient Laws of Ireland, which were codified around AD 400, gave a realistic description of the distance bees travel to visit flowers: as far as the crowing of a cock or the sound of a church bell can be heard.

Flight within 1 km of a hive provides a foraging area rather more than 3 km² (314 ha); a flight range of 2 km brings four times as much land within reach—over 12 km², 1 257 ha. Flight range is discussed further in 17.21; Ribbands' book (1953) has a chapter on foraging statistics.

Honey can only be made by bees after the raw materials have been collected from plants. If nectar is inaccessible to bees, or not sweet enough to attract them, or if the bees find a flower tiresome to collect from, because of its shape or behaviour, then it is unlikely that the nectar will be used for honey.

Figures 1.12/1–1.12/5 show how honeybees set about collecting nectar from some of the types of flower that are most important in honey

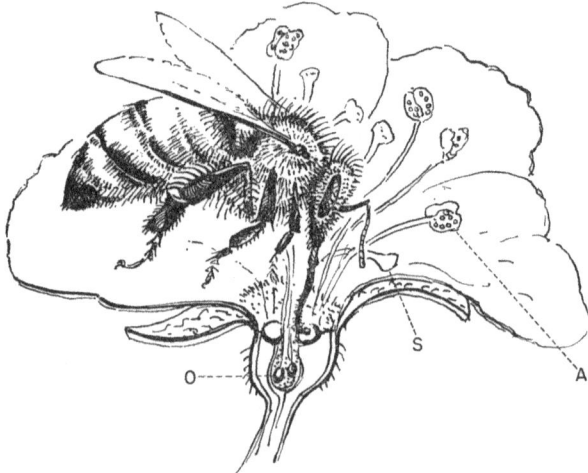

Figure 1.12/1   Section through a pear flower (*Pyrus communis*, Rosaceae) showing a honeybee sucking up nectar from the nectary. She stands on the flower petals to do so.   (Dorothy Hodges)
A = anther (presenting pollen)
S = stigma (which will receive pollen from this or another pear flower in the course of pollination)
O = ovary, not yet developed

production. The flower of Rosaceae (Figure 1.12/1) is cut in cross-section to show the bee's proboscis reaching down to the nectar on the nectary; in this position, the bee's head and thorax come into contact with the anthers and stamens of the flower, and the transfer of pollen is achieved.

Leguminosae (Figure 1.12/2) is one of the world's most important plant families for honey production, and white clover is a widespread source

Figure 1.12/2 Two honeybees collecting nectar from white clover (*Trifolium repens*, Leguminosae). The bee alights on a flower head and probes each floret in turn. (D. Hodges)

Figure 1.12/3 A honeybee collecting nectar from a lime flower (*Tilia*, Tiliaceae). She hangs upside down, and her tongue is thrust between the sepals where the nectar is secreted. (D. Hodges)

of good quality honey in both north and south temperate zones. Several of the limes (Tiliaceae) are among the most prolific nectar yielders; a bee usually works these flowers hanging upside down from them, as in Figure 1.12/3. Compositae and Scrophulariaceae (Figures 1.12/4 and 1.12/5 respectively—*see* Plate 1) are other important families. The bee pushes right into the toadflax (pollinating it in the process) to reach the nectar.

Some of the drawings show pollen foragers as well, with pollen loads already packed on their hind legs. The drawings have been made for this book by Dorothy Hodges, whose own book *The pollen loads of the honeybee* (1952) deals with this other part of the bee's foraging activity.

The foraging force of a colony consists of the older worker bees, and these are available for foraging only if there are enough young bees to feed the brood and keep the brood-nest temperature steady at 35° C (95° F). As well as nectar, the foragers must also collect sufficient pollen to provide protein for the developing brood. So the honey-getting capacity of a colony depends on the population of bees in it, and on the balance between different age groups. Ribbands (1953) discusses the subject in detail. In the active season, numbers for a strong colony might be something like this:

    1 queen
  300 drones
25 000 older workers, foraging for pollen and nectar
25 000 young workers in the hive; these rear the brood which might
          consist of:
 9 000 larvae requiring food
 6 000 eggs (from which future larvae will hatch)
20 000 older larvae sealed in cells, which need no attention except
          keeping warm.

A good beekeeper can recognize signs of balance or imbalance, and by suitable manipulations he gets his colonies at the right stage for the onset of each major nectar flow, so that they (and he) can get the maximum advantage from it.

An area of land may be under- or over-populated with bees at any one time in relation to the forage available. An area that will maintain fifty colonies during a three weeks' nectar flow—and yield surplus honey—may provide insufficient forage for the rest of the year. This fact has led to the practice of migratory beekeeping: the beekeeper moves his colonies to fresh forage when each flow starts. It is a very old practice—in Ancient Egypt, hives were moved down the Nile valley as the nectar flow advanced, on donkeys, and probably also by boat (Fraser, 1950, Armbruster, 1931). Throughout the world, beekeeping is becoming more and more dependent on migration. Current practice is summarized in 3.34.

Many countries could exploit more of their nectar sources by extended use of migration; studies of Polish conditions (Bornus, 1957, A. Demianowicz, 1957), and those in Czechoslovakia (Geisler, 1962), serve as examples. Over-population by bees can exist in many areas, including both towns and sites in open country to which bees are moved for a particular flow. In some of the suburbs round London there are about ten colonies of bees per 100 ha (four per 100 acres). It is the general experience amongst beekeepers there that the district is 'saturated' with bees, i.e. that if more colonies are introduced it is to the detriment of those present.

At the same time, under-populated, unexploited areas of bee forage are getting rarer every year. Access is usually the limiting factor, and as roads are built the bees are moved in (e.g. Mexico, 14.22). Most of the unexploited areas for bees are undeveloped lands. These are mainly in the tropics and subtropics: in South America (especially the Amazon basin), in various parts of Africa, and in Asia.

1.13 Characteristics of honey flows
Many laymen think of honey as accumulating continuously in the hive, the beekeeper being able to harvest it at his convenience. This is true only where there is a succession of plants which together flower and yield nectar continuously throughout the year. Such places are rare indeed; even areas of rich vegetation in the subtropics usually have an inactive period of 1–2 months. Elsewhere there is a more extended dearth period for the bees. In the temperate zones this is winter, when the temperature is too low for flowering and for bee flight. In the tropics it is a time of either drought (dry season) or excessive rainfall. Near the equator the season may be repeated every six months instead of annually.

Incoming nectar during a flow causes a rapid increase in the weight of a hive; during the night there may well be a loss in weight due to the evaporation of water as nectar is converted into honey. If a hive is weighed at the same time each day, the changes in weight provide a rough measure of the course of the nectar flow: examples have already been given in Figures 1.11/1 and 1.11/2.

Figure 1.13/1 shows daily weight changes of a hive during the three 'honey-getting' months May, June, and July, of twenty successive years. The records were made by Edwin Walker at Street, in the Vale of Glastonbury. By English standards the area was good for bees, one flow following another throughout the rather short honey season. In May there was fruit blossom, dandelion, and chestnut; instead of the 'June gap' there were flows from alder buckthorn (*Frangula alnus*) and charlock; clover and blackberry provided a third main flow in July. Edwin Walker's hives would have shown very little increase in weight from August to April; during the actual dearth period in winter the

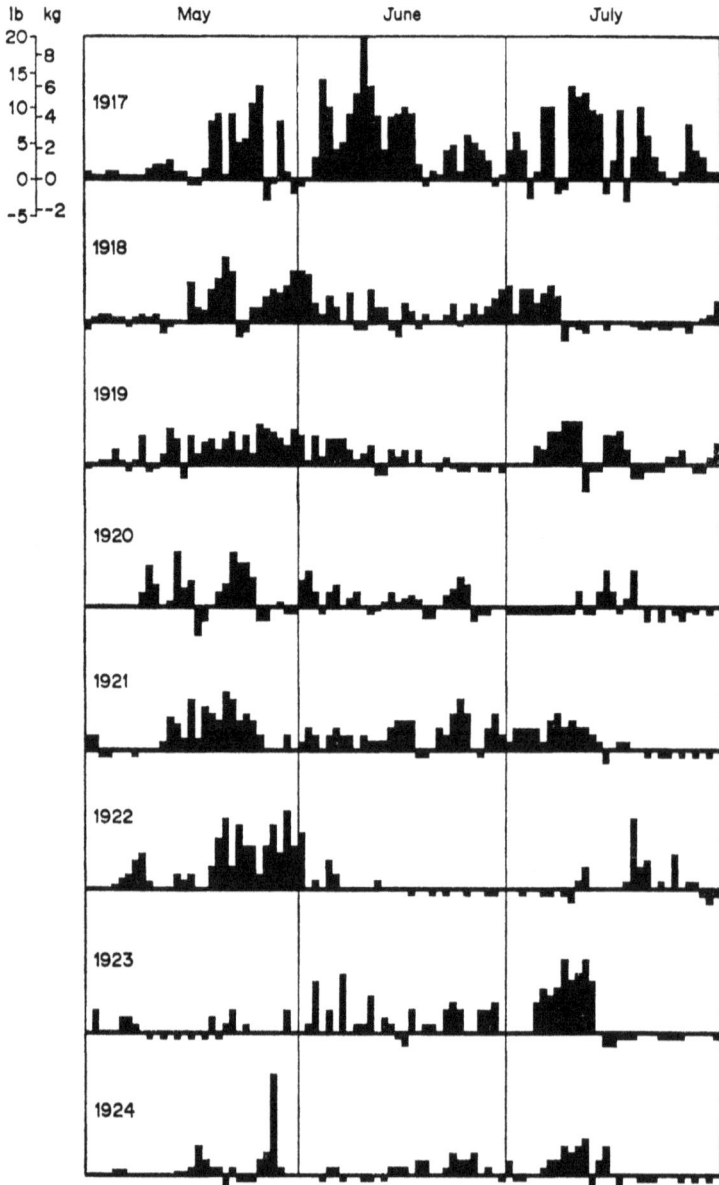

Figure 1.13/1   Daily gain or loss in weight (in pounds) of a hive at Street, Somerset, England, for twenty years [1 lb = 0·454 kg]. From unpublished records (1917–1936) by Edwin Walker in the Bee Research Association Library.

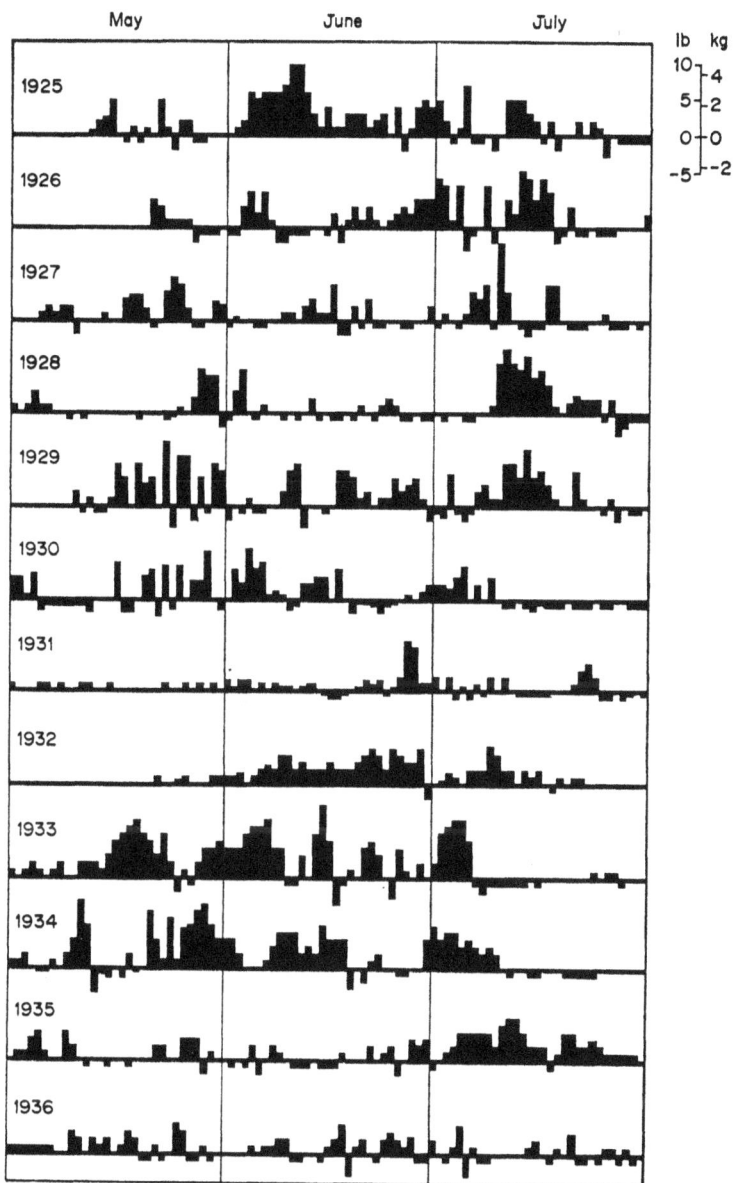

weight would have decreased, as stores were used up without chance of replacement. The year-to-year differences in Figure 1.13/1 show how dependent the honey yield is on the weather (1.14); the general decline in yield over the twenty years is due to reduction in bee forage, as explained in 1.11; the annual yields are quoted in 4.33.

Figure 1.13/2 shows a scale-hive record for a full season in Argentina, which is the world's largest exporter of honey. In winter and spring (June–November) there is no net gain. The summer flow starts in mid-November—from limes (*Tilia*), a wild marguerite (*Picris ichoiodes*), purple viper's bugloss (*Echium lycopsis*), white clover (*Trifolium repens*), etc. A lull in December (reminiscent of the 'June gap' familiar to English beekeepers) is followed by a larger flow from cardoon (*Cynara cardunculus*), spear thistle (*Cirsium vulgare*), and eucalypts. There is a final autumn flow from golden rod (*Solidago chinensis*) in March/April.

Figure 1.13/3 shows weekly weight changes in hives in Piracicaba, in the Brazilian State of São Paulo, 2 000 km north-east of La Plata and just within the Tropic of Capricorn. In this subtropical region the summer (January–March) is the rainy season and this is the period of dearth for the bees. Winter is the main honey-getting season, a wide variety of plants in bloom giving a continuous nectar flow for several months; autumn and spring also provide some nectar.

Honey production can nowadays be very profitable in certain north temperate regions with an intense short growing season, even though the winter is too long and too severe for bees to survive. 'Packages' of bees are flown in each spring from areas much further south, where spring comes early; the colonies made from these package bees develop very rapidly and are able to take advantage of the prolific short nectar flows. Figure 1.13/4 shows this type of season, as experienced in Manitoba, Canada. Virtually no weight gain of the hives is registered before July, and the peak of the flow is over before the end of the month. A minor flow may continue into August, after which the bees are destroyed and new packages brought in next April or May from California or Texas. All the flowering is concentrated into the short summer season, and the beekeeper may harvest 50, 100 or even up to 200 kg per hive. The Peace River district in northern Alberta and British Columbia is a notable new beekeeping region based on this system (Crane, 1966).

1.14   Effects of the weather

Neither the plant honey potential (1.11) nor the bees' foraging potential (1.12) can be realized in poor weather. No bees fly to collect nectar if the temperature is less than 12° C (53° F) or so. Plants cannot yield nectar unless the temperature is high enough: for bird cherry 8°, lime

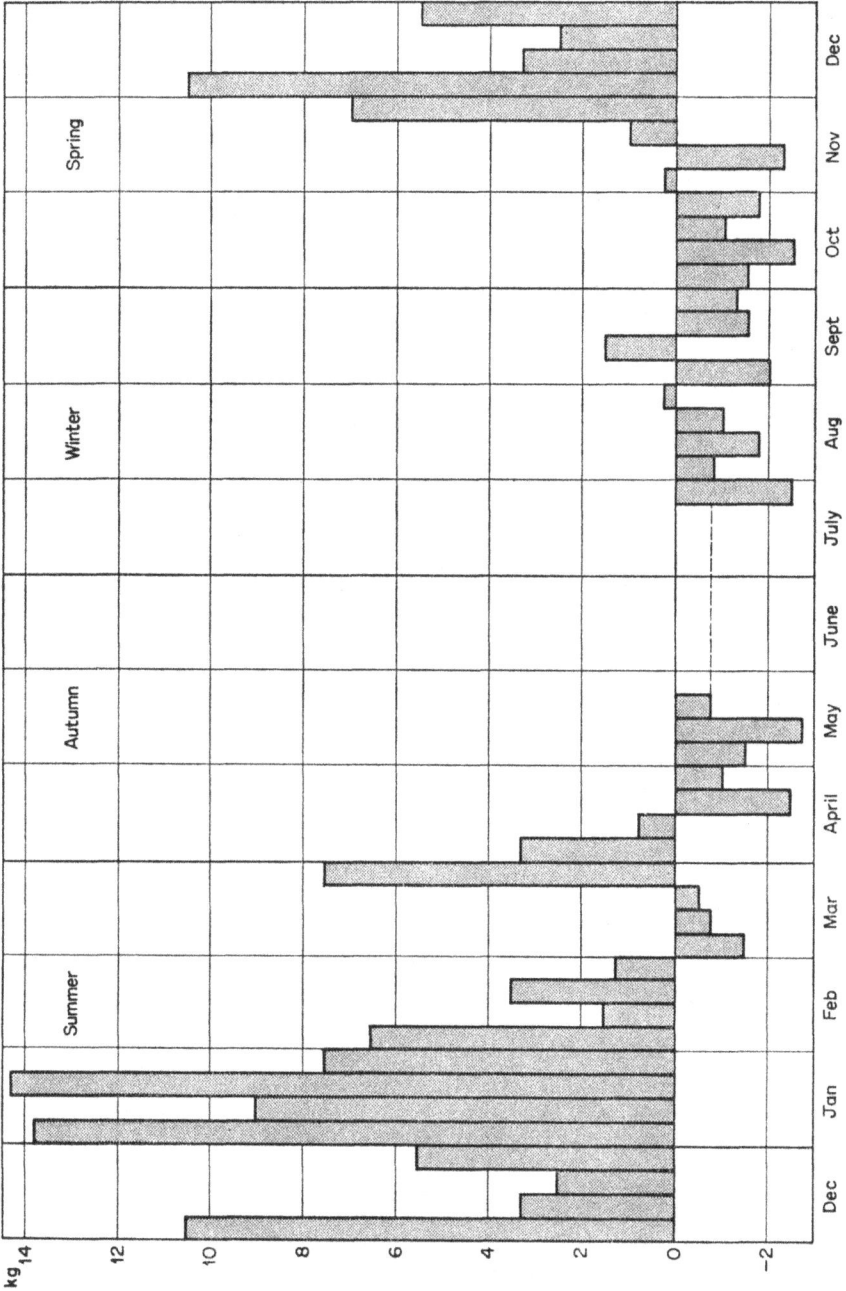

Figure 1.13/2 Weekly gain or loss in weight of a hive at the University of La Plata, Argentina, 1968–1969 (Cornejo *et al.*, 1971, and personal communication).

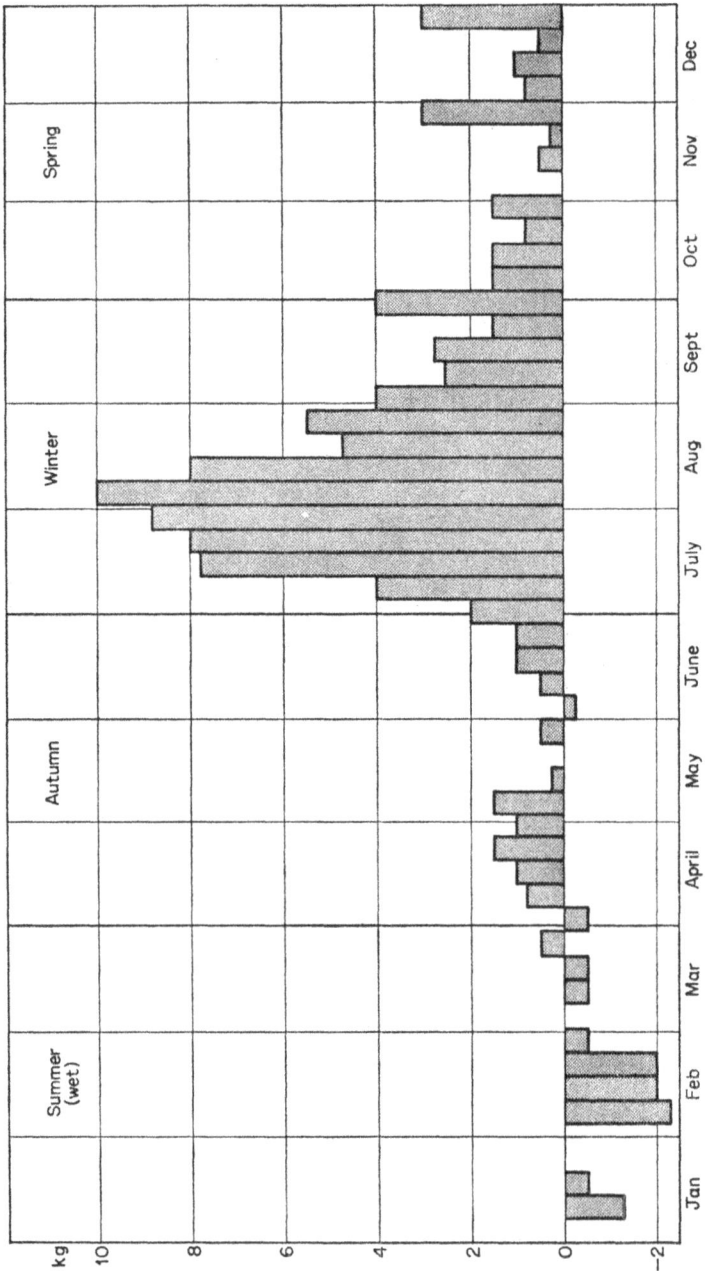

Figure 1.13/3 Average weekly gain or loss in weight (in kilograms) of three hives at Piracicaba, Brazil (Amaral, 1957).

and sainfoin 15°, white clover 23° C (47, 59, 73° F). Rain or high humidity can make nectar more dilute and give the bees much extra work, both in carrying the nectar home and in evaporating off the excess water. Few bees will fly in a wind of more than 25 km (15 miles) an hour.

Apart from damage by diseases and pests, and catastrophes such as fire, weather is the main factor underlying year-to-year variations in nectar flows from the same plant species.

These are very apparent in Figure 1.13/1, and the records show what an overriding effect weather can have on honey yields in England, with its variable oceanic climate; compare for instance 1934 and 1935. Bee-keeping may be much easier in a stable continental climate, although

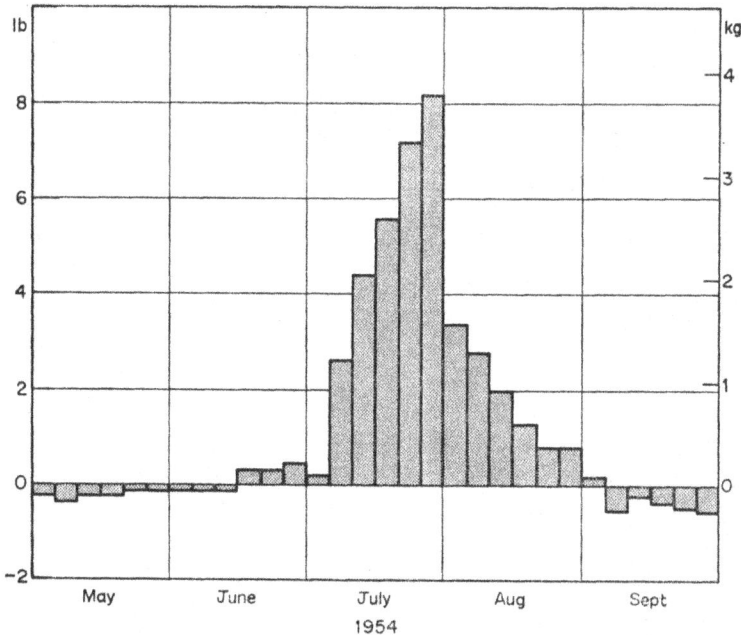

Figure 1.13/4 Daily gain or loss in weight (in pounds) of hives in Manitoba, Canada, 1954. The weights are five-day averages for twenty-one hives [1 lb = 0.454 kg] (Mitchener, 1955).

this also has its problems. A detailed study for the period 1918 to 1948 has been made in Kansas, and this report (Moffett & Parker, 1953) gives references to many earlier ones. The object of the Kansas study was to find the relationship between weather (rainfall, temperature, sunshine, etc.) and the amount of surplus nectar or honey stored by colonies of honeybees. No association was found between the nectar flows in successive seasons (nor is any such association apparent in Figure 1.13/1).

The average amounts of honey stored per hive varied from 3·8 to 10·3 lb in the two poorest years to 215 and 210 lb in the two best of the 31 years. The amount of honey stored in any one year was significantly $(P<0·01)$ correlated with the precipitation in June $(r = +0·590)$, and also with that during the previous winter $(r = +0·514)$. Higher April temperatures and lower May and June temperatures were associated with higher honey yields. The weight increase on any one day was related to the maximum temperature on the day, as set out below; $T(F)$ and $T(C)$ are the maximum temperatures in °F and °C; $W(p)$ and $W(k)$ are the average weight increases per hive per day in pounds and kilograms (the paper uses non-metric units); N is the number of days that fell within each temperature range.

$T(F)$	69 or less	70–79	80–89	90–99	100–109	110 or over
$W(p)$	− 0·65	0·42	2·81	3·33	0·99	− 0·13
$T(C)$	20 or less	21–26	27–32	32–37	38–43	44 or over
$W(k)$	− 0·29	0·19	1·27	1·51	0·45	− 0·06
N	23	154	737	932	401	48

The best temperatures were 80–100° F (27–37° C). Nectar secretion by plants flowering in cooler regions will be optimal at lower temperatures than those in Kansas. Moffett & Parker discuss in some detail the effects of temperature on nectar secretion, and also on bee flight, both of which are involved here, and which are subject to different sets of determining factors (*see also* Hambleton, 1925).

In Germany Koch has carried out extensive observations and calculations on the effects of weather on the nectar supply (1959), the seasonal cycle of nectar flows (1967), and the corresponding phenology of honeybee colonies (1957, 1961).

Discussions on weather can be continued for a very long time. This section will be closed with one example of abnormal weather affecting honey production. There was, unusually, a 'freeze' in Florida in December 1962 and this damaged many of the orange trees in the State. Flowering in March 1963 was variably affected but, even where trees bloomed freely, little or no nectar was produced. A graph showing hive weights during the citrus flow in one area (Robinson, 1964) shows gains of 52, 85, 61, −8 lb in 1960, 1961, 1962, 1963 respectively. The orange blossom honey flow failed completely as a result of the freeze.

1.15   Forecasting the honey flow
Attempts have been made to predict agricultural crops from very ancient times. Some of the mediaeval prognostications for honey harvests have recently been published (Crane, 1972a), and a manuscript of 1370

giving honey crop forecasts, in the Bodleian Library in Oxford, is reproduced in Figure 19.41/3. In this, honey is the only crop which is forecast each year, and it is for an abundant yield in six years out of seven—very much better than for livestock.

In recent years there have been many efforts to establish adequate scientific bases for predicting how good a particular honey flow will be and when it will start. This information would be especially valuable to migratory beekeepers (3.34), who need to know (a) when to move their bees to a new flow and, in some circumstances, (b) whether it is likely to be economically profitable to do so. There are direct and useful ways of dealing with the first problem, for a flow that is regularly profitable, such as that from acacia (*Robinia pseudoacacia*) in Hungary. There, scouts go out on motor cycles into the acacia forests when the flowers are expected to open; reports of the flowering are broadcast on the radio, and beekeepers and their families travel to the woodlands with their hives and camp out there while the flow lasts.

Prediction of the timing of a flow on a short-term basis may be aided by established phenological data, where these exist, since the *sequence* of flowering of different plants is more often stable than the actual dates. In 17 years' observation in the south of England, for instance, although the date on which Beauty of Bath apple came into flower varied by over a month, it usually followed about a week after Clapps Favourite pear. But even this interval varied between 3 and 14 days, and a report (Hodges, 1958) discusses the various hazards in trying to establish dates; it also gives a useful calendar of 105 flowering plants worked by bees, based on the 17 years' observations.

Variable weather can hasten or delay any nectar flow. A few weeks before a flow is due, the likely effects of *past* weather can be assessed for some flows, by painstaking observations which vary from plant to plant. Effects of *future* weather—including that during the flow— are inevitably less precisely predictable than the weather itself. So it is in general even more difficult to predict honey flows in oceanic climates than in continental climates where settled spells of weather are more regular. In Norway some success has been achieved in predicting the heather honey flow (*Calluna vulgaris*) from the temperatures in the preceding months. If the accumulated daily temperature above 16°C (60°F) between 20 June and 20 July (T) is less than 30°, it is probably not worth taking bees to the heather; if T is 30–40° the honey yield is likely to average about 10% of the maximum obtainable from the area; if T is over 60° the yield is likely to reach 50–80% of the maximum, i.e. a good crop. Details are given by Kierulf (1957) and Ukkelberg (1960).

A few other plants have been studied, from different points of view: buckwheat (*Fagopyrum*) by determining ascorbate oxidase activity

(Zauralov, 1966); and acacia and limes (*Robinia, Tilia*) by studying the floral primordia formed in autumn and spring respectively (Cîrnu, Tomescu & Sanduleac, 1965).

The long intervals between flowering of many eucalypts—often several years (*see* 1.51)—and the fact that the bud stage may last for a year or more if weather conditions are adverse, make flowering forecasts for the good honey-yielding species very desirable. In Victoria, 'budding statistics' from the Forests Commission are published regularly for the benefit of beekeepers (e.g. *Australian Bee Journal*, 1972), accompanied by the warning that they do not relate to the degree of flowering in any particular area. Some years ago Maclachlan tried to develop a more specific method, and in 1935 and 1936 he published a series of articles on the prediction of eucalypt honey crops by measuring their 'starch index'. Carbohydrate reserves are stored as starch in the sapwood of trees; this starch can be converted into sugar, and in theory there might be a direct relation between the starch content and the amount of sugar secreted during the next nectar flow. When Maclachlan died in 1945, some beekeepers in Victoria were using his starch-testing method for forecasting honey flows, whereas others declared it to be useless. Wykes undertook experiments in 1945 and 1946 to establish a reproducible method for starch estimation, and to compare the results obtained with subsequent nectar production. These experiments are described in an unpublished report (1947) in the Bee Research Association Library, which ends: 'Preliminary conclusions suggest that for a wide range of eucalypt species, a heavy starch content in the sapwood of well budded trees indicates the potentiality of a good honey flow, but that unfavourable climatic conditions during the flowering period may become a limiting factor, and modify this possibility. No reliable means of forecasting climatic conditions exists, but at the present stage of this investigation it appears that starch testing may provide a reliable basis for forecasting honey flows for the practical apiarist, and thus remove one of the uncertain factors of the Australian beekeeping industry.' No later work on the subject is known.

The honeydew flow (*see* 1.2 and 2.13) is determined by many complex factors, including the population dynamics of the insects producing the honeydew. These affect both the timing of the flow and the amount of honey that can be obtained from it. In Yugoslavia, observation stations were established in 1955, and with their aid a prediction service has been put into operation (Rihar, 1960, 1963). In Germany (D.D.R.) Scheurer made extensive observations on the distribution of the lachnids that produce honeydew, and of the wood-ants that visit the lachnids, and these studies are also used as a basis for predicting the honeydew flow (1967*a*, 1967*b*).

1.2 SOURCES OF HONEY APART FROM FLOWERS

Most honeys are derived from flower nectar, which is discussed in detail in 2.12. Some plants secrete nectar elsewhere than in flowers, including a few flowerless plants. Chapter 13 sets out the legal definitions of honey in eighteen countries, and in all of them the substance produced by bees from extrafloral nectar is also legally called honey. This makes sense, because there is no basic difference between floral and extrafloral nectaries (2.121). Until 1971 honey was defined more narrowly than this in Argentina: it had to be made by domestic bees *from flower nectar* (Argentina, 1953).

The formation and secretion of nectar by the nectary—wherever this is situated—is an active process in the cells of the glandular tissue. If plant tissue is cut, punctured or otherwise damaged, sap will tend to escape because of the pressure within the plant. One group of insects, the Rhynchota, are able to pierce some plant tissues and pump out minute amounts of the sap. Section 2.131 describes the process by which some of the fluid ingested by the insect is re-secreted—considerably altered in composition—as *honeydew*. If conditions are such that water evaporates quickly from the droplets of honeydew, these solidify into *manna*, which can be very sweet; the lerp secreted by some Psyllidae is composed of starch—it is the only known example of starch of animal origin (Basden, 1970).

Honeydew and manna are collected by bees; in some coniferous forests honeydew may provide the main honey flow, and a very prolific one. Section 2.13 gives up-to-date information; a more detailed survey of the whole subject is available in German (Kloft, Maurizio & Kaeser, 1965); Kloft has also written a summary in English (1963). Like floral and extra-floral nectar, honeydew is derived ultimately from the phloem sap of plants, but it differs from nectar in several ways: it is not actively secreted by the plant with selective transport of sugars, and it is processed by plant-sucking insects before the bees collect it. Also, a sticky coating of honeydew on deciduous leaves will often trap soot, dust or other particles falling on it from the air.

Honey produced from honeydew is explicitly included in the definition for honey in many countries; in other countries there is no explicit statement, but honeydew honey would probably be accepted as honey (13.68).

The phloem sap of plants is made accessible to bees in very large quantities when sugar cane is cut. For several days the sap rises in the stump, and is avidly collected by bees. In general the substance made from this sap by bees could legally be called honey. But a beekeeper in

Florida, visited by the author when he had small colonies of bees 'building up' on this sugar-cane flow, was emphatic that he would use it only for that purpose, and not for marketed honey. It is explained at the end of 2.221 that, during a very copious flow, the bees cannot process the incoming material as thoroughly as when the flow is less prolific. Apart from this aspect, the sap exuding from cut sugar cane would seem to satisfy the legislation of most if not all countries as a raw material for honey (Chapter 13).

In the north-east of North America, sugar maple trees (*Acer saccharum* and others) are tapped in February and March when the sap is rising, and the sap collected and concentrated to make maple syrup. In theory bees could collect this sap and make it into honey, but the sugar concentration ($1\frac{1}{2}$–5%) is not high enough to be attractive to them, and the temperature would in any case be too low for the bees to fly.

If the honeybee's mandibles were strong enough to pierce the skin of soft fruit, a further type of sweet material would be available for collection and conversion into honey. But they are not, so the bees can get juice from fruit only if it has been previously damaged. Caged bees given grapes as their sole source of food died of starvation. There is one early fig, with a crinkly porous skin, which bees can damage (Giordani, 1953).

Whatever the plant source, honey is essentially elaborated by bees from material collected while foraging. This material already undergoes changes in the bee's honey sac while she is flying back to the hive (2.21). What bees store in their combs from materials *fed in the hive* is not honey, either biologically or legally (13.67), unless the material fed is itself honey.

1.3   THE HONEY POTENTIAL OF DIFFERENT PLANTS SPECIES

Table 1.3/1 (Appendix 1, pages 48–55) gives the estimated 'honey potential' of 200 flowering plants (1.11). This is measured as the theoretical quantity of honey (in kilograms) that could be obtained in the course of a season from 1 hectare of land covered with the plant in question, which is nearly the same as the number of pounds of honey obtainable per acre. Optimal plant-growth conditions for nectar production are assumed, and an adequate force of foraging bees to collect all the nectar secreted, as well as suitable weather conditions. This gives a sensible basis for comparison, although in practice (1.12, 1.13, 1.14) these conditions are rarely all fulfilled.

In general this aspect of land use has received little attention except in the countries of eastern Europe, and most of the entries in Table 1.3/1 are derived from investigations there, especially those published by Glukhov (1955), Fedosov (1955) and Kuliev (1952) in the U.S.S.R.; Z. Demianowicz & Hlyń (1960), Demianowicz, Jabloński, Ostrowska & Szybowski (1963), Demianowicz & Jabloński (1966) and Demianowicz *et al.* (1960) in Poland; and Nyárády (1958) in Rumania. Nearly all the plant species represented in the more precisely derived Table 2.123/1 are included in the Table 1.3/1.

Plants are grouped into six classes, the honey potential for each class being as follows, measured in kilograms per hectare or in pounds per acre:

Class 1: 0–25   Class 4: 101–200
Class 2: 26–50   Class 5: 201–500
Class 3: 51–100   Class 6: over 500.

Where there are two entries for a plant, these represent independent estimates, often from different regions. Plants are grouped under their botanical families, and these are arranged in alphabetical order, as are the plant genera within each family. The common English name is given on the right.

Plants with the highest honey potential in Table 1.3/1

Labiatae	*Lamium album*
	Salvia officinalis
	Salvia verticillata
	Thymus vulgaris
Leguminosae	*Caragana arborescens*
(*see* Figure 1.12/2)	*Melilotus alba*
	Robinia pseudoacacia
Boraginaceae	*Anchusa officinalis*
	Echium vulgare
Compositae	none; 3 in Class 5
(*see* Figure 1.12/5)	
Tiliaceae	*Tilia caucasica*
(*see* Figure 1.12/3)	*Tilia cordata*
Aceraceae	*Acer campestre*
Asclepiadaceae	*Asclepias cornuti*
Dipsacaceae	*Cephalaria caucasica*
Hydrophyllaceae	*Phacelia tanacetifolia*
Onagraceae	*Chamaenerion angustifolium*
Scrophulariaceae	*Scrophularia nodosa*
(*see* Figure 1.12/4, Plate 1)	
Verbenaceae	*Vitex negundo incisa*

The 'top twenty' species are listed above. Outstanding families are Labiatae (with 4 species in Class 6 and 21 in Class 5) and Leguminosae (with 3 in Class 6 and 5 in Class 5). Next in importance, but a long way behind, are Boraginaceae and Compositae, both having 3 species in Classes 5 and 6 together. Species from 8 other families are also in Class 6.

The first five families listed are among the ten in Table 2.123/1 that give the highest sugar values (yielding more than 1 mg sugar per flower in 24 hours), which however do not take into account the length of the flowering period. Publications referred to on page 23 and in Table 2.123/1 indicate the methods used in obtaining the honey potential. Calculations are based on sugar value, length of flowering, and number of flowers per unit area of land, where possible supported by calculations from weight gains of hives during the flow in question.

It is hoped that the publication of tabulated results here will encourage botanists and bee scientists to investigate further plants, especially those belonging to families other than those in Table 1.3/1—which are all represented in Europe. When observations can be made on tropical species, and on the eucalypts, the relative importance of the plant families may well appear in a new light.

Jabłoński (1968) has shown that the intensity of honeybee visits to a plant can be used as a rough measure of its honey potential. He compared the daily number of bee visits to 1m² of different flowering plants with the calculated honey yield of the plants during the course of one day. On average, the number of bee visits was about the same as the honey yield in kilograms per hectare. On this basis the total number of bee visits to plants within 1m² during their whole flowering period would be a direct approximate measure of the honey potential of the plants in kilograms per hectare, as given in Table 1.3/1.

Jabłoński's analysis takes into account the considerable variation from year to year in the honey yields of the same plant species. He concludes that the direct source of these variations lies not so much in the difference in the amount of nectar secreted by individual flowers, or in the sugar concentration of the nectar, as in the density of the flowering— the number of flowers produced in a given area.

1.4 THE WORLD'S MOST IMPORTANT HONEY PLANTS

Table 1.3/1 gives information on the potential honey yield of plants on which the necessary measurements have been made. The present section deals with plants selected for their importance as sources of honey on a world scale; their characteristics and distribution, the characteristics of the honey from them, and their economic significance. The information is given in Table 1.4/1 (Appendix 2, pages 55–76). Plant families are listed in alphabetical order, as in Table 1.3/1; for each family the table gives the genera or species which are economically important honey yielders in one or more areas of the world. Some of these may rate fairly low in Table 1.3/1, but if grown in sufficient quantities much honey is obtained from them; many have never been assessed for Table 1.3/1.

The first line in each entry gives the botanical name and, where it exists, the common English name; then the type of plant: tree, shrub, climber, or perennial, biennial or annual (herbaceous) plant. Crop plants are generally noted as such. The geographical distribution is indicated, on a very broad basis only; that of cultivated crops changes from decade to decade, and that of indigenous and other wild flora may be reduced rapidly as new regions are opened up. (Figure 1.4/1 shows the main world vegetation types; Figure 1.4/2 the boundaries of the tropical region from the beekeeping point of view.) WHCR indicates an entry for the country named in the 'World Honey Crop Reports' that are published regularly in *Bee World*; since Report No. 3 (1966), reporters have been asked to name the main plant sources of the honey crops quoted. So an entry here 'WHCR Bulgaria' means that the plant in question was reported to be a main honey source in Bulgaria in one or more years since 1966. Entries HP 2, HP 6, and so on, give the honey potential class number (1–6) in Table 1.3/1.

The next line gives characteristics of the honey in liquid form: colour, density, viscosity (body), flavour, and aroma, any noteworthy aspects of granulation, and any other special features of the honey. Descriptions of the honey have been obtained largely from the country of origin; its flavour and quality, and even colour, may well be differently assessed elsewhere (*see* Chapter 13). Assessments which could not be obtained locally are derived from personal knowledge, or from that of specialists in the world honey market. The somewhat stereotyped descriptions— light amber, medium amber, etc.—are those in common use in the trade.

Appendix 2, with Section 1.5, is believed to represent the first comprehensive attempt to bring together information on the honey resources of the world. There have been many studies of honey plants of specific areas (1.5), and the present synopsis has been compiled largely from material extracted from these studies. In order to prepare a world atlas of honeys, comparable with H. Johnson's *World atlas of wine* (1971), much further detailed work would be needed, probably using computer techniques. The present table will, however, serve as a basis and any amendments or additions would be welcomed by the author.

E. B. Wedmore's book *A manual of bee-keeping for English-speaking bee-keepers* (1932, 1945) contained a 15-page list of honey sources, relating

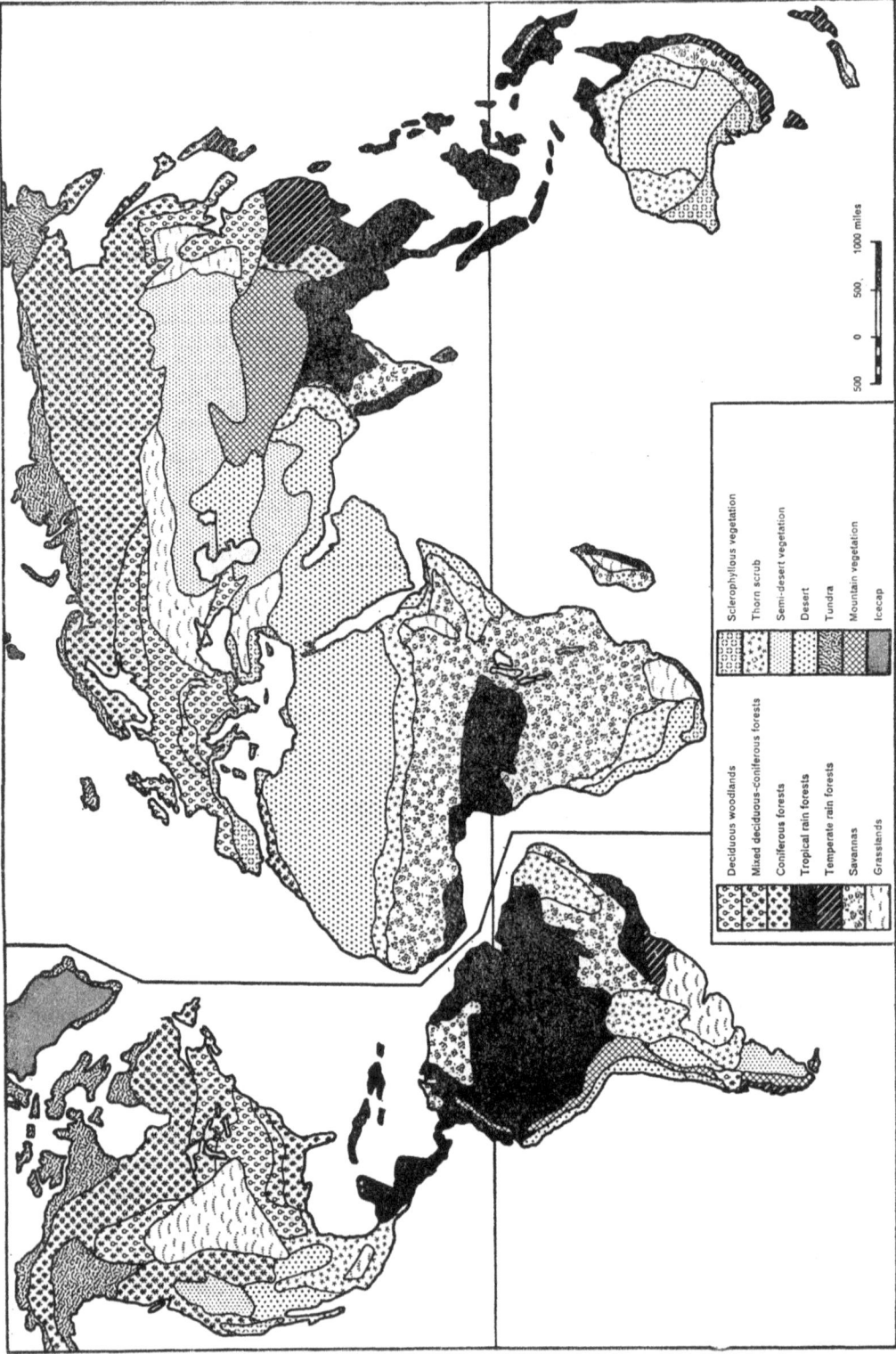

Deciduous woodlands
Mixed deciduous-coniferous forests
Coniferous forests
Tropical rain forests
Temperate rain forests
Savannas
Grasslands

Sclerophyllous vegetation
Thorn scrub
Semi-desert vegetation
Desert
Tundra
Mountain vegetation
Icecap

500   0   500   1000 miles

largely to English-speaking countries and with special emphasis on the British Isles. M. Yate Allen intended to follow his *European bee plants and their pollen* (1937) with a book on *World bee plants*, but the war prevented its publication. The manuscript material was presented to the Bee Research Association Library by A. G. Higgins in 1965, and it has been consulted in preparing this chapter. The basis of the proposed book was a series of articles (in English) in an Egyptian journal *The Bee Kingdom* between 1937 and 1939. Mr. Yate Allen, a Yorkshire parson, was one of the early amateurs who examined pollen grains in honey under a microscope. His 1939 manuscript includes the passage: 'Nobody has so far treated the subject in English, nor have any of the previous writers attempted to describe the pollen content of more than a few honeys outside those of their own country. In this work, through the kindness of friends who have supplied me with samples, I have been able

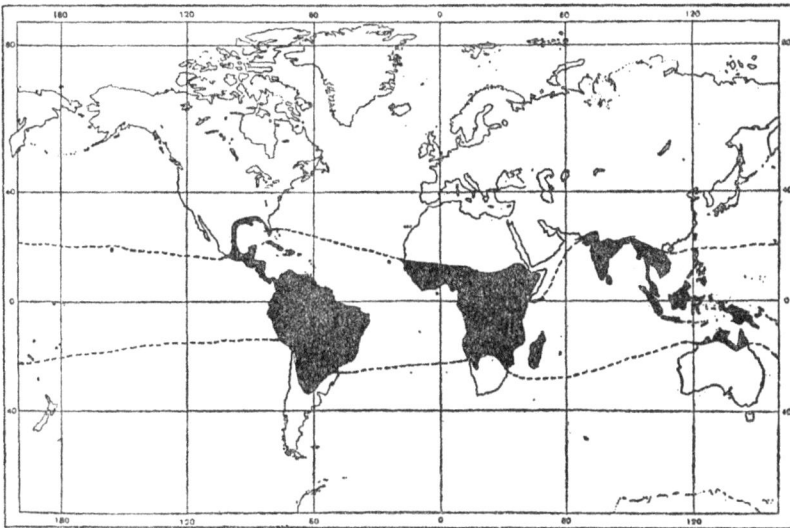

Figure 1.4/2 Boundaries of the tropical region (F. G. Smith, 1953).

to make what is often though erroneously called a Pollen Analysis of about 70 different samples of honey gathered in all the five continents, none of which is British, describing and illustrating the principal pollen grains found in each, and where possible giving the name and description of the plants or trees which produced them.' Yate Allen's pioneer efforts have been superseded by studies made with more precise methods and equipment, by scientists specializing in the subject, and this work is described in detail in Chapter 7.

1.5 INFORMATION ON HONEY PLANTS OF INDIVIDUAL COUNTRIES

1.51 Introduction

The map in Figure 1.4/1 shows the main types of world vegetation, and thus where a given plant is likely to be found. Outstanding honey-producing areas occur in many of these types of region. For instance: Peace River in Canada is cleared deciduous/coniferous forest; Yunnan in China is temperate rain forest and Yucatán in Mexico is tropical rain forest; Western Australia has sclerophyllous (drought-tolerant) vegetation, and parts of eastern Australia and Argentina savanna-type. Such maps therefore do not serve to identify honey-producing regions, nor do maps of floristic regions of the world such as Takhtajan's (1969).

While certain plant families include many good honey plants (1.3), the existence of a family or genus in an area does not guarantee the presence of a prolific nectar-yielding species. Moreover a species may yield much nectar in one area and little in another where conditions are different (2.125).

Many books and bulletins on the honey flora of different parts of the world exist. Most of these have been compiled on a national basis, although the distribution of honey plants depends on soil and climate rather than on political boundaries.

European countries are the best documented, and the use of pollen analysis (Chapter 7) for identifying honey sources has also been most extensive there. Many individual states and provinces in North America are well documented in regional beekeeping bulletins. Several extensive books have been published in the U.S.S.R., which pays great attention to utilizing its national resources of bee forage. Japan has been fairly well studied.

As we progress southwards towards the Equator, we enter less well known territory, the least known and potentially the most important being China (see 14.25). Studies are now well under way in India, and in the eastern Mediterranean, and some work has been done in Africa both north and south of the Sahara. We owe much to the late Dr. G. S. Ordetx for his studies of Central American countries. Argentina, which surpasses even Mexico as a honey exporter, is perhaps most in need of attention; other countries in South America are no better studied, but are less important. New Zealand is quite well documented. Australia occupies a unique position, in that its beekeepers must be first and foremost botanists to learn how to identify each eucalypt and to recognize the clues that indicate when it will come into flower; this may be once in 2, 4, or even 8 years, and buds may hang unopened for a year or more if conditions are adverse. So there are detailed descriptions of honey sources for many of the separate areas of vegetation in Australia.

1.52   Guide to published information on honey sources

A proper account of the world's honey sources would occupy a complete book; the most useful provision that can be made here is a guide to the information that is currently available. The list below has been compiled with the aid of all the resources of the Bee Research Association, and with few exceptions the publications are in the B.R.A. Library. The list should be of special use to scientists and technicians concerned with identifying the sources of imported honeys. It is also hoped that its publication here will stimulate new observations where they are especially needed. Enquiries made in compiling the list have already initiated pre-liminary observations in a number of countries; some have now been published (Crane, 1973).

This guide to published information on honey sources is arranged by continent as follows:

Europe, including all U.S.S.R.	page 30
Asia	page 35
Africa	page 37
North America	page 41
Central America	page 43
South America	page 45
Oceania	page 46

Countries are listed in alphabetical order within continents. Publications for each country are roughly in order of usefulness. For well documented countries the selection has been fairly rigorous; occasionally an important older publication is included to complete the record. Books and bulletins on beekeeping which have a chapter on honey plants are included where specialist publications are lacking. Some countries are so little known that any clue is valued, even if only a beekeeping traveller's report. The entry 'No information' means that no publication has been traced, institutions in the country and the embassy in London having been asked for information, but without success.

Titles are quoted in English, and the language is indicated for all publications not in English. Many publications include Latin names of plants, so unfamiliarity with a language should not deter those who need the information from pursuing it. P indicates a study based on pollen analysis of honey; most of the other publications are based on observa-tions in the field, by the author quoted or by others who provided him with information. K indicates a useful key to plants or to pollen grains, but not specifically related to honey plants. To save space, references are somewhat abbreviated. For publications marked R, fuller information may be obtained from the list of references in the Bibliography (page 489). Figures such as 328/67 refer to an English summary of the publication in *Apicultural Abstracts* (No. 328 in 1967).

EUROPE, INCLUDING ALL U.S.S.R.

GENERAL

R P *The book of bee plants*. A. Maurizio & I. Grafl (Munich, 1969)
German [396/70

R P *Flowers, nectar, pollen, honey*. A. Maurizio (Nürnberg, 1960a)
German [217/61

R P *Pollen of European bee plants*. A. Maurizio & J. Louveaux (Paris,
1965) French [652/65

R P *Photographic atlas for pollen analysis of honey*. J. Louveaux (Paris,
1970) French [1026/71

 P *European bee plants and their pollen*. M. Y. Allen (Alexandria, 1937)

R P Determination of the origin of honey. Vol. I–V. E. Zander (1935–
1951) German; *see* page 554 for details
No information is available for the following countries:
Andorra
Monaco
San Marino

ALBANIA

Beekeeping in Albania. J. Svoboda *Včelařství* 1: 118–119 (1953)
Czech [217/54

AUSTRIA

R P Upper Austrian honey. F. Ruttner *Bienenvater* 77: 82–90 (1956)
German [220/57

R P Styrian honeys. A. Fossel *Bienenvater* 77: 156–163 (1956) German
[303/56

Forest flow and forest flow observations in Austria. F. Ruttner
Bienenvater 81: 196–203 (1960) German [174/61

BELGIUM

Flowers worked by hive bees in the Liège region (1945–1959)
J. Leclerq *Bull. Inst. agron. Gembloux* 29: 79–90 (1961) French
[596/64

R P Study of bee forage in Belgium by pollen analysis. N. Martens,
O. van Laere & C. Pelerents *Biol. Jaarb.* 32: 292–325 (1964)
Flemish [489/66

BRITAIN

Plants and beekeeping. F. N. Howes (London, 1945)

 P *Survey of British honey sources*. A. S. C. Deans (Rep. B.R.A. 142,
1957) [195/65

R   A calendar of bee plants. D. Hodges *Bee Wld* 39: 63–70 (1958)
[337/59

British bee plants. A. F. Harwood (Foxton, 1947)
Survey of heather areas in Scotland. S.B.A. Research Committee
Scot. Beekpr 26: 180–182 (1950) [286/57
Bee-plants and their honey, Parts I–IX. M. Grieve *Bee Wld* 4, 5
(1922–23)
P *Nectar-producing plants and their pollen*. G. Hayes (London, 1925)

BULGARIA
Bee plants. V. Petkov (Sofia, 1973) Bulgarian [418/74

CYPRUS no information

CZECHOSLOVAKIA
Beekeeping. B. Tomšík *et al.* (Prague, 1953) Czech [242/43
Beekeeping encyclopaedia. J. Svoboda (Prague, 1956) Czech [104/59

DENMARK
P Studies on the pollen content of Danish honey samples. O. Hammer,
E. G. Jørgensen & V. M. Mikkelsen *Tidsskr. Planteavl* 52:
293–350 (1948) Danish

P Phenological studies on some important nectar plants. B. Fredskild
Tidsskr. Planteavl 61: 133–148 (1957) Danish [290/58
R P Studies on the pollen content of honey samples from Danish heather
regions. P. Wolthers *Tidsskr. Planteavl* 58: 683–721 (1955)
Danish [65/58

FINLAND
Beekeeper's manual. L. Mali & K. J. Laaksonen (Helsinki, 1961)
Finnish
R P Composition and source of origin of Finnish honey. E. Martimo
Maataloust. Aikakausk. 17: 157–169 (1945) Finnish

FRANCE
Medicinal plants, nectar plants, useful and harmful plants. G. Bonnier
(Paris, 1920?) French
R P A study of French honeys by pollen analysis. J. Louveaux *XVI Int.
beekeep. Congr. prelim. sci. Meet.* (1956) [260/56

GERMANY, EAST (D.D.R.)
*Statistical evaluation of results of a questionnaire sent to beekeepers on
bee forage*. G. Pritsch (Berlin, 1957) German [265/61
R P Bee forage at various localities with light soil, with special reference
to its investigation by means of the pollen analysis of honey.

G. Pritsch *Arch. Geflügelz. Kleintierk.* 7: 184–247, 282–310 (1958)
German   ⌊218/61

Improvement of bee forage. G. Pritsch (Berlin, 1959) German [169/61

GERMANY, WEST (B.R.D.)

Beekeeping handbook. Vol. VII: Bee pasture. U. Berner & H. Müller
(Stuttgart, 2nd ed., 1967) German   [140/68

Practical advice on the improvement of bee pasture. J. Mayer
(Freilassing, 1950) German   [68/50

Index of bee plants. O. Dengg (Leipzig, 1953) German   [315/57

R P Pollen analysis of honeys of the B.R.D. J. Evenius *Dt. Bienenwirt.*
11: 76–80 (1960) German   [139/61

R P Determination of the origin of honey. Vol. 1–5. E. Zander (1935–
1951) German; *see* page 554 for details

GREECE

Beekeeping: modern intensive methods. N. J. Nicolaidis (Athens,
3rd ed., 1959) Greek   [355/59

HUNGARY

Bees and flowers. G. Lengyel (1943)

R *Plants providing bee forage.* A. Nyárády (Bucharest, 1958) Hungarian   [56/61

Among the bees. Z. Örösi-Pál (Budapest, 5th ed., 1957) Hungarian
[282/58

ICELAND no information (no bees are kept there)

IRELAND

A survey of beekeeping in Ireland. T. N. Hillyard & J. Markham
(Dublin, 1968)   [309L/70

ITALY

Honey plants. L. Fossati (Rome, 1971) Italian

First national beekeeping enquiry (pp. 83–170). A. Zappi-Recordati
(Rome, 1930–1937) Italian

Beekeeping. G. Canestrini (Milan, 15th ed., 1955) Italian   [401/57

Modern beekeeping. A. Zappi-Recordati (Rome, 1955) Italian
[132/56

see also Vorwohl (1972), Battaglini & Ricciardelli (1971, 1972) in
Bibliography

LUXEMBURG

Bee forage in forests. C. Kirpach *Luxemb. Bienenztg* 83: 21–23 (1968)
German

P The pollen spectrum of Luxemburg honeys. A. Maurizio *Luxemb.*
Bienenztg 86(1/2): 1–44 (1971) German   [495/72

R P The pollen spectrum of honeys from Luxembourg. A. Maurizio
Apidologie 2(3): 221–237 (1971) French   [496/72

MALTA
  Beekeeping in the Maltese islands. K. Stevens *Glean. Bee Cult.*
  97(2): 102–108, 119 (1969)   [74L/70
NETHERLANDS
  The comprehensive bee book. R. P. Groenveld (Groningen, 1961)
    Dutch   [403/62
NORWAY
  Manual of beekeeping. R. Lunder (Oslo, 2nd ed., 1947) Norwegian
POLAND
  Nectar-bearing plants. Z. Demianowicz (Warsaw, 1953), Polish
                                                                [155/55
  The honey yield of the main honey plants in Polish conditions.
  Z. Demianowicz *et al. Pszczel. Zesz. Nauk.* 4: 87–104 (1960);
  7: 95–111 (1963) Polish   [853, 854/64
  Characteristics of the nectar flow and honey yield in Poland
  (1950–1963). L. Bornus & M. Gromisz *Pszczel. Zesz. Nauk.*
  8: 1–19 (1964) Polish   [779/65
PORTUGAL
  P Pollen analysis of some honey samples. J. A. Martins d'Alte *Publ.*
    Inst. Bot. Sampaio No. 7: 179 pages (1951) Portuguese   [80/54
RUMANIA
  R *Plants providing bee forage.* A. Nyárády (Bucharest, 1958) Hungarian
                                                                [56/61
  R P Pollen analysis of the honeys of the Rumanian People's Republic.
    C. Pelimon *Annls Abeille* 3: 339–347 (1960) French   [333/61
SCANDINAVIA (general)
  K *Atlas of the distribution of vascular plants in N.W. Europe.* E. Hultén
    (Stockholm, 1950) Swedish   [26/53
  K *An introduction to a Scandinavian pollen flora.* G. Erdtman,
    B. Berglund & J. Praglowski (Stockholm, 1961) English   [7/63
  K *An introduction to a Scandinavian pollen flora.* Volume II.
    G. Erdtman, J. Praglowski & S. Nilsson (Stockholm, 1963)
                                                                [454/64
SPAIN
  Nectar plants. F. N. Howes [translation of *Plants and beekeeping*]
    (Barcelona, 1953) Spanish   [69/54
  P Study of the pollen spectrum of some Spanish honeys. J. Louveaux
    & P. Vergeron *Annls Abeille* 7: 329–347 (1964) French   [376/67
  R P Chemical composition and pollen spectrum of Spanish honeys.
    B. S. Pérez & A. T. Rodríguez *An. Bromat.* 22(4): 377–406 (1970)
    Spanish   [494/72

SWEDEN
  Bee plants. A. Hansson (Orebro, 1968) Swedish
R P Pollen analytical investigations on Swedish honey. R. Lunder
  Medd. Växtskyddanst. No. 45 (1945) Swedish

SWITZERLAND
  P Quantitative pollen analysis of Swiss honey. A. Maurizio *Schweiz.*
  Bienenztg 62(11): 642–646 (1939) German
  P Pollen analytical observations 6–9. A. Maurizio *Ber. Schweiz. bot.*
  Ges. 51: 77–95 (1940) German
  P Pollen analytical observations 10–12. A. Maurizio *Schweiz.*
  Bienenztg 65(11): 524–534 (1942) German
  P Types of honey occurring in Wallis [Valais]. A. Maurizio *Muri-*
  thienne (64): (1946/47)
R P Pollen analysis research on honey and pollen loads. A. Maurizio
  Beih. Schweiz. Bienenztg 2(18): 320–441 (1949*a*, 1949*b*) German
                                                                [176/50
R P Honey types in Italian Switzerland. A Maurizio *Riv. Svizz. Apic.*
  41: 20–26 (1958*a*) Italian   [279/59
  Nature and landscape. Vol. 1: Bee pasture. C. Brodbeck (Basel,
  1950) German   [62/50

TURKEY *see* under Asia

U.S.S.R.
R   *Bee plants.* M. M. Glukhov (Moscow, 6th ed., 1955) Russian [415/57
R   *The study of plants yielding nectar and pollen.* A. M. Kuliev (Moscow,
  1952) Russian   [94/54
  Honey sources and the development of beekeeping in the central regions
  of the U.S.S.R. A. M. Kovalev (Moscow, 1959) Russian [761L/65
  The utilization and development of bee forage in Latvian SSR.
  K. Balode (Riga, 1957) Latvian   [93/60
  Bee plants of the Ukraine. P. M. Bergegovii (Kiev, 1959) Ukrainian
  Bee plants of Tomsk Region. N. N. Kartashova (Tomsk, 1955)
  Russian   [583/63
  Bee forage sources of Northern Kazakhstan. S. G. Min'kov (Alma-Ata,
  1956) Russian   [125L/64
  Bee plants of Khabarovsk Region and their use. N. V. Usenko
  (Khabarovsk, 1956)   [805L/63

YUGOSLAVIA
  Migratory beekeeping. F. Šimić (Zagreb, 1957) Croatian   [158/60
R P The pollen spectrum of imported Yugoslav honeys. A. Maurizio
  Z. Bienenforsch. 5: 8–22 (1960*b*) German   [464/61

ASIA

No information is available for the following countries:

Bangladesh	Jordan	Singapore
Bhutan	Laos	Syria
Burma	Mongolian Republic	Thailand
Cambodia	Sabah	United Arab Emirates
Hong Kong	Saudi Arabia	Vietnam, N. and S.
Indonesia	Sikkim	Yemen

AFGHANISTAN

R   Honey sources in some tropical and subtropical countries. E. Crane
     Bee Wld 54: 177–186 (1973)   [875/73

CEYLON *see* Sri Lanka

CHINA (People's Republic)

R   A beekeeping journey in the People's Republic of China. H.
     Oschmann *Arch. Geflügelz. Kleintierk.* 10: 235–255 (1961) German
                                                                   [769/64
R P The pollen spectrum of Chinese honeys. E. Focke *Z. Bienenforsch.*
     9: 196–206 (1968) German   [807/69
     Entomological excerpts from southeastern China (Fukien Province).
     C. R. Kellogg (Claremont, Calif., 1968)   [476/68
     see also Taiwan

INDIA

    Code for conservation and maintenance of honey bees. Indian
    Standards Institution AFDC 11 (1017)
R   *Beekeeping in India.* S. Singh (New Delhi, 1962)   [291/64
  P A pollen analytical study of Indian honeys. P. K. K. Nair *J. Indian
     Bot. Soc.* 43: 179–191 (1964)
R P *A pollen study of major honey yielding plants of Mahabaleshwar hills.*
     G. B. Deodikar & C. V. Thakar (Poona, 1953)   [185/55
     Floral calendar of major and minor bee forage plants in Mahaba-
     leshwar Hills (Western Ghats). C. V. Thakar, V. V. Diwan, &
     S. R. Salvi *Indian Bee J.* 24:35–48 (1962)   [582/63
R   Physico-chemical composition of major unifloral honeys from
     Mahabaleshwar (Western Ghats). R. P. Phadke *Indian Bee J.*
     24: 59–65 (1962)   [907/64
     Bee flora of Karnatak and Kerala. Khadi & Village Industries
     Commission *Indian Bee J.* 21: 90–92 (1959)   [85/62
     Bee flora of Northern India. N. Kohli *Indian Bee J.* 20:113–118
     (1958)   [84/62

Some important honey plants of the Punjab (India). S. Singh.
Rep. Ia St. Apiar. for 1948: 34–42 (1949)
see also Kashmir, Nepal

IRAN

R   Honey sources in some tropical and subtropical countries. E. Crane
    Bee Wld 54: 177–186 (1973)   [875/73
R P Comparative qualitative studies on Iranian honeys. S. Gassparian &
    G. Vorwohl. *Apidologie* 5: 177–190 (1974) German

IRAQ

The land of Iraq and its bees. O. Morgenthaler *Südwestdtsch. Imker*
10: 304–310 (1958) German   [106/59

ISRAEL

Beekeeping. A. Ben-Niryah (Tel Aviv, 1949) Hebrew   [146/52
Acclimatization of woody plants in Israel. I. Gindel (Tel Aviv, 1956)
Hebrew   [348/57
Nectaries of honey plants in Israel. A. Fahn (Tel Aviv, 1948)
Hebrew   [65/52

JAPAN

The principal honey plants in Japan . . . I. Okada *Anim. Husb.* 12:
506–510 (1958) Japanese   [156/59
New beekeeping. Y. Tokuda (Tokyo, 1958) Japanese   [143/59
Beekeeping of the future. T. Inoue (Tokyo, 1951) Japanese   [8/60
P   Studies on the identification of pollens from bee plants. S.
    Hisamichi *Jap. J. Pharmacogn.* 13(2): 71–124 (1959) German
                                                                   [132/63
R K *Pollen grains of Japan.* M. Ikuse (Tokyo, 1956) Japanese & English
                                                                   [161/58

KASHMIR

Bees in Kashmir. A. M. Shah *Glean. Bee Cult.* 87: 424–425, 431
(1959)   [6/61

KOREA, N. and S.

Beekeeping. Sang Yul Cho & Young Tai Cho (Seoul, 1955) Korean
                                                                   [222/59
Melliferous plants . . . of Korea. Beekeepers' Association of the
Korean P.D.R. *XX Int. Beekeep. Congr.*: 286–293 (1965)

LEBANON

Beekeeping practised from immemorial times in Lebanon. R.
Yazbeck *Rev. franc. Apic.* 3: 597–599 (1953) French   [116/55

MALAYSIA

Honey sources in some tropical and subtropical countries. Part 2. E. Crane *Bee Wld* 56: *in press* (1975)

NEPAL

Beekeeping in Central Nepal. S. R. Wadhi *Indian Bee J.* 23: 54–55 (1961)

PAKISTAN

R A contribution to bee-flora of Pakistan. A Latif, A. Qayyum, & Manzoor-ul-Haq. *Pakist. J. sci. Res.* 10: 67–71 (1958)

R Honey sources in some tropical and subtropical countries. E. Crane *Bee Wld* 54: 177–186 (1973)   [875/73

PHILIPPINES

Beekeeping in the Philippines. R. A. Morse & F. M. Laigo *Fm Bull. Univ. Philippines Coll. Agric.* No. 27 (1968)

SRI LANKA

Some bee plants of Ceylon. A. W. Kannangara *Bee Wld* 21: 94–96 (1940)

Beekeeping for beginners. C. Drieberg & St. L. H. de Zylva *Bull. Dep. Agric. Ceylon* No. 92 (1941)

TAIWAN

Beekeeping. Fan Tsung Deh (Taipeh, 1959) Chinese   [282/61

Beekeeping in Formosa. Fan Tsung Deh *Bee Wld* 33: 150–151 (1952)
                                                          [260/53

TURKEY

Modern beekeeping. K. Senocak (Ankara, 1956) Turkish   [230/57

Studies on the honey bee and beekeeping in Turkey. F. S. Bodenheimer (Istanbul, 1942) Turkish & English

U.S.S.R. *see* under Europe

AFRICA

No information is available for the following countries:

Central African Republic	Ivory Coast	Réunion
Chad	Lesotho	Sahara
Dahomey	Liberia	Sierra Leone
Equatorial Guinea	Mali	Sudan
Gabon	Mauritania	Swaziland
Gambia	Niger	Togo
Ghana	Portuguese Guinea	Upper Volta

GENERAL

R   *Beekeeping in the tropics.* F. G. Smith (London, 1960)   [362/60
R P Pollen and spores of tropical Africa No. 16. Travaux et Documents
     de Géographie Tropicale (1974) French 282 pp.

ALGERIA

Beekeeping in French North Africa. C. Paradeau *XIV Int. Beekeep.*
Congr. Paper 19 (1951)

ANGOLA

Prospects and research in beekeeping in Angola. J. F. Rosário
Nunes & G. C. Tordo (Lisbon, 1960) Portuguese   [595/62

AZORES

Beekeeping in São Miguel. V. C. Paixão *Abelhas* 9: 34, 61, 90,
110–111, 122–123 (1966) Portuguese   [475L/68, 655L/69

BOTSWANA

R   Honey sources in some tropical and subtropical countries. E. Crane
    Bee Wld 54: 177–186 (1973)   [875/73

BURUNDI *see* under Congo (Brazzaville)

CAMEROON

R   Honey sources in some tropical and subtropical countries. E Crane
    Bee Wld 54: 177–186 (1973)   [875/73

CANARIES

Beekeeping in the Canary Islands. C. R. Templer *Bee Wld* 38: 184
(1957)   [256/59

CONGO (Brazzaville)

Beekeeping in the Belgian Congo and in Ruanda-Urundi. L. Dubois &
E. Collart (Brussels, 1950) French   [9/58

CONGO (Kinshasa) *see* Zaire

EGYPT, ARAB REPUBLIC OF

Pollinators of the chief sources of nectar and pollen grain plants, in
Egypt. A. K. Wafa & S. H. Ibrahim *Bull Soc. ent. Egypte*
43: 133–154 (1959)   [221/64

ETHIOPIA

R   Honey sources in some tropical and subtropical countries. E. Crane
    Bee Wld 54: 177–186 (1973)   [875/73

GUINEA

Life and behaviour of bees. M. Mathis (Paris, 1951) French   [7/53

KENYA

Kenya bee keeping pilot project (Oxfam). Kenya Ministry of Agriculture (1967, 1969)
See also references under Tanzania

LIBYAN ARAB REPUBLIC

Introduction of modern beekeeping to Cyrenaica. O. Brittan *Bee Craft* 37: 145–146 (1955); 38: 4–5 (1956) [214/58

MADAGASCAR *see* Malagasy Republic

MALAGASY REPUBLIC

Estimation of production and improvements which can be achieved in Madagascar. M. Douhet *XX Int. Beekeep. Congr.*: 638–646 (1965)
Beekeeping in Madagascar. L. Partiot *Abeille Fr.* 46: 17–19 (1967) French
'Mother of honey', the bee of Madagascar. P. Latique *Climats* (1954) French [155/57

MALAWI

Beeswax and honey production—the Nyasaland potential. J. S. Sheriff (Zomba, 1963) [69/64

MAURITIUS

R Honey sources in some tropical and subtropical countries. E. Crane *Bee Wld* 54: 177–186 (1973) [875/73

MOROCCO

Present state of apiculture in Morocco and its development programme. E. C. Barbier *XXII Int. Beekeep. Congr.*: 372–374 (1969)
Report on beekeeping in Morocco. P. Haccour *XX Int. Beekeep. Congr.*: 688–689 (1965)
R Honey sources in some tropical and subtropical countries. E. Crane *Bee Wld* 54: 177–186 (1973) [875/73

MOZAMBIQUE

Mocambique. R. Guy *S. Afr. Bee J.* 43(5): 11–19 (1971)
R Honey sources in some tropical and subtropical countries. E. Crane *Bee Wld* 54: 177–186 (1973) [875/73

NIGERIA

R Honey sources in some tropical and subtropical countries. E. Crane *Bee Wld* 54: 177–186 (1973) [875/73

RHODESIA

Bulletin on Eucalyptus *spp.* P. Papadopoulo (Salisbury, 1966) [146/6ε

Rhodesian indigenous trees, shrubs and wild flowers: cultivated crops,
   trees, flowers. P. Papadopoulo (Salisbury, 1966)   [147/68
Rhodesian indigenous trees and shrubs. P. Papadopoulo (Salisbury,
   1970)   [708L/72)

RWANDA *see also* under Congo (Brazzaville)
   Study of the natural beekeeping environment for setting up apiaries in
      Bugesera-Mayaga. R. Bauduin (Kigali: Ministère de l'Agriculture
      et de l'Élevage République Rwandaise, 1966) French   [654/72

SENEGAL
   Beekeeping in Senegal. J. Linder (Jerusalem, 1967) French   [65/69
   Production and commercialization of beekeeping products in
      Senegal. M. Douhet *Santé Abeille* (16): 124–130 (1970) French
                                                             [81L/72

SEYCHELLES
   Bee-keeping in the Seychelles. R. E. M. Silberrad *Br. Bee J.*
      97: 18–19 (1969)

SOMALIA
   R   Honey sources in some tropical and subtropical countries. E. Crane
         Bee Wld 54: 177–186 (1973)   [875/73

SOUTH AFRICA
   Beekeeping in South Africa. R. H. Anderson, B. Buys, & M. F.
      Johannsmeier *Bull. Dep. agric. tech. Serv.* No. 394 (1973)
   About South Africa honey flora. W. F. Crisp *S. Afr. Bee J.* 32(1): 1,
      3, 5; (3) 12–13; (5): 7 (1957)   [87/62
   Maculate aloes as nectar producers in the Transvaal and its
      implications. G. P. Beyleveld *S. Afr. Bee J.* 39(5): 10–12 (1967)
                                                              [362/68
   Nectar and pollen producing trees and plants of the Transvaal.
      G. P. Beyleveld *S. Afr. Bee J.* 40(4): 10–11 (1968)   [650L/70
   Detailed list of nectar and pollen producing plants of economic
      importance in the Transvaal. G. P. Beyleveld *S. Afr. Bee J.*
      40(6): 13–14 (1968)   [651L/70
   Eucalyptus species suitable for the production of honey. E. E. M.
      Loock *Bull. Dep. For., Pretoria* No. 46 (1970)   [920L/72

SOUTH WEST AFRICA
   Honey sources in some tropical and subtropical countries. Part 2.
      E. Crane *Bee Wld* 56: *in press* (1975)

TANZANIA
   P   *Bee botany of Tanganyika.* F. G. Smith (University of Aberdeen
         (*thesis*), 1956)   [268/56
   R   *Beekeeping in the tropics.* F. G. Smith (London, 1960)   [362/60

Bee botany in East Africa. F. G. Smith *E. Afr. agric. J.* 23: 119–126
(1957)   [95/60
see also Zanzibar

TUNISIA

Tunisian apiculture. K. Hicheri & M. Bouderbala *XXII Int.
Beekeep. Congr.* 440–443 (1969)
Life and behaviour of bees. M. Mathis (Paris, 1951) French   [7/53

UGANDA

A survey of beekeeping in Uganda. E. Roberts *Bee Wld* 52(2): 57–67
(1971)   [320/73

ZAIRE

Beekeeping in the Belgian Congo and in Ruanda-Urundi. L. Dubois &
E. Collart (Brussels, 1950) French   [9/58

ZAMBIA

Bark-hive beekeeping in Zambia. W. D. Holmes *Bull. Forest Dep.,
Repub. Zambia* No. 2 (1965)   [675/69
*Beekeeping in Northern Rhodesia: its prospects and recommendations
for its development.* F. G. Smith (N.D.B. thesis, 1959)

ZANZIBAR

R   Honey sources in some tropical and subtropical countries. E. Crane
Bee Wld 54: 177–186 (1973)   [875/73

NORTH AMERICA

CANADA

R P Pollen analysis of some Canadian honeys. J. Louveaux *Z. Bienen-
forsch.* 8: 195–202 (1966) German   [566/66
Nectar and pollen producing plants in Manitoba. A. V. Mitchener
Sci. Agric. 28: 475–480 (1948)   [202/51
See also under 'U.S.A., General'. (Manitoba is the only Province
from which special studies have been published.)

GREENLAND

The Greenland experiment. N. C. Bastholm *Tidsskr. Biavl* 84:
147–149, 163–164 (1950) Danish   [48/54

U.S.A.: GENERAL

R   *American honey plants.* F. C. Pellett (New York, 4th ed., 1947)
Honey plants manual. H. B. Lovell (Medina, Ohio, 1956)   [417/57
Let's talk about honey plants. H. B. Lovell. This series of monthly
articles in *Gleanings in Bee Culture* ran for eleven years: *see*
137/56; 416/57; 121, 232/59; 94, 370/61; 271/62; 148L–151L/68

Honey and pollen plants of the United States. E. Oertel *Circ. U.S. Dep. Agric.* No. 554 (1939)

Beekeeping regions in the U.S.A. An (editorial) monthly series which ran for six years in *Gleanings in Bee Culture* also covered honey plants; *see* 70/56; 274/57; 148/58; 223/59

U.S.A.: NORTH EAST STATES

Bee culture in Maine. O. B. Griffin. *Quart. Bull. Me Dep. Agric.* No. 17 (1918)

Honey and pollen plants of Massachusetts. F. R. Shaw *Spec. Circ. Mass. Ext. Serv.* No. 27 (1950)   [241/53

Woody honey plants for roadside planting in New Jersey. W. C. Morrison *Circ. N.J. Dep. Agric.* No. 403 (1957)   [48/59

Pennsylvania beekeeping. W. W. Clarke, Jr. & E. J. Anderson *Ext. Circ. Pa St. Coll. Agric.* No. 472 (1965)

U.S.A.: SOUTH EAST STATES

Some honey plants of Florida. L. E. Arnold *Bull. Fla Univ. agric. Exp. Sta.* No. 548 (1954)   [270/56

Honeybee plants of south Florida. J. F. Morton *Proc. Fla St. hort. Soc.* 77: 415–436 (1964)   [921L/71

R P The microscopic spectrum of some Florida honeys. G. Vorwohl. *Apidologie* 1: 233–269 (1970) German   [793/72

Beekeeping in Virginia. J. M. Grayson & J. O. Rowell *Bull. Va agric. Ext. Serv.* No. 178 (1950)   [112/53

U.S.A.: SOUTH CENTRAL STATES

Beekeeping in Alabama. C. C. Baskin & G. H. Blake *Circ. Auburn Univ. co-op. Ext. Serv.* P–64 (1967)

R   Nectar flow and pollen yield in southwestern Arkansas, 1945–1951. V. C. Thompson *Rep. Ser. Ark. agric. Exp. Sta.* No. 94 (1960)
[371/61

Honey and pollen plants of Louisiana. E. Oertel *Bees* 3(6): 4–5; (7): 7–8; (10): 6–8 (1949)   [126/52

R P A melissopalynological study of 54 Louisiana (U.S.A.) honeys. M. H. Lieux. *Rev. Palaeobot. Palynol.* 13: 95–124 (1972)

[474/74

U.S.A.: NORTH CENTRAL STATES

Illinois honey and pollen plants. V. G. Milum *Contr. Dep. Hort. Univ. Ill.* 3rd ed. (1957)   [294/59

Honey plants of Iowa. L. H. Pammel, C. M. King *et al.* (1930) *Bull. Ia geol. Survey* No. 7 (1930)

Basic beekeeping [Michigan]. E. C. Martin *Ext. Bull. Mich. State Univ.* E625 (1968)

Bee lines [North Carolina]. F. B. Meacham, H. E. Scott, & J. F. Greene *Circ. N.C. agric. Ext. Serv.* No. 334 (1967)
Wisconsin honey, production and marketing. P. D. Weber *Spec. Bull. Wis. Dep. Agric.* No. 61 (1956)   [224/59
Beekeeping in Wisconsin. H. J. Rahmlow *Circ. Wis. Univ. coop. Ext. Progm* No. 659 (1968)

U.S.A.: WEST STATES

Nectar and pollen plants of Colorado. W. T. Wilson, J. O. Moffett, & H. D. Harrington *Bull. Colo. Univ. Exp. Sta.* No. 503S (1958)
[295/59
Bee culture in Kansas. R. L. Parker *Bull. Kans. agric. Exp. Sta.* No. 357 (1953)   [259/54
A preliminary list of the honey-producing plants of Nebraska. C. E. Bessey *Bull. agric. Exp. Sta., Neb.* 7: 141–152 (1895)

U.S.A.: PACIFIC STATES

Beginning in beekeeping. W. Stanger *Leafl. Calif. agric. Exp. Sta. Ext. Serv.* No. 183 (1965)
Fundamentals of California beekeeping. W. Stanger (ed.) *Man. Univ. Calif. agric. Ext. Serv.* No. 42 (1971)
Nectar and pollen plants of Utah. W. P. Nye *Monograph Ser. Utah St. Univ.* 18(3) (1971)
Beekeeping [Washington]. C. A. Johansen *Publ. co-op. ext. Serv. Wash. St. Univ.* PNW 79 (1966)

CENTRAL AMERICA

No information is available for the following countries:

El Salvador
Haiti
Panama

GENERAL

Bee plants of tropical America. G. S. Ordetx (Havana, 1952) Spanish
[240/52
Beekeeping in the tropics. G. S. Ordetx & D. Espina Pérez (Mexico, D.F., 1966) Spanish   [53/67
Beekeeping in the Caribbean. H. Addleman *Glean. Bee Cult.* 90: 80–83, 122 (1962)

COSTA RICA

Beekeeping in Costa Rica. E. J. Dyce *Am. Bee J.* 93: 296–298 (1953)
[2/54

Beekeeping in Costa Rica. C. L. Calvo *Gac. Colmen.* 32: 144, 146, 148–149 (1970) Spanish   [655L/72

CUBA

Bee plants of Cuba. G. S. Ordetx *Rev. Agric., La Habana* 27: 5–160 (1944) Spanish   [27/52

DOMINICAN REPUBLIC

Study of the bee botany of the Dominican Republic. G. S. Ordetx (Santo Domingo, 1964) Spanish   [553/65

GUATEMALA

R   Beekeeping in Guatemala. C. W. Elmenhorst *Bee Wld* 33: 93–96 (1952)   [261/53

HONDURAS

Bee plants of Honduras. G. S. Ordetx (Tegucigalpa, 1963) Spanish
[334/64

MARTINIQUE

Beekeeping in Martinique. Anonymous *Apiculteur algér.* 4: 399–400, 419–420 (1957) French   [358/59

MEXICO

P. 311–344 (by G. S. Ordetx) from: *Modern beekeeping.* P. Aragon Leiva (1958) Spanish   [225/62

R   *Beekeeping encyclopaedia.* Vol. II. A. Wulfrath & J. J. Speck (Mexico, D.F., 1958) Spanish   [403/59

Beekeeping. J. F. Martinez López (Mérida, Yucatán, 1951) Spanish
[342/57

NICARAGUA

Report on the apicultural resources of Nicaragua. G. S. Ordetx (Managua, 1963) Spanish   [552/65

PUERTO RICO

Porto Rican beekeeping. E. F. Phillips *Bull. Porto Rico agric. exp. Stn.* No. 15 (1914)

WEST INDIES

Bees and beekeeping in Jamaica. G. P. Chapman, C. D. Frankson, & S. C. Jay *Bee Wld* 51: 173–181 (1970)   [356/72

Honey sources in some tropical and subtropical countries. Part 2. E. Crane *Bee Wld* 56: *in press* (1975)

Beekeeping at the Abbey Mount St. Benedict, Trinidad, West Indies. C. Cully *Am. Bee J.* 108:474 (1968)

SOUTH AMERICA

No information is available for the following countries:

Belize (British Honduras) French and Dutch Guiana
Bolivia Peru
Colombia

ARGENTINA

Economic prospects for apiculture in Argentina (the honey market).
A. A. Coscia *Instituto Nacional de Tecnologia Agropecuaria
Informe Tecnico* No. 13 (1963) Spanish
Contribution to the study of bee plants in the Argentine. M. Medici
Argentina Min. Agric. misc. Pub. No. 181 (1948) Spanish
Bee plants. A. de Ariño *Gac. Colm.* 16 (188): 2–10 (1954) Spanish
[294/55
Characterization of honeys from the province of Buenos Aires.
A. M. Gamero, L. G. Cornejo, & R. Tomasevich *Producción Anim.*
2(1): 1–34 (1969) Spanish [200/72
Characterization of honeys from the province of Buenos Aires
Zone II. A. M. Gamero, L. G. Cornejo, & E. Schminke *Producción
Anim.* 2(4): 133–158 (1971) Spanish
Characterization of honeys from the southern zone of the province
of Buenos Aires. A. M. Gamero, L. G. Cornejo, & E. Schminke
Producción Anim. 3(1): 1–26 (1972) Spanish

BRAZIL

Manual of beekeeping (pp. 180–212) ed. J. M. F. de Camargo (San
Paulo, 1972) Portuguese [322/73
R P Main pollen types found in honey samples. Preliminary note.
C. F. de O. Santos *Rev. Agric., Piracicaba* 36: 93–96 (1961)
Portuguese [908/64
Nectaries of some bee plants. C. F. de O. Santos *Universidade de
São Paulo (thesis),* (1954) Portuguese [418/57
P *Morphology and taxonomic value of pollen of the principal bee plants.*
C. F. de O. Santos (*Escola Superior de Agricultura Piracicaba
(thesis),* 1961) Portuguese [807/63
Scientific and practical beekeeping. W. E. Kerr & E. Amaral (San
Paulo, 1960) Portuguese [227/62
All depends on the flowers. W. E. Kerr *Guia Rural*: 23–25 (1966/67)
Portuguese
Bee plants of Curitiba. R. Braga *Bull. Univ. Paraná, Bot.* (2): 1–11
(1961) Portuguese [519/68
R P Pollen spectra of some Brazilian honeys. O. M. Barth *Z. Bienen-
forsch.* 9: 410–419 (1969) German [1012/70

R P Microscopic analysis of some honey samples. 1. Dominant pollers
O. M. Barth *Anais Acad. bras. Cienc.* 42: 351–366 (1970).
Portuguese; also (1970b, 1971a, 1971b) in Bibliography   [795/72

CHILE
Conditions and methods of management in South Chile. K.-H.
Franz *Bienenvater* 81: 47–49, 82–84 (1960) German   [162/61
K *Pollen and spores of Chile.* C. J. Heusser (Tucson, Arizona, 1971)
[923/71

ECUADOR
R   Honey sources in some tropical and subtropical countries. E. Crane
Bee Wld 54: 177–186 (1973)   [875/73

GUYANA
R   Honey sources in some tropical and subtropical countries. E. Crane
Bee Wld 54: 177–186 (1973)   [875/73

PARAGUAY
List of important nectar and pollen plants in Primavera, Paraguay.
W. Braun *Dusenia* 61–67 (1954) German   [276/55
Development of beekeeping in Paraguay. L. A. Ibarra *Gac. Colmen.*
30: 178–181 (1968) Spanish   [344L/70

URUGUAY
Bee forage of Uruguay *Mundo de las Abejas* (26): 30–31 (1974)
Spanish
Beekeeping in Uruguay. F. Rodriguez Ycart *Glean. Bee Cult.* 87:
711–715 (1959)   [117/61

VENEZUELA
R   Honey plants of Venezuela. M. Stejskal *Turrialba* 21(1):119–120
(1971) Spanish
Common plants of Venezuela. L. Schnee (Maracay, 1960) Spanish

OCEANIA

AUSTRALIA: NEW SOUTH WALES
The honey industry in New South Wales. R. E. Cooke-Yarborough
Rev. Market agr. Econ., Sydney 31: 3–37 (1963)   [772/64

AUSTRALIA: QUEENSLAND
The honey flora of South-Eastern Queensland. S. T. Blake & C. Roff
(Brisbane, 1958)   [59/60
Honey flora of coastal central Queensland. C. Roff *Adv. Leafl. Qd
Dep. Agric.* No. 733 (1963)

AUSTRALIA: TASMANIA
Beekeeping in Tasmania. T. D. Raphael & D. G. Cunningham (Hobart, 1954)   [270/55

AUSTRALIA: VICTORIA
Honey flora in Victoria. Victoria Dep. Agriculture (5th ed., 1949)

AUSTRALIA: WESTERN AUSTRALIA
Honey plants in Western Australia. F. G. Smith *Bull. Dep. Agric. West. Aust.* No. 3618 (1969)   [143/71
Commercial bee-keeping. 2. Honey flora of Western Australia. R. S. Coleman *Bull. Dep. Agric. W. Aust.* No. 3038 (1962)
к Trees of Western Australia, no. 1-109. C. A. Gardner Series in *J. Dep. Agric. W. Aust.* (1952-1966)   [293/53, 272/62, 675/66

FIJI
Rural industry: bee keeping. G. B. Gregory *Agric. J. Dep. Agric. Fiji* 31: 42-43 (1961)

HAWAII
Rehabilitation of the beekeeping industry in Hawaii. J. E. Eckert (Hawaii, 1951)   [158/53
Fundamentals of beekeeping in Hawaii. J. E. Eckert & H. A. Bess *Ext. Bull. Univ. Hawaii* No. 55 (1952)
A brief survey of Hawaiian beekeeping. E. F. Phillips *Bull. U.S. Bur. Ent.* No. 75: 43-58 (1909)

NEW ZEALAND
Handbook of New Zealand nectar and pollen sources. R. S. Walsh *Upper Hutt, N.Z.* (1967)   [731/68
P Pollen in honey and bee loads. W. F. Harris & D. W. Filmer *N.Z.J. Sci. Tech.* A 30: 178-187 (1948)

APPENDIX 1

Table 1.3/1.
Potential honey yields of 200 honey plants

This table gives the estimated 'honey potential' of 200 flowering plants measured as the maximum quantity of honey (in kilograms) that could be obtained in the course of a season from 1 hectare of land (or in pounds obtainable per acre, which is nearly equivalent). Optimal growing conditions are assumed, and an adequate force of foraging bees to collect all the nectar secreted. (*See* Section 1.3, pages 22-24.)

Class 1 = 0–25 kg/ha   Class 4 = 101–200 kg/ha
Class 2 = 26–50   Class 5 = 201–500
Class 3 = 51–100   Class 6 = over 500

Many of the entries are based on sources in the U.S.S.R., especially
M. M. Glukhov (1955), N. F. Fedosov (1955), A. M. Kuliev (1952);
Poland, especially Z. Demianowicz (1960, 1963); Rumania, especially
A. Nyárády (1958).

Plants in Appendix 1 and Appendix 2 are listed in alphabetical order
of family, and within each family in alphabetical order of genus. Where
possible, authorities for plant names have been obtained from:

Flora Europaea ed. T. G. Tutin *et al.* (Cambridge: University Press)
    Vol. 1 (1964); Vol. 2 (1968)
Manual of cultivated plants L. H. Bailey (New York: Macmillan) 1949,
    reprinted 1971

I am indebted to the staff of the Royal Botanic Gardens, Kew, for help
with plants not described in any publications accessible to me. Names
whose authority could not be established are quoted as in the source
publication.

	Class						
Plant species	1	2	3	4	5	6	*Common name*
Aceraceae							
Acer campestre L.						x	common maple
Acer platanoides L.				x			Norway maple
Araliaceae							
Hedera helix L.					x		common ivy
Asclepiadaceae							
Asclepias syriaca L.						x	common milkweed
Boraginaceae							
Anchusa officinalis L.			x			x	bugloss
Borago officinalis L.				x			borage
Cynoglossum officinale L.				x			hound's tongue
Echium vulgare L.						x	viper's bugloss
Symphytum asperum Lepech.					x		rough comfrey
Symphytum caucasicum				x			Caucasian comfrey
Caprifoliaceae							
Lonicera caucasica	x						Caucasian honeysuckle
Lonicera fragrantissima	x						yellow jasmine
Lind. & Paxt.							honeysuckle

Plant species	Class						Common name
	1	2	3	4	5	6	
Lonicera iberica	x	.	.	.	.	.	Iberian honeysuckle
Lonicera xylosteum L.	x	.	.	.	.	.	fly honeysuckle
Sambucus ebulus L.	.	x	.	.	.	.	danewort
Viburnum opulus L.	.	x	.	.	.	.	guelder rose

Compositae

Arctium lappa L.	.	.	.	.	x	.	great burdock
Aster amellus	.	x	.	.	.	.	Italian star-wort
Aster tripolium L.	.	x	.	.	.	.	sea aster
Carduus hamulosus	.	x	.	.	.	.	hooked thistle
Centaurea cyanus L.	.	.	x	.	.	.	cornflower
Centaurea iberica	.	.	x	.	.	.	Iberian cornflower
Centaurea jacea L.	.	.	.	x	.	.	brown-rayed knapweed
Cichorium intybus L.	.	.	x	.	.	.	chicory
Cirsium ciliatum	.	.	x	.	.	.	hairy thistle
Echinops cummutatus	.	.	.	x	.	.	
Echinops sphaerocephalus L.	.	.	.	.	x	.	globe-thistle
Helianthus annuus L.	.	x	.	.	.	.	common sunflower
Solidago L.	.	.	.	x	.	.	golden rod
Solidago gigantea Aiton	.	.	.	.	x	.	
Taraxacum officinale Weber	.	.	.	x	.	.	dandelion

Cruciferae

Barbarea vulgaris R. Br.	.	.	x	.	.	.	winter cress; yellow rocket
Brassica napus L. subsp. *oleifera* DC.	.	.	.	x	x	.	rape, coleseed
Brassica oleracea L.	.	.	x	.	.	.	wild cabbage
Eruca vesicaria subsp. *sativa* Miller	x	.	.	.	.	.	salad rocket
Sinapis alba L.	x	.	x	.	.	.	white mustard
Sinapis arvensis L.	.	.	.	x	x	.	charlock
Isatis tinctoria L.	.	.	.	x	.	.	woad
Raphanus raphanistrum L.	.	x	.	.	.	.	wild radish

Cucurbitaceae

Citrullus vulgaris Schrad.	x	.	.	.	.	.	water melon
Cucumis melo L.	.	x	.	.	.	.	melon
Cucumis sativus L.	.	x	.	.	.	.	cucumber
Cucurbita maxima Duchesne	.	.	x	.	.	.	giant pumpkin
Cucurbita pepo L.	.	x	.	.	.	.	pumpkin/squash/marrow

H—3

Plant species	Class 1 2 3 4 5 6	Common name
Dipsacaceae		
Cephalaria caucasica	 x	
Dipsacus fullonum L. ssp. *sylvestris* Hudson	. . . x . .	wild teasel; fuller's teasel
Dipsacus strigosus	. . . x . .	
Scabiosa bipinnata	. . . x . .	
Scabiosa caucasica Bieb.	. . . . x .	Caucasian scabious
Ericaceae		
Calluna vulgaris (L.) Hull	. . . x . .	ling heather
Fagaceae		
Castanea sativa Miller	. x	sweet chestnut
Geraniaceae		
Geranium pratense L.	. . x . . .	meadow crane's bill
Grossulariaceae		
Ribes rubrum L.	. . . x . .	red currant
Ribes uva-crispa L.	. . x . . .	gooseberry
Hippocastanaceae		
Aesculus carnea Hayne	. . . x . .	red horse chestnut
Hydrangeaceae		
Philadelphus caucasicus	. . x . . .	Caucasian syringa
Hydrophyllaceae		
Phacelia tanacetifolia Bentham	. . . . x x	phacelia
Labiatae		
Ajuga genevensis L.	. . . . x .	erect bugle; gout ivy
Ajuga orientalis	. . . x . .	oriental bugle
Ajuga reptans L.	. . . . x .	bugle
Ballota nigra L.	. . x . . .	black horehound
Ballota ruderalis	. . . x . .	
Clinopodium vulgare L.	. x	wild basil
Dracocephalum L.	. . . . x .	dragon's head
Dracocephalum moldavicum L.	. . . . x .	Moldavian balm
Galeopsis angustifolia Hoffm.	x	narrow-leaved hemp-nettle

Plant species	Class 1 2 3 4 5 6	Common name
Glechoma hederacea L.	. . x . . .	ground ivy
Hyssopus officinalis L.	. . . x x .	hyssop
Lamium album L.	. . . x . x	white dead-nettle
Lamium maculatum L.	. . . x . .	spotted dead-nettle
Lamium purpureum L.	. x	red dead-nettle
Lavandula	. . . x . .	lavender
Leonurus cardiaca L.	. . . . x .	motherwort
Marrubium vulgare L.	. x . . x .	white horehound
Melissa officinalis L.	. x . x . .	balm
Mentha longifolia (L.) Hudson	. . . . x .	horse-mint
Nepeta cataria L.	. . . x x .	catmint
Nepeta grandiflora	. . . x x .	
Nepeta mussinii Sprengel	. . . . x .	
Nepeta nuda	. . . x . .	
Nepeta transcaucasica	. . . . x .	
Nepeta zangezura	. . . . x .	
Ocimum basilicum L.	. . x . . .	sweet basil
Origanum vulgare L.	. . . x . .	marjoram
Perilla L.	. x	perilla
Phlomis pungens	. . . x . .	
Phlomis tuberosa	. . . x . .	
Prunella laciniata (L.)L.	. . x . . .	
Prunella vulgaris L.	. . . . x .	self-heal
Rosmarinus officinalis L.	. . . x . .	rosemary
Salvia nemorosa L.	. . . . x .	woodland sage
Salvia officinalis L.	. . . . x x	sage
Salvia sclarea L.	. . . x . .	clary
Salvia verbenaca L.	. . . x . .	English clary; wild sage
Salvia verticillata L.	 x	whorl-flowered clary
Salvia virgata	. . . x . .	
Satureia vulgaris	. . x . . .	savory
Stachys annua (L.)L.	. . . x . .	woundwort
Stachys germanica L.	. . . . x .	downy woundwort
Stachys iberica	. . x . . .	Iberian hedge-nettle
Stachys olympica Poir.	. . . . x .	lamb's ear
Stachys palustris	. . . . x .	marsh woundwort
Stachys sylvatica L.	. . . x . .	hedge woundwort
Teucrium chamaedrys L.	. x	wall germander
Teucrium orientale	. . x . . .	oriental germander
Teucrium polium L.	. . . x . .	pennyroyal germander

Plant species	Class						Common name
	1	2	3	4	5	6	
Thymus kotschyanus	.	.	.	.	x	.	
Thymus pulegioides L.	.	.	.	x	.	.	larger wild thyme
Thymus rariflorus	.	.	.	.	x	.	
Thymus serpyllum L.	.	.	.	x	.	.	wild thyme
Thymus vulgaris L.	.	.	.	.	.	x	common thyme
Ziziphora tenuior	.	x	.	.	.	.	
Leguminosae							
Astragalus alpinus L.	.	x	.	.	.	.	alpine milk-vetch
Astragalus cancellatus	x	.	.	.	.	.	
Astragalus stevenianus	.	.	x	.	.	.	
Caragana arborescens Lam.	.	.	.	.	.	x	pea tree
Cercis siliquastrum L.	.	.	x	.	.	.	Judas tree
Cicer arietinum L.	x	.	.	.	.	.	chick pea
Gleditschia Clayton	.	.	.	.	x	.	honey-locust
Gleditschia triacanthos	.	.	x	.	.	.	honey-locust
Lathyrus digitatus (Bieb.) Fiori	.	.	.	.	x	.	
Lathyrus hirsutus L.	.	x	.	.	.	.	hairy vetchling
Lathyrus miniatus	.	.	.	x	.	.	
Lathyrus pallescens (Bieb.) C. Koch	.	.	.	x	.	.	
Lathyrus sativus L.	.	x	.	.	.	.	chickling-vetch
Lotus corniculatus L.	.	x	.	x	.	.	birdsfoot-trefoil
Medicago sativa L.	.	.	.	.	x	.	lucerne; alfalfa
Melilotus alba Medicus	.	.	.	.	x	x	white melilot
Melilotus officinalis (L.) Pallas	.	.	.	x	.	.	common melilot
Onobrychis arenaria (Kit.) DC	.	.	.	x	.	.	
Onobrychis cyri Grossh.	.	.	.	x	.	.	
Onobrychis radiata (Desf.) Bieb.	.	x	.	.	.	.	
Onobrychis viciifolia Scop.	.	.	x	x	.	.	sainfoin
Robinia pseudoacacia L.	.	.	.	.	.	x	acacia
Sophora japonica L.	.	.	.	x	.	.	sophora
Trifolium campestre Schreber	.	x	.	.	.	.	hop trefoil
Trifolium hybridum L.	.	.	.	x	.	.	alsike clover
Trifolium pratense L.	.	.	.	.	x	.	red clover
Trifolium repens L.	.	.	x	x	.	.	white clover
Trifolium resupinatum L.	.	.	.	x	.	.	Persian clover
Vicia cracca L.	.	.	x	.	.	.	tufted vetch
Vicia faba L.	.	.	x	.	.	.	field bean
Wistaria sinensis (Sims) Sweet	.	x	.	.	.	.	Chinese kidney-bean tree

Plant species	Class 1 2 3 4 5 6	Common name
Liliaceae		
Allium cepa L.	. . x . . .	onion
Linaceae		
Linum bienne Miller	x	pale flax
Lythraceae		
Lythrum salicaria L.	. . . . x .	purple loosestrife
Malvaceae		
Althaea officinalis L.	. . x x . .	marsh mallow
Althaea rosea (L.)Cav.	. . . x . .	hollyhock
Gossypium	. . . x . .	cotton
Gossypium hirsutum L.	x	upland cotton
Hibiscus cannabinus L.	. x	Deccan hemp
Lavatera trimestris L.	. . x . . .	
Malva sylvestris L.	. x	common mallow
Onagraceae		
Chamaenerion angustifolium (L.) Scop.	. . . . x x	rosebay willowherb
Epilobium hirsutum L.	. . x . . .	great hairy willowherb
Pedaliaceae		
Sesamum orientale L.	. x	sesame
Pinaceae		
Picea abies (L.) Karsten	. . . x x .	Norway spruce
Plumbaginaceae		
Limonium vulgare Miller	. x	sea lavender
Statice gmelini	. x	
Polemoniaceae		
Polemonium caeruleum L.	. . x . . .	Jacob's ladder
Polygonaceae		
Fagopyrum esculentum Moench.	. . x . x .	buckwheat
Polygonum baldschuanicum Regel	. . . x . .	
Rhamnaceae		
Frangula alnus Miller	. . x . . .	alder buckthorn

Plant species	Class 1 2 3 4 5 6	Common name
Rosaceae		
Crataegus L.	. x	hawthorn
Crataegus kyrtostyla Fingerh.	. x	
Cydonia oblonga Miller	x	quince
Malus orientalis	x	
Malus sylvestris Miller	. x	crab apple
Prunus dulcis (Miller) D. A. Webb = *Prunus amygdalus* Batsch	x	almond
Prunus armeniaca L.	. x	apricot
Prunus avium L.	. x	wild cherry
Prunus cerasifera Ehrh.	. x	cherry-plum
Prunus cerasus L.	. x	sour cherry
Prunus domestica L.	. x	plum
Prunus spinosa L.	x	blackthorn
Pyrus caucasica Fedorov	x	Caucasian pear
Pyrus communis L.	x	pear
Rubus caesius L.	x	dewberry
Rubus idaeus L.	. . x . . .	raspberry
Rubus sanguineus Friv.	. . x . . .	
Sorbus aucuparia L.	. x	rowan; mountain ash
Sorbus graeca (Spach) Kotschy	. x	
Rutaceae		
Ruta graveolens L.	. . . . x .	rue
Salicaceae		
Salix L.	. . . x . .	willow
Saxifragaceae		
Deutzia scabra Thunb.	x	
Scrophulariaceae		
Digitalis purpurea L.	. . . . x .	foxglove
Scrophularia nodosa L.	 x	figwort
Solanaceae		
Nicotiana rustica L.	. x	tobacco
Nicotiana tabacum L.	. x	tobacco
Tiliaceae		
Tilia caucasica Rupr.	 x	
Tilia cordata Miller	. . . x . x	small-leaved lime

Plant species	Class 1 2 3 4 5 6	Common name
Umbelliferae		
Coriandrum sativum L.	. . . . x .	coriander
Eryngium campestre L.	. . . x . .	field eryngo
Heracleum sphondylium L.	. . . x . .	hogweed
Verbenaceae		
Vitex negundo incisa Clarke	 x	
Vitaceae		
Parthenocissus tricuspidata (Siebold & Zucc.) Planchon = *Ampelopsis veitchii* Lynch	. . . . x .	

APPENDIX 2

Table 1.4/1.
Descriptions of honey and world distribution, for 150 important plant genera and species

Plants known to be major sources of the world's honey are arranged in alphabetical order of plant family (*see* note to Appendix 1, pages 47–48); *see also* Section 1.4, pages 24–25.

The details given are in two parts:

(1) Botanical name [common name]; type of plant, world distribution and importance. WHCR refers to entries in World Honey Crop Reports (*Bee World*, 1966–1974). HP 1 to HP 6 indicates the 'honey potential' class as set out in Table 1.3/1, pages 47–55.

(2) Characteristics of the honey (in liquid form): colour, density, viscosity (body), flavour and aroma, natural granulation, and any special features.

Acanthaceae
Dyschoriste Nees spp.—annuals/perennials in tropical African wooded grassland
  honey very light; fine flavour
Hypoestes Soland spp.—shrubs native to tropical and southern Africa and Madagascar, in wooded grassland
  honey very light, fine flavour

Aceraceae
Acer L. spp. [maple, etc.] in Europe, especially *A. pseudoplatanus* [sycamore]—trees common in most cool temperate regions of the

world as indigenous and/or introduced species; WHCR England and Wales; HP 4/6

honey pale amber, sometimes greenish; unremarkable flavour and aroma; slow, fine granulation

Amaryllidaceae

Agave L. spp. [agave]; *A. sisalana* Perrine = sisal—tropical and sub-tropical Africa, America, Asia; some species cultivated; plant flowers after 6–100 years, in various regions yielding much honey; plant then dies

honey dark (or light), poor quality, unpleasant flavour

Anacardiaceae

Anacardium occidentale L. [cashew nut]—tree native to W. Indies, cultivated also in India and other tropical areas

honey characteristics not known

Lannea spp.—trees and shrubs in open forests and woodlands of tropical Africa

honey white to light amber, excellent flavour

Mangifera indica L. [mango]—tree from S.E. Asia; cultivated also in India, Africa, C. America, etc.

honey amber, dense; delicious flavour; heavy early morning nectar flow

Pistacia vera L. [pistachio]—tree grown in Mauritius and elsewhere in the tropics, and in the Mediterranean region

honey dark, second grade

Rhus L. spp. [sumac, etc.]—shrubs/trees distributed over both N. and S. temperate regions

honey colour varies according to species; good quality; some have bitter flavour, but this disappears after storage

Aquifoliaceae

Ilex glabra [gallberry] (also other spp., holly, etc.)—shrub in south-east U.S.A.; other spp. elsewhere

honey very light, heavy body; mild delicious flavour; slow to granulate; other spp. with 'pleasant twang'

Asclepiadaceae

Asclepias syriaca and spp. [milkweed, silkweed]—perennials, native to America and S. Africa, naturalized in temperate regions of U.S.S.R. and elsewhere; HP 6

honey very light; good quality; said to retain the aroma of the flowers; may take years to granulate

Balsaminaceae
Impatiens glandulifera Royle and spp. [balsam (Himalayan and others)]—
annuals in tropical Asia and Africa and elsewhere, including Europe,
U.S.S.R.
honey light, sweet, no noticeable aroma

Berberidaceae
Berberis L. spp. [barberry]—shrubs widespread in north and south
temperate zones, and mountains elsewhere
honey light amber to amber; good flavour (said to be similar to
that of *Ranunculus*)

Bombacaceae
Durio zibethinus Murr. [duryon]—tree native to Indo-Malaya region
honey characteristics not known

Boraginaceae
Borago officinalis L. [borage]—annual/perennial of European provenance,
now widely distributed in temperate regions from U.S.S.R. to New
Zealand; HP 4
honey whitish with yellow-grey tint
Echium lycopsis L. [purple viper's bugloss, Patterson's curse, Salvation
Jane]—biennial, distribution as *E. vulgare*
honey as *E. vulgare*
Echium vulgare L. [viper's bugloss, blueweed]—biennial European weed
naturalized widely in temperate regions, including rest of U.S.S.R.,
N. Africa, Australia, New Zealand; HP 6
honey white to light golden, delicate flavour

Cactaceae
Opuntia Miller (200 spp.) [prickly pear, etc.]—large shrubby cacti whose
natural range extends from Utah to Patagonia, and some spp. have
become troublesome weeds in Australia, etc.
honey of *O. engelmanni* Salm.-Dyck. light amber, high viscosity, strong
flavour; said to granulate in large crystals in clear liquid

Combretaceae
Combretum trothae Engl. and Diels.; *guienzii* Sond.; *farinosum* H.B.K.
[chupamiel, Spanish]; also 250 other spp. of shrubs in tropical
African scrub; C. America, especially Honduras; flowers + honey of
C. farinosum used in folk medicine for eye affections
honey of *C. trothae* in Tanzania light amber; excellent quality

Compositae

Arctotheca calendula (L.) H. Levyns [cape weed]—herb widespread in agricultural areas in south-west of W. Australia

honey light amber, rather yellow; medium flavour

Aster L. spp. [aster]—mostly perennials, many species in N. America and other temperate regions; HP 2

honey light to medium amber; characteristic aroma; *A. tripolium* L. has a salty flavour

Bidens L. spp. [Spanish needle, etc.]—annuals/perennials, widely distributed in warm and temperate regions

honey amber, dense; strong aromatic flavour, sometimes disagreeable

Calea urticifolia (Miller) DC. and spp. [jalacate (Spanish)]—tree in Central America

honey excellent

Carduus spp. [thistles]—annuals/biennials/perennials; see *Cirsium*; HP 2

honey see *Cirsium*

Carthamus tinctorius L. [safflower]—annual, native of Asia, now widely grown as an oil-crop there, and in N. America, etc.

honey dark, strong unpleasant flavour and aroma

Centaurea L. (600 spp.) [knapweed, cornflower, etc.]—chiefly of Mediterranean origin, now widely distributed throughout the temperate zones; WHCR Israel, U.S.S.R.; HP 3/4

honey light amber, thin; some with sharp flavour and characteristic bitter after-taste; soft granulation

Cirsium Miller spp. [thistles]—annuals/biennials/perennials, European weeds widely naturalized in uncultivated areas of cool temperate regions, including e.g. U.S.S.R. and New Zealand; WHCR Argentina; HP 3

honey light; good flavour, extra sweet, no pronounced aroma

Cynara cardunculus L. [cardoon]—thistle-like plant of Mediterranean region, widespread in Argentina, etc.; WHCR Argentina

honey light, mild, good quality

Guizotia abyssinica Cass. [niger]—annual tropical African oilseed crop, grown also in India and elsewhere

honey characteristics not known

Helianthus annuus L. [sunflower]—annual of N. American origin, widely grown as oilseed crop in warm areas of all continents; one of the few plants said to be worth growing for honey production; WHCR Bulgaria, Morocco, U.S.S.R., Argentina; HP 2

honey variously reported as: (*a*) egg-yolk yellow; characteristic flavour, strongly aromatic, reminiscent of dandelion; (*b*) dark, and unpleasant flavour; (*c*) mild

Senecio jacobaea L. [ragwort], also 1 300 other spp.—perennial weed of

neglected land throughout many parts of temperate zones, including New Zealand; *S. palustris* (L.) Hooker in N. European marshes
honey light amber, often with bitter flavour and strong rank aroma
Solidago spp. [golden rod]—very widely distributed in Europe, and in N. America where nearly all 80 spp. are native; HP 4/5
honey deep golden yellow, thick; pronounced flavour, strong aroma; granulates fairly quickly
Taraxacum officinale Weber [dandelion]—widespread perennial weed in cool temperate regions, especially N. America, Central Europe, U.S.S.R., but also other temperate regions; so dense as to provide an important honey flow in many areas; WHCR Finland, Sweden, Argentina; HP 4
honey intense golden yellow; sharp flavour, pronounced aroma; coarse hard rapid granulation
Vernonia Schreb. spp. [ironweed]—annuals/perennials in tropical upland areas of Africa, S. America, Asia, usually wooded grassland
honey very light, fine flavour
Viguiera grammatoglossa DC. [acahual]—herb yielding one of Mexico's export honeys
honey light amber, exquisite flavour, butter-like consistency
Viguiera helianthoides H.B.K. [romerillo (Spanish), tah (Maya)]—herb of C. America, in subtropical lowland rain forest, also dry areas; important in Yucatán
honey light, medium quality; pronounced flavour and aroma; granulates rapidly

Convolvulaceae
Ipomoea L. spp. [campanilla, bell-flower, morning glory, aguinaldo (Spanish)]—climbers in C. America (especially Cuba) and elsewhere in warmer parts of the world; from Yucatán (Mexico) *I. sidaefolia*
honey pearly white, often thin-bodied, distinct but pleasant flavour and aroma
Rivea corymbosa (L.) Hall. [aguinaldo blanco (Spanish)]—perennial in tropical and subtropical America
honey water-white (low mineral content), said to be the lightest honey in the world; delicate flavour

Cruciferae
Brassica juncea (L.) Czern. [Indian mustard, Chinese mustard]—tropical annual oilseed crop, especially in U.S.S.R., China, India, Pakistan
honey: see *Brassica napus*
Brassica napus L. subsp. *oleifera* DC. [summer/winter/swede rape]—

very widely grown annual oilseed crop in temperate zones of the world; WHCR Germany, Netherlands, Poland, Scandinavia; HP 4/5

honey water-white; flavour sweet, with aroma varying from almost none to rather unpleasant characteristic scent; granulates very rapidly, sometimes in the combs. Some other brassicas give darker honeys (amber), of unpleasant flavour

Brassica rapa (B. campestris L.) var. *oleifera* [(winter) rape = turnip rape] —widely grown biennial oilseed crop in (temperate) zones with a short growing season, including U.S.S.R., Canada, China; WHCR England and Wales, Finland, Sweden, Japan; HP 3

honey: see *Brassica napus*

Brassica rapa (B. campestris L.) var. *sarson* [sarson]—tropical oilseed crop, in India, Pakistan, etc.

honey: see *Brassica napus*

Brassica rapa L. *(B. campestris* L.) var. *toria* [toria]—tropical oilseed crop, in China, India, Pakistan, etc.

honey: see *Brassica napus*

Sinapis alba L., *Brassica nigra* (L.) Koch [white, black mustard]— widespread crop in temperate zones; HP 1/3

honey: see *Brassica napus*

Sinapis arvensis L. [charlock, wild mustard]—annual widespread weed throughout the temperate zones of the world; HP 4/5

honey: see *Brassica napus*

Cucurbitaceae

Cucumis L. spp. [cucumber, melon, etc.]—climbers widely cultivated in warm and temperate regions; WHCR Morocco; HP 2

honey from cucumber light amber

Cunoniaceae

Weinmannia racemosa L. [kamahi]—tree in New Zealand (only); WHCR New Zealand

honey extra light amber, bitter flavour dominant in blends, and increases with age; granulation coarse and uneven

Weinmannia silvicola Sol. ex A. Cunn. [towai, tawhero]—tree in New Zealand (only)

honey extra light amber, similar to *W. racemosa* but flavour considerably better

Cyrillaceae

Cyrilla racemiflora L., *Cliftonia ligustrina* [ti-ti]—shrubs in eastern N. America to West Indies

honey amber, strong harsh flavour

Ebenaceae
Diospyros virginiana L. and spp. [persimmon etc.]—trees/shrubs in tropical and cooler parts of Asia, N.C. and S. America, some spp. cultivated
> *honey* characteristics not known

Ericaceae
Calluna vulgaris (L.) Hull, only species [ling heather]—low shrub confined to moors and open woodlands with acid soils in N. and W. Europe, western Siberia, and extreme north-east of N. America, also New Zealand; WHCR Denmark, Netherlands, Norway, Poland, Portugal, Sweden, U.S.S.R.; HP 4
> *honey* light, dark or reddish brown; pronounced and characteristic flavour and aroma; thixotropic

Erica L. spp. [heaths] e.g. *E. carnea* in Alps, *E. arborea*, *E. umbellata* L., *E. vagans* L. in Mediterranean region; shrubs in Africa and Europe (470 of the 500 spp. native to S. Africa); WHCR Portugal
> *honeys* light/dark amber; some with strong or tart flavour

Erica cinerea L. [bell heather]—low shrub in heathland near western seaboard of Europe
> *honey* brownish 'port-wine' colour; characteristic flavour

Oxydendrum arboreum DC., only species [sourwood]—tree in E. and S.E. United States
> *honey* very light, delicious flavour and aroma; granulates fairly rapidly

Rhododendron ferrugineum L., *R. hirsutum* L. [Alpine rose]—shrubs growing throughout the higher Alps, also Pyrenees and Apennines
> *honey* has high enzyme content; very light and mild

Rhododendron ponticum L. [rhododendron]—shrub in Asia Minor, introduced elsewhere
> *honey* contains andromedotoxin when fresh, poisonous to humans, and a cause of illness in Xenophon's troops in 40 BC

Vaccinium L. spp. [blueberry, huckleberry, cranberry, whortleberry, bilberry etc.]—shrubs in boggy ground in northern parts of N. America, Europe, and Asia; some species cultivated
> *honey* white (some spp. amber), good body, mild flavour; may granulate hard

Eucryphiaceae
Eucryphia Cav. spp. [leatherwood]—tree in Australia (Tasmania) only
> *honey* light amber, distinctive flavour (reminiscent of almond), considered by some to be the finest of all honeys; granulated honey hard, and can be packed in small paper-wrapped cubes

Euphorbiaceae

Croton L. (600 spp.) [croton]—trees/shrubs native to equatorial cloud
forest, widespread in warm parts of the world; several spp.
cultivated for oil, bark, etc.; WHCR Kenya, Tanzania
> *honey* of some species dark amber, strong flavoured

Ricinus communis L. [castor-oil plant]—shrub native to tropical Africa,
widely grown in India, also C. America
> *honey* dark amber; strong flavour; probably from extrafloral nectaries

Fagaceae

Castanea sativa Miller [sweet chestnut]—native to Mediterranean area,
introduced e.g. W. and S. Europe, Korea; in N. America other
species; WHCR Belgium; HP 2
> *honey* light/dark amber, often reddish; flavour pronounced and may be
> bitter (N. America); aroma reminiscent of the flowers; granulates
> slowly and finely

Quercus L. spp. [oak]—trees widespread in N. temperate regions, India,
Malaya, Pacific coasts, etc.
> *honey:* important source of honeydew honey (see under Pinaceae)

Hippocastanaceae

Aesculus L. spp. [horse chestnut]—trees/shrubs, several species in
Europe, Asia, and N. America (buckeye) yield honey; some species
cause poisoning in cattle and bees; WHCR Japan; HP 4
> *honey* reported light, and also very dark; dense, medium quality

Hydrophyllaceae

Phacelia Juss. (*tanacetifolia* Bentham and spp.) [phacelia]—annual
native to California, introduced to Europe in 1832; rapid growth and
high nectar yield make it a worthwhile catch crop (sown for its
honey yield) in U.S.S.R. and neighbouring countries; WHCR
U.S.S.R.
> *honey* amber, flows freely, granulates quickly

Labiatae

Dracocephalum moldavicum L. [Moldavian balm]—annual from E.
Siberia, introduced to Europe in 1600s; HP 5
> *honey* characteristics not known

Lavandula spica L. [lavender]—shrubs in and around the whole
Mediterranean region and in U.S.S.R., etc.; the hybrid *L. spica* ×
L. latifolia (lavandin, French) grown for perfumery; WHCR
Bulgaria, France, Spain, Morocco
> *honey* golden/dark amber; flavour very highly regarded; may have high
> water and sucrose contents; granulated honey almost as smooth as
> butter; lavandin honey very light

Lavandula stoechas L. [French lavender]—shrub in S.W. Europe, widely introduced e.g. Australia; WHCR France, Portugal, Morocco, Argentina

honey: see *L. spica*

Mentha L. spp. [mint] including *spicata*, *piperita*, and (N.Z.) *viridis*, *pulegium*—perennials widely distributed over most of the world outside the tropics; WHCR Bulgaria; HP 5

honey amber (*M. viridis* light amber), distinctive sharp aroma; fine granulation; honey from *M. aquatica* unusual in containing vitamin C (1·6 mg/g)

Nepeta spp. [catmint, etc.]—annuals/perennials widely distributed in N. hemisphere outside tropics; HP 4/5

honey has piquant flavour, and is a good addition to mild honeys; granulates smoothly

Ocimum L. spp. [*O. basilicum* L. is sweet basil]—sub-shrubs in wooded grassland of tropical Africa and other warm parts of the world; HP 3

honey very light; fine flavour

Origanum L. *vulgare* [marjoram]—perennial in Europe (especially Mediterranean), also U.S.S.R., India, etc.; HP 4

honey fine quality, distinct flavour; see also *Thymus*

Rosmarinus officinalis L. [rosemary]—shrub in Mediterranean region; WHCR France, Spain, Morocco; HP 4

honey exceedingly fine and delicious; the famous Narbonne honey (France) was probably from rosemary; see Figure 12.1/1

Salvia L. (550 spp.) [sage, etc.]—shrubs, etc. widely dispersed over warm and temperate regions; HP 4/6

honey water-white, heavy body; very fine quality, many with mild characteristic flavour; slow to granulate

Satureia L. spp. [savory]—shrubs and perennials in Mediterranean region; *S. thymbra* nectar may contain 85% sugar; HP 3

honey: see *Thymus*

Stachys annua (L.) L. [woundwort]—180 other species; annual in C. and S. Europe and Asia, including U.S.S.R.; WHCR U.S.S.R.; HP 4

honey very light amber; soft fine granulation

Teucrium scorodonia L. [wood sage]—perennial throughout most of Europe

honey light

Thymus vulgaris L. and 150 spp. [thyme]—many species of shrubs native to Mediterranean region; widespread in Europe, U.S.S.R., New Zealand (Otago), and (*T. serpyllum* L.) in a few parts of N. America; WHCR Portugal, Spain, Morocco; HP 6

honey golden amber, strongly aromatic, rich in enzymes; the famous Hymettus honey from Greece probably mostly from thyme/savory/marjoram (*Thymus/Satureia/Origanum*)

Lauraceae

Persea americana Miller [avocado]—tree grown extensively in tropical and subtropical America, Israel, etc.
  honey dark, heavy body

Leguminosae

Acacia L. spp. [wattle, etc.]—many species of trees and shrubs scattered over warm dry regions; WHCR Kenya, Tanzania
  honey: in parts of southern U.S.A., several species (catsclaws) give very light, heavy bodied honeys of delicious flavour that granulate slowly; in central Tanzania honey is light to extra light amber; in Australia, where the species are most abundant, they are less useful for honey production

Arachis (*hypogaea* L. and spp.) [ground nut, peanut, earth nut]—annual oilseed crop from S. America widely grown in warm regions; its reported importance as a honey plant is questioned
  honey fairly light; thick; characteristic mild flavour

Astragalus L. spp. [milk vetch]—mostly annuals or perennials, in N. temperate regions especially Asia, e.g. China, Japan, Iran; some species can cause poisoning in bees; WHCR Japan; HP 1/3
  honey from some spp. water-white; quality varies according to species

Brachystegia Bentham spp. with *Julbernardia*, these trees are the main honey source in widespread areas of central African open forest and woodland [miombo]; WHCR Rhodesia, Tanzania
  honey medium amber (e.g. W. Tanzania), extra light amber (e.g. copper belt of Zambia); dense, medium to strong but pleasant flavour; slow to granulate, coarse crystals

Caragana arborescens Lam. [(Siberian) pea tree]—tree/shrub in U.S.S.R. (Siberia), Manchuria, also introduced to Canada; HP 6
  honey light, fine quality

Dalbergia L. spp. [rosewood, sissoo, etc.]—trees/climbers in India, Africa (Nigeria, Tanzania), C. America, probably other tropical forest regions
  honey dark amber, strong flavour

Gleditschia Clayton spp. [honey locust]—trees in warm areas of Asia, America, U.S.S.R.; HP 3/5
  honey characterics not known

Glycine max (L.) Merr. [soya-bean]—annual oilseed crop in China, Japan, N. America (introduced 1854); now increasingly grown for protein content of seeds
  honey light, rather thin; peculiar unpleasant flavour

Haematoxylon campechianum L. [logwood, campeche]—tree in tropical

America, e.g. Guyana, also introduced, e.g. Mauritius; main export honey of Jamaica

honey light, delicious flavour, superb quality

Hedysarum coronarium L. [sulla]—perennial of European origin; fodder crop in Mediterranean region and southern U.S.A.

honey very light, mild flavour

Julbernardia globiflora, paniculata and spp. [julbernardia, mua (Tanzania)] *see Brachystegia*; WHCR Rhodesia, Tanzania

honey extra light amber, dense, medium flavour, excellent quality; slow to granulate (coarsely)

Lespedeza spp. [lespedeza]—shrubs native to N. America, Asia (Japan, Korea, etc.), and Australia

honey of some spp. bright golden or dark, with fine delicate flavour

Lotus corniculatus L. [bird's-foot trefoil]—perennial drought-resistant forage crop widespread in temperate zones, including N. America, Europe, U.S.S.R., China; *L. ulginosus* Schkuhr in N. Europe; WHCR Belgium; HP 2/4

honey light, medium/good quality

Medicago L. (*sativa* L. and spp.) [lucerne (= alfalfa)]—perennial grown extensively as fodder crop in nearly all temperate regions; does not yield nectar well in all conditions; WHCR Bulgaria, Italy, Argentina, U.S.A.; HP 6

honey light, mild, granulates rapidly; often mixed with other sources

Melilotus Miller (*alba* Medicus and other spp.) [white (yellow, etc.) sweet clover, melilot]—biennial/annual native to Central Asia; grown for over 2 000 years round the Mediterranean, and now widespread as a fodder crop in cool temperate regions from U.S.S.R. to New Zealand, but less grown than in 1920s; most important for honey in N. America; WHCR Belgium, U.S.S.R.; HP 4/6

honey white or light greenish yellow; good quality, delicate flavour (like cinnamon or vanilla)

Onobrychis viciifolia Scop. [sainfoin]—widespread perennial fodder crop in cool temperate regions, probable source of the famous Gâtinais honey (France); important in Iran; WHCR Bulgaria, U.S.S.R.; HP 3/4

honey pale/light amber, very sweet, with more pronounced flavour than other legumes; plant yields nectar at lowish temperatures

Ornithopus sativus Brot. [seradella]—annual grown as fodder plant in some temperate and subtropical areas, e.g. U.S.S.R., S. Europe, N. Africa

honey light in colour, good quality

Phaseolus L. spp. [bean (runner, black-eyed, lima, etc.)]—annuals/ perennials widely distributed in warmer (and some temperate) regions, many species being grown as food crop *honey* light; mild, undistinguished flavour; granulates rapidly (some honeys dark amber)

Piscidia piscipula L. [ha'bin (Maya), Jamaica dogwood]—tree in C. America, **W.** Indies; important source in Yucatán (Mexico)
honey characteristics not known

Prosopis (*glandulosa* Torr. and spp.) [mesquite, also *P. juliflora* in C. America, Hawaii]—tree native to C. America, widespread in semi-desert areas of N., C., and S. America, Hawaii, some spp. in India
honey light, good quality, granulates quickly

Psoralea pinnata L. [blue pine weed, taylorina]—shrub native to S. Africa, also in New Zealand, Australia (WA)
honey light to extra light amber, good body, distinctive flavour said to be reminiscent of desiccated coconut; slow coarse granulation

Robinia pseudoacacia L. (also other spp.) [robinia, acacia, black locust]—tree native to N. America, widely planted (and naturalized) there and in poor dry soils of Europe (e.g. in and around the Danube basin), U.S.S.R., China, Korea, Iran, New Zealand. In Rumania *R. p. decasniana* (thornless) is the best honey yielder, up to 1 500 kg/ha; WHCR U.S.S.R.; HP 6
honey water-white, heavy body, sweet; fine flavour, little aroma; granulates slowly (several years); high fructose and low enzyme contents

Sophora japonica L. [pagoda tree]—tree native to China and Korea, introduced to U.S.S.R., Europe, U.S.A., and elsewhere; HP 4
honey from *S. secundiflora* in C. America has excellent flavour and aroma

Tamarindus indica L. (only species) [tamarind]—tree in tropical Africa and Asia (India, Mauritius, etc.), also C. America
honey dark, second-grade

Trifolium alexandrinium L. [Egyptian clover, barseem, berseem]—crop grown especially in Egypt, India, and Pakistan
honey probably similar to *T. pratense*

Trifolium hybridum L. [alsike clover]—perennial native to Sweden; widespread crop at higher latitudes of temperate regions of Europe and U.S.S.R., less now in N. America; HP 4
honey white, mild, good flavour, granulates rapidly

Trifolium incarnatum L. [crimson clover]—annual grown in more southerly parts of N. America, Europe, U.S.S.R., also N.Z.
honey very light amber, good quality, similar to *T. repens*

Trifolium pratense L. [red clover]—perennial whose long corolla makes nectar difficult to reach; widespread pasture and fodder crop in Europe (to 71°N), U.S.S.R., N. America, N. Africa, New Zealand, etc.; tetraploid varieties increasingly used; WHCR Germany, Italy, Poland, U.S.S.R.; HP 5 *honey* water-white, mild, good flavour; granulates rapidly

Trifolium repens L. [white clover]—perennial pasture crop widespread in cool temperate regions, and a major honey source in e.g. Europe, U.S.S.R., Canada, U.S.A., New Zealand; Ireland, Netherlands, Sweden, Japan, Argentina; WHCR Belgium, Denmark, New Zealand; HP 3/4

honey light, mild; granulation uniform, generally slow

Trifolium resupinatum L. [Persian clover]—annual cultivated as catch crop in Europe, N. America and elsewhere; HP 4

honey light

Vicia L. spp. [vetches]—150 spp. of annuals/perennials native to north temperate regions and S. America; many grown as fodder plants; HP 3

honey from some spp. dark amber, flavour mild but stronger than *Trifolium*

Vicia faba L. (see also *Phaseolus*) [field bean, broad bean]—annual crop widely grown throughout temperate regions, especially now for its high protein yield; WHCR Egypt; HP 3

honey light (dark if with honeydew); pleasant mild flavour; inclined to granulate rather quickly and coarsely

Vicia villosa Roth [hairy vetch]—annual grown as fodder crop in N. America, Europe, etc.

honey light, heavy body, mild pleasant flavour, granulates readily

Liliaceae

Allium L. spp. [onion, leek, garlic, etc.]—bulbs widely cultivated in N. Hemisphere; HP 3

honey light amber; 'oniony' aroma which disappears after storage

Aloe spp. [aloe]—most of the 200 species of these shrubby plants are indigenous to Africa and adjacent islands; a few in the Saponariae group are very high honey yielders, e.g. in Transvaal up to 230 kg per plant but bees reported to become aggressive; HP 6

honey very light, almost no flavour or aroma; granulates immediately after extraction

Asparagus officinalis L. [asparagus]—perennial cultivated in Europe, N. America, etc., also widely naturalized

honey light/medium amber; characteristic rather acid taste

Lythraceae

Lythrum salicaria L. [purple loosestrife]—perennial in damp places, spreading throughout north temperate regions (and Australia); HP 5

honey reported very dark, strong flavour—also light, good flavour; wax and cappings golden yellow

Magnoliaceae

Liriodendron tulipifera [tulip tree, tulip poplar]—tree in eastern N. America, introduced to U.S.S.R. and elsewhere
  honey dark brown, heavy body; delicious mild quince-like flavour, exceptionally high nectar yield per flower (1·64 g)

Malvaceae

Gossypium hirsutum L. and spp. [cotton]—shrub cultivated in tropical/ subtropical parts of U.S.S.R., India, Egypt, S. America, and U.S.A., etc.; WHCR Bulgaria, Egypt, Morocco, Israel, U.S.A.; HP 1/4
  honey light; good quality but may be thin-bodied; pleasing flavour (but on dry sandy soils may be dark and strong flavour); extrafloral nectaries important

Musaceae

Musa L. spp. [banana, plantain, etc.]—large perennials cultivated in many tropical regions and yielding nectar copiously
  honey said to be dark and thick; flavour not very attractive

Myrtaceae

Eucalyptus spp. [common names are confused, and vary from State to State in Australia, where the 500 + species are native]. In Australia, eucalypts comprise three-quarters or more of the total indigenous vegetation; they are now cultivated in nearly all countries except northern Europe and Canada, and are important in the whole Mediterranean region, as well as warm temperate parts of Africa, Asia, and the Americas; WHCR Portugal, Spain, Kenya, Morocco, Rhodesia, Argentina, Australia
  honey varies widely from species to species (see those listed below); many flower only once in 2–8 years, but give copious yields
Eucalyptus albens Miq. ex Bentham [white box]—tree in Australia (NSW, Vic, SA, Qd)
  honey almost water-white, fairly dense, excellent flavour, rapid fine granulation
Eucalyptus calophylla R. Br. ex Lindl. [marri]—tree in Australia (WA), where it is the most dependable honey source
  honey light amber, excellent flavour, granulates finely
Eucalyptus camaldulensis Dehnh. [(river) red gum]—tree in Australia (SA, Qd, Vic, WA)
  honey medium amber, dense, pleasant mild woody flavour, granulates fairly slowly with large brown grains
Eucalyptus citriodora Hook. (replaces *E. maculata* Hook. in some areas) [lemon-scented gum]—tree in Australia (Qd, SA), Africa, etc.
  honey medium amber

Eucalyptus diversicolor F. Muell. [karri]—tree in forests of W. Australia only; these enormous trees give one of the most intense flows known, yielding 200 kg honey per hive, but only once in 4–8 years; WHCR Australia

honey extra light amber, finest quality, mild, characteristic flavour, granulates readily but fairly coarsely

Eucalyptus grandis Hill ex Marden—a widely introduced species (tree) which is probably a better honey producer in some African countries (where it exceeds *E. paniculata*) than in Australia

honey medium-dark, granulates readily

Eucalyptus hemiphloia F. Muell. ex Bentham [grey box]—tree in Australia (Vic, Qd, NSW)

honey medium amber, liable to darken during storage; density often low; pleasant flavour; may granulate rapidly

Eucalyptus leucoxylon F. Muell. [blue gum and other names]—tree in Australia (SA, Vic)

honey good quality, pale straw colour, clear, good body; mild flavour, granulates rather quickly

Eucalyptus loxophleba Bentham [York gum]—tree in W. Australia

honey light to medium, varies from season to season; pleasant flavour

Eucalyptus maculata Hook. [spotted gum]—tree in Australia (NSW, Qd), Africa

honey medium amber, strong but not unpleasant flavour

Eucalyptus marginata Sm. [jarrah]—tree in W. Australia

honey dense, quality variable, pleasant nutty flavour near coast, poorer quality inland; characteristic slight frothiness; rarely granulates

Eucalyptus melliodora A. Cunn. ex Schau. [yellow box]—tree in Australia (especially NSW, Vic, Qd); prolific yielder (said to be the best honey-producing plant in the world)

honey usually extra light, dense; good quality; sweet cloying flavour; very slow to granulate

Eucalyptus obliqua L'Herit. [stringy bark]—tree in Australia (NSW, Vic, Tas)

honey not best quality; medium/dark amber

Eucalyptus paniculata Sm. [grey ironbark]—tree in Australia (NSW, Vic), Africa, prolific yielder, comparable with *E. melliodora*

honey light, medium density; pleasant aroma, excellent flavour; slow, fine granulation

Eucalyptus platypus Hook. [moort]—tree in W. Australia, four-year flowering cycle, but may give copious flow for 2 months

honey first class

Eucalyptus robusta Sm. [swamp messmate]—tree in Australia (NSW, Qd), Mauritius *honey* dark amber, reasonable density, pleasant flavour

Eucalyptus rudis Endl. [flooded gum]—tree in W. Australia with two-year flowering cycle

honey first class, light amber, pleasant flavour

Eucalyptus siderophloia Benth. [broad-leaved ironbark]—tree in Australia (NSW)

honey golden, medium density; sweet mild flavour; coarse brown granulation

Eucalyptus sideroxylon A. Cunn. ex Benth. [mugga, red ironbark]—tree in Australia (NSW, Qd, Vic)

honey light amber; good density; fine rapid granulation

Eucalyptus tereticornis Sm. [forest red gum]—tree in Australia (Vic), Mauritius, etc.

honey first grade, dark, strongly flavoured

Eucalyptus viminalis Labill. [manna gum, white gum]—tree in Australia (Vic, Tas), New Zealand

honey clear amber, not very dense, medium flavour, granulates rather readily

Eucalyptus wandoo Blakely [wandoo]—tree in W. Australia

honey first-class, amber to light amber; good density, mild flavour; granulates medium/light creamy colour; yields of 90 kg per hive common

Leptospermum scoparium J. R. et G. Forst. and spp. [manuka, ti tree, tea tree]—*L. scoparium* in New Zealand, but most of the 35 species of shrubs are native to Australia; WHCR New Zealand

honey from *L. scoparium* light amber, from prolific nectar flow; thixotropic (like *Calluna* honey), with distinctive flavour best obtained from comb honey

Melaleuca leucadendron L. [cajeput]—tree in Australia (native to Qd); widespread in warm temperate regions, C. America, tropical Asia; copious nectar flow

honey amber, mild but distinct flavour and considerable, even penetrating, aroma; also reported unpalatable

Metrosideros excelsa Sol. ex Gaertn. [pohutukawa]—tree in New Zealand (Auckland); WHCR New Zealand

honey water-white; unique salty flavour, greatly valued; granulates (coarsely) in a few days

Metrosideros umbellata Cav./*robusta* A. Cunn. [rata]—trees in New Zealand (WHCR)

honey water-white, delicate distinctive flavour; fine, silky (rapid) granulation

Myrtus communis L. [myrtle]—shrub in Mediterranean area and S. Europe to W. Asia

honey light

Tristania conferta (and spp.) [scrub box]—tree in Australia (NSW, Qd)
honey light amber, reasonable density, good slightly aromatic flavour, granulates quickly with hardish white grain

Nyssaceae

Nyssa (aquatica and spp.) [tupelo]—tree in Florida; other spp. in S. Asia
honey from *N. aquatica* white, fine mild flavour, does not granulate if pure (high fructose content), hence used in pharmaceutical industry; by law it can be specified as tupelo honey when sold

Onagraceae

Chamaenerion angustifolium (L.) Scop. [rosebay willowherb, fireweed]—
perennial widespread in cool temperate regions, in open woodland especially where recently cleared by fire (Europe, N. America, U.S.S.R., etc.); WHCR England/Wales, Finland, U.S.S.R.; HP 5/6
honey white, from very long nectar flow; high quality, fine flavour; granulates finely

Palmaceae

Cocos nucifera L. [coconut palm]—tree in C. America, E. Africa, Ceylon, Philippines, and many other tropical areas, especially islands
honey amber (water-white when pure?); good quality; reliable information needed

Roystonea (regia H.B.K. and spp.) [royal palms]—trees native to tropical America, planted there and in other subtropical regions
honey from *R. regia* light amber, characteristic delicious flavour and aroma

Serenoa repens (Bartr.) Small (*serrulata*), also other spp. [saw palmetto]—
shrub in south-east U.S.A.
honey light amber, thick and waxy; pronounced but delicious flavour

Pedaliaceae

Sesamum orientale L. [sesame]—annual tropical oilseed crop, grown in China, India, Africa, C. America, etc.; HP 2
honey very light amber, good quality

Pinaceae

Abies alba Miller [silver fir]
Abies bornmuellerana Mattfeld
Larix decidua Miller [larch]
Picea abies (L.) Karsten [spruce]
Pinus halepensis Miller
Pinus mugo Turra [mountain pine]

Pinus nigra Arnold [Austrian pine]

Pinus sylvestris L. [Scots pine]

and other species

> trees in forests of central Europe (where these species are known to be a main honey source), probably yield honey in U.S.A. (e.g. Calif., Colo.), and in coniferous forests of other continents where the summer temperatures are high enough; adequate studies have not yet been made; WHCR Austria, Bulgaria, Greece, Poland; HP 5
>
> *honeydew honey* from these and other conifers is light to dark amber, *A. alba* being the darkest, with a greenish tinge; heavy bodied, strong flavour; slow to granulate

Polygonaceae

Antigonon leptopus Hook. and Arn. [coral vine]—climber in C. and S. America; one of the few plants which may be economically worth growing for its honey production alone

> *honey* light; good flavour, marked characteristic aroma

Erigonum fasciculatum Benth. [wild buckwheat]—shrub in Western U.S.A.

> *honey* deep light amber; heavy body; good flavour; granulates readily; commonly marketed as 'Californian sage honey'

Fagopyrum esculentum Moench [buckwheat]—annual grain crop in (north) temperate regions, grown less than formerly: China, Korea, N. America; E. and S. Europe and a few places in W. Europe; HP 3/5

> *honey* very dark brown; very strong characteristic flavour and aroma; used in N. Europe for making traditional honey cakes, and by Jews in many countries for their honey wine

Gymnopodium antigonoides (Robinson) Blake [dzidzilché (Maya)]—tree in C. American subtropical lowland rain forest; most important honey plant in Yucatán (Mexico)

> *honey* light amber, delicious aroma and flavour

Polygonum L. spp. [bistort, heartsease, etc.]—annuals/perennials of world distribution, especially in temperate regions; HP 4

> *honey* light amber; pronounced unpleasant flavour; very white cappings

Proteaceae

Banksia spp. [banksia]—shrubs/trees in Australia (NSW, Qd, Vic)

> *honey* often dark and not of table quality

Banksia menziesii, *B. prionotes* [Menzies, orange banksia]—small trees in W. Australia

> *honey* extra light amber, mild flavour, excellent quality; *B. menziesii* may contain 10–12% sucrose, occasionally more

Dryandra sessilis, also other spp. [parrot bush]—shrub in W. Australia

> *honey* light to extra light amber, darkening rather quickly in storage,

HMF (hydroxymethylfurfuraldehyde) content also increasing; characteristic flavour variously appreciated

Grevillea (*robusta* A. Cunn. and 230 spp.) [grevillea]—shrubs/trees native to Australian upland (cloud) forest, widespread in WA; also cultivated in many tropical and subtropical parts of Africa and America

honey from some species very light and mild, but most medium amber; in Tanzania, *G. robusta* honey dark amber, high density, pronounced flavour

Knightea excelsa R. Br. [rewarewa, New Zealand honeysuckle]—tree in New Zealand

honey medium/dark amber, heavy body, rich distinctive (slightly burnt) flavour, slow coarse granulation

Protea spp. [protea]—shrubs/trees in wooded grassland in S. and tropical Africa, giving copious nectar yield

honey characteristics not known

Rhamnaceae

Rhamnus L. spp., *Frangula* Miller spp. [buckthorn, alder buckthorn, etc.]—trees/shrubs in C. and S. Europe, U.S.S.R., and other temperate regions; *R. alaternus* L. in Spain; HP 3

honey amber (*R. purshiana* and other spp. dark), thick; some species with fine flavour and aroma; granulates quickly

Rosaceae

Crataegus L. (*oxycantha* Thuill. and spp.) [hawthorn]—trees now widely distributed in both temperate zones (e.g. Europe, U.S.S.R., New Zealand); WHCR England and Wales, Ireland; HP 2

honey light/dark amber; fine nutty flavour

Eriobotrya japonica Lindl. [loquat]—tree in China (Fukien Province), India, and other parts of tropical Asia/America

honey amber, agreeable flavour

Malus. Mill./*Pyrus* L./*Prunus* L. spp. [pome and stone tree fruit (apple, pear, plum, cherry, etc.)]—orchard trees widespread in temperate regions; WHCR Belgium, Finland, Japan; HP 1/2

honey light, excellent delicate flavour and fine aroma, said to granulate quickly, with soft fine grain

Prunus see *Malus*

Rubus fruticosus L. [blackberry, bramble]—climbing shrub, (secondary) honey source widespread in most cool temperate regions; WHCR Ireland

honey light; good density; medium flavour, rather similar to clover; slow to granulate

Rubus idaeus L. [raspberry]—shrub cultivated in N. and C. Europe, Canada, U.S.S.R., etc.; WHCR Bulgaria, Scandinavia, U.S.S.R.; HP 3
  honey light, very fine mild flavour and aroma

Rubiaceae

Calycophyllum candidissimum (Vahl.) DC. [dagame (Spanish)]—tree in C. America
  honey light, sometimes water-white; good density; fine flavour
Coffea (arabica L. and spp.) [coffee]—shrub in upland tropical forest areas of Africa, S./C. America, India, etc.; WHCR El Salvador, Tanzania
  honey light; characteristic flavour; mixed coffee-orange honey very highly valued

Rutaceae

Citrus aurantium L. [orange]—tree widely planted in citrus groves in warm regions of the world; WHCR Italy, Portugal, Spain, Israel, Morocco, Rhodesia, U.S.A.
  honey very light, heavy body; flavour delicious; diastase commonly absent
Citrus L., other spp. [lemon, grapefruit, lime, citron, etc.]—trees distributed as *C. aurantium* L. but more localized; WHCR Israel
  honey from lemon light amber; acid flavour, aroma characteristic of the plant

Salicaceae

Salix L. (500 spp.) [willow]—trees/shrubs widely distributed, especially in N. temperate parts of Asia, America, and Europe; HP 4
  honey light amber; mild flavour, fine aroma

Sapindaceae

Euphoria longan (Lour.) Steud. [longan, lengeng]—tree native to and cultivated in S.E. Asia; a main honey source of many areas there, e.g. Taiwan and parts of China; introduced to C. America
  honey characteristics not known
Nephelium litchi Comb. = *Litchi chinensis* Sonn. [litchi, lychee]—tree is native to S. China; widely grown there and e.g. Singapore, Mauritius, where it is a major honey source
  honey excellent, fine flavour
Sapindus mukorossi Gaertn., possibly other spp. [soapnut]—tree in tropical Asia (India, China, etc.)
  honey water-white, mild flavour (also described as light golden, good aroma and flavour)

Scrophulariaceae
Scrophularia L. spp. [figwort]—perennials widespread in N. temperate zone including U.S.S.R., some as weeds; HP 6
honey light, excellent quality, little aroma

Simarubaceae
Ailanthus altissima (Miller) Swingle [tree of heaven]—tree in S.E. Asia, introduced elsewhere
honey greenish brown; flavour initially poor but after a few weeks becomes very fine, with aroma reminiscent of muscat wines; granulates finely (only after several months)

Sterculiaceae
Dombeya rotundifolia (Hochst.) Planch. and spp. [dombeya]—tree native to African equatorial cloud forest, wooded grassland and scrub; introduced to S. America, India, etc.; same family as cocoa (*Theobroma cacao* L.); *Dombeya* WHCR Kenya, Tanzania
honey extra light amber, fine flavour

Tamaricaceae
Tamarix L. spp. [tamarisk]—shrubs indigenous to Mediterranean region and northern parts of Old World, introduced to N. America
honey dark brown, minty aroma

Tiliaceae
Tilia L. spp. [lime, linden, basswood]—trees widespread in cool temperate regions, especially N. and S.E. Europe, China, Japan, U.S.S.R.; *T. americana* [basswood] in N. America; large trees can give heavy crops; many species are available, which together give a long succession of honey flows, in many regions from honeydew also; WHCR Bulgaria, Finland, Germany, Netherlands, Poland; HP 4/6
honey light, somewhat greenish; density often low; pronounced characteristic flavour and aroma; granulates smoothly and rapidly
Triumfetta rhomboidaea Jacq. and spp. [triumfetta]—herbs/shrubs in equatorial cloud forest of Africa, C. America
honey dark amber, strong flavour

Umbelliferae
Anthriscus cerefolium (L.) Hoffm. [cow parsley]—biennial weed of pastures in Europe, also e.g. Siberia, N. Africa, and introduced to N. America
honey light amber, harsh flavour

Daucus carota L. [carrot]—wild and cultivated biennial in Europe,
N. Africa, W. Asia, America
honey light amber

Foeniculum vulgare Miller [fennel]—perennial native to Mediterranean
region, naturalized in most temperate countries
honey light amber

Heracleum sphondylium L. [hogweed, cow parsnip]—biennial weed of
pasture land; Europe S. of 61°, W. and N. Asia, western N. Africa,
introduced to N. America; HP 4
honey light/dark amber, harsh flavour, rich in diastase

Verbenaceae

Avicennia nitida Jacq. [black mangrove]—tree in C. America
honey very light, thin bodied (ferments readily), mild, good flavour

Citharexylum spp. [fiddlewood, etc.]—trees/shrubs in subtropical parts of
America, heavy honey yielders
honey characteristics not known

Lippia spp. [carpet grass, etc.]—shrubs native to S., C., N. America (and
Africa), many very long-flowering; introduced elsewhere, e.g.
citriodora in U.S.S.R.
honey: repens, light amber, heavy body, similar to lucerne; *reptans*,
clear amber, good quality; *virgata*, delicious, aromatic

Vitex negundo (*incisa* Clarke)—*V. negundo* [bannah (India)] is shrub
native to S.E. Asia; used in China for baskets, also cultivated elsewhere (*V. n. incisa* in U.S.A., HP 6)
honey characteristics not known

CHAPTER 2

HOW BEES MAKE HONEY

by Dr. Anna Maurizio

FORMERLY RESEARCH WORKER AT FEDERAL BEE RESEARCH DEPARTMENT,
LIEBEFELD-BERN, SWITZERLAND

> So smell the Breath about the hives
> When well the work of honey thrives;
> And all the busie Factours come
> Laden with wax and honey home.
> ROBERT HERRICK (1591–1674),
> TO ANNE SOAME

2.1 THE RAW MATERIALS OF HONEY

In temperate climates honey is produced mainly from the nectar of flowers and from honeydew. In other climates the bees may use other raw materials as well, such as nectar from extrafloral nectaries, or sap from sugar cane and some other plants. Here we shall consider chiefly floral nectar and honeydew, since these are the most important raw materials of honey. Both originate from the phloem sap of higher plants.

2.11 Phloem sap: the basic raw material

The sieve tubes—in conifers the sieve cells—form a transport system throughout the plant, in which the nutritive substances, dissolved in water, move under considerable pressure (20–40 atmospheres). In this system the phloem sap removes assimilates on the one hand and, on the other, furnishes all parts of the plant with necessary nutrients.

Much research has been done in recent years on the composition of the phloem sap, mainly of deciduous trees and herbaceous plants. Little work has yet been done on phloem sap of conifers (von Dehn, 1961, Eschrich, 1961, 1963a, 1963b, Mittler, 1953, 1957, 1958a, Peel & Weatherley, 1959, Wanner, 1953, Ziegler, 1956, 1962, 1968a, Ziegler & Kluge, 1962, Ziegler & Mittler, 1959, G. Zimmermann, 1961, M. Zimmermann, 1957, 1958a, 1958b, 1960, 1961a, 1961b, 1964). Phloem sap is a clear, usually colourless liquid, which is sometimes fluorescent; its dry weight varies between 5 and 30% (average 15–25%) of the total; its ash content is 1–3% of the dry weight. The pH is 7·3–8·6, slightly on the alkaline side of neutral.

Among the constituents of phloem sap, the carbohydrates (sugars) predominate, forming up to 90% of the dry weight. The plants so far examined can be divided into three groups according to the sugars found in their phloem sap. In the first group are plants whose phloem sap sugar is mostly or entirely sucrose. This group includes the majority of plant families, including Leguminosae and Coniferae. The second group consists of plant families whose phloem sap contains, besides sucrose, fairly large amounts of oligosaccharides. These oligosaccharides belong to the 'raffinose series' and are characterized by the presence of one or several galactose molecules attached to the sucrose molecule (stachyose, verbascose). In this group are species of Oleaceae, Bignoniaceae, Verbenaceae, Combretaceae, Myrtaceae, and Onagraceae. The third group consists of plants whose phloem sap, in addition to the sugars already mentioned, contains a fair amount of sugar alcohols (mannitol, sorbitol). Certain Oleaceae (*Fraxinus*, *Syringa*) and Rosaceae (*Prunus serotina*, *Malus sylvestris*; Ziegler, 1968a, M. Zimmermann, 1957, 1960, 1961a) belong to this group.

The carbohydrate content of phloem sap is usually around 10–30%, except for the Cucurbitaceae, in which the sugar content is under 1% of the fresh weight. In a few plants sugar phosphates and monosaccharides have been found in the phloem sap (*Tilia tomentosa*, *Centaurea scabiosa*, *Campanula rapunculoides*, *Cirsium arvense*; von Dehn, 1961, Eschrich, 1961; *see also* Ziegler, 1968a). The sugar content of phloem sap shows consistent changes throughout the day and throughout the season; within the plant itself the concentration decreases from the tip to the base.

Phloem sap contains other substances as well as carbohydrates: nitrogen compounds, fats, organic acids, nucleic acids, vitamins, and minerals. These substances occur only in very small amounts, however, compared with the carbohydrates.

The nitrogen content of phloem sap rises to a peak in spring, at the time when the plants come into leaf, and again in autumn when the leaves change colour. It comprises some proteins, but for the most part amino acids and amides, of which glutamic acid and glutamin, and aspartic acid and asparagine, are the most common (von Dehn, 1961, Mittler, 1958a, Ziegler, 1956, G. Zimmermann, 1961). The composition varies between different species, and even within one species it fluctuates with the season. For example, in the phloem sap of some species of trees, urea derivatives have been found (allantoic acid, allantoin, citrulline, in *Acer*, *Platanus*, *Aesculus*, *Betula*, *Alnus*, *Carpinus*, *Juglans*); other trees contain putrescin, and many Leguminosae canavenin (*Robinia*; Ziegler & Schnabel, 1961). Cucurbitaceae are an exception in this respect also:

two-thirds of the dry weight of their phloem sap consists of nitrogen compounds (protein, aspartic acid, citrulline; Eschrich, 1963a, 1963b, Ziegler, 1968a).

Fats have been found only in the phloem sap of *Robinia pseudoacacia* and *Tilia platyphyllos* (0·13 and 0·54% of the dry weight; Kluge, 1964). Small amounts of organic acids (citric acid, vinous acid, oxalic acid, fumaric acid, malonic acid, malic acid, and gluconic acid, among others) occur in the phloem sap of *Robinia pseudoacacia, Salix viminalis,* and *Tilia platyphyllos* (Kluge, 1964, Peel & Weatherley, 1959). Nucleic acids, again in very small quantities, have been found in the phloem sap of *Robinia pseudoacacia* and *Tilia platyphyllos* (Ziegler, 1968a, Ziegler & Kluge, 1962). The vitamin content of phloem sap has been well established: an examination of 37 species of trees and herbaceous plants showed considerable amounts of thiamin, pantothenic acid, nicotinic acid, *meso*-inositol, vitamin C, and usually also pyridoxine; riboflavin, biotin and folic acid occurred more rarely and in smaller quantities (Ziegler & Ziegler, 1962). The mineral spectrum of phloem sap is characterized by the predominance of potassium and the almost complete absence of calcium; only very small quantities of other minerals, such as sodium, magnesium, phosphates, nitrates, and trace elements, have been detected (Ziegler, 1968a, which gives further references).

2.12 Nectar

Nectar is an aqueous, sugar-containing secretion of plant glands called nectaries. It is not possible in this book to give detailed information about the structure and function of nectaries; only the most important facts concerning nectar as a raw material of honey can be dealt with. Further information can be obtained from the comprehensive references given by Lüttge (1969), Schnepf (1969), and Ziegler (1968b).

2.121 *Nectaries*

Nectaries can be found on any parts of a plant which are above ground, and they occur in ferns as well as in flowering plants. According to their position on the plant we distinguish between floral (or nuptial) nectaries, situated in the region of the flowers, and extrafloral (or extranuptial) nectaries in other parts of the plant. Floral nectaries may be situated on the axis of the flower, on sepals or petals, on stamens, or on carpels. Extrafloral nectaries may occur on the cotyledons, on the trunk, on leaves, stipules and bracts, or on petioles, etc. There is no basic difference between floral and extrafloral nectaries in either structure or function (Figures 2.121/1 overleaf and 2.121/2, Plate 1). In both we distinguish between undifferentiated and differentiated structures, the latter being subdivided into histoid and organoid nectaries. Undifferentiated nectaries are hardly distinguishable with the naked eye from the surrounding plant tissues. Differentiated nectaries occur in various shapes—flat, hollow, or like scales, discs or hairs. As a rule, differentiated nectaries have a glandular tissue consisting characteristically of small cells rich in plasma;

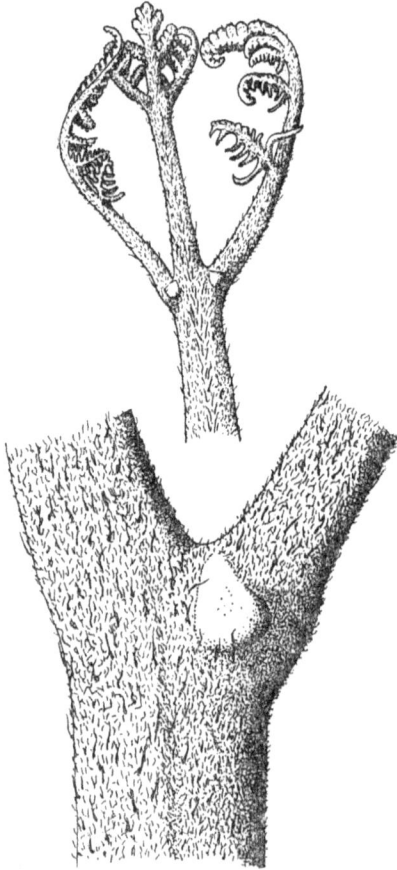

Figure 2.121/1   Extrafloral nectaries on a young frond of the fern *Pteridium aquilinum.*
Above: position of the nectaries on the frond
Below: a nectary (magnified)
               (after Schremmer, 1969)

the nectaries themselves are often clearly visible, being strikingly shaped and coloured organs (Fahn, 1953, Frey-Wyssling & Häusermann, 1960, Schremmer, 1969, Ziegler, 1968b, J. G. Zimmermann, 1932).

The innervation of the nectaries is characteristic for individual genera, sometimes for whole families. Nectaries are commonly served by the neighbouring vascular bundles, but it is not unusual for a special system of phloem branches to lead to the secretory tissue (Figures 2.121/3 and 2.121/4). The sugar content of nectar depends largely on the type of innervation of the nectaries: nectaries plentifully supplied with phloem sap, i.e. those directly connected to the phloem (sieve tube) part of the vascular bundles, produce nectar with a higher sugar content than those whose innervation depends mainly on the xylem (woody part) of the vascular bundles (Agthe, 1951, Frei, 1955, Frey-Wyssling & Agthe, 1950).

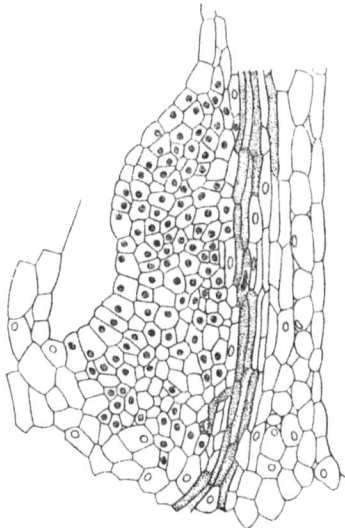

Figure 2.121/3 Longitudinal section through the nectary of white clover (*Trifolium repens*), phloem tissue on the right.
(after Frei, 1955)

So far there have been several different opinions on the significance of nectar secretion for the plant, and on its mechanism (*see* Lüttge, 1959, Schnepf, 1969, Shuel, 1966, Ziegler, 1968b). According to one of these opinions 'nectar is secreted phloem sap', the main purpose of the nectaries being to regulate the supply of sap in the plant (Frey-Wyssling & Agthe, 1950). During their development the young parts of the plant (stalks, leaves, flowers) are plentifully supplied with sap for their formation, and when growth comes to a halt this supply continues for a time. But the sap is no longer (or only partially) used, so it begins to accumulate, and is secreted by the nectaries in the form of nectar. The nectaries thus serve as 'sap valves' to regulate sap pressure within the plant. This interpretation has now been modified, and today it is believed that selective transport takes place in the nectary, water and sugar being the main substances secreted (Lüttge, 1961, 1964, 1969, Schnepf, 1969, Ziegler, 1956, 1968b, Ziegler & Lüttge, 1960). This leads to a reduction in the osmotic pressure within the vascular system of the plant, so that the current of formative sap is directed to the points where it is needed. On this basis the nectaries are to be regarded as 'sugar valves', which regulate

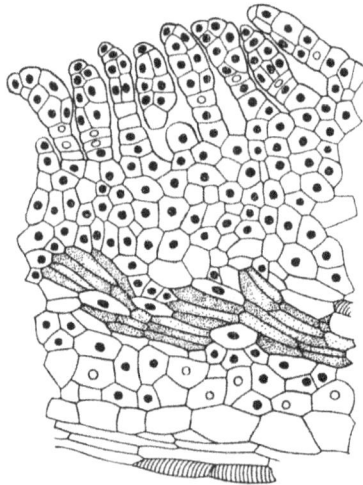

Figure 2.121/4   Longitudinal section through the hair nectary of *Tilia tomentosa*; phloem tissue below.

(after Frei, 1955)

the sugar content (and through it the osmotic pressure) in the vascular system. This view would account for the fact that nectaries are often situated outside the flower area, and occur also in non-flowering plants. As a secondary function, the nectaries in the flower area have come to attract pollinating insects to visit them.

The process of nectar secretion in the nectary is complex and not yet entirely understood. It is, however, certainly not a passive process of filtration under pressure from the sieve tubes; it is an active secretion by the glandular tissue. Water containing the dissolved assimilates is certainly conveyed to the vicinity of the nectaries by the transport system, but the formation and secretion of nectar is an active process in the cells of the glandular tissue. The nature of this process is still largely obscure. Probably the sequence proceeds by phosphorylation (enzymatic linking of the sugars to phosphorus, followed by their splitting off). Whether during nectar secretion only the sugars are secreted and other substances retained, or whether all substances are first secreted and nitrogen compounds, phosphates, and minerals are subsequently resorbed by the gland tissue, is a question still unanswered. The latter possibility is made to seem likely by experiments where substances (sucrose, glutamic acid, phosphates, sulphates, calcium) placed on nectaries have been resorbed (Lüttge, 1961, 1962b, 1969, Pedersen et al., 1958, Shuel, 1961, 1967, 1970, Ziegler, 1968b, Ziegler & Lüttge, 1960).

2.122 *Chemical composition of nectar*

Nectar consists in the main of an aqueous solution of various sugars. Other substances—such as nitrogen compounds, minerals, organic acids, vitamins, pigments, and aromatic substances—are present only in small amounts. The ash content is 0·023–0·45%. As a rule, nectar shows an acid or neutral reaction (pH 2·7–6·4), more rarely alkaline (pH up to 9·1). The following vitamins have been found in nectar: thiamin, riboflavin, pyridoxine, nicotinic acid, pantothenic acid, folic acid, biotin, *meso*-inositol and ascorbic acid. Apart from ascorbic acid (vitamin C), which occurs in appreciable quantities in the nectar—and honey—of certain plants, the vitamin content is low. The nitrogen contribution, especially amino acids and amides, is related to the degree of differentiation of the nectaries, nectaries with the most primitive structure being richest in nitrogen compounds (Beutler, 1930, 1953, Hahn, 1970, Lüttge, 1961, 1962a, 1962b, 1964, 1969, Mostowska, 1965, Ziegler, 1968b, Ziegler, Lüttge & Lüttge, 1964). Substances which prevent pollen germination have also been discovered in nectar (Gonnet, 1969). A few nectars contain substances harmful to bees, to humans, or to both (*see* Chapter 5 and 2.132).

The dry substance of nectar consists chiefly of a mixture of sugars. The total sugar content varies considerably (from 5 to 80%), and there are great differences in the sugars present and in their proportions. Nectars from about a thousand plant species have so far been examined, and in these three groups can be distinguished, according to their sugar spectrum: (1) nectars in which sucrose predominates or (in extreme examples) is the only sugar present; (2) nectars containing roughly equal amounts of sucrose, glucose, and fructose; (3) nectars in which sucrose is present only in small quantities (or is completely absent), and the two hexoses glucose and fructose predominate. There are also differences in the proportion of the two hexoses; they are frequently present in equal amounts (especially in groups 1 and 2), but sometimes one is greatly predominant, especially in group 3. Usually these latter nectars contain more fructose than glucose, and the fructose:glucose ratio may be as high as 28; nectars with glucose predominating are more rare, and the fructose:glucose ratio does not fall below 0·7.

The proportion in which the three main sugars occur is characteristic of certain species, sometimes of entire plant families. For example, the nectar of *Rhododendron ferrugineum* (Alpine rose), and of several cultivated *Rhododendron* species, contains almost exclusively sucrose, whereas sucrose is almost completely absent in the nectar of most Cruciferae. The nectar of *Castanea sativa*, *Robinia pseudoacacia*, *Trifolium pratense*, and many of the Labiatae, contains much more fructose than glucose, whereas glucose predominates in the nectar of *Taraxacum officinale* and of *Brassica napus*. The nectar of most Leguminosae is characterized by the equal proportions of the three main sugars (Barbier, 1963, Beutler, 1930, Fahn, 1949, Furgala *et al.*, 1958, Maurizio, 1959b, Percival, 1961, Wykes, 1952a).

Since the phloem sap contains chiefly sucrose, and no hexoses, the question arises how these simple sugars originate in nectar. They are partly produced by enzymatic hydrolysis of sucrose—new, complex types of sugar being simultaneously constituted through group transference. The presence of transglucosidases and transfructosidases in the nectar of *Robinia* and *Impatiens holstii* has been established (M. Zimmermann, 1953, 1954). Moreover, experiments with excised nectaries maintained on artificial nutrients have shown that, separated from the plant, nectaries can continue to secrete nectar for a time: not only can they break down sucrose into glucose and fructose, they can also convert hexoses into sucrose and higher-molecular sugars (Agthe, 1951, Frey-Wyssling *et al.*, 1954, Matile, 1956, Shuel, 1956, Ziegler, 1955, 1956, 1968b; *see also* 2.132). On the other hand, it is also possible that such sugars are produced by the action of micro-organisms in the nectar.

The fact that the sugar spectrum of nectar is not uniform may be of practical importance in beekeeping. Honeybees—and bumble bees (Pouvreau, 1974)—*prefer* mixtures of different sugars to solutions of any one of the components at the same concentration (Jamieson & Austin, 1958, Wykes, 1951, 1952b). It is possible that, in their choice of sources of food, bees are influenced not only by the quantity and concentration of the nectar, but also by its sugar spectrum (*see* 2.132). The choice of a nectar source is likely to be determined to some extent also by its aromatic content (Loper & Waller, 1969, 1970).

2.123 *Evaluation of factors influencing nectar production*

The nectar production of a plant is usually measured by the quantity (in milligrams) and sugar concentration (%) of nectar secreted by one flower in 24 hours. From this the 'sugar value' is obtained: the amount of sugar (in milligrams) secreted per flower per 24 hours (Table 2.123/1). The quantity and sugar concentration of nectar fluctuate considerably according to external circumstances; on the other hand the sugar value has proved to be constant and characteristic for individual plant species. It therefore provides a useful basis for comparing different species and their importance as sources of honey. Another measure sometimes used to characterize nectar yield is the 'honey potential', the quantity of honey (in kilograms) that could theoretically be obtained in the course of a season from 1 hectare of ground covered with the plant in question (*see* 1.3). This value is calculated from the sugar value, the average number of flowers per plant, and the average number of plants per unit area. It is important when an estimate is needed of the prospective honey yield of large areas of a uniform crop (e.g. rape, buckwheat, sunflower), or of wild flowers like heather; also for the proper siting of migratory colonies of bees (Demianowicz *et al.*, 1960, 1963, Jabloński, 1968, Kropáčová & Haslbachová, 1970).

Table 2.123/1
Sugar values of different plants (in milligrams of sugar produced per flower per 24 hours).
Compare Table 1.3/1, which also includes common names of the plants.

Plant	Sugar value	Author
Boraginaceae		
Borago officinalis L.	1·30	Be
,, ,,	1·10	Bo
,, ,,	0·2–4·9	D *et al.* 1960
Cynoglossum officinale L.	0·38–1·3	D *et al.* 1963
Echium vulgare L.	1·64	Bo
,, ,,	0·09–1·3	D *et al.* 1960
Compositae		
Centaurea jacea L. (single flower head)	0·46–6·1	DH
Centaurea hyalolepis Boiss.	0·07	F
Helianthus annuus L.	0·12	Bo
,, ,,	0·30	BS
,, ,,	0·12	F
Cruciferae		
Brassica napus L. emend Metzger	0·79	Bo
,, ,,	0·50	BS
,, ,,	0·4–0·8	D
, ,,	0·03–2·1	D *et al.* 1960
,, ,,	0·53	HM

Plant	Sugar value	Author
Eruca sativa Miller	0·02–1·1	D *et al.* 1960
Sinapis alba L.	0·4	Bo
,, ,,	0·007–0·2	D *et al.* 1960
Sinapis arvensis L.	0·05	Bo
,, ,,	0·05–0·38	D *et al.* 1963
Raphanus raphanistrum L.	0·04–0·56	D *et al.* 1963
Ericaceae		
Calluna vulgaris (L.) Hull.	0·12	Bo
Hippocastanaceae		
Aesculus hippocastanum L.	1·10	Be
,, ,,	2·08	Bo
Hydrophyllaceae		
Phacelia tanacetfolia Bentham	0·069–0·98	D *et al.* 1960
Labiatae		
Lavandula spica L.	0·26	Bo
,, ,,	0·09	K
Lav. spica × *Lav. latifolia* Medicus	0·23–1·0	Ba
Dracocephalum moldavicum L	0·18–0·54	D *et al.* 1960
,, ,,	0·1–0·75	Sz
Lamium album L.	0·12–0·14	D *et al.* 1960
,, ,,	0·69–0·72	G
Lamium amplexicaule L.	0·06	G
Lamium galeobdolon (L.) L.	0·72–0·89	G
Lamium garganicum	0·88	G
Lamium maculatum L.	0·10–0·13	D *et al.* 1960
,, ,,,	0·32	G
Lamium purpureum L.	0·07	G
Salvia farinacea Bentham	0·33	F
Salvia leucantha Cav.	4·38	F
Salvia nemorosa L.	0·08–0·28	D *et al.* 1960
,, ,,	0·05–0·25	J
Salvia officinalis L.	0·3–3·3	D *et al.* 1960
,, ,,	0·37–1·56	J
Salvia pratensis L.	0·60	M
Salvia sclarea L.	0·3–1·18	D *et al.* 1963
Salvia splendens Sello ex Nees	0·70	M
Salvia triloba L.	0·63	F
Salvia verticillata L.	0·06–0·58	D *et al.* 1963

Plant	Sugar value	Author
Thymus capitatus (L.) LK et Hoffm.	0·09	F
Thymus pulegioides L.	0·02–0·08	D *et al.* 1963
Thymus serpyllum L.	0·02–0·12	D *et al.* 1963
Thymus vulgaris L.	0·01–0·09	D *et al.* 1963
Leguminosae		
Lotus corniculatus L.	0·08	BS
Medicago sativa L.	0·04–0·2	D *et al.* 1963
Melilotus alba Medicus	0·001–0·06	D *et al.* 1963
Melilotus officinalis (L.) Pallas	0·0005–0·004	D *et al.* 1963
Onobrychis viciifolia Scop.	0·24	BS
,, ,,	0·2–0·3	D *et al.* 1963
,, ,,	0·01–0·15	KH
Robinia pseudoacacia L.	0·22–2·3	D *et al.* 1960
,, ,,	1·36	H
Sophora japonica L.	0·3–1·0	HS
Trifolium hybridum L.	0·01	M
Trifolium incarnatum L.	0·07	BS
,, ,,	0·003	M
Trifolium pratense L.	0·19	Bo
,, ,,	0·08	BS
,, ,,	0·03	M
,, ,,	0·061–0·14	S
Trifolium repens L.	0·14	Bo
,, ,,	0·04	BS
,, ,,	0·02–0·1	D *et al.* 1960
,, ,,	0·012	M
Liliaceae		
Allium schoenoprasum L.	0·37–0·48	KK
Onagraceae		
Chamaenerion angustifolium (L.) Scop.	0·62	Bo
,, ,,	0·46–4·26	D *et al.* 1960
,, ,,	0·02	Ke
Polygonaceae		
Fagopyrum esculentum Moench	0·10	BS
,, ,,	0·29–2·68	D *et al.* 1960
,, ,,	0·05–0·19	DR
Rosaceae		
Malus sylvestris Miller	1·37	Bo

Plant	Sugar value	Author
Malus sylvestris Miller	0·70	BS
,, ,,	0·03–1·94	E 1940
,, ,,	0·28–1·37	F
Prunus armeniaca L.	0·31–0·84	E 1940
Prunus avium (L.) L.	0·50	Be
,, ,,	1·50	Bo
,, ,,	0·08–0·79	E 1940
Prunus cerasus L.	1·27	Bo
,, ,,	1·20	BS
,, ,,	0·15–1·31	E 1940
Prunus domestica L.	0·13–1·47	E 1940
,, ,,	0·25	F
Prunus persica (L.) Batsch.	0·54–1·38	E 1940
Pyrus communis L.	0·09	Bo
,, ,,	0·30	BS
,, ,,	0·05–0·16	E 1940
,, ,,	0·16	F
Rubus idaeus L.	3·80	Bo
,, ,,	1·00	BS
,, ,,	0·18–1·13	E 1940
,, ,,	1·8–8·1	D *et al.* 1960

Saxifragaceae

Ribes sylvestre (Lam.) Mert & Koch	0·70	Bo
,, ,,	0·04–0·5	E 1940
Ribes uva-crispa L.	4·11	Bo
,, ,,	1·0	BS
,, ,,	0·13–5·41	E 1940

Tiliaceae

Tilia cordata Miller	0·9	Be
,, ,,	0·35	BW
,, ,,	3·06	Bo
,, ,,	0·01–2·52	D *et al.* 1960
,, ,,	0·03–2·58	DH
,, ,,	0·1–3·57	E 1936/38
Tilia euchlora C. Koch	0·42	BW
,, ,,	0·3–2·9	D *et al.* 1960
,, ,,	0·17–1·98	DH
,, ,,	0·12–2·31	E 1936/38

Plant	Sugar value	Author
Tilia platyphyllos Scop.	0·82	BW
,, ,,	0·31–2·85	D *et al.* 1960
,, ,,	5·4	DH
,, ,,	0·16–7·70	E 1936/38
,, ,,	1·10	K
Tilia tomentosa Moench	0·82	BW
,, ,,	0·07–5·0	E 1936/38
Tilia vulgaris Hayne	1·04	BW

AUTHORS (see Bibliography)

Ba = Barbier, 1963
Be = Beutler, 1930
Bo = Boetius, 1948
BS = Beutler & Schöntag, 1940
BW = Beutler & Wahl, 1936
D = Z. Demianowicz, 1968a
D *et al.* = Demianowicz *et al.*, 1960, 1963
DH = Demianowicz & Hlyń, 1960
DJ = Demianowicz & Jabloński, 1966
DR = Demianowicz & Ruszkowska, 1959
E = Ewert, 1936, 1938, 1940
F = Fahn, 1949

G = Gulyas, 1967
H = Haragsim, 1974
HM = Hasler & Maurizio, 1949, 1950
HS = Haragsim & Slavikova, 1968
J = Jabloński, 1968
Ke = Kleber, 1935
KH = Kropáčová & Haslbachová, 1970
KK = Kropáčová & Kropáč, 1968
M = Maurizio, 1954a
S = Švec, 1968
Sz = Szklanowska, 1965

2.124 *Nectar production: influence of internal factors*

The production of nectar, and of sugar, is affected by a number of factors inherent in the plant itself. As well as the innervation mentioned above, these include the size of the flower and of the nectary surface, the age and maturity of the flower, the position of the flower on the plant, and the species, variety or cultivated race to which the plant belongs (cf. summaries Beutler, 1953, Maurizio, 1960c, Maurizio & Grafl, 1969, Shuel, 1964, 1970). Relationships between flower size, nectary surface, and nectar production, for example, have been established for *Phacelia* and various species of *Lamium* and *Citrus* (Fahn, 1949, Gulyas, 1967, Zimna, 1959). It is likely that differences in the nectar and sugar secretion of diploid and polyploid forms of the same species depend partly on differences in the nectary surface (Bond, 1968, Maurizio, 1954a, 1958c). Considerable differences in nectar and sugar production have been established between species, varieties, cultivated races, and clones, e.g. for rape, cherry, plum, apple, orange, lime, willow, lucerne, red and white clover, buckwheat, *Phacelia*, and cultivated bilberries (Bailey *et al.*, 1954, Bilozorova, 1964, Brewer & Dobson, 1969, Z. Demianowicz, 1957, 1968a, Demianowicz & Hlyń, 1960, Demianowicz & Ruszkowska, 1959, Fahn, 1949, Kropáčová & Laitová, 1965, Kropáčová & Nedbalová, 1970, Mommers, 1966, Oertel, 1956, Pankiw & Bolton, 1965, Pedersen, 1953,

1958, Shaw *et al.*, 1956, Shuel, 1952, Szklanowska, 1957, 1965, 1967, Vansell, 1939). Differences have been found between cultures of chives 1, 2, and 3 years old (Kropáčová & Kropáč, 1968).

The influence of the position of the flower on the plant is less clearly understood. Flowers at the top of *Tilia platyphyllos* secrete less nectar than those on lower branches, but the sugar content is higher at the top, and the sugar value of the two groups is nearly the same (Beutler & Wahl, 1936). Experiments with *Brassica napus* and *Rubus fruticosus* gave similar results (Beutler, 1953, Hasler & Maurizio, 1949, Percival, 1946). The diameter of the branch on which the flower is situated seems to have a more definite effect (Percival, 1946). Flowers of the male and female forms of plants show certain differences in their nectar secretion: male flowers of the banana (*Musa paradisiaca*) secrete 4–5 times as much nectar and sugar as female ones, and male willow flowers (*Salix* sp.) produce more nectar and sugar than female flowers; with cucumber and marrow (*Cucumis, Cucurbita*) the reverse is true, the female flowers producing 3–4 times as much nectar and sugar as the male (Beutler, 1953, Fahn, 1949, Kropáčová & Nedbalová, 1970). Differences in sugar concentration have been found between male-sterile and male-fertile flowers of onion (*Allium cepa*) (Lederhouse *et al.*, 1968).

The age and condition of flowers have important effects on the secretion of nectar. Sometimes secretion starts already in the closed bud; as a rule, it increases from the opening of the flower until its full development and then decreases as it fades. Usually secretion ceases after fertilization, though sometimes very old flowers secrete large quantities of nectar containing little sugar. If pollination and fertilization do not occur (because of rain or because the flower has been enclosed), nectar secretion may go on for quite a long time. Visits by insects, or artificial withdrawal of the nectar, stimulate secretion (Barbier, 1963, Beutler, 1953, Boetius, 1948, Ewert, 1938, Fahn, 1949, Zimna, 1959, where further references are given). Towards the end of the period of secretion some of the secreted nectar is re-absorbed (Agthe, 1951, Boetius, 1948, Bonnier, 1879, Lüttge, 1961, 1964, Pedersen *et al.* 1958, Shuel, 1961, Ziegler, 1955, 1968b, Ziegler & Lüttge, 1960). Flowers with several stages of maturity often show differences between the male and female stage. For example, the flowers of lime trees and *Phacelia* species secrete twice or several times as much nectar and sugar at the female stage as at the male stage of development (Demianowicz & Hlyń, 1960, Szklanowska, 1957, 1965, 1967).

2.125   *Nectar production: influence of external factors*

External factors influence nectar production substantially: for instance soil humidity; type of soil and use of fertilizers; temperature and wind;

time of day and of the year; length of day and sunshine. In addition to the general climatic conditions, the microclimate of the immediate vicinity of the flower is likely to be significant (Büdel, 1956, 1957, 1959).

If the relative humidity is high, nectar is generally secreted in large amounts but contains little sugar; in dry air less is secreted, but the sugar concentration is high (Barbier, 1951, Beutler, 1930, 1953, Boetius, 1948, Dietz, 1966, Fahn, 1949, Hasler & Maurizio, 1949, Huber, 1956, Kropáčová, 1963, Szklanowska, 1957, Zimna, 1959, where further references may be found). This phenomenon is due to the hygroscopic effect of the sugar contained in the nectar: nectar absorbs more water from a saturated atmosphere than from dry air (Beutler, 1930, Shuel, 1952). This applies chiefly to flowers in which the nectar is exposed (Beutler, 1930, Fahn, 1949, Park, 1929). It has been found, however, that in hot climates (Israel, Morocco) nectar is more concentrated at high than at lower relative humidities (Barbier, 1951, Fahn, 1949).

The nature of the soil, its moisture content, and the use of fertilizers, can affect nectar secretion in various ways—observations have shown that some crops (lucerne, mustard) produce a particularly good nectar flow on limestone soil and others (white clover, buckwheat) on sandy soil or marl (Beutler, 1953, Bonnier, 1879, Ewert, 1936, Johnson, 1946, Shuel & Pedersen, 1953). Precise experimental methods have revealed that aeration and moisture of the soil, are the decisive factors: poor aeration and lack of moisture reduce nectar production. An increase in secretion was obtained with a soil saturation of 45–75%, with an optimum at 60% (Beutler, 1953, von Czarnowski, 1952, 1953a, Percival, 1946, Shuel & Shivas, 1953). It seems that the combined action of temperature, aeration, and saturation of the soil with water increases assimilation and, in consequence, the production of nectar.

In considering the effect of fertilizers on nectar secretion, two points have to be kept in mind: the influence on secretion by individual flowers, and the indirect effects on the plant as a whole—the number of flowers per plant and per unit area. In general, a comprehensive fertilizer programme favour flowers formation and prolonged flowering; both of these are reduced by the absence of certain substances, with the result that the nectar yield per unit area is lessened. Numerous, and to some extent contradictory, observations have been made on the direct influence of fertilizers on nectar secretion. There is general agreement that an excessive supply of nitrogen has an adverse effect, whereas secretion is increased by an addition of potassium as a fertilizer (Beutler, 1953, Beutler & Schöntag, 1940, von Czarnowski, 1953b, Demianowicz & Ruszkowska, 1959, Ewert, 1935, 1936, Hasler & Maurizio, 1949, 1950, Ryle, 1954, Schöntag, 1953, Shaw et al., 1957, Shuel, 1955, 1957, 1964, Shuel & Pedersen, 1953, Kropáčová, 1963, Szklanowska, 1965, 1967).

The addition of calcium and magnesium had a positive effect with red clover, but has been found ineffective with rape and *Phacelia*, as were additional treatments with phosphorus and boron (Hasler & Maurizio, 1949, 1950, Ryle, 1954, Szklanowska, 1965).

Temperature, length of day, and amount of insolation are important factors in nectar secretion. Secretion does not start below a certain temperature, the critical threshold varying according to plant species. The nature of the process is probably enzymatic, and needs further clarification (Barbier, 1951, Beutler, 1953, Boetius, 1948, Fahn, 1949, Shuel, 1952, 1970, Vansell, 1940). The relationship is less clear with shrubs and trees (blackberry, lime) because woody plants can draw on previously formed reserves for the carbohydrates used in nectar production (Maclachlan, 1940, Maksymiuk, 1960, Percival, 1946). This applies also to plants which flower before the leaves appear.

Normally nectar is not secreted at a constant rate. With most plants there is not only a difference between day and night, but a characteristic rhythm of secretion throughout the 24 hours (Beutler, 1930, Kleber, 1935). Usually nectar secreted during the night contains more water than nectar secreted in daytime, probably because of the higher humidity of the air. During the day, the rate of nectar secretion may remain constant while the sugar content varies (*Borago, Teucrium*) or, conversely, the quantity may fluctuate while the sugar concentration remains the same (*Lythrum, Salvia*). In some plants, the peaks of nectar volume and of nectar sugar content are reached at the same time (*Helianthus*); in others, the two peaks occur at different times (*Tilia, Chamaenerion*). For many plants the optimal time for nectar secretion (maximum quantity and sugar content) is during the morning (*Helianthus, Hyssopus, Origanum, Salvia*); more rarely it is in the afternoon (*Tilia*).

The foraging of bees is adjusted to the daily plant rhythm: the bees show a marked sense of time (Kleber, 1935). Their visits to a certain species of flowers reach a maximum in the peak period of nectar secretion, i.e. when the flowers offer the greatest quantity of the most highly concentrated nectar. When a plant has two separate peak periods, the curve plotted for bee visits will also show two peaks (Figure 2.125/1). During the intervening periods of lower nectar secretion, the number of visiting bees decreased (Kleber, 1935).

2.13 Honeydew

2.131 *The origin of honeydew, and honeydew producers*

Certain plant-sucking insects excrete a substance containing sugar which is called honeydew. All honeydew-producing insects belong to the order Rhynchota and are characterized by the structure of their mouthparts.

These are adapted for sucking, with four 'pricking bristles', which can move against each other and enable the insect to puncture plant tissues. The two central bristles each have two grooves; placed closely together

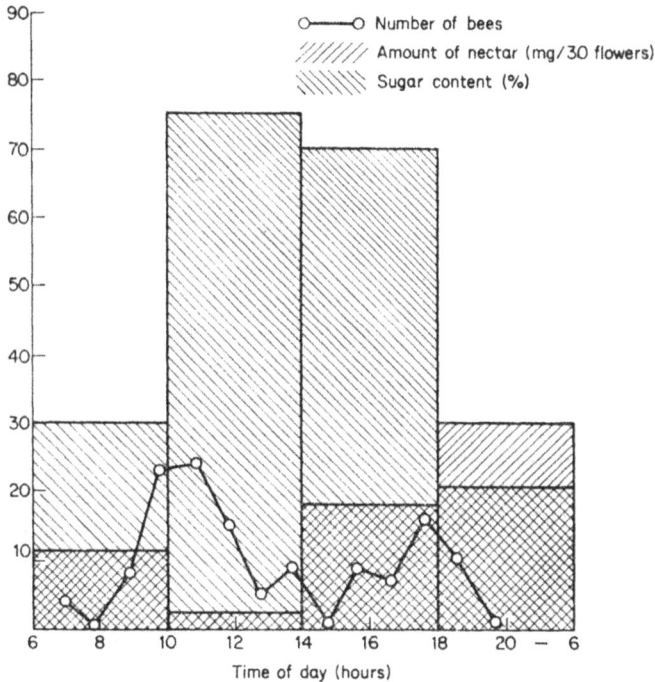

Figure 2.125/1 Diurnal distribution of nectar secretion by *Tilia cordata*, and of honeybee visits to the flowers. (drawing by Schmitz, after Kleber, 1935)

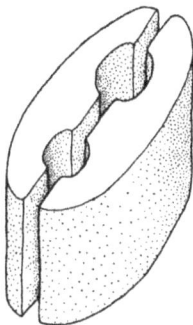

Figure 2.131/1 Cross section through the two central bristles of the sucking apparatus of Rhynchota, showing the (wider) feeding tube and the (narrower) salival tube. (after Kloft, 1965)

they form a double tube (Figure 2.131/1). Through the wider of these, the feeding tube, the plant sap passes into the gut of the insect, partly by means of the pressure within the plant itself, and partly by the insect's active pumping. Through the narrower (salival) tube, saliva containing enzymes is introduced through the puncture into the plant tissue. All the bristles are enclosed in the lower lip, as in a sheath, the whole forming the sucking apparatus of the insect.

The individual groups of Rhynchota get their food from different parts of plant tissue; the most prolific producers of honeydew are those which suck from the sieve tubes. They puncture the sieve tubes of the vascular bundles, and derive their food directly from the phloem sap. This group, the most important for honey production, includes a large number of Coccina, all the Aleyrodina and Psyllina, and most of the Aphidina and Cicadina (cf. Kloft, 1965).

The plant sap ingested through the sucking apparatus passes into the gut, which in plant-sucking insects often contains so-called filter chambers, parts of the hindgut which are attached directly to the front part of the foregut, and allow any excess liquid from the ingested food to pass directly into the hindgut, by-passing the midgut. Such a large quantity of food is ingested that honeydew production can be very prolific. An adult nymph of *Tuberolachnus salignus*, for example, secretes up to $1 \cdot 9$ μl an hour, corresponding to a liquid intake one-third of its own weight; with young larvae the liquid intake per hour may greatly exceed their body weight (Kloft, 1965, Mittler, 1957, 1958a, 1958b).

Honeydew is secreted in small droplets, which fall on to the surface of leaves and twigs, and are there collected by bees and other insects (Figure 2.131/2—see Plate 2). The plant-sucking insects skilfully avoid getting their own bodies sticky from the honeydew, which quickly solidifies in the air (Kunkel, 1969, 1972). Newly secreted it is colourless; it becomes light brown only when exposed to the air. On average, it contains 5–18% of dry substance, its relative density is $1 \cdot 0$–$1 \cdot 3$, and its pH $5 \cdot 1$–$7 \cdot 9$. The droplets solidify quickly through evaporation, the dry substance rising to 35–50% or more of the total. In its passage through the gut of the insect, the plant sap has undergone considerable physico-chemical changes, so the secreted honeydew differs greatly from the original phloem sap (Auclair, 1958, 1959, 1963, Ehrhardt, 1961, 1962, 1963, Kloft, 1958, 1960a, 1960b, 1960c, 1962, 1963, 1965, Kloft & Ehrhardt, 1962, Maltais & Auclair, 1962, Maurizio, 1965a, Michel, 1942, Mittler, 1957, 1962, Munk, 1969, Pourtallier, 1962, where further references are given).

In recent years a great deal of research has been done on the feeding, digestion, and metabolism of plant-sucking insects and the mechanism of honeydew production. This has shown among other things that, contrary to earlier views, the insects do not absorb mainly nitrogen from the phloem sap, leaving the carbohydrates to be excreted unchanged in the honeydew. The total nitrogen content of honeydew is often only slightly less than that of the phloem sap; also, the carbohydrate spectrum is changed considerably during the passage through the insect. The situation is further complicated by the fact that the plant-sucking insect is host to endosymbionts capable of synthesizing sterins, vitamins, and amino acids, which thus aid its metabolism. This symbiosis with micro-organisms enables the insect to make up for the fluctuations (or absence) of certain substances in the phloem sap (Ehrhardt, 1969, where further references are given).

2.132 Chemical composition of honeydew

Honeydew always contains enzymes derived from the secretions of the salivary glands and the gut of plant-sucking insects. So far, invertases, diastase, a peptidase, and a proteinase have been found (Duspiva, 1953, 1954a, 1954b, Ehrhardt, 1962, Geinitz, 1930, Kloft, 1960b, Srivastava & Auclair, 1962a, 1962b, 1963, Wolf & Ewart, 1955b).

The nitrogen content of honeydew amounts to 0·2–1·8% of the dry substance; 70–90% of it are amino acids and amides. In general the amino acid content of honeydew shows seasonal fluctuations corresponding to those of the amino acid level in the phloem sap of the host plants; there are, however, certain regular differences in the composition of the two substances. The honeydew of some plant-sucking insects has been found to contain as many as 22 amino acids and amides, among them aspartic acid and asparagine, glutamic acid and glutamin, cystine, serine, glycine, threonine, alanine, tyrosine, valine, leucine and *iso*leucine, phenylalanine, histidine, lysine, and proline. Several of these do not occur at all in the phloem sap of the host plants (von Dehn, 1961, Ehrhardt, 1962, Gray, 1952, Maltais & Auclair, 1962, Mittler, 1958a, 1958b, Tamaki, 1964a, 1968).

Honeydew always contains organic acids, especially citric, more rarely malic, succinic, and fumaric acids (Gray, 1952, Lamb, 1959). Carbohydrates comprise 90–95% of the dry substance of honeydew. In the main they consist of various sugars, often including some that are not present in the phloem sap but are synthesized during the passage through the insect, by the action of enzymes from the gut and the salivary glands. The action of these enzymes, and their range of activity, vary with different plant-sucking insects, and even in the same species there are differences between the action of enzymes in the saliva and of those in the gut. As a rule, both enzymes break down sucrose, maltose, and trehalose, the enzymes of the gut being the more effective. Melezitose, raffinose, melibiose, and starch are less affected, and often by the enzymes of the gut only. The inverting enzymes of plant-sucking insects not only convert complex into simple sugars; they also synthesize new sugars by means of group transference, linking one or several molecules of fructose or glucose to the sugars already present. Dependent upon the properties of the active enzymes, the transference can take place either as transglucosidation or as transfructosidation, producing di- or trisaccharides or higher-molecular oligosaccharides; free glucose or fructose is produced at the same time. For example, the enzyme of the gut of many plant-sucking insects acts as a transglucosidase which, in breaking down sucrose, produces two characteristic trisaccharides: melezitose and fructomaltose (erlose) (Bacon & Dickinson, 1957, Duspiva, 1953, 1954a, 1954b, Ehrhardt, 1962, Ewart & Metcalf, 1956, Gray & Fraenkel, 1953, 1954, Haragsim, 1963, 1966, Maurizio, 1965a, Pourtallier, 1962, Smith & Srivastava, 1957, Stephen, 1959, Wolf & Ewart, 1955a, 1955b, where further references are given; cf. also 2.122).

Several groups of plant-sucking insects can be distinguished by the sugar spectrum of their honeydew. The simplest honeydew contains only the three main sugars (sucrose, glucose, fructose). Often these are accompanied by maltose, trehalose, melezitose, fructomaltose, and a number of higher-molecular oligosaccharides. The two trisaccharides (melezitose and fructomaltose) are not mutually exclusive, but one or the other is usually present in a much larger quantity. Occasionally further sugars are found in honeydew, such as raffinose, melibiose, mannose, rhamnose, stachyose; some of these sugars are harmful to bees (Basden, 1965, 1966, 1968, Geissler & Steche, 1962, Sols et al., 1960, Staudenmayer, 1939, Tamaki, 1964b, Vogel, 1931). The 'manna' or 'lerps' found on some types of Eucalyptus and Angophora, for example, may contain up to 80% raffinose (Basden, 1965).

In certain cases the sugars in honeydew are replaced or accompanied by sugar alcohols (dulcitol, sorbitol, inositol, ribitol). Sugar phosphates have occasionally been found (Duspiva, 1954a, Gray & Fraenkel, 1954, Hackmann & Trikojus, 1952).

As with nectar, the carbohydrate spectrum of honeydew may affect the honey yield, since not all sugars are equally attractive to bees. Honeydew rich in sugar alcohols is disregarded by many insects, among them the honeybee (von Frisch, 1934, Jamieson & Austin, 1958, Vogel, 1931, Wykes, 1951, 1952b).

Nectar, as well as honeydew (and honey), may contain substances injurious to bees and to human beings. Bees foraging exclusively on *Rhododendron ponticum*, indigenous to Asia Minor, produce poisonous honey, a fact known in antiquity and thoroughly investigated in recent years. The active substance, andromedotoxin-acetylandromedol, has been detected in the nectar of several Ericaceae, e.g. *Rhododendron ponticum, Kalmia latifolia, Ledum palustre*, and *Tripetaleia paniculata* (Gruch, 1957, Maurizio, 1968c, Plugge, 1891, Poltev, 1956, Tokuda & Sumita, 1925, White & Riethof, 1959). In New Zealand, honey produced from the honeydew flow on *Coriaria arborea* has been found to have a toxic effect. The honeydew of this plant—secreted by the insect *Scolypopa australis*—contains the active substances picrotoxin, tutin, and mellitoxin (Harris & Filmer, 1947, Palmer-Jones, 1947*a*, 1947*b*, Paterson, 1947, Sutherland & Palmer-Jones, 1947*a*, 1947*b*).

2.2 THE ELABORATION OF HONEY

2.21 Collection of raw materials by the honeybee

The honeybee ingests solid and liquid foods through its mouthparts, from which they are conveyed into the honey sac and the gut. These organs, in particular the mouthparts and the honey sac, are differently developed in the three castes of honeybee (queen, worker, drone). General descriptions of the anatomy and function of the mouthparts and the gut will be found in the literature (Chauvin, 1968*a*, Dade, 1962, Leuenberger & Morgenthaler, 1954, Morgenthaler, 1953, Snodgrass, 1949, 1956, Weber, 1966, Zander, 1951); here only the anatomy of the worker will be considered, in so far as it relates to the ingestion and processing of the raw materials of honey.

2.211 *The oral cavity, mouthparts, and honey sac of the worker honeybee*

The oral cavity consists of two parts: the preoral cavity (cibarium) and the oral cavity proper. The preoral cavity is bounded by the scutellum and the labrum (*Lm*) in front, and by the prementum (*Prmt*) at the back; at the sides are the labium (*Lb*) and the maxillae. The oral cavity proper forms a groove across and behind it, between the base of the mouthparts, leading to the pharynx, and behind it, between the base of the mouthparts, leading to the pharynx, the oesophagus (*Oe*) and the honey sac (Figure 2.211/1A).

The mouthparts of the honeybee consist of four parts: the single upper lip or labrum (*Lm*), paired mandibulae and maxillae, single labium (*Lb*); Figure 2.211/1A. Attached to the base of the maxillae are three branches, called lacinia or lamina interna (*lcl*), galea or lamina externa (*Ga*), and maxillary palp (*MxPlp*). The base of the labium is formed by the triangular postmentum (*Pmt*) and the elongated prementum (*Prmt*). Attached to the prementum is the narrow hairy glossa (*Gls*), with two short paraglossae (*Pgl*), and two labial palps (*LbPlp*). At the tip of the glossa is the labellum (*Fbl*), shaped like a spoon and equipped with hairs, which facilitates the intake of liquid food (Figure 2.211/B).

The most important part of the mouthparts is the proboscis, a tube bounded by the galeae of the maxillae and the labial palps. The glossa

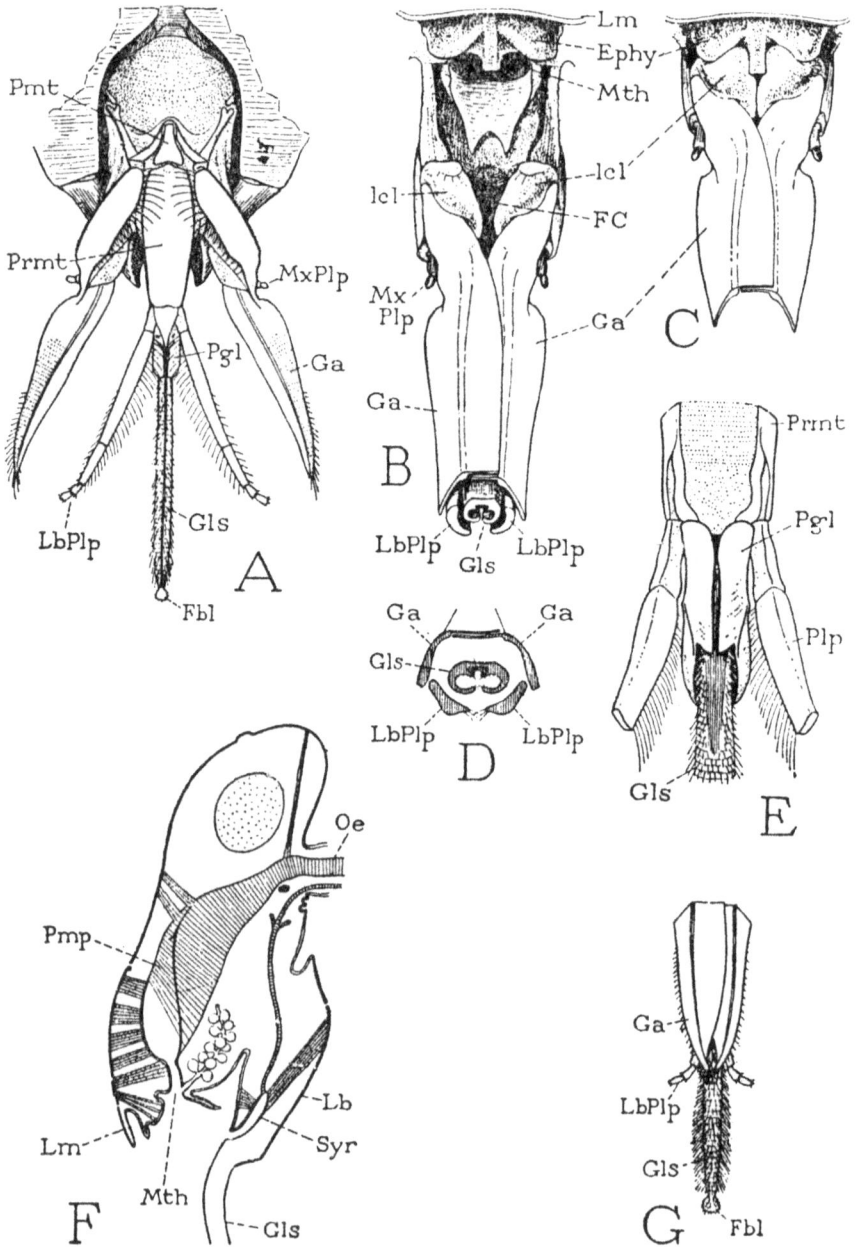

Figure 2.211/1 Mouthparts of the worker honeybee.   (after Snodgrass, 1949)
A proboscis, seen from behind, with its parts artificially spread out
B front view of base of proboscis pulled down from head, exposing mouth
   (Mth) and deep food channel (FC) leading up to mouth

moves back and forth inside this tube (Figure 2.211/1, B-F). In contrast to those of other insects, the honeybee's proboscis is not permanently closed. It functions through the interlocking of the mouthparts concerned and the action of the pharyngeal muscles, which together produce an airtight closure of the proboscis, enabling the bee to pump liquid food into the pharynx and the honey sac. Round the glossa the closure of the proboscis is effected by the action of the maxillary galeae and the labial palps, which close like a sheath. There is, however, a gap higher up, and while the bee sucks with the proboscis extended this is closed by the epipharynx (*Ephy*); *see* Figure 2.211/1, B, C, D. When the proboscis is retracted, this gap remains open and leads to the glossal groove, a smooth shallow groove at the base of the glossa through which the bee can pump back the contents of the honey sac and transfer it to another bee. The functioning of the proboscis is ensured by strong muscles, which provide the power for both ingestion and regurgitation of liquids. At rest, the proboscis is folded back by the tension of the muscles and placed on the head. Two muscles running across from the labium to the rear wall of the head enable the bee to extend the proboscis into its working position. The length of the proboscis in worker bees varies in different genetic races, from 5·9 to 7·1 mm.

The liquid ingested through the proboscis passes through the pharynx and oesophagus into the honey sac (Figure 2.211/2). Anatomically this pear-shaped sac is a widening of the oesophagus rather than a part of the gut, and its elastic walls are of a structure similar to those of the oesophagus. When filled to capacity, the honey sac holds 50–60 μl and weighs 40–70 mg. The weight of the full honey sac may amount to more than 90% of the body weight of the bee (Armbruster, 1921*a*, Betts, 1929, Chauvin, 1968*a*, von Frisch, 1934, Gontarski, 1935, Snodgrass, 1956, Wells & Giacchino, 1968).

The honey sac is not a digestive organ, but a collecting chamber, whose function is to receive and transport liquid foods. It is closed off from the ventriculus—the next part of the alimentary canal—by the pro-

---

C   same, food channel closed by base of proboscis brought up against the mouth
D   cross section through middle of proboscis
E   base of tongue, paraglossae, labial palpi (*LbPlp*) and distal end of labial prementum (*Prmt*), anterior view
F   lengthwise vertical section of head showing sucking pump (*Pmp*), salivary syringe (*Syr*), and associated structures
G   distal part of proboscis with tongue protruded
Key
Ephy, epipharynx; *Fbl*, labellum; *FC*, food channel on base of proboscis; *Ga*, galea; *Gls*, tongue (glossa); *Lb*, labium; *LbPlp*, labial palps; *lcl*, lacinial lobe of maxilla; *Lm*, labrum; *Mth*, mouth; *MxPlp*, maxillary palp; *Oe*, oesophagus; *Pgl*, paraglossa; *Pmt*, postmentum; *Prmt*, prementum.   (after Snodgrass, 1949)

ventriculus, which is anatomically still part of the foregut. The proventriculus consists of a funnel-shaped 'valve' protruding into the honey
sac and a thin stem which connects it with the midgut. The proventricular
valve is made up of four triangular flaps whose inner edge has a chitinous
rim from which bristles protrude at an angle into the proventriculus.
When the flaps are closed these bristles close its cross-shaped opening.
Under the grooves, i.e. between adjacent flaps, there are four pouches
(Figure 2.211/2).

Figure 2.211/2   Details of the proventriculus. Drawn by Hodges from
Zander (1951), Bailey (1952).

   A Proventricular valve viewed from the honey sac
   B Longitudinal cross-section through the proventriculus with
      pollen grains in the pouches
   F fold   *H* fringe of hairs
   L chitin layer   *p* pouch with pollen grains (*right*) and
                                 dislodged pollen mass (*left*)

The function of the proventriculus is to regulate the passage of food
from the honey sac to the midgut, and to retain the liquid collection in
the honey sac while it is being transported to the hive. It is thus the
proventricular valve which provides the access to the actual digestive
tract (ventriculus). Moreover, small solid components in the honey sac,
such as pollen grains, or nosema and foul brood spores, are filtered off
by the proventriculus and moved on, first into the four pouches, and then
(in small packets) into the midgut. In this way the pollen content of
nectar, for instance, is reduced to a half or a third of its original amount
within 15 minutes after ingestion (Bailey, 1952, 1954, Maurizio, 1949*b*,
Sturtevant & Revell, 1953, Todd & Vansell, 1942, Whitcomb & Wilson,
1929).

2.212 The glands in the head and thorax

In the head and thorax of the worker honeybee, a number of glands are connected to the oral cavity. Those important for processing and ripening honey are the pharyngeal or hypopharyngeal glands, the postcerebral, thoracic, and labial glands, and possibly also the mandibular glands (Figure 2.211/3).

The hypopharyngeal glands lie in pairs in the front part of the head cavity. They consist of two tubes, about 16 mm long and studded with about a thousand rounded glandular lobes, which open separately into the back wall of the pharynx immediately behind the oral cavity. The function of these glands is twofold: they secrete part of the food needed for brood rearing, 'bee milk', and they also produce a secretion rich in enzymes (diastase, invertase, glucose oxidase) used in elaborating honey. The invertase activity of the hypopharyngeal glands is dependent on the age and physiological condition of the bee. In the undeveloped glands of newly emerged workers the activity is very low; it is higher in the (still young) nurse bees whose glands are fully developed and which secrete brood food; activity reaches a maximum in the foraging bee about four weeks old, whose glands have already degenerated. In winter, bees' invertase activity is much reduced, and it does not increase again until

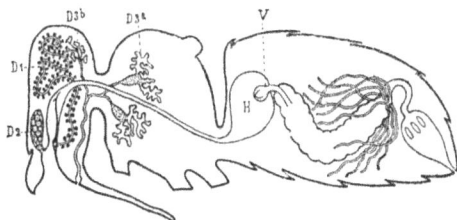

Figure 2.211/3 Digestive apparatus, and head and thoracic glands, of the worker honeybee. (after Leuenberger & Morgenthaler, 1952)
D1 pharyngeal gland, D2 mandibular gland, D3a post-cerebral gland, D3b labial gland, H honey sac, V proventriculus

brood rearing starts in spring (Gauhe, 1940, Gontarski, 1954, Gontarski & Hoffman, 1963, Kosmin & Komarov, 1932, Kratky, 1931, Maurizio, 1954b, 1961, 1962a, 1962b, 1968b, Örösi-Pál, 1968, Simpson, 1960, 1962, Simpson, Riedel & Wilding, 1968, Soudek, 1927, Wahl, 1963, Zander & Weiss, 1964).

The salivary gland of the head and thorax—both paired—are situated respectively at the back of the head cavity and in the anterior ventral part of the thorax. They consist of a large number of pouches, elongated (thoracic glands) or pear-shaped (postcerebral glands), which drain through a common channel at the base of the glossa. Most scientists who have studied the secretion from these glands believe that it provides the liquid needed to dissolve solid food and does not contain any enzymes to break down carbohydrates (Kosmin & Kosmarov, 1932, Kratky, 1931, Örösi-Pál, 1968, Simpson, 1960, 1962). An isolated contrary view has been expressed that this secretion does contain diastase and invertase (Inglesent, 1940).

The mandibular glands are pairs of pouches opening outside the oral cavity, at the front inner base of the mandibles. Their secretion for the most part contributes to the brood food; it has however been discovered in the honey sac contents, and there is a possibility that it may contribute to the ripening of honey (Örösi-Pál, 1968).

2.22   Processing the material collected

2.221   *Receiving the raw materials*

The raw sugar-containing materials (nectar, honeydew or other liquids) are carried into the hive by the foraging bee in her honey sac. The raw material is already mixed with saliva, and received its first dilution when it was ingested by the forager, so the contents of the honey sac generally has a lower sugar concentration than the original raw material. The saliva which is added comes from the hypopharyngeal and salivary glands. By her ability to dilute the material collected, the bee is enabled to ingest highly concentrated raw materials and ripe honey. Admixture of saliva even makes it possible for the honeybee to ingest dry sugar (Free & Durrant, 1966, Gontarski & Hoffmann, 1963, Örösi-Pál, 1968, Simpson, 1960, 1962, 1963, 1964).

When she enters the hive, the forager passes on the contents of her honey sac to one or several 'house' bees. She opens her mandibles with her proboscis retracted, and a drop of liquid appears at the base of the glossa. The house bee extends her proboscis fully, and sucks up the drop. Experiments with sugar solutions marked with dyes or radioactive tracers have shown that the food received is quickly passed on from one bee to another; according to some authors, drones are also included in this food chain. The speed of food transmission depends on various factors, such as temperature, the age of the bees, their race, the strength of the colony, and the supply of raw material (Delvert-Salleron, 1963, Gösswald & Kloft, 1961, Hoffmann, 1966, Kloft, 1969, Lecomte, 1961, Lensky, 1961, Nixon & Ribbands, 1952, Oertel *et al.*, 1953, Park, 1923, Pershad, 1967). The duration of the processing, and the number of bees taking part in it, depend on the strength of the colony and the supply of raw material. During a strong nectar flow the partly-ripened honey is stored in the cells of the comb immediately, or after only a few transferences from bee to bee. During a moderate or weak flow the food is passed on to and by many bees before it is stored. The greater the number of bees in the food chain, the richer the ripe honey will be in their secretions and hence in enzymes.

2.222 Ripening the honey

As collected, the raw materials of honey contain too much water to allow storage: much of the water must be evaporated, and this is done during the ripening process. For a long time it was believed that the reduction of the water content of the collected material begins in the honey sac during the bee's return flight. It was thought that the walls of the honey sac were semi-permeable and that part of the water passed from the honey sac into the haemolymph and thence through the Malpighian tubules into the rectum, from which it was excreted in flight (Brünnich, 1940). We now know that the walls of the honey sac are impermeable to water, that the collected material brought into the hive usually contains more water than the raw material ingested, and that the liquid discharged by bees in flight has a different physiological significance (Oertel et al., 1951, Park, 1932, 1949, Pasedach-Pöverlein, 1941). So it is accepted that the concentration and ripening of honey takes place only in the hive. There are two phases of this process; the bees are actively involved in the first phase, but only passively in the second, during which excess water evaporates from the honey stored (Park, 1925, 1927, 1928, 1933, 1949).

A bee actively engaged in processing nectar into honey pumps out the contents of her honey sac into a flat drop on the underside of the proboscis; she then draws it up again (Figures 2.222/1 and 2.222/2—see Plate 2), the process being repeated rapidly for 15–20 minutes. The liquid is thereby exposed, in a thin film, to the warm, dry air of the hive, and loses some of its water. In this way the bees produce half-ripened honey, containing about 50–60% (maximum 70%) of dry substance. The second, passive, phase of honey ripening follows. The bees deposit the half-ripened honey in small droplets on the cell walls, or in a thin film on the cell floor. As a rule, $\frac{1}{4}$ to $\frac{1}{3}$ of the cell is filled; but during a strong flow, or if there is lack of space, $\frac{1}{2}$ or $\frac{3}{4}$ of each cell is filled straight away. Normally, when the honey is nearly ripe, the bees move it again, and the cells are then filled to $\frac{3}{4}$ of their capacity. The final ripening takes 1–3 days, depending on the water content when the honey is first put into the cells, the level to which the cells are filled, the amount of air movement achieved (which itself depends on the strength of the colony, and hive ventilation), and temperature and relative humidity (Gontarski & Hoffmann, 1963, Park, 1927, 1928, 1933, 1949, Rheinhardt, 1939,

Woźnica, 1966). In experiments where the combs in a hive were covered with wire netting (allowing the bees no access to them), a 60% sugar solution was ripened within 48 hours, and a 20% solution in 72 hours (Figures 2.222/3 and 2.222/4). If only ¼ of the cells are filled, the honey ripens faster than if they are ¾ full. Additional ventilation provided in the hive can speed up the ripening process by 1–4 days; inadequate ventilation may delay it up to 21 days (Park, 1925, 1929, 1933, 1949, Rheinhardt,

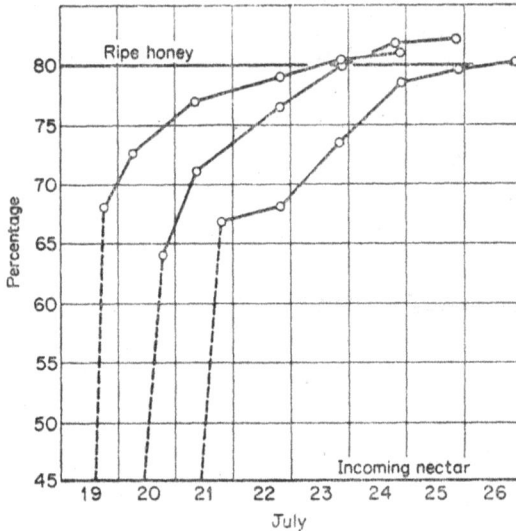

Figure 2.222/3 Honey ripening in the cells. Evaporation of the honey stored in cells until it is ripe (water content 20%).

(after Park, 1925, 1949)

1939). Only when the honey is ripe, i.e. when it contains 20% water or less, do the bees fill the cells completely and seal them with an airtight wax capping. This airtight seal prevents subsequent absorption of water from the surrounding air and thus the risk of fermentation. If temperatures are low and the relative humidity high, for instance in damp winter climates, there is still a risk of fermentation in the stored honey (cf. Chapter 9).

2.223 *Chemical changes during the ripening process*

While water is being evaporated from it, ripening honey undergoes chemical alterations, which chiefly affect the carbohydrates. The action of inverting enzymes changes the sugar spectrum of the original raw material into the form characteristic of ripe honey. Most of the enzymes derive from the secretions of the bees, but certain amounts of invertases have been found in the raw materials themselves. Honeydew is especially rich in invertases (Duspiva, 1953, 1954a, 1954b, Ehrhardt, 1962, Frey-Wyssling et al., 1954, Kloft, 1960b, Maurizio, 1965a, 1968a, Shuel, 1956, Rinaudo et al., 1973, M. Zimmermann, 1953, 1954; cf. 2.122 and 2.132, where further references are given).

The raw material contains varying amounts of sucrose, and often of other sugars, according to its origin (see 2.122 and 2.132). During the ripening process these sugars are broken down by enzymatic action into the simple sugars, glucose and fructose. Simultaneously with this breakdown, a synthesis takes place of sugars that have been newly discovered and are little known. These sugars are absent from the raw materials, but are characteristic of the sugar spectrum of honey (Duisberg, 1967, Maurizio, 1962c, 1965a, 1968a, and Chapter 9, where further references are given).

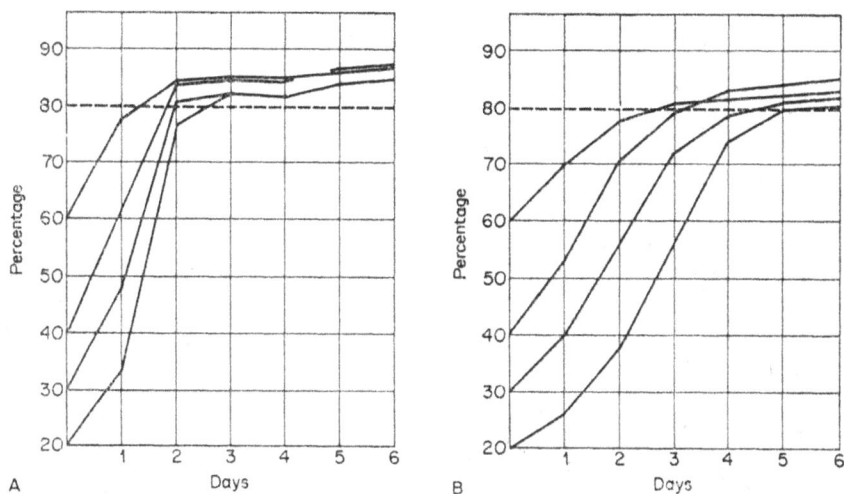

Figure 2.222/4 Honey ripening in combs where the cells are one-quarter (A) and three-quarters (B) full. The combs were in the hive, but inaccessible to the bees.   (after Park, 1925, 1949)

The sugar spectrum of honey is thus to be interpreted as the final result of the action of a number of plant and animal invertases on the sugars of the original raw materials. Usually enzymes derived from the bees predominate; in extreme examples, however, the sugars and enzymes of the raw materials may appear in the sugar spectrum of the ripe honey. For example honeys from *Robinia* and *Tupelo* are rich in fructose; honeys from rape and dandelion are rich in glucose; and honeys from larch and spruce honeydew contain melizitose (cf. 2.122 and 2.132 for further references).

CHAPTER 3

THE BEEKEEPER'S PART IN PRODUCING HONEY

by Dr. Francis G. Smith

SENIOR APICULTURIST, DEPARTMENT OF AGRICULTURE,
WESTERN AUSTRALIA

Yes, we shift, boys, shift, for there isn't the slightest doubt
The buds are dripping nectar in the country further out.
To where there's glowing prospect we will take a little load
And we will go site hunting on the Beeman's Road.

B. R. W.,
THE BEE-MAN'S ROAD (1962)

3.1 BASIC METHODS

3.11 Honey hunting

The most elementary method of producing honey is hunting for the nests of bees in forests. This form of honey production is also the most ancient, and today it is still of great importance in tropical Africa and southern Asia (*see* 4.25, 4.26, 17.2, 19.21).

In collecting honey from wild nests, the bees usually are driven off with the smoke from a bundle of smouldering twigs, and the combs cut out. Very often the brood combs are eaten with the rest. Nests of the African honeybee *Apis mellifera adansonii*, and of *Apis dorsata* in Asia, are usually robbed at night, as there is then less danger of the honey hunter being stung. The disturbed bees fly to the light of the burning twigs and die.

3.12 Simple-hive beekeeping

In some tropical countries there are very large numbers of simple hives, made of hollowed logs, of bark which may be protected with straw, of woven cane plastered with dung, or of coiled straw; or they may be gourds, earthenware pots, dried mud cylinders or, in more sophisticated areas, wooden boxes. Even in Europe, straw hives (skeps) are still used in fairly large numbers in north Germany and the Netherlands, and cork or wooden box hives in Spain and Portugal.

Beekeeping with these simple hives is based on the swarming instinct of bees. Empty hives are made attractive to the bees by smearing hot beeswax or propolis inside and near the entrance. The beekeeper places the hives where swarms are likely to occupy them e.g. in the crown of a tree. Less frequently, swarms that have issued and clustered are put into the hives by the beekeeper. In Central and South America and parts of Africa honey is also obtained from 'stingless bees' (Meliponinae); the complete nest of these bees is transferred into the hive. (*See* 17.3.)

The honey crop is collected after the main flow, when the hives are known to be full of honey. The methods of collection are very much the same as those from nests in hollow trees. In dry areas, where no water is available to the bees for several weeks or even months of the year, all combs are taken, because the bees would otherwise migrate as soon as water supplies failed. In areas with permanent water, there is a tendency to leave the brood combs and some food for the bees, so that the colony has a chance of survival.

In European simple-hive beekeeping, the crop is taken from only some of the hives. The bees are 'driven' from the hives to be robbed into the hives containing colonies which are being preserved.

Production from each simple hive is low, probably not more than 7 kg honey a year, because swarming keeps the size of the colony small, and the bees have to rebuild their combs each year. But as the hives cost nothing but the time spent in making them, each beekeeper can own several hundred. In Africa it is estimated that some 50 000 tons of honey or more are produced each year from simple hives and wild nests.

A step forward from the simple hives, in which the bees build combs in any way they fancy, is the use of bar hives. These may be made of any simple materials but they are opened from the top. Across the top of the hive are placed parallel bars of wood from which the bees are persuaded to build their combs, by shaping the underside of the bars or baiting them with wax or strips of comb. When the bees build their combs suspended from these bars, any one comb can be removed from the hive more easily without damaging the nest, particularly if the hive is made wider at the top than at the bottom. These hives are 'movable-comb' hives, but the comb is not in a frame.

3.13 Frame-hive beekeeping

Hives in which 'movable frames' of wood are used for the combs produce most of the world's honey supplies. The frames can be removed for inspection (Figure 3.13/1—*see* Plate 3) and extraction of the honey without damaging the combs or killing the bees. After extraction the combs can be used again by the bees, which is a considerable economy.

The frame hive is capable of vertical expansion and contraction by the addition or removal of boxes, each of which contains eight to ten frames. These hives are usually kept 30–100 together in an apiary (Figure 3.13/2—see Plate 3); a greater number would normally 'overstock' the foraging area; *see also* 4.4.

Although frame hives can be transported with bees in them very easily, most beekeeping is done with the hives in one place throughout the year. Under favourable conditions crops up to 100 kg per hive can be obtained from static beekeeping, but 15–20 kg is more usual. The greatest production is obtained by moving apiaries from place to place, as the various plants come into flower in different parts of the country. Migratory honey production has reached its highest development in Australia; here the best commercial honey producers consistently average 200 kg per hive each year, and a record of 350 kg per hive from a holding of 450 hives has been achieved, *see also* 4.24.

3.2 MODERN METHODS

3.21 Single-queen methods

The majority of beekeepers operate their hives with one queen to each colony, as is normal in nature. Commercial honey producers replace their queens at least every two years, or every year in some regions with long active seasons. This is necessary because egg production tends to fall off in the third year, and because swarming becomes more common after the first year of a queen's life.

Some beekeepers confine the queen and her brood to one or two hive boxes (chambers). Others permit the queen to lay anywhere in the hive. The former group may use one single box with deeper frames for the brood nest, and shallower boxes for the storage of honey; this practice is to be found among the most successful and extensive commercial honey producers. But the majority use the same standard depth of box throughout the hive, and the same design of frame for brood rearing and for honey storage. A wire grid which permits worker bees to pass through, but which stops the larger queen (a queen excluder), is used to confine her to the brood chamber.

3.22 Two-queen and multiple-queen methods

A small proportion of beekeepers operate their hives with two or more queens during the most active period of colony development, reducing again to one queen during the main honey flow. (This is most effective where there is only one honey flow in the year.) Each additional queen is separated from her neighbour by two queen excluders and one or more honey chambers (supers). The progeny of the queens combine into one strong foraging force during the honey flow.

Other methods use two or more queens throughout the year, housed in adjoining chambers which are separated from each other by a solid wooden wall, or by queen excluders which provide worker-access to neutral ground in the form of a common honey storage chamber. In the second type, additional honey supers are added on top of the central chamber, leaving each of the brood chambers readily accessible on each side of the hive.

3.23   Mechanization in the apiary

In addition to motor trucks and power-driven extracting equipment, beekeepers are using machines in the apiary to assist them to load and unload hives when 'migrating' with their hives from one flow to another, and to convey boxes of honey combs from the hives on to the truck or into the extracting plant.

Some migratory beekeepers use motorized barrows to load hives. These barrows run up a ramp on to the truck, and some have a small fork-lift incorporated to raise the hive for stacking.

Loaders mounted on the trucks are becoming increasingly common among the migratory commercial honey producers, and they make light work of moving apiaries. Boom loaders handle one or two hives at a time, and gantry loaders handle a number of hives on pallets. Some producers use tractors with fork-lifts for loading hives. These same tractors are also useful for clearing apiary sites and making fire breaks.

3.24   Pollination and other services

Bees are kept mainly for the production of honey, but beeswax is a major by-product. It is obtained from old and damaged combs and from the cappings with which the bees cover the cells full of honey. These cappings must be cut off by the beekeeper before the honey can be extracted. There is a very wide range of uses for beeswax; much goes into cosmetics and pharmaceutical products.

In regions where large areas of fruit or seed crops are grown, bee-keepers may hire out their apiaries for pollination purposes. This work may be additional to the use of the apiaries for honey production at other seasons. In important fruit and seed areas, however, the apiaries may be used so extensively on pollination work that there is very little or no production of honey. Such apiaries are moved from one orchard or field to another, in the same way that honey producers follow the honey flows in Australia (see 3.34).

Bees may also be kept for the production of royal jelly, the food produced by worker bees for feeding to queens in the larval stage. This is at present used for cosmetic and pharmaceutical purposes. Pollen may be 'trapped' and harvested, venom extracted, and propolis collected, all for pharmaceutical purposes.

3.3 HIVE MANAGEMENT

3.31 Essentials of management

There are phases of hive management which are common to beekeeping operations in any part of the world.

The first is ensuring that the foraging force of bees in the colony is built up at the right time for the collection of nectar; the second, providing space for the storage and ripening of the nectar into honey by the bees; third, removing the honey from the hive and extracting it from the combs; fourth, preparing the colony to withstand any period of dearth, whether due to excessive cold or heat, or excessive drought or rainfall. The first of these dearth periods is the temperate-climate winter, and the other three occur in various parts of the tropics and subtropics.

3.32 The build-up

During the first phase the maximum production of brood is required. The colonies are provided with vigorous young queens and adequate comb space within the hive for the queen to lay her eggs. The beekeeper also ensures that there is sufficient food for raising the brood. This is usually in the form of honey left over from the previous honey flow, together with fresh supplies of pollen from plants flowering within range of the apiary. When there is a dearth of pollen, the bees may be given other protein foods in place of pollen, or pollen which has been stored from an earlier season.

Empty hives may be stocked at the start of the active season with 'package bees'—2 or 3 lb (about 1 kg) of bees with a queen. Package bees may also be used to reinforce the colonies during the build-up.

Where the build-up is prolonged, the colonies may reach full strength prematurely and divide into two by swarming; this is their natural method of reproduction. Some beekeepers seek to overcome this problem by dividing their colonies before swarming is likely to occur, and reuniting the two parts at the beginning of the honey flow. The progeny of the additional queen then adds to the foraging force.

It is customary to raise new queens during the build-up period, since this is the most suitable part of the colony's reproductive phase. This usually occurs when most pollen is available; colonies also rear drones then, and these are needed for mating with the new queens.

3.33 The honey flow

Beekeepers strive to have the colonies at the peak of their strength at the beginning of the honey flow. At this time queen excluders may be used to restrict the queens to a smaller part of the hive than during the build-up period. Brood is now required only to maintain colony strength during the flow; it is unprofitable for the bees to use the flow for building up a larger population, since this will not be productive once the flow is over.

Extra boxes or supers are added to the hives for the storage and ripening of the nectar into honey, either on top of the existing boxes or immediately above the brood box and under the full or nearly full boxes.

Where there is only one flow, supers are added as required until the flow ends. But where there may be more than one flow, or the flow is prolonged, full supers are removed and extracted as soon as they are ready, and returned to the hives.

3.34 Migratory beekeeping

Where honey flows occur at different places at different times, it is possible for the beekeeper to move his bees from flow to flow to harvest a series of honey crops. In many countries, where the same plants flower and yield nectar each year, the annual programme of migrations is predictable and can be planned ahead. In other countries, such as Australia, the main honey-producing plants may flower only in alternate years or at longer intervals, because the period between flower bud formation to seed fall may be several years. The migratory beekeeper requires a detailed knowledge of his honey plants, and must observe their behaviour closely over a wide area (for instance at different altitudes, on different soils), to plan his migrations. Moves may involve shifting a single truck load of hives 200 km; alternatively thousands of hives may be taken 500 km or more, and mistakes can be very costly.

In cool climates the bees can be shut in their hives for the journey, provided the hives are adequately ventilated by travelling screens. In hot climates many colonies can be killed if they are shut in the hives. So methods have been devised for loading and transporting hives with their entrances left open. The hives are loaded on trucks in the evening, and the trucks left in the apiary until dusk to collect all the flying bees. The journey takes place during the night, and if a halt is made for a meal, the truck engine is kept running: this prevents the bees from crawling out of the hives and over the load; provided it is dark, the bees will not fly. If the journey cannot be completed in one night, the truck is parked in the shade of the trees during the day, water being provided for the bees, and the journey is resumed at dusk. On arrival at the new apiary site, the truck is parked in the apiary, and the hives are unloaded as soon as it is light enough for the bees to fly. (*See* Figure 3.34/1—Plate 4.)

3.35 Harvesting

When the honey supers are full, and the honey is ripe, the bees are cleared from the combs. Many commercial beekeepers use chemical repellants for this work. The liquid is sprinkled lightly on absorbent material mounted in a frame which fits on the top of the full supers, and drives the bees down into the lower boxes in a few minutes. The most effective repellant in warm and hot climates is pure carbolic acid (phenol) dissolved in water. Benzaldehyde, introduced more recently, has proved particularly useful in cool weather.

An alternative method is the use of a 'bee escape' board. This is a frame covered with metal gauze or plywood, in which is set a 'bee escape'; this acts like a valve, allowing bees to pass through in one direction only. Often several escapes are mounted in one board. The escape board is placed under the supers of each hive on one day, and by the next day the bees have passed down from the supers into the lower box. Escape boards are used mostly by beekeepers whose apiaries are close to their base, and for whom two visits to each apiary cause little additional expense.

The old method of taking each comb of honey from the hive individually, and brushing the bees off it, is still practiced by some commercial beekeepers who do not use queen excluders. It can be done only while the honey flow is still in progress, otherwise it leads to bees robbing from each others' hives.

Some beekeepers have recently adopted the use of a 'bee blower', basically a vacuum cleaner in reverse, for removing the bees from honey supers. The blower is powered by a small petrol engine, and is either carried on the beekeeper's back or dragged around the apiary on skids or wheels. The honey super to be removed is stood on end, or on a special stand, and the bees are blown with a jet of air from between the combs. The bee blower is quick in operation, but subjects the beekeeper to constant noise, and gives him additional weight to move around the apiary. It is possible that in time these disadvantages will be overcome by the use of silent engines and light-weight units.

When the honey flow is followed by a period of dearth, enough honey is left in the hives as food for the bees until the next flowering season. But if another flow follows on, even if it is only a light flow, some commercial beekeepers 'clean out the hives', robbing the bees of all their stores.

If the honey flow has been a failure, and the bees have insufficient stores to see them through the period of dearth, the colonies are fed with sugar syrup.

3.36 Period of dearth

In regions where there is a very long cold winter, some beekeepers kill their colonies of bees after the honey flow, and stock the hives again in the spring with package bees obtained from warm regions. This is considered by some to be more profitable than trying to keep the bees through the winter; it releases more of the honey for sale, since none need be left in the hives as winter stores. But in most beekeeping areas the colonies are maintained during the periods of dearth, whether the dearth be the result of a cold winter, heavy rains, or drought. By reducing the number of boxes, the hives are contracted to a size that the bees can look after, and the surplus boxes of combs are stored. Queen excluders are usually removed.

In order to survive the period of dearth, the hives must contain adequate food stores, be well ventilated, protected from pests, and left undisturbed in a suitable apiary site.

3.4 EXTRACTION OF HONEY

3.41 Extracting plants

Beekeepers who operate their apiaries within 100 km or so of their base bring their full honey supers home for extracting. Such beekeepers may upon occasion travel further, to high-yielding and profitable honey flows. Migratory beekeepers (*see* 3.34), who frequently operate at a greater distance from their base, use mobile extracting plants in specially built caravans.

3.42 Migratory extracting plants

The mobile plants have the same basic essentials as the central plant, but lack facilities for storage, and have to be of compact design. Because the mobile plant is operated at the apiary site, the supers are extracted as soon as they are removed from the hives, and returned to them at once. The honey does not have time to become cold, and there is no need to store more boxes than those being uncapped and extracted at any one time.

Some beekeepers have small settling tanks mounted either on the caravan or on the back of the truck alongside. The honey is run into bulk containers while extraction is in progress.

A mobile plant has a boiler that provides steam for the uncapping knife and cappings melter, and a small petrol engine to drive the extractor and honey pump. The more modern mobile plants are fitted with a hot-water circulating system operated by bottled gas. This is more economical in fuel and water consumption than a steam boiler; it permits better temperature control, reduces or eliminates heat damage to the honey, and provides more pleasant working conditions for the operator. In the latest mobile plants, extractors, honey pumps, and even compact uncapping machines are being run by electric motors, power being produced by a portable generator situated at a distance from the extracting van so that the operator is not subjected to engine noise. The provision of

H—5

electricity also permits the installation of many conveniences in the honey producer's living van.

Using a semi-radial extractor taking nine or twelve frames, if one man takes honey off the hives while another uncaps the combs and loads the extractor, about a ton of honey can be extracted in a day. Such a unit will service 300–600 hives run on intensive migratory lines, in apiaries of ninety hives or more.

3.43   Central extracting plants

The central honey extracting plant can be equipped with better facilities for working, bigger machines, and efficient systems for filtering, settling, and storing the honey. It is suited to any type of operation, from domestic to really large-scale commercial, and can be used to extract anything up to 3 or 4 tons of honey a day, or more.

Honey extraction in general is dealt with in Chapter 9.

3.5   SOURCES OF INFORMATION

Many of the publications listed in 1.5 provide information on beekeeping, and some others are listed below, almost all written in English.

For Europe, a series of articles in *Bee World*, 'Facts about beekeeping in . . .' provides information on Czechoslovakia (Svoboda, 1958), Denmark (Johnsen, 1954), England (Hillyard, 1965), France (Borneck, 1959), Greece (Nicolaidis, 1955), Italy (Zappi-Recordati, 1956), Netherlands (Campbell, 1953), Yugoslavia (Kulinčević, 1959). There are many publications in Russian on beekeeping in the U.S.S.R.; Glushkov (1959) and Crane (1963a) have written articles in English.

The hive and the honey bee edited by Grout (1963) deals well with North American beekeeping; articles by Crane (1954, 1957, 1966) and Jamieson (1958) are also informative. Central American beekeeping is important but less well documented; Mexico (Wulfrath & Speck, 1955, 1958; Willson, 1953) and Guatemala (Elmenhorst, 1952) are among the countries that have been described. The same is true of South America; Vitez (1965) has written on Argentina.

In Asia, India is dealt with by Muttoo (1944, 1956) and Singh (1962), China briefly by Crane (1960), Israel by Kalman (1962), and the Philippines by Morse & Laigo (1969). More has been written about Africa —from Seyffert (1930) to Irvine (1957) and Guy (1972); F. G. Smith has published a book (1960) and many articles and pamphlets (e.g. 1953, 1958a, 1962).

Most of Australia has been covered by the 'Facts . . .' series: Cunningham (1961), Langridge (1961), Masterman (1961), Roff (1962), and F. G. Smith (1964); also New Zealand (Cook, 1967).

CHAPTER 4

THE WORLD'S HONEY PRODUCTION
by Dr. Eva Crane
DIRECTOR, BEE RESEARCH ASSOCIATION

While Honey is in Every Flower, no doubt
It takes a Bee to get the Honey out.
ARTHUR GUITERMAN,
A POET'S PROVERBS (1924)

4.1 INTRODUCTION

Chapter 1 is concerned with the plant sources that honey is derived from in different parts of the world. Most of this honey is used up in maintaining the colonies of bees throughout the year, and the bee-keepers' harvest can only be the honey that is surplus to the bees' requirements; this is likely to be somewhere between 5% and 40% of the total. The term 'honey production' is, however, most commonly used to mean the beekeepers' harvest, and as such it is the subject of the present chapter. Relationships between the amount of honey the bees produce and the amount the beekeeper harvests are discussed in more detail in 4.4. Except for Table 4.1/1, information on the history of honey production will be found in Chapter 19.

Table 4.1/1
Estimated percentage of the world's honey crop (from species of *Apis*) harvested in different continents at 250-year intervals

Period (AD)	1470	1720	1970
Europe	30?	30	25
Asia	35?	35	22
Africa	35?	30	13
New World	0	5	40
Total	100	100	100

Table 4.1/1 shows what an adaptable insect the honeybee is. Although it is an Old World species, once it was taken to the New World it throve there, foraging on both indigenous and introduced flowering plants. Some 40% of the world's annual honey crop is now produced in the New World, and the relative importance of the Old World continents has become consistently less during the past 500 years; *see also* Table 4.27/1. In the future, regions which still have large unexploited areas (South America, Africa, and Asia) are likely to produce an even higher proportion of the world's honey crop, and Europe a correspondingly lower one.

Almost all the world's honey harvest is produced from one species of honeybee—*Apis mellifera*. Only in Asia are there other honeybee species (Chapter 17), and it is unlikely that they account for more than 10% of the world's honey crop. Perhaps 2% is produced from non-*Apis* bees, all of them tropical species and most living in South and Central America; Chapter 17 discusses these also.

4.11   Description of Tables in 4.2
The Tables 4.21/1 to 4.27/1 give estimates for the honey production of individual countries, together with other data which are needed for a proper understanding of the honey figures:

Table 4.21/1 Europe and all U.S.S.R.
Table 4.22/1 North America
Table 4.23/1 Central and South America
Table 4.24/1 Oceania
Table 4.25/1 Asia
Table 4.26/1 Africa
Table 4.27/1 World summary

Countries for which no figures are available are listed at the foot of the appropriate Table, with their areas. Where applicable, combined estimates for such countries are added to the Table and included in the continental totals.

Each Table gives the following information for the individual countries that are listed in alphabetical order in Column 1.

COLUMN 2: *A*. AREA IN 1 000 SQ KM

This is taken from the *Statesman's Yearbook* (Paxton, 1970); all countries larger than 100 000 sq km are included, and a few smaller ones that have some special point of interest or are needed to complete a group.

COLUMN 3: *B*. NUMBER OF BEEKEEPERS IN 1 000S

This is a figure less frequently available than either the number of colonies (*C*, Column 4), or the honey production (*H*, Column 5). The FAO

Statistical Yearbooks quote *H* only, and World Honey Crop Reports in *Bee World* give *H* and *C* only. Sometimes *B* is restricted to those in a beekeeping organization. *B* is usually a primary figure, but where it is not available it may be calculated from the total number of colonies (*C*) and an estimate of the average holding per beekeeper (*C/B*).

COLUMN 4: *C*. NUMBER OF COLONIES IN 1 000S

This refers to colonies of honeybees living in hives. It does not include unoccupied hives—of which there may be as many again in some countries of tropical Africa. Nor does it include colonies of honeybees of any species *living wild*, from which honey is collected; a major part of the honey crop may come from such colonies in some countries of southern Asia. Movable-frame hives and fixed-comb hives are both included; the code letter *f* indicates that a large proportion are fixed-comb hives.

COLUMN 5: *H*. ANNUAL HONEY PRODUCTION IN TONS

The unit is the metric ton (tonne) except where the official figures are quoted in long tons (2 240 lb); the two tons differ only by 2%.

The numbers of beekeepers *B* and of their colonies *C* (Columns 3 and 4) show important long-term changes, but they are fairly steady from one year to the next. The honey production of a country, on the other hand, fluctuates greatly because of the weather (*see* e.g. Figure 1.13/1, Table 4.33/1). So a 10-year or other long-term average is quoted wherever possible. Details are given in 'Notes to Tables'.

COLUMN 6: *H/C*. AVERAGE YIELD PER COLONY IN KILOGRAMS

This is usually calculated from *H* and *C*, and is subject to most of the variations affecting the total honey production of a country (*H*). But, being unrelated to the country's size, it gives a useful measure of honey productivity.

COLUMN 7: *C/B*. AVERAGE NUMBER OF COLONIES PER BEEKEEPER

This is usually calculated from *C* and *B*. It can be a useful indicator of the profitability of beekeeping.

COLUMN 8: *C/A*. AVERAGE COLONY DENSITY (PER SQ KM OR 100 HA)

This 'bee population density' is an indicator of the importance of beekeeping as a pursuit over the country as a whole—whether for cultural or commercial reasons—and can be useful in pollination considerations. In countries which have large areas of desert or other barren land, the parts which do support vegetation may have a much higher bee population than the national average; Egypt is an extreme case. In the agricultural atlas published by Skibbe (1958), the colony densities (Crane, 1958) are calculated per unit area of agricultural land; here they are related to the total area.

COLUMN 9: H/P. AVERAGE HONEY CONSUMPTION PER CAPITA (IN 100G)

This is quoted from published sources where these exist. Otherwise it is calculated from the population given in the *Statesman's Yearbook* (Paxton, 1970) and the annual honey production, imports, and exports. This calculation may be misleading, for instance if stocks are held over from a good year to the next; nevertheless, there is no reason to doubt the large differences between different continents that are shown in Table 4.27/1.

COLUMN 10: S/P. AVERAGE SUGAR CONSUMPTION PER CAPITA (IN KILOGRAMS)

These figures are taken from Table 164 of the *U.N. Statistical Yearbook* (1970); most relate to 1969, a few to 1968. No attempt has been made to assess their validity. The figures must be multiplied by ten to get a direct comparison between sugar consumption and the honey consumption in Column 9.

CONTINENTAL TOTALS

The totals for each continent have been calculated as follows:

For B, C, H, the total of the entries is used; in Table 4.25/1 and 4.26/1 it is augmented by a proportionate amount for the 'no-information' areas.

For H/C, the total (i.e. average) is the total of the entries for H divided by the total of the entries for C, considering all those countries which have entries for both H and C. It is a weighted mean from data given.

C/B is treated on a similar basis, and also C/A (but all countries have an A entry).

For H/P and S/P the total (i.e. average) is not always directly calculable, and the U.N. totals are inconsistent with national figures. Non-calculated assessments from figures given take into account which countries have large populations that dominate the situation.

4.12 General comments on Tables in 4.2

Most published estimates for honey production (*see* 4.28) are far from complete, and do not mention countries for which no data are given; it is not easy for a reader to assess the importance of these missing countries, or even to identify them, and he is likely to assume—incorrectly

—that the sum of the figures quoted is a meaningful total. Many of the officially published figures underestimate the true honey production, and the amount of any bias may be extremely difficult to determine. A better estimate is more likely to emerge from discussions with government and non-government beekeeping specialists in the country concerned than from consulting published figures alone (*see also* 4.28). One cause of underestimation is simply lack of information about some of the apiaries in an area: hives are mostly kept in secluded places. Owners of only a few hives may be intentionally omitted; 'unorganized' beekeepers (i.e. those not in the official beekeepers' organization) may be excluded; returns may refer only to movable-frame hives, ignoring a greater number of fixed-comb hives. Even if information is provided about known omissions in an original document, it may be omitted when the figures are quoted elsewhere.

Figures for a country's exports and imports (Chapter 14) are likely to be closer to reality, since it is easier to ascertain what goes through ports of exit and entry than what is taken from each hive, or how many hives there are. And honey is not such a valuable commodity that it would be shipped as contraband. Where a country exports virtually all its honey and imports none, the export figure gives a useful lower approximation to the honey production—provided no honey is kept in stock until the next year.

Source material for the Tables, and earlier published and unpublished statistics, are discussed after the Tables have been presented and discussed, in 4.28.

4.2 PRESENT WORLD HONEY PRODUCTION

Notes to Tables 4.21/1–4.27/1; further details in 4.11.

Column 1	*f*	entries under 5 and 6 low because fixed-comb hives form an appreciable proportion of the total
Column 3	( )	estimates that are especially uncertain
and others	+	figures known to be an underestimate
	D	recent estimates from various sources
	E	estimates obtained in the late 1950s, most derived directly from the country concerned
Column 3	↑	beekeeping has since increased
	↓	beekeeping has since decreased
Column 4, 5, 6	W	from World Honey Crop Reports (*Bee World* 1965–1972), derived directly from the country concerned
	71	and similar entries—the figure quoted is for the year indicated

Column 5, 6   *F*   From *F.A.O. Production Yearbook*, Table 131
                        (1969)
                U   from U.S.D.A. publications (*see* Bibliography)
                × 10   recent 10-year average
                × 5   recent 5-year average
Column 6, 7   A few entries appear inconsistent with data
                        quoted in Columns 3–5; this is because they are
                        derived from data for different periods, or from
                        different sources
Column 10   Except for N. America, continental means are
                        quoted from the same source as national means,
                        but they do not always appear to be consistent
                        with them. *Note that all entries here* must be
                        multiplied by 10 for comparison with Column 9.

4.21   Europe and U.S.S.R.

The U.S.S.R. is such a large country that it is treated as a separate
'continent' in comparative statements, instead of being combined with
the rest of Europe and Asia. It is dealt with at the end of the present
section, and at the end of Table 4.21/1.

The most notable general feature of Europe from the point of view of
honey production (Table 4.21/1) is the high population density of colonies
of bees (C/A). The average is 2·8 per sq km, 7 times as high as in Africa
or U.S.S.R.—the runners up—and 50 times as high as in Australia. This
is partly linked with a high human population density, but not entirely.
Many countries in Asia that are as densely populated have far fewer
hives of bees; three equivalent pairs will serve as examples: Belgium has
10 times the hive density of Taiwan; Switzerland 400 times that of India;
Czechoslovakia 900 times that of Pakistan. Czechoslovakia, Greece,
and Switzerland have the highest hive densities (C/A) in the world.

The root cause of the high concentration of bees in Europe would seem
to be the long tradition of beekeeping there (Chapter 19). It is certainly
not high honey yields, for the average European yield per hive is only a
third or a quarter as much as in the New World continents.

Of the individual countries of Europe, West Germany and France are
the largest honey producers (H), Spain and Poland coming next. These
four large countries all produce over 10 000 tons a year. Each of them has
around a million colonies of bees or more (C), as do Czechoslovakia,
Greece, and Rumania.

Sweden, with a very low population density of humans and of bees,
comes out as having the highest honey yield per colony (H/C), probably
because of the long summer days and intense short growing season.

Table 4.21/1
Honey production in Europe and U.S.S.R.

See 4.2 (pages 119-120) for explanation of headings to columns 1-10 and other code letters etc.

1 Country	2 A (1 000 km²)	3 B (1 000s)	4 C (1 000s)	5 H (tons)	6 H/C (kg)	7 C/B	8 C/A (/km²)	9 H/P (100 g)	10 S/P (kg)
Albania†	29	9 E	85 E	350 E	4 E	7 E	3	2·9 E	
Austria	84	43 ↑ E	460 W 71	6 000 W ×10	13 W ×10	10 E	6	6 E	36
Belgium	34	12 E	120 W 71	960 W ×10	8 W ×10	8 E	4	3 E	43
Bulgaria	111	80 ↑ E	760 W 71	4 252 F ×5	6 ×5	6 E	7 E	7 E	47
Czechoslovakia	128	125 ↑ E	1 124 W 70	6 572 W 70	6 W 70	7 E	9	3 E	46
Denmark	43	(10) D	(100) W 68	(2 050) F ×5	(20) F ×5	(10) D	(2)	6 E	51
Finland	305	5 E	29 W 71	470 W ×10	16 W ×10	5 E	0·1	2 E	42
France	552	(100) D	977+ D 66	13 000 W ×10	13 W ×10	(15) E	1·7	4 E	38
Germany, E.	108	50 ↓ E	520 W 67	4 000 W ×10	8 W ×10	16 E	5	5 E	37
Germany, W.	249	160 ↓ E	1 150 W 69	13 500 W ×10	12 W ×10	10 E	4	10 E	40

1 Country	2 A (1 000 km²)	3 B (1 000s)	4 C (1 000s)	5 H (tons)	6 H/C (kg)	7 C/B	8 C/A (/km²)	9 H/P (100 g)	10 S/P (kg)
Greece†	132	47↑ *E*	990 *W* 71	8 452 *F* × 5	8 × 5	11 *E*	8	4 *E*	20
Hungary	93	50↑ *E*	600 *W* 67	7 000 *W* × 10	12 *W* × 10	10 *E*	7	1 *E*	41
Iceland	103	0	0	0	0	0	0	(0·2) *E*	
Irish Republic	69	3 *E*	11 *W* 71	200 *W* × 10	19 *W* × 10	5 *E*	0·2	(0·2) *E*	63
Italy	301	80 *E*	750 *W* 69	6 800 *W* × 10	9 *W* × 10	8 *E*	2·5	1·5 *E*	29
Luxembourg *see* Belgium									
Netherlands	34	15↓ *E*	60 *W* 69	444 *F* × 5	7 × 5	7 *E*	1·8	20 *E*	69
Norway	324	9 *E*	55 *W* 71	950 *W* × 10	17 *W* × 10	7 *E*	0·2	2 *E*	45
Poland	313	180↑ *E*	1 300 *W* 70	10 500 *W* × 10	8 *W* × 10	7 *E*	4	3·8 *E*	43
Portugal†	92	60 *E*	548 *W* 71	2 400 *F* × 5	4 × 5	8 *E*	6	2·3 *E*	23
Rumania	237	20↑ *E*	976 *W* 69	8 700 *W* × 10	9 *W* × 10	30 *E*	4	5·3 *E*	21
Spain	505	80↑ *E*	975 *W* 71	12 400 *W* × 10	13 *W* × 10	10 *E*	2	3·4 *E*	27
Sweden	411	(22) *E*	69+↓ *W* 70	(2 400) *F* × 5	26 × 5	(5) *E*	0·2	3·1 *E*	42

1 Country	2 A (1 000 km²)	3 B (1 000s)	4 C (1 000s)	5 H (tons)	6 H/C (kg)	7 C/B	8 C/A (/km²)	9 H/P (100 g)	10 S/P (kg)
Switzerland	41	37 E	339 E	1 920 U ×10	6 E	9 E	8	5·5 E	55
United Kingdom	231	90↓ E	160 W 71	3 340 F ×5	17 ×10	4 E	0·7	2·5 E	53
Yugoslavia	256	80 E	825 W 68	4 020 F ×5	5 W 68	9 E	4	2·2 E	28
TOTAL*	4 682	1 367	12 983	120 680	9	9·5	2·8	4	36
U.S.S.R. (Europe+Asia)	22 400	(1 150) E	10 000 W 70	103 000 U ×10	11 E	(8) E	0·4	5 E	45

* The total area is 6% larger (4 929), but the unadjusted figures are more representative.

In many countries of Europe, the urge to keep bees does not necessarily depend on getting financial profit from them. Central Europe is probably the greatest traditional stronghold of beekeeping in the world, and in parts of this region hives are kept packed together in a bee house, and the average beekeeper's holding (C/B) is more than where hives are free-standing in the open, as elsewhere in the world. The bee-house tradition is more or less confined to areas where German is the primary or secondary language (Germany, Austria, German-speaking Switzerland, Czechoslovakia, Slovenia in Yugoslavia).

Widespread sources of honey in Europe are clover and rape, with heather also on the Atlantic seaboard. In Central Europe the acacia or locust (*Robinia pseudoacacia*) is perhaps the most notable. A native of North America, it thrives on sandy soil and has been planted in large areas of Hungary, Rumania, and Yugoslavia, where it provides the main export honey (14.26). Various species of lime (*Tilia*) are important also in Rumania and Poland. Honeydew from coniferous forests is predominant in many parts of Austria and Germany. Orange honey is notable in Spain, rosemary there and also in France, where lavender is another specialty honey. Greek thyme honey is famous.

Much detailed information on European honey sources is referred to in 1.52. Each country, and each area within it, has its own characteristic honey flora and beekeeping conditions—hot or cool in summer, cold or mild in winter, wet or dry, high or low in altitude, sheltered or windy, open land or urban development. The honey yield, its sources, and the capabilities of the local bees, all vary from place to place. The honey yields quoted in the Table 4.21/1 and subsequent Tables are averages of results depending on the intimate relationships between bees and the plants within their flight range, throughout the area in question.

The whole U.S.S.R. (final entry in Table 4.21/1) is 5 times as large as the rest of Europe. The best honey-producing regions are in the large Asiatic section (*see* e.g. Glushkov, 1959, Crane, 1963*a*). The total number of colonies is rather less than in the rest of Europe, and these are much more sparsely scattered, at an average of only 0·4 per sq km; this is about the same as in the United States. The total honey production is also about the same as in the States—but there the honey is produced from less than half as many hives. Honey sources over this vast Union of Republics are many and varied (*see* 1.52); limes, sunflower, buckwheat, willow-herb (fireweed), raspberry, and cotton are among the most important.

4.22 North America

North America is a continent where modern frame-hive beekeeping is even more universal than in Europe. The two countries concerned (Table 4.22/1) are about the same size, but the United States has 10 times as

Table 4.22/1

Honey production in North America

See 4.2 (pages 119-120) for explanation of headings to columns 1–10 and other code letters etc.

1 Country	2 A (1 000 km²)	3 B (1 000s)	4 C (1 000s)	5 H (tons)	6 H/C (kg)	7 C/B	8 C/A (/km²)	9 H/P (100 g)	10 S/P (kg)
Canada	9 976	10 D 65	408 W 70	18 200 W × 10	57 × 10	41 E	0·04	10 E	51
United States	9 660	425 E	4 346 W 71	107 000 W × 10	25 × 10	13 E	0·4	6 E	48
Greenland	2 176	no beekeeping; excluded from totals below							
TOTAL	19 636	435	4 754	125 200	26	14	0·24	7	49

many colonies, 10 times the colony density (and 10 times the human population); it has 40 times as many beekeepers, but they tend to have fewer hives each than in Canada. Canada's high honey productivity (57 kg per colony) is a notable example of the honey yields obtainable at high latitudes; they are here harvested with package bees raised much further south. Honey consumption is higher than in Europe or the U.S.S.R., partly because honey can be produced more cheaply; the sugar consumption is also higher, but not by such a large factor.

Canada is a country of commercial honey production, for home consumption and export; the average holding of 41 hives per beekeeper covers a range from several thousand down to less than 10 among the hobbyists. The U.S.A. also has large holdings, up to 10 000 or more; but in the east especially there is also a considerable amount of hobby beekeeping, and this brings the national average down.

Honeys from white clover, sweet clover, and a mixture of these with alfalfa, are widely produced in Canada and some northern parts of the U.S.A.; these honeys are usually described as 'clover'. Basswood (*Tilia*) is also common to both countries. British Columbia produces more aromatic honeys. Further south in the United States, cotton, and citrus (usually marketed as orange) are important, and in California *Erigonum fasciculatum* (wild buckwheat, Californian sage). There is a host of other sources, to which reference can be found in 1.52.

4.23   Central and South America

We now leave the comparatively well documented regions of the world, returning to them in only one other section (4.24). But future expansion of honey production is likely to be greatest in the less developed, less well documented regions, which are thus in need of special attention (Table 4.23/1).

Central America is dominated by Mexico, whose dramatic rise as a honey-producing country is described in 14.22. In several other countries the honey yield per hive is also higher than in the United States. In spite of this high honey productivity, the honey *consumption* is very much smaller than in Europe or North America.

Central and South America between them produce about as much honey as the U.S.S.R. or the U.S.A. But, unlike the two last countries, those of Latin America export most of what they produce, and in so doing they dominate the world honey trade (14.2). The average holding is 12½ colonies per beekeeper, about the same as in the United States, and the yield per colony rather higher—but without all the expertise that has been applied to North American beekeeping.

Argentina is the greatest honey producer in South America, with Brazil and Chile a long way behind. Colonies are more thinly spread than

Table 4.23/1

Honey production in Central and South America

See 4.2 (pages 119-120) for explanation of headings to columns 1-10 and other code letters etc.

1 Country	2 A (1 000 km²)	3 B (1 000s)	4 C (1 000s)	5 H (tons)	6 H/C (kg)	7 C/B	8 C/A (/km²)	9 H/P (100 g)	10 S/P (kg)
CENTRAL									
British Honduras	23			190 F × 5					
Costa Rica	51	(0·2) E	10 E	360 E	36 E	(50) E	0·2 E	0·8	
Cuba	115	10 E	216 E	7 833⁺ U × 4	23	21 E	2 E	0 E	77
Dominican Republic	48			312 F × 3					32
Guatemala	109	5 E	118 E	2 260 U × 10	19 × 10	23 E	1·1 E	1·4 E	23
Honduras	112 no information								
Jamaica	11	0·2 D 68	20 D 68	(378) F × 5	19 × 5	100	2		
Mexico†	1 967	110 E	(1 000) E	30 000 U × 10	30 × 10	(10) E	(0·5) E	0·3 E	38
Nicaragua	148 no information								
Salvador	20		20 E	1 060 F × 5	50		1 E		
SOUTH									
Argentina	2 809	(28)↑ E	800 W 71	23 300 W × 10	30 × 10	(19) E	0·3	3 E	37

1 Country	2 A (1 000 km²)	3 B (1 000s)	4 C (1 000s)	5 H (tons)	6 H/C (kg)	7 C/B	8 C/A (/km²)	9 H/P (100 g)	10 S/P (kg)
Bolivia	1 099			(1 160) F×5					22
Brazil	8 512	(15)	(440) D 68	7 386 W×10	(17) ×10	(30)	(0·05)	0·8	38
Chile[f]	742	(8) ↑ E	417	6 030 U×10	(14)	(18) E	0·6	5	36
Colombia	1 139	2 E	30 E	850 E	27 E	15 E	0·02 E	0·6 E	25
Ecuador	271	no information							21
Guiana, Netherlands (Surinam)	181								
Guyana	210			40 F×5					
Paraguay	407	(5) E	(75) E			(15) E			
Peru	1 285								27
Uruguay	187	4 E	(73) E	780 F×5	(11) E	18 E	(0·4) E	2·7 E	42
Venezuela	912								39
(Total)*	16 069	182	3 124	79 177					
Other countries	4 716	53	917	23 237					
TOTAL	20 785	235	4 041	102 414	25†	17†	0·19†	1	42

* only countries with entries for A, B, C, and H   † calculated from countries referred to under *

in the U.S.A., and very much more so than in Europe. Sugar consumption is around half that in continents considered previously; Cuba, with sugar as its main crop, is an exception—it tops the world for sugar consumption. Throughout Latin America honey consumption is low; honey is a source of income rather than something to eat at home.

White clover and lucerne (alfalfa) are among the main honey sources now exploited in temperate South America, as well as the thistle-like cardoon, *Cynara cardunculus*. In tropical regions the sources are much less known, and indeed it is here (for instance in the Amazon basin) that the honey resources are barely exploited at all, let alone known and evaluated. In various parts of Central America the following are especially important: dzidzilché, tah, campeche, Jamaica dogwood, campanilla, royal palms. Their botanical names are *Gymnopodium antigonoides*, *Viguiera helanthoides*, *Haematoxylon campechianum*, *Piscidia piscipula*, *Ipomoea* spp. and *Rivea corymbosa*, *Roystonea* spp. Further information is given in 1.52, 14.21, and 14.22.

4.24 Oceania

This term is used in preference to Australasia as being more comprehensive. It includes Australia and New Zealand, and a host of islands to the south and east of Asia proper which, however, contribute only 7% to the total area and are unimportant in the present context, so they are omitted from Table 4.24/1. Oceania is nearly twice as large as Europe (Table 4.21/1), but less than half the size of any other continental unit.

New Zealand honey production is not unlike that in other productive temperate regions. The native flora yields some most interesting honeys, and a few that have objectionable properties. But the bulk of the honey comes from white clover; it is eaten at home, and exported (14.24).

Australia has several unusual or unique features from the beekeeping point of view. The large proportion of the country that is desert gives it the lowest overall hive density of any land mass in the world. Yet it is a country of large holdings and high honey yields. The average number of hives per beekeeper (83) would be higher still if it were not for the immigrants from Europe since the last war, who brought with them their tradition of hobby beekeeping, with ten hives or less.

The bulk of Australian honey comes from indigenous species of eucalypt. Many of these produce nectar copiously, but some have other less convenient characteristics from the beekeeper's point of view. Some yield no pollen, and most of them flower only at biennial or less frequent intervals. Stands of different trees may provide rich honey flows in succession, which a beekeeper can harvest by moving his hives from one flow to the next. Very large annual crops can thus be obtained—up to say 300 kg per hive—but a low price for honey may set a limit to the amount

Table 4.24/1
Honey production in Oceania

See 4.2 (pages 119-120) for explanation of headings to columns 1-10 and other code letters etc.

1 Country	2 A (1 000 km²)	3 B (1 000s)	4 C (1 000s)	5 H (tons)	6 H/C (kg)	7 C/B	8 C/A (/km²)	9 H/P (100 g)	10 S/P (kg)
Australia	7 687	5·5 W 70	493 W 71	18 100 W×10	37	83	0·06	5 E	56
New Zealand	269	6·6 E	191 W 71	5 750 W×10	30	29 E	0·7	5 E	53
TOTAL*	7 956	12·1	684	23 850	35	56	0·09	5	57

* The total area is 7% larger (8 510), to which Papua New Guinea contributes 460, but the unadjusted figures are considered to be more representative.

it is worth paying out in transport costs. The classic migration story relates how two beekeepers once moved 1 600 hives nearly 4 000 km, by train, from New South Wales to Western Australia to work the karri (*Eucalyptus diversicolor*). The trees bloomed from December 1948 until July 1949; during this time some 15 000 hives were in the karri forest, and there were recorded instances of 25 kg of honey being harvested from the same hive every 14 days. But after one flowering it may be 4, 8 or 12 years before the karri forest flowers again.

4.25 Asia

Here, again, is a completely different region: another Old World continent, with indigenous honeybees—but four separate species, three of which occur nowhere else. The fourth (*Apis mellifera*) has been introduced into many areas from Europe, and it is also native to parts of Asia around the Black and Caspian Seas. The respective parts played by these bees in honey production are explained elsewhere (17.2).

Statistics are hard to come by, and a number of those in Table 4.25/1 are very imprecise. Apart from notable exceptions such as Israel, Japan, and Taiwan, and possibly parts of China (People's Republic), honey productivity is low in Asia. Reasons for this include the wide use of fixed-comb hives and the accompanying lack of control over the bees; lack of knowledge and of mechanization which would allow migratory beekeeping; the generally low productivity of *Apis cerana*; the difficulties of harvesting honey from the other two species, *dorsata* and *florea*, which build a single comb in the open and cannot be kept in ordinary hives (17.2); finally, the many enemies of bees in tropical regions. In our present circuit of the world, Asia is the first continent discussed which has tropical regions with honeybees *that evolved there*—and with enemies of honeybees that evolved there too. Africa (4.26) is the only other such continent.

The People's Republic of China, three times as large as India and twice the size of Europe, is coming more and more to dominate honey production in Asia—as Mexico and Argentina dominate the situation in Central and South America respectively. But the development in China is more recent (14.25), and is probably still gathering momentum; no full assessment of the conditions is available, and so the future is very difficult to predict—except that further expansion is likely. The present national average yield per hive (11 kg) is depressed by the many fixed-comb hives still remaining, and the present estimated total production is not much more than twice that of Japan, a country only 4% as large. Israel and Taiwan are two tiny countries which show something of the honey-producing potential of parts of Asia.

In general, sugar is a much less important food in Asia than in any

Table 4.25/1
Honey production in Asia

See 4.2 (pages 119-120) for explanation of headings to columns 1-10 and other code letters etc. and page 134 for *Note*

1 Country	2 A (1 000 km²)	3 B (1 000s)	4 C (1 000s)	5 H (tons)	6 H/C (kg)	7 C/B	8 C/A (/km²)	9 H/P (100 g)	10 S/P (kg)
Bangladesh figures included with Pakistan									
Cambodia	181			140 E +					7
Ceylon see Sri Lanka									
Chinese People's Republic[f]	9 597	(300) E	(1 400) E	18 100 U×4	(11) E	(10) E	(0·1) E	0·02 E	4
Cyprus	9			(206) F×5					
India[f]	3 054	(40) E	(170) E	1 400 E	(4) E	(4) E	(0·05) E	0·03	6
Indonesia[f]	1 904		20 E	(10) E	0·5 E		(0·01) E		7
Iran[f]	1622		(500)	(2 000) E	(4) E				
Iraq[f]	438	2·3 E	17 E	125 E	7 E	8 E	0·04 E	0·3 E	36
Israel	21	0·8 E	47 W 71	1 950 U×8	40 U×8	50 E		♦ E	57
Japan	370	18↑ E	248 W 71	7 350 W×10	30	7 E	0·7	2·3 (0·02E)	26
Jordan[f]	98	0·8 E	8 E	12 E	15 E	10 E	0·08 E	0·04 E	26

1 Country	2 A (1 000 km²)	3 B (1 000s)	4 C (1 000s)	5 H (tons)	6 H/C (kg)	7 C/B	8 C/A (/km²)	9 H/P (100 g)	10 S/P (kg)
Korea, N.	122								11
Korea, S.†	98	0·3 ↑ E	101 F	570 F×5	6 E	12 E	0·04		6
Lebanon†	10	1·3 E	15 E	120 F×5	8			0·7 E	
Malaysia	334								39
Pakistan†	944	(3) E	(11) E	150 E	4 E	(4) E	(0·01) E	0·02 E	5
Philippines	299								18
Sri Lanka†	65	(1) E	(4) E	25 E	(4) E	(4) E			25
Syria†	186	(8) E	64 E	230 F×5	(4) E	(8) E	(0·35) E	0·5 E	20
Taiwan	36	0·36 E	14 E	260 F×5	28 E	38 E	0·4 E		15
Thailand	514								9
Turkey†	781	(132) E	(1 300) E	10 750 U×8	8 ×8	(10) E	1·7	3 E	19

U.S.S.R. included in Table 4.21/1

1 Country	2 A (1 000 km²)	3 B (1 000s)	4 C (1 000s)	5 H (tons)	6 H/C (kg)	7 C/B	8 C/A (/km²)	9 H/P (100 g)	10 S/P (kg)
Vietnam, S.	172								7
(Total)*	15 698	508	3 399	41 042					
Other countries	12 198	394	2 641	31 891					
TOTAL	27 896	902	6 040	72 933	12†	7†	0·2†	(0·04)	7

* only countries with entries for A, B, C and H † calculated from countries referred to under *

Note to Table 4.25/1: In many parts of south-east Asia and the Indian subcontinent, honey is collected from wild colonies of *Apis cerana*, *Apis dorsata*, and *Apis florea*; this is included in Column 5 (only).

Countries for which no information is available, with their areas in 1 000 sq km:

Afghanistan	658	Oman	212
Burma	678	Saudi Arabia	2 400
Laos	238	South Yemen	160
Sarawak	121	Vietnam, N.	159
Mongolia	1 565	Yemen	195
Nepal	141		
		Total	6 527

other continent, and the same is true of honey. Habits are changing, however (*see* 14.4.10). In the next ten years the honey situation in Asia is likely to change markedly, and figures may perhaps be available to give a more complete and reliable picture of the position then.

With Asiatic U.S.S.R., the continent stretches from the Arctic to the Equator and beyond. Even without the northern Soviet area, the plants from which honey is harvested are legion, and in many areas have still not been documented. The exported honey from China includes much from limes (*Tilia*) and litchi (*Litchi chinensis*); in Israel citrus is the main source. Such information as is available can be traced through the lists in 1.52.

4.26 Africa

The final continent to be considered includes the largest tropical land area on earth. It is an Old World region where the most widespread indigenous honeybee is the subspecies *adansonii* of the Eurpoean bee *Apis mellifera*. No honeybee except *Apis mellifera* lives there, but there are several sub-groups (Smith, 1961). In much of Africa there is a strong beekeeping tradition, as in Europe. But the marketed product in tropical Africa has not been honey but beeswax, which is easier to transport where there are no roads. Most of the honey is used for fermentation, in the production of honey beer (Seyffert, 1930); *see also* 16.43, 19.23.

North of the Sahara, beekeeping is not dissimilar from that in other Mediterranean countries. At the southern extremity of Africa normal modern beekeeping is practised, except that one local subspecies (*A. m. capensis*) has some unusual characteristics (Smith, 1961).

In countries between the Sahara and Kalahari deserts, *A. m. adansonii* thrives. It is prolific, building strong colonies which swarm freely; the swarms occupy any available cavity of a suitable size. Beekeeping is based on setting up empty bait hives into which swarms move; this is true even in modern commercial systems (Guy, 1971). Combs are built quickly, temperatures being high enough to encourage wax production by the bees. A 'traditional' beekeeper might have up to several hundred hives, hung in trees as a protection against their many enemies, half the hives being empty, awaiting the next occupants. At the right season most or all of the combs are cut out, using smoke and working at night. There is a crude extraction of honey by straining and pressing, and the wax is melted and made into blocks for transport to a marketing point. The honey is likely to be sold to the local beer maker, or used domestically for the same purpose (16.43).

Entries for these countries in Table 4.26/1 are extremely vague: statistics for most of the columns have never been collected, and many Africans dislike counting things, considering it unlucky to do so. In the wax-producing countries the most reliable figures are those for *beeswax*

Table 4.26/1
Honey production in Africa

See 4.2 (pages 119–120) for explanation of headings to columns 1–10 and other code letters etc. and page 138 for Note

1 Country	2 A (1 000 km²)	3 B (1 000s)	4 C (1 000s)	5 H (tons)	6 H/C (kg)	7 C/B	8 C/A (/km²)	9 H/P (100 g)	10 S/P (kg)
Algeria^t	2 467	(35) E	(220) E	(700) E	(3) E	(6) E	(0·3) E	(1·0) E	18
Angola^t	1 247	(250) E	(1 000) E	(6 000) E	(6) E	(4) E	(0·7) E	(14)	
Central African Rep.^t	625	(160)	(630)	(3 800) F×5	(6)	(4)	(0·01)		
Chad^t	1 284	(13) E	(52) E 40	(300) E 40	(6) E	(4) E	(0·04) E		
Congo	332	figures included with Zaire							
Congolese Republic *see* Zaire									
Ethiopia^t (including Eritrea)	1 222	(200) E	(800) E	(4 800) E	(6) E	(4)	(0·7) E	(2·5) E	
Kenya^t	583	(50) E	(200) E	(1 200) E	(6) E	(4) E	(0·4) E	(1·4) E	15
Libya^t	2 100	(1·6) E	(8) E	(50) F×5	(6) ×5	(5)	(0·005) E		
Malagasy Republic (formerly Madagascar)^t	594	(150) E	(1 700)	10 000 F×2	(6) E	(4) E	(1·0) E	(8) E	
Malawi^t	94	(2)	(8)	50 F 61	(6)	(4)	(0·1)		
Morocco^t	500	(27) D	(400) D	4 500 W 71	(11)	(15)	(0·8)		26

1 Country	2 A (1 000 km²)	3 B (1 000s)	4 C (1 000s)	5 H (tons)	6 H/C (kg)	7 C/B	8 C/A (/km²)	9 H/P (100 g)	10 S/P (kg)
Nigeria[f]	924	(100) E	(700) E	(2 800) E	(4) E	(7) E	(0·8) E	(0·3) E	
Senegal[f]	197	(5)	(20) E 40	(110) E 40	(6) E 40	(4)	(0·1) E		
South Africa	1 221	(4) E	41 E	(380) F × 5	9	(10) E	(0·03) E	0·2 E	40
Sudan[f]	2 500	(50) E	(200) E	(1 200) E	(6) E	(4) E	(0·1) E	(1·3) E	15
Tanzania[f]	940	(400) E	1 700	10 380 W × 10	(6)	(4)	(2)	(9) E	
Tunisia[f]	164	(10) E	(220) E	(1 200) E	(6) E	(20) E	(1·3) E		
Uganda	236								15
U.A.R. (Egypt)[f]	1 000	(110) E	688 W 65	4 190 F × 5	(6) × 5	(6) E	(0·7) E		15
Zaire[f]	2 345	(40) E	(160) E	(1 000) E	(6) E	(4) E	(0·07) E	(0·6) E	
(Total)	20 574	1 608	7 647	52 660					
Other countries	11 740	918	4 363	30 047					
TOTAL	32 314	2 526	12 010	82 708	(6)	(6)	(0·4)	(2·6)	11

Note to Table 4.26/1: Countries for which no information is available, with their areas in 1 000 sq km:

Botswana	575	Mozambique	785
Cameroon	474	Niger	1 187
Dahomey	113	Rhodesia	391
Gabon	267	Sierra Leone	73
Ghana	239	Somalia	700
Guinea	246	S.W. Africa	823
Ivory Coast	323	Sahara	2 172
Liberia	111	Upper Volta	274
Mali	1 204	Zambia	752
Mauritania	1 031		
Total			11 740

exports. From these the total amount of beeswax produced has been estimated, and the number of occupied hives is calculated on the basis of one for every $\frac{1}{2}$ kg of beeswax produced. The honey production is estimated at 6 kg per hive, i.e. a ratio of 12 to 1 for honey and beeswax yields. The average holding is taken as 4 occupied hives per beekeeper; individual ownership may vary from 1 to 1 000, and half the hives are likely to be empty. Honey consumption is calculated on the assumption that all available is collected, and that none is exported. This basis was suggested by Dr. F. G. Smith in the late 1950s; in view of the changes taking place in tropical Africa, adjustments might well be necessary for any future calculations.

Accepting the limitations of the figures used, which are not contradicted by such independent checks as could be made, the following facts can be deduced from Table 4.26/1. The important honey-producing countries in tropical Africa are Tanzania, Malagasy Republic, Angola, and Ethiopia. North of the Sahara, Morocco and Egypt produce most honey. The average honey 'consumption' *per capita* is much higher than in any other group of countries except the highly developed ones, but as some 80% of the honey in tropical Africa is used for brewing, comparatively little is eaten as such. The beer is normally drunk by men only, so their consumption will be 2–3 times that quoted in Column 9.

There are, and have been, many beekeeping development programmes in Africa, some of which are successful in a number of ways. Most aim to increase the output of marketable honey by improving methods of collection and processing, as well as by getting bee management on a more rational basis—though not necessarily by use of precision-made frame hives. Traditional beekeeping is on the wane in many parts (e.g. Brokensha *et al.*, 1972). So, as with Asia, the next ten years may well see many changes in this rich honey-producing region, which has in the past been remarkable as the world's main source of beeswax. The fertile regions of Africa north of the Sahara are also capable of much development, by use of efficient beekeeping methods and by migratory beekeeping as and when roads are made. Citrus and eucalypts are two of the main honey plants in the north, and also in South Africa, which has other prolific sources including aloes. In the tropical areas *Brachystegia* and *Julbernardia* are among the trees that are good honey producers, and they also provide sites for hanging hives. Further details of the rich and varied honey flora can be obtained from 1.52, and general details from Smith (1953, 1960).

4.27 World summary

Table 4.27/1 gives the averages and totals for the continents covered separately in Tables 4.21/1–4.26/1. Some of the component figures are

Table 4.27/1
Honey production: world summary

See 4.2 for explanation of headings to columns 1-10 and other code letters etc. The additional column 11 gives the approximate amount of honey produced annually (kg) per beekeeper. Entries relating to the Old World are in roman type and those relating to the New World are in italics.

1 Continent	Table	2 A (1 000 km²)	3 B (1 000s)	4 C (1 000s)	5 H (tons)	6 H/C (kg)	7 C/B	8 C/A (/km²)	9 H/P (100 g)	10 S/P (kg)	11 H/B (kg)
Europe*	4.21/1	4 682	1 367	12 983	120 680	9	9.5	2.8	4	36	90
U.S.S.R.	4.21/1	22 400	1 150	10 000	103 000	11	8	0.4	5	45	90
N. America	*4.22/1*	*19 636*	*435*	*4 754*	*125 200*	*26*	*14*	*0.24*	*7*	*49*	*300*
S. & C. America	*4.23/1*	*20 785*	*235*	*4 000*	*102 000*	*25*	*17*	*0.19*	*1*	*42*	*50*
Oceania	*4.24/1*	*7 956*	*12*	*684*	*23 850*	*35*	*56*	*0.09*	*5*	*57*	*2 000*
*Asia**	*4.25/1*	*27 896*	*900*	*6 000*	*73 000*	*12*	*7*	*0.2*	*0.04*	*7*	*70*
Africa	*4.26/1*	*32 315*	*2 526*	*12 000*	*82 700*	*6*	*6*	*0.4*	*2.6*	*11*	*30*
TOTAL		135 670	6 625	50 421	630 430	12.5	8	0.4	1.7	20	95
Old World % of total		87 293 64%	5 943 90%	40 983 81%	379 380 60%	9	7	0.5	1.3	15	64
New World % of total		*48 377 36%*	*682 10%*	*9 438 19%*	*251 050 40%*	*27*	*14*	*0.2*	*3.8*	*46*	*370*
New World/Old World		×0.4	×3	×3	×3	×3	×2	×0.4	×3	×3	×6

* excluding U.S.S.R.

frail, but estimates are complete in that they have been adjusted to include notional figures for countries without statistics of their own; there are no hidden gaps.

The world production of honey fluctuates from year to year; the present average is assessed at rather more than 600 000 tons. This honey is harvested from 50 million colonies of bees, by about 4½ million bee-keepers. The estimated human population in the world is about 3 500 million, and the number of individual honeybees will be about 500 times as great as this. A summary published in 1963 (Crane, 1963) gives the annual honey production as 500 000 tons, the number of colonies as 40–45 million and the number of beekeepers as 5 million. The increase in honey production and the number of colonies, and the slight decrease in the number of beekeepers who manage them, are in accordance with general experience.

The final entries in Table 4.27/1 have been added to bring out what is perhaps the most striking feature of world honey production: the high productivity of the New World (which had no indigenous honeybees) compared with the Old World (where they evolved). With only 18% of the hives of bees, the New World produces 40% of the honey, exporting a substantial amount of this to the Old World. The average honey pro-duction per colony is 27 kg for the New World as a whole, three times as much as the Old World average of 9 kg.

The number of colonies managed by any one beekeeper may be any-thing between 1 to over 1 000; the average is twice as high in the New World (14, compared with 7 in the Old World). The colony density in Europe is relatively so high that it raises the Old World average to 1 per 2 sq km, compared with 1 per 5 sq km in the New World. Honey consumption is highest in the advanced countries, and three times as high in the New World as in the Old. Tropical Africa holds a special position, the honey being used for fermentation into beer.

On a world basis, the amount of honey *produced per beekeeper* is perhaps the most striking figure, and it is included as Column 11 of Table 4.27/1. The highest productions are achieved where both plenty of land and mechanized equipment are available. The New World outstrips the Old by a factor of nearly 6. Asia and Africa (and South and Central America) still have unexploited land, and when mechanization can be applied effectively to beekeeping in these continents, their *per capita* production could rise steeply. It is less likely to do so in Europe, where shortage of land (and bee forage) is more likely to be the limiting factor.

The evidence shows why honey is a world industry, and it points to the establishment of honey as a connoisseur's food, although there are still some regions of the world where it is one of the subsistence foods. There is at present a world shortage of honey, and part of the increasing

Production

30 000 tons

4 000 tons

4 tons

1 mm³ = 4 tons

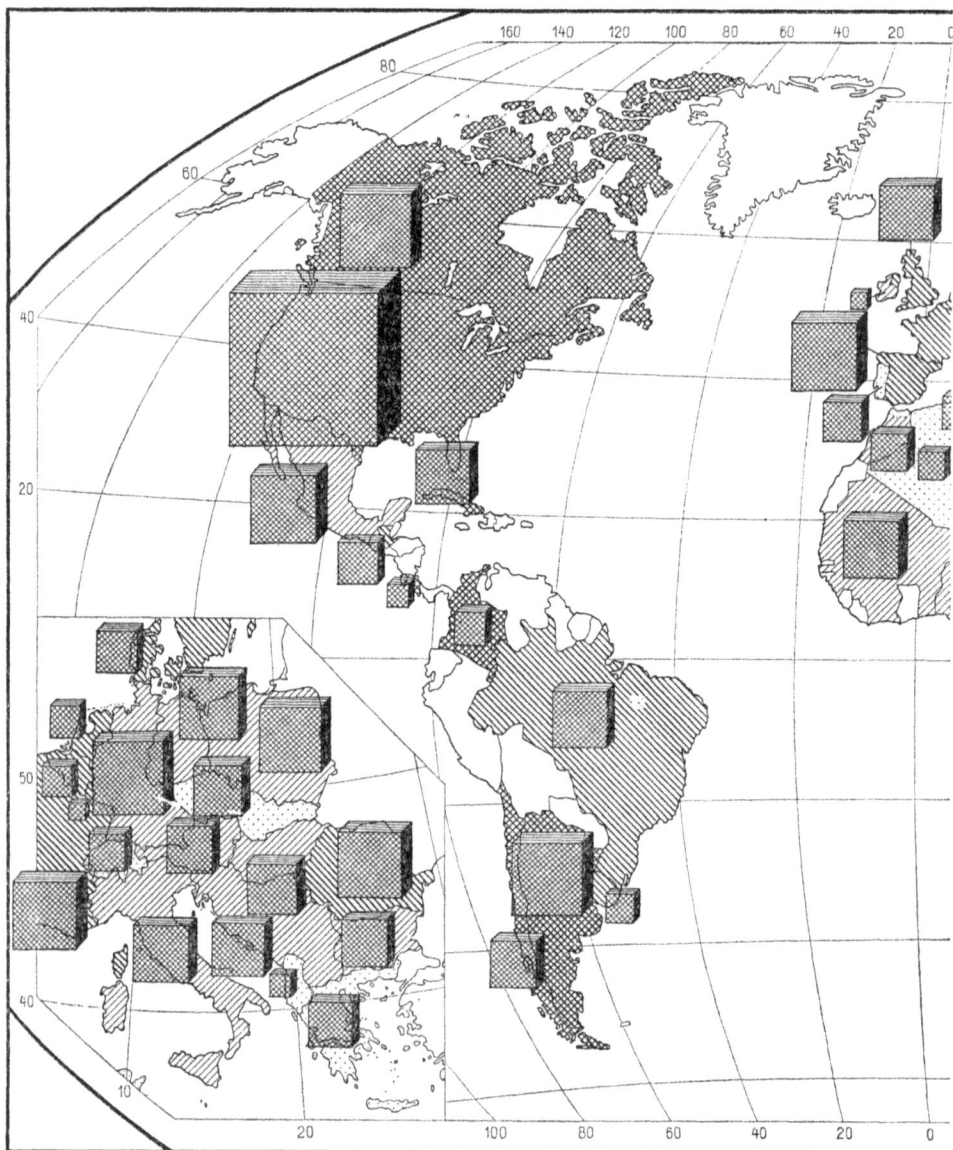

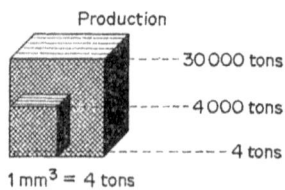

Yield per colony (kg)

⬚ <6   ⬚ 10-20
⬚ 6-10   ⬚ >20

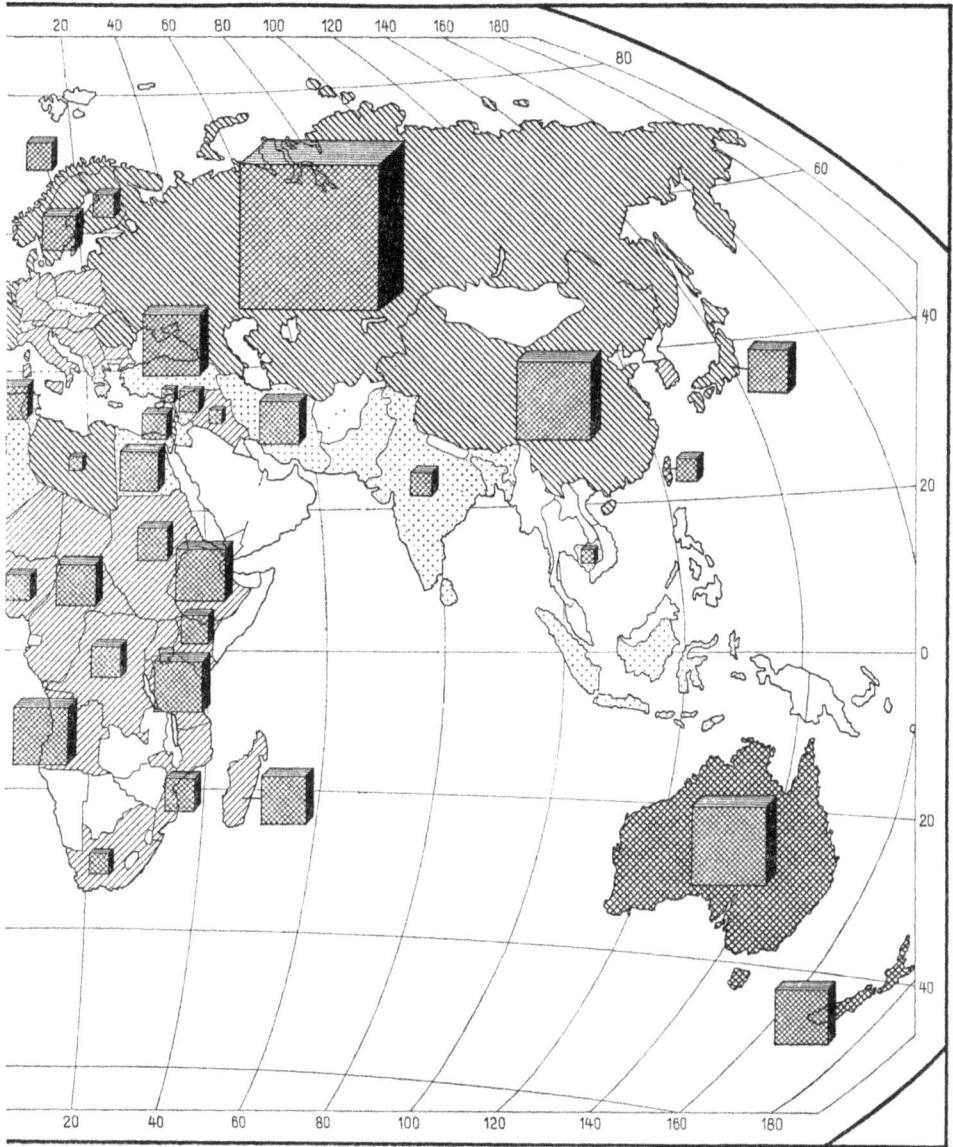

Figure 4.28/1   Total annual honey production, and yield per colony, in different countries.
(Crane, in Skibbe (1958))

demand is from young people interested in 'natural' foods. Honey straight from the comb comes near to this ideal; the more extensively honey is processed—especially by heat treatment (*see* 9.31)—the less 'natural' it becomes. So we are now seeing a move away from the bulk blending and processing which has been successful in establishing a stable honey trade in the past decades. Many customers are willing to pay more for honeys of specific types, with their own special flavours and aromas, than for a standardized product. Analogy can perhaps be drawn from the cheese trade, where favoured varieties are clearly distinguished from the 'mousetrap' kind and command higher prices.

4.28   Source material past and present

The data presented in Tables 4.21/1 to 4.27/1 are an outcome of nearly twenty years' collection of statistics on beekeeping and honey production. A preliminary paper on the subject was read at the XVI International Beekeeping Congress in Vienna in 1956, but not published. Tentative data were contributed to the *World agricultural atlas* edited by Skibbe (1958) in the form of two maps. One (Figure 4.28/1) showed the total national honey yields in tons (H) and the average honey yield in kilograms per colony (H/C). The other (Figure 4.28/2) showed the number of hives of honeybees per 100 hectares (i.e. per sq km) of agricultural land in each country. For some countries, such as Czechoslovakia and Hungary, this is nearly the total area, and the hive density shown in the map is not very different from C/A in 4.2. For others with large areas of desert, tundra, or bare mountains, the hive density on the map is much higher; for Egypt it is even 40 times as high.

Selected summarized data were later included in a chapter on world beekeeping (Crane, 1963), which concluded: 'The present world production of honey is nearly 500 000 tons, the work of 40 to 45 million colonies of bees in the hands of perhaps 5 million beekeepers. The estimated human population of the world is about 3 000 million, and the number of individual honeybees will be about 500 times as great as this.'

No other collections of data are known which try to assess the world situation. F.A.O. publishes figures for 'Bee hives' and 'Honey production' in its *Production Yearbooks*, with figures for certain earlier years. In 1969 there were entries for 57 countries, compared with 81 in Tables 4.2. The most useful source of information is the series of *Foreign Agricultural Circulars* published by the United States Department of Agriculture from 1963 onwards. Data on honey production and numbers of hives are also included in some of the publications giving information on world honey trade (Chapter 14). A memorandum issued by the Centre National du Commerce Extérieur, France (1956) is one example.

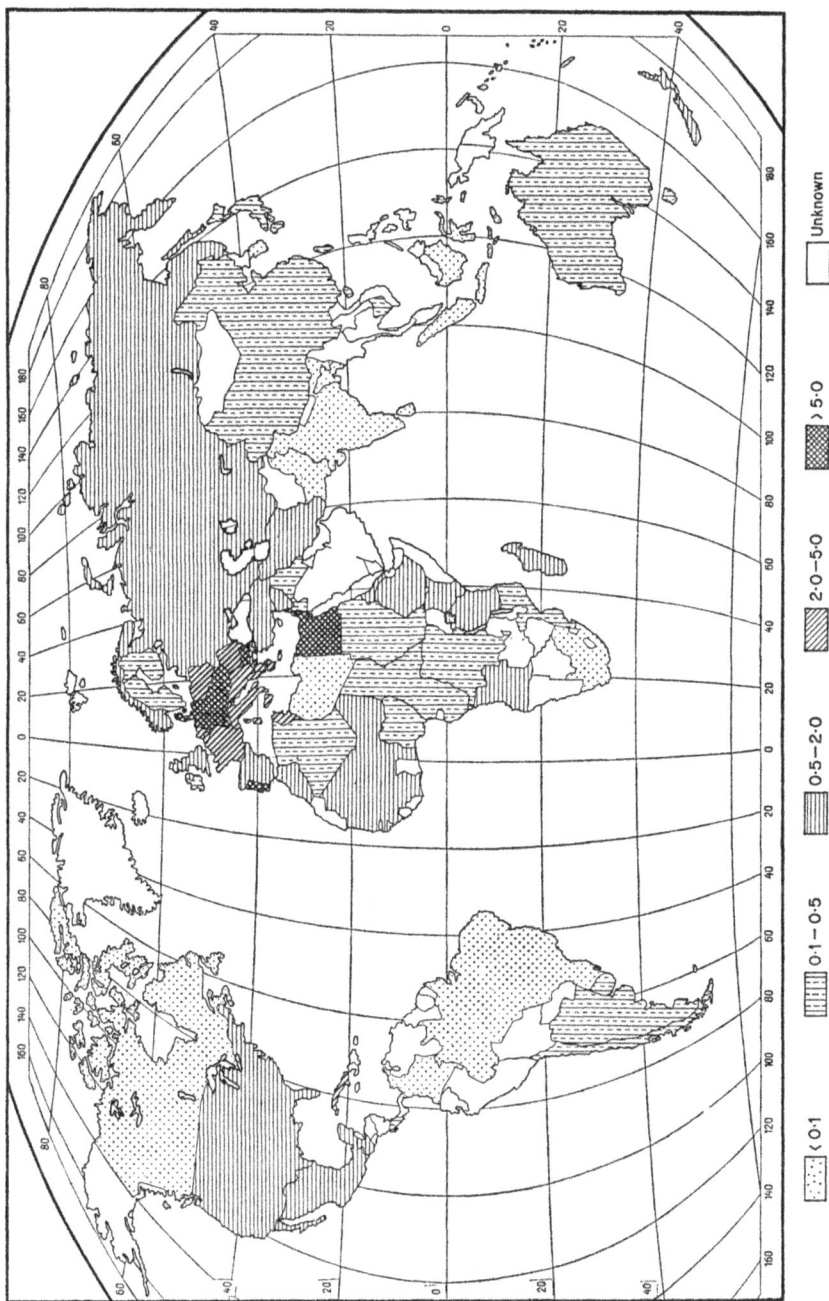

Figure 4.28/2 Number of hives per sq km of agricultural land, in different countries.   (Crane, in Skibbe (1958)).

Going back to earlier years, a report was published in 1940 by Schubring, using 'all available statistics' for hives, honey and beeswax production, imports, and exports. He gives figures for one or more years for about 20 European countries and 15 others—the countries differ in the different tables, which relate to any available years in the 1930s. There is useful material here, but no attempt seems to have been made to obtain figures from countries without official statistics, nor to consider the validity of those that are included. Some of the statements therefore create a false impression; for instance the only African country for which beeswax production is given is Algeria, yet its output is insignificant compared with that of the major wax-producing countries in tropical Africa.

Another paper published in 1940, by Guglielmo, is on honey and wax in industry and trade; it gives figures for a small number of countries, mostly in the 1920s.

Alphandéry's *Traité complet d'apiculture* (1931, pp. 503–544) included a section 'L'apiculture dans le monde' which has much interesting information, especially about some less well known countries; a certain number of figures are given, but there is no assessment of their validity.

In 1912 Dr. P. E. Spielmann tried to collect information from many countries on honey production.* Some of the information brought to light is most interesting, but many of the countries approached were unable to provide data. The Spielmann papers are valuable source material because the date and origin of each is known. By contrast, lists have been published from time to time in the newspapers and bee-keeping journals of different countries with no indication of the source or date of the material. They may have an appearance of authenticity which is quite unwarranted, in view of the known difficulties in obtaining valid data. Moreover, it is rarely made clear that a large number of countries are omitted from the lists. Import and export figures are much more satisfactory in this respect, because lack of data is likely to imply absence of trade.

4.3   THE CHANGING PATTERN OF WORLD HONEY PRODUCTION

Any change in the vegetation or climate of a region can change its potential honey production, as can the introduction of bees with markedly different honey-getting abilities. The latitude effect is discussed at the end of Section 1.13.

---

* He was also involved in proposed legislation on bee diseases and on the sale of mead, in the United Kingdom. In 1955 he presented his papers to the Bee Research Association, where they are in safe keeping until opportunity arises for a study to be made of them.

4.31 Change of land use

Rather little 'virgin land' exists in the world today, but striking changes in honey productivity can occur when an area is settled for the first time.

Where coniferous forests are cleared for agricultural use in northern Canada and Siberia (U.S.S.R.), the cleared land requires nitrogen, which can profitably be provided by growing legumes for seed. At these latitudes the clear air and long hours of summer daylight give high insolation, and even with the short growing season both seed yields and honey yields are excellent. In the Peace River area of northern Alberta and British Columbia the average honey yield per colony is around 70–100 kg, although the beekeeping season lasts only $4\frac{1}{2}$ months—from mid-April when package bees are received until the end of August when they are killed. If planting is left too late for legumes, rape is grown for seed instead, and this also yields well, although the honey is not of such good quality as that from the clovers; see Figure 1.13/4.

The favoured position of these areas for honey production will, however, cease if and when legume seed production is no longer the most profitable form of agriculture.

Most of the uncleared land of the world lies in the tropics and subtropics. Much of this is good, and some extremely good, for honey production. In these areas, clearing is likely to decrease the honey productivity of the land. But in the process of opening up, other good honey country that is more remote may become accessible for the first time. There are for instance still large areas in northern Argentina that could yield excellent honey crops if they were accessible. In many jungle regions, such as Yucatán in eastern Mexico, access is still the limiting factor in honey production. The same is true of the province of Yunan in southern China, another subtropical area with a very high honey potential (Oschmann, 1961). There are many more examples, some well known and others hardly suspected yet.

Clearing land for agriculture can lead to a heavy loss in honey production where indigenous trees and shrubs are good honey yielders. In Australia, for instance, the native eucalypts give a far higher yield than any crop likely to succeed them. Where the land is farmed, it is not farmed *for* honey production—the bees have to gather what they can from plants which have some other economic use. This may be crop production (e.g. cotton, seed legumes), grazing (clovers, meadows), conservation of birds (heather), land stabilization (*Robinia*) and so on. There are some infertile regions where the economic return from honey production is actually higher than any other, and clearing the natural ground cover for grazing reduces the financial return per hectare. Parts of the sandy coastal plains of Western Australia probably come into this category.

In some areas, expansion of built-up areas can increase honey productivity, because of intensive garden cultivation and amenity planting.

4.32 Agricultural practices

Where unproductive land is brought under cultivation—a desert irrigated or water drained—there can be nothing but gain in honey production. New Dutch polders provide an interesting succession of honey sources in their initial years.

Increasing the scale on which crops are grown can be detrimental to honey production, where it reduces the variety of forage available to maintain colonies all through the year. This lack of variety of bee plants reaches its limit in large monocultures surrounded by land with no vegetation, as in California for example. If the crop grown is a good nectar yielder, and the pest control programme is safe for bees, then very large honey crops can be harvested by moving bees into the area at the right time; the colonies are taken to forage elsewhere at other periods.

Many of the pests that damage crops are insects, and many of the insecticides developed since the Second World War for pest control are poisonous to bees as well as to the pests they are designed to kill. A number of countries have passed through a very bad period, when there was little understanding on the part of the crop growers or spraying contractors that bees on the flowering crop must not be killed because they were pollinating it. The situation is now somewhat improved, through increased knowledge and education, coupled with the current emphasis on selectivity in pest control—especially through biological methods—but continuous watchfulness is needed, and some countries are only now passing through the phase of indiscriminate insect killing. Recent summaries of the position are given by Johansen (1966), Anderson & Atkins (1968), and Beran (1970).

4.33 Economic developments

The honey productivity of an area can only be fully utilized (by man) if the density of honeybee colonies is adequate. This depends on access (4.31), and on the willingness and ability of beekeepers to operate bees in the area—which usually depend on the adequacy of the financial reward. Capital investment is needed, and the income obtainable must offer sufficient inducement. Where there is a strong beekeeping tradition, the threshold 'inducement level' is likely to be lower than elsewhere. As a sideline spare-time occupation, beekeeping can be operated at a lower inducement level than is possible where it is a full-time occupation and labour must be paid for. But the incentive of a good harvest is still very important. Where there has been a general decline in beekeeping (as in England, for instance, Table 4.33/1), a good harvest one year may prevent a decrease (or even lead to an increase) in the number of bee-keepers the following year. In Gloucestershire (Table 4.33/1), the number of colonies dropped from 9 103 to 7 535 after the poor 1956 harvest, then rose to 7 778 after the better 1957 yield. But 1958 was another poor year, and in 1959 there were only 6 324 colonies. After the very good 1961 harvest the number of colonies increased to 7 792 in 1962 —but the yield that year was very low, and in 1963 only 6 062 were left.

The decrease in honey production over the past twenty years in many European countries is mostly due to the drop in the number of colonies (and of beekeepers), but there is also evidence of a general downward trend in the honey yield per hive. The figures for Gloucestershire from 1945 to 1971 (Table 4.33/1) are probably fairly representative. The honey yields from 1917 to 1936 from a single scale hive at Street in Somerset, a county adjoining Gloucestershire, are interesting for comparison— although Street seems to have been an unusually good honey area. These are given below; they relate to Edwin Walker's scale hive (*see* Figure 1.13/1):

1917	140 kg	1922	36 kg	1927	15 kg	1932	31 kg
1918	48 kg	1923	34 kg	1928	20 kg	1933	66 kg
1919	43 kg	1924	6 kg	1929	39 kg	1934	54 kg
1920	16 kg	1925	49 kg	1930	7 kg	1935	28 kg
1921	51 kg	1926	25 kg	1931	4 kg	1936	16 kg

Honey production is fairly stable but slightly on the increase in the U.S.S.R. In parts of South and Central America there have been dramatic increases as existing potentials were newly exploited, and the same is now true of the People's Republic of China (14.25). In general the efficiency of beekeeping has increased, so that one beekeeper can manage more hives, and *with the same conditions* the honey yield per hive has also increased. But in many parts the conditions have not remained static. Pesticides take their toll of bees, and farming managed on efficient and economic lines reduces the honey-yielding forage: weeds are fewer, forage legumes are harvested rapidly before they flower, and hedges and 'waste land' are cleared up. In the United States, for instance, the increased efficiency of management during the past forty years has only just cancelled out the losses due to various factors, including those mentioned. As a result, the average yield per hive has remained more or less unchanged. The problem is also discussed in Section 1.1.

Table 4.33/1
The decrease in beekeeping since the end of the last war in England and Wales, with representative honey yields for one county

Columns 2, 3, and 4 are derived from 'Returns of application for sugar for autumn feeding' for 1945–1953, and thereafter from *Bee health and beekeeping in England and Wales* (Ministry of Agriculture, Fisheries, and Food). Column 5 is derived from the Crop Reports published annually in the *Yearbook of the Gloucestershire Beekeepers' Association*. It gives the average yield of surplus honey per hive, from reports on 5–10% of the colonies in the county. Early figures relate to the second of the two years.

	England and Wales		*Gloucestershire*	
Year	*No. Beekeepers*	*No. Colonies*	*No. Colonies*	*Honey Yield (kg)*
1945/46	73 369	369 387	11 932	8
1946/47	83 370	418 543	11 665	17
1947/48	76 155	368 290	13 214	11
1948/49	87 078	464 399	13 043	25
1949/50	—	—	—	—
1950/51	84 481	443 446	13 346	23
1951/52	74 020	376 778	11 746	14
1953	80 567	396 247	—	5
1954	—	—	—	2
1955	71 853	305 507	11 130	25
1956	56 925	266 142	9 103	8
1957	48 002	220 599	7 535	23
1958	48 140	233 452	7 778	5
1959	48 562	223 387	6 324	30
1960	47 760	219 545	6 208	12
1961	48 089	222 726	6 425	34
1962	47 923	228 496	7 792	9
1963	44 485	192 652	6 062	5
1964	44 458	201 564	5 994	17
1965	42 895	198 777	5 269	6
1966	39 122	182 254	4 812	8
1967	36 811	161 519	4 634	21
1968	36 436	165 737	4 185	9
1969	33 418	158 510	4 438	18
1970	32 699	152 608	4 718	14
1971	32 025	158 219	4 795	23

4.4 HONEY PRODUCTION FROM THE VIEWPOINT OF ENVIRONMENTAL BIOLOGY

'Honey production' can have several different meanings, relating to both bees and beekeepers, and these are now discussed. Lower-case letters are used here, referring to a single colony, to prevent confusion with letters used in 4.1 and 4.2.

The amount of honey removed by the beekeeper during and after the active season (h) is his harvest from the hive. He leaves in the hive enough food to last the bees through the dearth period or, if not, he augments it by feeding sugar (honey-equivalent f, including sugar fed at any other time). If c denotes the colony's consumption of honey during the year, its 'total income' (t) taken from the colony's environmental resources, together with food provided (f), must at least balance $c+h$, i.e. the colony's own consumption plus the beekeeper's harvest. If $t+f$ is greater than $c+h$, there is a reserve of stores to start the next season. If the reverse is true, the net amount of food available ($t+f-h$) would be less than the colony's requirement for consumption (c), and the colony would starve in the dearth period—unless kept going by a reserve from earlier years. In those countries where there is a very high honey yield per hive, beekeepers usually do not undertake the expense and work of feeding sugar, and $h=t-c$. In the U.S.S.R. and some neighbouring countries, the honey production of a colony may be counted as *all* the honey stores in the hive at the end of the active season, plus any removed earlier, but corrected for any sugar fed. This is $h+c_d-f$, where c_d is the consumption allowed for the dearth period; it may be considerably higher than h, especially if h is small. The amount c_d is fairly easily recognized by the beekeeper, since it is normally present in the hive when the dearth period starts. But in fact the amount used in the active season (c_a) is far larger, and is often the largest item in the colony's balance sheet for the year.

It is instructive to see how h, t, c, and f vary in different parts of the world. In Table 4.4/1, Britain, U.S.A., and Western Australia are taken as examples of regions with low, medium, and high honey yields, and N.W. Canada refers to the package-bee industry (3.32), where bees are introduced only for the short summer season—*see* 1.13.

In Britain the honey harvested (h) is 10% of the total (t) produced by the bees; in a bad year it might be under 5%. In good honey country this 'efficiency' may rise to 30–40%, or in exceptional circumstances it might be 50%. This difference helps to explain why European honey is in general more expensive than honey imported from countries with high honey yields: utilization of nectar resources is only a third or quarter as efficient. The use of package bees is especially interesting

Table 4.4/1
Annual honey harvest (h) taken from a colony of bees, and its food consumption during the year (c), in relation to the food provided by the beekeeper (f) and the total honey-equivalent of food collected by the colony during the year (t), all in kg

See text for further details

Type of honey yield	low	medium	high	high
Example	Britain	U.S.A.	W. Australia	package bees
h	15	40	100	100
c	145	200	200	180
				$c_a = 130$
				$c_d = 50$*
f	10	—	—	10
$t = h + c - f$	150	240	300	270
h/t	10%	17%	33%	37%

* This is used in the production of the package bees in southern U.S.A.

from the present point of view since it takes advantage of two environments. Bees are produced at low latitudes, with a long season and early spring; honey is produced at a high latitude, with a short summer giving intense growth and nectar production. The bees do not have to survive a dearth period, since they are killed before it starts.

In Tables in Section 4.2, the 'honey production' H should theoretically be the number of colonies $\times h$, but Section 4.12 explains something of its frailty; where the entry under H is known to be too low, this is noted.

In many countries, published statistics on honey production relate not to h, however, but to the average amount of honey per colony put on the market (m). This may be close to h in countries with high honey yields where each beekeeper has several hundred colonies and sells all his honey. But m is likely to be much less than h in countries where bee-keeping is mainly a hobby, and many beekeepers have only enough hives to provide honey for their families and friends. The honey production per colony quoted in government statistics may be much lower than h for other reasons, discussed in 4.12.

SECTION 2

CHARACTERISTICS OF HONEY

CHAPTER 5

COMPOSITION OF HONEY

by Dr. Jonathan W. White, Jr.

CHIEF, PLANT PRODUCTS LABORATORY,
UNITED STATES DEPARTMENT OF AGRICULTURE

What is sweeter than honey?

JUDGES 14, 18,

NEW ENGLISH BIBLE

5.1 INTRODUCTION

Man is twice indebted to the honeybee for searching earth's fields and forests for their treasured sweetness—for the honey itself, with its variety of appeals to our senses, and for the increased crop yields resulting from inadvertent pollination of the flowers the foraging bee visits. Literally vital to the honeybee colony, honey is to us simply a desirable and delectable variety in our diet. It is a twice-stolen sweet (and stolen sweets are said to be the best)—taken by the bee from the flower (and from the owner of the plot upon which it grows)—and then stolen from the bees by their keepers.

Honey as it is found in the hive is a truly remarkable material. Prepared by the bees from the natural sugar solutions we know as nectar, it is changed from an easily spoiled, thin, sweet liquid to a stable high-density, high-energy food. By inverting the sucrose in the nectar, the bee increases the attainable density of the final product, and thus raises the efficiency of the process in terms of caloric density. At the same time, the resistance of the stored product to spoilage by micro-organisms is greatly increased, because of the higher osmotic pressure attained. A price is paid, however, in terms of the increased tendency of the honey to absorb atmospheric moisture, with consequent liability to fermentation when yeast levels are sufficiently high.

5.2 AVERAGE COMPOSITION OF HONEY*

The composition of a particular honey sample will depend upon two general factors: most important, the composition of the nectar(s) whence

* Discussion in this chapter has been restricted to honey from nectar (floral and extrafloral); honeydew has been excluded because of space limitations, but much useful information on honeydew and honeydew honey is included in Chapter 2.

it originates; of less importance, certain external factors. As detailed in
Chapter 2, nectars from different plants vary widely in the identity and
concentrations of their constituent sugars; in fact, honey types are
ascribed to plant sources by flavour or gross composition alone. Weather
or climatic conditions and beekeeper practices in removing and extracting
honey may affect composition to a minor extent.

Two considerations make the concept of an 'average composition' of
honey somewhat uncertain—the degree to which the analytical methods
used to establish the individual values actually reflect the true composition of the sample, and secondly, the considerable variation in
composition encountered among honeys because they represent different
floral types. This latter factor may be compounded by general differences
from one area or country to another in nectar sources available. Hence,
the average composition given in Table 5.2/1 of honey from the United
States would probably differ from similar values for honeys from regions
of differing climate, topography or agricultural pattern.

Table 5.2/1
Average composition of U.S. honey and ranges of values
(White, Riethof, Subers & Kushnir, 1962)*

Component	Average	Standard Deviation	Range
Moisture	17·2	1·5	13·4 −22·9
Laevulose	38·2	2·1	27·2 −44·3
Dextrose	31·3	3·0	22·0 −40·7
Sucrose	1·3	0·9	0·2 − 7·6
'Maltose'†	7·3	2·1	2·7 −16·0
Higher sugars	1·5	1·0	0·1 − 8·5
Free acid (as gluconic)	0·43	0·16	0·13 − 0·92
Lactone (as gluconolactone)	0.14	0·07	0·0 − 0·37
Total acid (as gluconic)	0·57	0·20	0·17 − 1·17
Ash	0·169	0·15	0·020− 1·028
Nitrogen	0·041	0·026	0·000− 0·133
pH	3·91		3·42 − 6·10
Diastase value	20·8	9·8	2·1 −61·2

* Values as percentage of honey, except the last two entries
† Reducing disaccharides calculated as maltose

Table 5.2/1 shows the average values, with standard deviation and
range, obtained in the analysis of 490 samples of U.S. honey representing
3 single floral types and 93 blends of 'known' composition (White *et al.*,

1962). The extreme values found among the samples are shown in the column marked 'range'. A better idea of variability may be obtained by use of the standard deviation (S) in the table. With normal distribution of values, about two-thirds of the samples will fall within $\pm 1S$ of the average value shown; 95% will be found in the interval from $2S$ less than the average to the average plus $2S$. Thus, although in this study the extreme values for laevulose content were 27·2 and 44·3%, 95% of the samples were within the range 34·0 to 42·4% laevulose.

Table 5.2/2 shows average composition of honey samples of various floral types (White et al., 1962). This table will provide an idea of the variability among different kinds of honey. The high dextrose content of fast-granulating cotton honey contrasts with the low values for non-granulating tupelo honey.

The data in Tables 5.2/1 and 5.2/2 were obtained by the most advanced analytical methods, with adequate care that dextrose and laevulose values are accurate and not influenced by other more complex reducing sugars in the sample (White et al., 1962). Many published data on sugar composition of honey (particularly older reports) should be viewed with caution. The nature of the sugar complex of honey is such that results can be misleading unless either a suitable pre-analysis separation of sugar classes is made, or highly specific analytical methods are used. Pourtallier (1962) recognized the inadequacy of the non-specific methods previously used and proposed a modified procedure. It is not clear whether all sources of potential error have been taken into account in his methods. He has examined the application of thin-layer chromatography (1964) and gas-liquid chromatography (1967) to carbohydrate analysis of honey. Echigo (1970) has also reported the use of gas-liquid chromatography. Results on a single sample differed considerably from those by ion-exchange chromatography. Siddiqui (1970) referred to analysis of 95 Canadian samples by a paper-chromatographic procedure but gives only average values and ranges for fructose, glucose, and oligosaccharides, with no information on sample identity or origin. His claim that results by this procedure are 'equally as good as, if not better than, those afforded by other available methods' is not supported by any data and seems highly unlikely.

In Table 5.2/3 are shown representative analytical values for honeys from various areas over the world. The fragmentary nature of the data is apparent. One may generalize that dextrose content is somewhat lower than laevulose content, that moisture content is usually between 15 and 21%, sucrose about 1–3%, and ash between 0·09 and 0·33%. Comparison of data shown in different entries must be made with knowledge of the analytical methods used and of their shortcomings; among the older methods of honey analysis, variability due to methods

Table 5.2/2
Average composition of specific honey types
(White et al., 1962)[1]

Type	No. samples	Dextrose	Laevulose	Sucrose	Maltose[2]	Higher sugars	Total acidity[3]	Ash	Nitrogen	Water
Alfalfa, Lucerne *Medicago sativa*	23	33·4	39·1	2·6	6·0	0·9	0·53	0·093	0·033	16·5
Aster *Aster* spp.	5	31·3	37·5	0·8	8·4	1·0	0·44	0·302	0·043	17·4
Basswood *Tilia americana*	3	31·6	37·9	1·2	6·9	1·4	0·46	0·084	0·022	17·4
Blackberry *Rubus* spp.	3	25·9	37·6	1·3	11·3	2·5	0·57	0·399	0·055	16·4
Buckwheat *Fagopyrum esculentum*	5	29·5	35·3	0·8	7·6	2·3	0·82	0·224	0·064	18·3
Wild buckwheat *Eriogonum fasciculatum*	4	30·5	39·7	0·8	7·2	0·8	0·63	0·136	0·054	16·3
White clover *Trifolium repens*	12	30·7	38·4	1·0	7·3	1·6	0·62	0·156	0·046	17·9
Sweet clover *Melilotus* spp.	8	31·0	37·9	1·4	7·7	1·4	0·52	0·084	0·038	17·7
Cotton *Gossypium hirsutum*	10	36·7	39·3	1·1	4·9	0·5	0·58	0·339	0·037	16·1
Fireweed *Chamaenerion angustifolium*	3	30·7	39·8	1·3	7·1	2·1	0·52	0·108	0·032	16·0
Gallberry *Ilex glabra*	6	30·1	38·9	0·7	7·7	1·2	0·40	0·163	0·028	17·1

	n									
Goldenrod *Solidago* spp.	3	33·1	39·6	0·5	6·6	0·6	0·43	0·263	0·045	17·0
Black locust *Robinia pseudoacacia*	3	28·0	40·7	1·0	8·4	1·9	0·31	0·052	0·018	17·3
Mesquite *Prosopis glandulosa*	3	36·9	40·4	0·9	5·4	0·3	0·32	0·129	0·012	15·5
Orange/Grapefruit *Citrus* spp.	13	32·0	38·9	2·8	7·2	1·4	0·59	0·073	0·014	16·5
Loosestrife *Lythrum salicaria*	3	29·9	37·7	0·6	8·1	2·3	0·57	0·125	0·044	18·3
Sage *Salvia* spp.	3	28·2	40·4	1·1	7·4	2·4	0·57	0·108	0·037	16·0
Sourwood *Oxydendrum arboreum*	3	24·6	39·8	0·9	11·8	2·5	0·33	0·230	0·020	17·1
Star thistle *Centaurea solstitialis*	4	31·1	36·9	2·3	6·9	2·7	0·81	0·097	0·055	15·9
Tulip tree *Liriodendron tulipifera*	4	25·8	34·6	0·7	11·6	3·0	0·84	0·460	0·076	17·6
Tupelo *Nyssa ogeche*	5	25·9	43·3	1·2	8·0	1·1	0·72	0·128	0·046	18·2
Hairy vetch *Vicia villosa*	9	30·6	38·2	2·0	7·8	2·1	0·45	0·056	0·030	16·3

[1] All values as percentages
[2] Footnote †, Table 5.2/1
[3] As gluconic acid

can exceed variability due to difference in samples (White, Ricciuti & Maher, 1952). Carbohydrate values obtained by White & Maher's (1954) selective adsorption method are those of Austin (1958), Anderson & Perold (1964), Gryuner & Arinkina (1970), and da Silva Ferreira (1970).

Other factors influencing the composition of honey are the period and conditions of storage. Early data on this subject were reviewed, and storage changes in carbohydrates, acidity, and diastase were quantified, by White et al. (1961, 1962). Gonnet (1965) has also examined storage changes in honey, verifying in general the findings of White et al.

5.3 CARBOHYDRATES

Table 5.2/1 shows that by far the largest portion of the dry matter in honey consists of sugars. In the main, the sugars are responsible for much of the physical nature of honey, its viscosity, hygroscopicity, granulation properties, energy values, and so on.

5.31 Monosaccharides and disaccharides

In nearly all honey types, laevulose (fructose) predominates; a few honeys —such as rape (*Brassica napus*), dandelion (*Taraxacum officinale*), blue curls (*Trichostema lanceolatum*)—appear to contain more dextrose (glucose) than laevulose. These two sugars together account for 85–95% of honey carbohydrates. More complex sugars (oligosaccharides) made up of two or more molecules of glucose and fructose constitute the remainder, except for a trace of polysaccharide. Recent research in the United States, Japan, and Canada has shown that at least eleven disaccharides are present in honey in addition to sucrose. Most of these sugars are quite rare, and their recovery from honey was the first isolation from natural material. White & Hoban (1959) separated the sugars and, by use of the infra-red spectra of the free sugar and of its acetate, identified maltose, *iso*maltose, nigerose, turanose, and maltulose. Watanabe & Aso (1960) crystallized the acetates to identify maltose, *iso*maltose, nigerose, and kojibiose, and tentatively identified leucrose. Further confirmation of the occurrence of all of these sugars (except leucrose) was reported by Siddiqui & Furgala (1967). They attained a more rigorous identification by isolation and characterization of crystal-line sugars (sucrose, turanose) or crystalline acetates (*iso*maltose, kojibiose, maltose, nigerose). In addition, neotrehalose, gentiobiose, and laminaribiose were identified (as acetates). Tentative identification of *iso*maltulose and maltulose was also reported.

5.32 Tri- and higher saccharides

In their definitive study of honey carbohydrates, Siddiqui & Furgala (1968a) have reported the isolation and identification of eleven oligo-

Table 5.2/3
Composition of honeys from different countries[1]

Country	No. samples	Water	Total reducing sugars	Dextrose	Laevulose	Sucrose	Maltose	Dextrin	Ash	Nitrogen	Free acidity[2]
EUROPE											
Britain	13	18·9 15·9–23·4		34·6 30·0–36·8	39·8 36·1–44·4						
Bulgaria	190	21·3	71·7						0·15		0·68
Italy (Sicily)			70·9–77·2			1·6 1·6–2·1					0·42–1·24
Netherlands	41			30·5 20·4–39·5	41·5 38·1–53·9		4·6 1·8–7·5	2·3 0·9–4·0			
Portugal	10	18·1 16·7–19·8		32·2 27·2–34·6	36·2 31·0–38·2	0·96 0·70–1·15	6·68 5·13–8·92	1·90 1·04–3·56			0·64 0·45–0·82
Rumania	257	16·5		34·0	38·4	3·1		3·8	0·17		
Spain	23	17·3 14·3–21·6	75·6	28·4 21·2–36·8	36·5 31·9–41·7	0·9 0·4–2·3	8·2 2·4–14·9	1·1 0·4–3·5	0·18 0·040–0·75	0·050 0·014–0·089	0·45 0·23–1·20
U.S.S.R. (1970)	10	19·3 17·7–23·6		32·5 28·7–36·7	34·9 31·5–37·5	1·43 0·0–4·75	4·32 2·26–6·84	1·26 0·10–2·57	0·20 0·035–0·34	0·066 0·015–0·152	
U.S.S.R. (1963)	217	18·6 15–23	73·8 62·7–84·4	35·9 26·4–44·4	37·4 21·7–49·7	2·11 0–10·3			0·16 0·01–0·59		0·57 0·09–1·20
Yugoslavia	43	17·7 13·8–20·8	73·0 65·5–79·1	35·2 26·1–43·2	37·3 30·1–44·9	1·91 0·15–4·70			0·25 0·07–0·70		0·38 0·04–0·85
ASIA											
India (Mahabaleshwar)[3]	3[4]	17·1		35·1	41·2	2·74		1·5	0·13	0·065	0·23
India[3]	12	19·2 16·2–22·1		35·7 34·2–39·2	39·3 36·8–40·5	0·60 0·3–1·0			0·10 0·03–0·46		0·45 0·25–1·25

Country	No. samples	Water	Total reducing sugars	Dextrose	Laevulose	Sucrose	Maltose	Dextrin	Ash	Nitrogen	Free acidity[2]
Japan (Aso et al., 1960)	15	20·5 15·8–26·2	69·2 60·5–76·1	32·6 22·2–38·6	36·0 30·0–48·5	2·83 1·0–5·8					
Japan (Arai et al., 1960)	30	20·4 13·8–25·4		29·8 20·4–38·7	40·8 30·4–46·0				0·11 0·03–0·46		
Pakistan[3]	15	14·3–18·6		39·0–53·8	27·7–34·2	1·9–2·75			0·11–0·32		0·30–0·68
AFRICA											
Angola	4	19·3 13·4–25·0		33·9 32·1–35·0	36·4 34·2–38·5	0·86 0·15–1·50	6·48 4·56–7·79	0·98 0·52–1·32			0·53 0·22–1·03
Mozambique	4	18·7 17·4–21·8		32·0 28·6–35·3	36·2 33·6–38·2	1·10 0·65–1·60	6·51 5·10–7·58	1·76 0·82–3·16			0·81 0·32–1·71
Portuguese Guinea	4	19·4 16·2–20·4		31·2 28·5–34·4	38·3 35·6–40·8	1·06 0·45–2·20	6·36 3·55–8·80	0·80 0·16–1·48			0·98 0·61–1·51
São Tomé + Principe[5]	4	22·9 19·0–24·6		31·0 28·5–32·4	34·8 33·9–36·2	0·61 0·10–1·00	5·97 5·13–7·34	1·19 0·72–1·36			0·92 0·67–1·40
South Africa	66	16·2 13·9–18·8		31·5 22·3–39·4	35·5 22·3–40·1	0·54 0·0–6·24	5·4 2·1–10·0	0·50 0·2–3·3	0·33 0·03–0·94	0·043 0·018–0·13	
AMERICA											
Argentina (1945)	58			34·3 29·4–37·2	40·9 37·7–44·9						
Argentina (1930)	16	15·4	74·0			0·72			0·093		0·17
Canada	40	17·5[6]		33·8 30·8–37·4	38·8 35·4–40·7	1·2 0·02–3·4	6·1 4·0–11·1	1·3 0·5–2·8			

Chile	10	16·0 14·7–17·9	79·2 73·7–80·7	32·7 30·1–35·0	44·1[7] 39·1–47·0	2·5 0·9–4·4	0·15	0·54	0·53
Uruguay	32	17·3	67·3			4·9			
OCEANIA									
Australia	99	15·6	73·5	30·2	43·3[8]	2·5	0·17	0·054	0·30
New Zealand	21	17·5 16·2–19·1		36·2 32·4–40·2	40·0 38·4–42·0	2·8 1·5–4·8	0·18 0·04–0·39	0·040 0·023–0·077	0·32 0·13–0·59

Britain—Marshall & Norman (1938)
Bulgaria—Zoneff (1927)
Italy (Sicily)—Sorges (1933)
Netherlands—van Voorst (1941)
Portugal—da Silva Ferreira (1970)
Rumania—Pelimon & Baculinschi (1955)
Spain—Pérez & Rodríguez (1970)
U.S.S.R.—Gryuner & Arinkina (1970)
U.S.S.R.—Chudakov (1963)
Yugoslavia—Čermagič et al. (1964)
India (Mahabaleshwar)—Phadke (1962)
India—Giri (1938)
Japan—Aso, Watanabe & Yamao (1960)
Japan—Arai et al. (1960)

Pakistan—Latif et al. (1956)
Angola—da Silva Ferreira (1970)
Mozambique—da Silva Ferreira (1970)
Portuguese Guinea—da Silva Ferreira (1970)
São Tomé+Principe—da Silva Ferreira (1970)
South Africa—Anderson & Perold (1964)
Argentina—Ugarte & Karman (1945)
Argentina—Ceriotti & Delpino (1930)
Canada—Austin (1958)
Chile—Masson & Schmidt-Hebbel (1963)
Uruguay—Bertullo & Lembo (1943)
Australia—Chandler et al. (1974)
New Zealand—Thomson (1936)

[1] Single values are averages, others are ranges. All values are given as percentages.
[2] Calculated as gluconic acid
[3] From *Apis cerana indica*; see 17.22
[4] These three types represent 80% of the area's production
[5] Islands off West African coast, near Equator
[6] All Canadian analyses are calculated to 17·5% moisture
[7] Includes reducing disaccharides, *see* text
[8] Calculated as difference between total reducing sugars and glucose

saccharides, largely by rigorous procedures. Earlier, Goldschmidt & Burkert (1955a) had inferred the presence of melezitose, erlose, kestose, raffinose, and dextrantriose, using paper-chromatographic behaviour and colour reactions for identification. Such procedures alone are not acceptable for absolute identification of sugars: some other physical properties of the sugar or a derivative are needed also. Siddiqui points this out in discussing the alleged presence of raffinose reported by five investigators, which he could not confirm. The carbohydrates reported by Siddiqui & Furgala are: 1-kestose, melezitose, 6^G-α-glucosylsucrose, panose, *iso*-maltotriose, erlose, 3 a-*iso*maltosylglucose, *iso*panose, maltotriose, *iso*-maltotetraose, *iso*maltopentaose and two not identified. One of the latter was later characterized as o-α-D-glycopyranosyl-(1→4)-O-[α-D-gluco-pyranosyl-(1→2)-D-glucose] and given the trivial name centose (Siddiqui & Furgala, 1968b); it was estimated to constitute 0·0018% of the honey sample. Siddiqui & Furgala (1968a) have speculated on the origin of the various oligosaccharides they found in honey; they concluded that the enzymatic production of these materials cannot be explained from our present knowledge of plant, bee, and other insect enzymes.

There are many reports, especially in the earlier literature, of honey 'dextrin'. This has been considered an ill-defined, higher molecular-weight carbohydrate material found in honey, and usually estimated by precipitation with alcohol. This property provided the misleading name, since dextrins (partial hydrolysis products of starch) have similar properties. Rychlik & Fedorowska (1962a) have proposed a direct method for dextrin determination, in which sugars are first removed by a preliminary fermentation by baker's yeast. Most of the disaccharides listed above are unfermentable, so they would be measured as 'dextrin' by this procedure—which does not measure 'dextrin', whatever that may be, but rather unfermentable sugars. Many years ago, Barschall (1908) noted that the apparent molecular weight of the dextrin of conifer honey (honeydew honey) corresponded to a trisaccharide; Fellenberg & Ruffy (1933) reported that the dextrins of floral honey were also in this molecular-weight range. With the advent of paper chromatography, it was easily demonstrated that the higher sugars of honey differ from starch dextrins in containing fructose. In fact, the addition of starch syrup to honey can be demonstrated by the presence of a series of glucose saccharides (White, 1959a). Helvey (1953), investigating the colloids of honey, reported three components, two of protein nature and one a polysaccharide of molecular weight about 9 000. Siddiqui (1965) isolated a polysaccharide representing 0·002% of honey and showed it to be a highly branched arabogalactomannan (molar proportion 1·0:2·04:4·04) of molecular weight less than 10 000. General structural features were outlined.

5.33   Changes in carbohydrates with time

In modern honey analysis, all the reducing disaccharides are measured together and reported (Tables 5.2/1, 5.2/2) as 'maltose', the best-known member. The amount of such sugars in honey appears to be a function of the period since the honey was ripened, and the storage conditions since that time. Täufel & Müller (1953) suggested that, since they found no minor sugars in pollen, the source of those in honey might be by secondary conversion (enzymes in acid solution) in honey. Later (Täufel & Müller, 1957), however, using both classical methods of quantitative analysis and paper chromatography, they concluded that significant changes in sugar composition of honey do not occur in storage. It has since been shown, using modern analytical methods and statistical treatment (White, Riethof & Kushnir, 1961), that storing honey for two years at 'room temperature' brings about a 69% increase in 'maltose' at the expense of dextrose plus laevulose, which decreased to 86% of its initial value in this period. Earlier, Austin (1958) had proposed that the 'maltose' content of honey would depend to some extent upon apiary management, and on storage temperature and moisture content of honey. The report of Chudakov (1963) that glucose content increase in 8-months' storage (statistically evaluated) must be discounted because of the inadequate analytical procedures used, which would not distinguish between aldose monosaccharides and oligosaccharides. Kalimi & Sohonie (1964b) confirmed the increase of higher sugars and decrease of monosaccharides during honey storage at 28–30°C (83–86°F) for 6–12 months.

The increase in the oligosaccharides in honey is caused by two mechanisms: enzyme activity and acid reversion. The sucrose-splitting enzyme present in honey is in reality a transglucosylase (White & Maher, 1953a, White & Kushnir, 1967b), which synthesizes several of these sugars when it splits sucrose.

Although many of these disaccharides can be split by honey invertase to their constituent monosaccharides under proper conditions, the low water concentration of honey appears to favour a moderate accumulation of the disaccharides, since the free monosaccharides are formed by transfer of the glucosyl residue to water. When solutions of monosaccharides (dextrose, laevulose) remain in concentrated solution in the presence of acids, disaccharides and other carbohydrates are formed (Pigman & Goepp, 1948, page 434). The extent to which these two processes continue is probably limited. An analysis of a honey sample 36 years old (White, Riethof & Kushnir, 1961) showed 16·4% of 'maltose'. The failure of Täufel and Müller to detect these changes was probably due to inadequate analytical procedures.

5.34   Effects of complexity on analysis for sugars

The accurate analysis of honey for individual sugars, even dextrose and laevulose alone, is not simple. The presence of the minor sugars described above introduces errors unless they are removed before analysis, or specific methods of analysis are used. This means that caution must also be exercised when literature reports of the sugar composition of honey, particularly laevulose and dextrose values, are compared. For example, Masson & Schmidt-Hebbel (1963) reported an average dextrose content for 10 samples of Chilean honey as 32·7% (a reasonable value), but the laevulose content was said to be 44·1%. This is higher than the average reported (White *et al.*, 1962) even for tupelo honey, the U.S. honey of highest laevulose content. Although the specific glucose oxidase method was used for dextrose in the Chilean honey (and hence a reasonably good dextrose value obtained), laevulose was estimated simply by subtracting this value from the total amount of reducing sugars, which includes the reducing disaccharides. Such a 'laevulose value' would be considerably higher than the actual laevulose content. The laevulose value should be corrected by subtracting the contribution of the reducing disaccharides, which were not measured; estimating this to be 5–7%, the laevulose value would become 37–39% which appears reasonable.

Another pitfall in honey analysis has been described by White, Ricciuti & Maher (1952): determination of dextrose by a non-specific method for aldose sugar determination, and subsequent estimation of laevulose as the difference between total reducing sugars and the erroneously high 'dextrose' value. The presence of reducing disaccharide aldoses, as listed above, gives too high an apparent dextrose value, and too low an apparent laevulose value. An example of this error is seen in the analysis of 10 Japanese honeys (Watanabe, Motomura & Aso, 1961). Averages of 36·95% dextrose and 36·08% laevulose were reported, which can be compared with other values for Japanese honey in Table 5.2/3.

Further analytical data are needed on the true laevulose and dextrose contents of honey types found elsewhere in the world than in the U.S. and Canada, using suitable analytical methods with adequate controls. Chandler (1974) does not report true laevulose in Australian honeys.

5.4   THE ACIDS OF HONEY

The flavour of honey results from the blending of many 'notes', not the least being a slight tartness or acidity. Not only does this honey characteristic add its note to honey flavour; the level of acidity of honey contributes to its stability towards micro-organisms. In fact, it was early thought that the last act of the bees in ripening honey was to add formic acid to preserve the honey (König, cited by Browne, 1908). As we shall see later, the bees in effect do increase the acidity of honey during ripening; this perhaps achieves the same end, but not in so direct a manner.

The acidity of honey can be examined in two ways. The kinds of acids present and their relative or total amounts can be considered; or the effect of the acids and other materials which affect acidity, such as minerals, can be expressed in terms of the concentration in the honey of hydrogen ions, which all the acids have in common. A brief review of both aspects is given below.

5.41 Identity of the acids

The complexity of honey extends to the number of acids present. Perhaps largely because newer methods have extended the range of possible investigations, much of our knowledge of the honey acids has been obtained within the past twenty years.

Formic acid, once thought to be 'the' acid of honey, was early recognized (Farnsteiner, 1908, Heiduschka & Kaufmann, 1911, Merl, 1914) to be but a minor component. The following acids have been identified in honey by unequivocal procedures:

acetic	(Stinson *et al.*, 1960)
butyric	(Stinson *et al.*, 1960)
citric	(Nelson & Mottern, 1931; Goldschmidt & Burkert, 1955*b*, Stinson *et al.*, 1960)
formic	(Vogel, 1882, cited by Farnsteiner, 1908)
gluconic	(Stinson *et al.*, 1960)
lactic	(Stinson *et al.*, 1960)
maleic	(Goldschmidt & Burkert, 1955*b*)
malic	(Hilger, 1904, Nelson & Mottern, 1931, Goldschmidt & Burkert, 1955*b*, Stinson *et al.*, 1960)
oxalic	(von Philipsborn, 1952)
pyroglutamic	(Stinson *et al.*, 1960)
succinic	(Nelson & Mottern, 1931, Stinson *et al.*, 1960)

The following acids have been identified in honey without rigorous proof of their identity, and it is considered that they are probably present:

glycollic	(Maeda *et al.*, 1962)
α-ketoglutaric	(Maeda *et al.*, 1962)
pyruvic	(Maeda *et al.*, 1962)
tartaric	(Heiduschka & Kaufmann, 1913; Vavruch, 1952)
2- or 3-phosphoglyceric acid	(Subers *et al.*, 1966)

α- or β-glycerophosphate   (Subers *et al.*, 1966)
glucose-6-phosphate   (Subers *et al.*, 1966)

Many of the acids listed in the first group have been reported by other investigators besides those cited, but without rigorous proof of identity.

It is now known (Stinson, Subers, Petty & White, 1960; Maeda *et al.*, 1962) that gluconic acid is present in honey in considerable excess over all other acids; it is produced by the action of an enzyme in honey upon the dextrose in it. The various amino acids in honey are dealt with in Section 5.6. Except for gluconic acid, the sources of the various honey acids are not known. Many of the acids are intermediates in the Krebs cycle of biological oxidation and are of widespread occurrence; they may be present already in the nectar.

The identification of gluconic acid in honey provides an explanation of a difficulty long encountered by analysts seeking to measure the total amount of the various acids in honey. This is done by titration with alkali, and an indistinct or fading endpoint is often encountered, which leads to uncertainty or error in the measurement. Gluconic acid exists in solution in equilibrium with its lactone, or internal ester, which does not have an acid function. The proportion present in each form is governed by several factors. The titration of the total acidity of honey can easily be done by a modified procedure (White *et al.*, 1958). Values in Tables 5.2/1 and 5.2/2 for total acidity were determined by this method, as were total acid values reported by da Silva Ferreira (1970) for Portuguese and African honeys.

Since inorganic ions such as phosphate, chloride, and sulphate are known in honey, we may also consider the corresponding acids to be honey constituents.

5.42 Active acidity

All of these acids have in common the dissociation in aqueous solution to provide protons or hydrogen ions. It is to these that much of the 'sourness' and other characteristics of acids are ascribed. A measure of the total concentration of hydrogen ions provides information on the strength of acidity and allows comparisons between materials. This is expressed in the logarithmic *p*H scale, in which *p*H 1 (0·1-molar hydrogen ion concentration) is about the acidity of a dilute solution of an acid like hydrochloric; *p*H 7 represents neutrality. On this scale values for honey fall in the range 3·2 to about 4·5, averaging about 3·9. This value is affected somewhat by the amounts of the various acids present, but mostly by the mineral content—calcium, sodium, potassium, and other ash constituents (Section 5.5). Honeys rich in ash generally show high *p*H values.

Chudakov (1964a, 1964b) has examined the buffer mechanism of the establishment of pH in honey.

5.5 MINERALS

The scientific literature on honey ash falls into three subject categories— amounts of total ash, amounts of the principal constituents, and the identities of minor metallic constituents, which often appear in extremely minute amounts.

Reference to Table 5.2/1 shows that the average ash of U.S. honey is 0·17%, with a range 0·02–1·03%. Table 5.2/3 shows a similar average for other honeys. It is of interest that many years ago in Germany, honeys containing less than 0·1% ash were viewed with suspicion (Schwarz, 1908, Utz, 1908a). A survey of U.S. honeys (White et al., 1962) showed 193 of 492 authentic samples had less than this amount. On the other hand, although the U.S. (advisory) standard permits a maximum of 0·25% ash, 103 of the samples exceeded this.

Tables 5.5/1, 5.5/2, and 5.5/3 summarize the literature on the mineral composition of honey. The data of Table 5.5/1 resulted from the research of Schuette and his colleagues at the University of Wisconsin on United States honeys. Table 5.5/2 gives a summary of data available on honeys from elsewhere. These data were recalculated as far as possible to be comparable to those in Table 5.5/1. Rather than present data on both bases (which is not available for all honeys listed), the results are shown as the percentage of ash for several honeys where the ash content of the honey is not stated in the report, and in parts per million for others, the ash content being given where available. A summary of qualitative (spectrographic) assays of honey minerals appears in Table 5.5/3. Elements listed in Table 5.5/2 were usually reported, but are not included in Table 5.5/3.

5.6 PROTEINS AND AMINO ACIDS

Although it has been known for many years that honey contains protein materials, little is known of their characteristics. The occurrence of protein in honey was used in two ways to demonstrate adulteration of commercial honey. Lund (1909) precipitated proteinaceous material with a tannin solution and attempted to set limits for the volume of the precipitate for genuine honeys. This was later modified (Lund, 1910) using phosphotungstic acid (Voerman & Bakker, 1911) or alcohol (Laxa, 1923). Another means of using honey proteins for this purpose developed from the work of Langer (1903), who demonstrated that serum from an animal immunized with buckwheat honey protein material gave a

Table 5.5/1
Mineral elements of honey*

Mineral element	Honey colour	No. samples	As percentage of ash		As parts per million of honey	
			Range	Average	Range	Average
Potassium (K)	light	13	23·0 –70·8	35·30	100–588	205
	dark	18	2·0 –61·6	33·00	115–4733	1 676
Sodium (Na)	light	13	0·96– 9·26	3·59	6–35	18
	dark	18	0·20–11·20	4·68	9–400	76
Calcium (Ca)	light	14	3·54–13·00	8·77	23–68	49
	dark	21	0·46– 7·30	3·57	5–266	51
Calcium as lime (CaO)	light	14	4·95–18·19	12·27	32–95	69
	dark	21	0·64–10·21	5·00	7–372	71
Magnesium (Mg)	light	14	1·00– 9·24	3·42	11–56	19
	dark	21	0·66–11·47	2·77	7–126	35
Iron (Fe)	light	10	—	—	1·20–4·80	2·40
	dark	6	—	—	0·70–33·50	9·40
Copper (Cu)	light	10	—	—	0·14–0·70	0·29
	dark	6	—	—	0·35–1·04	0·56
Manganese (Mn)	light	10	—	—	0·17–0·44	0·30
	dark	10	—	—	0·46–9·53	4·09
Chlorine (Cl)	light	10	4·52–13·21	10·20	23–75	52
	dark	13	2·26–14·46	9·67	48–201	113
Phosphorus (P)	light	14	1·03– 9·55	6·37	23–50	35
	dark	21	0·84– 6·67	3·67	27–58	47
Sulphur (S)	light	10	5·77–16·24	11·49	36–108	58
	dark	13	2·67–14·36	7·98	56–126	100
Silica (SiO_2)	light	10	0·58– 2·23	1·60	7–12	9
	dark	10	0·17– 1·79	1·00	5–28	14
Silica, crude	light	14	1·60– 7·07	3·86	14–36	22
	dark	21	1·03– 5·82	2·87	13–72	36

* Schuette *et al.* (1932, 1937, 1938, 1939)

Table 5·5/2
Inorganic constituents of honey

Country honey type	Germany blossom[1]	U.S.S.R. raspberry-fireweed[2]	U.S.S.R. buck-wheat[2]	Yugoslavia[11]	Australia eucalyptus[4]	Sweden[5] clover	Sweden[5] heather	Austria[6]	Italy[12]	S. Africa floral	Hungary[13] acacia	Australia[14] eucalyptus	Australia[14] clover	India[17]
No. samples	—	1	1	43	1 each of 4 spp	2	3	4	39	17	12	1	1	—
	Percentage of ash			*Parts per million parts honey*										
Total ash	25·4-42·2			2 500 / 700-7 000	1 830-5 830	890-1 952	4 554-4 902		2 010 / 370-7 800	270-9 380	790	1 940	1 130	—
Potassium	254-422		44·7		548-1 785	318-826	2 090-2 280	502-2 130		141-2 945		1 032	367	
Sodium	4·1-7·4		2·3		37-200	36-65	85-121			34-806		17	186	18·2-45·4
Calcium	1·5-5·7		2·6		107-214	41-56	80-129	128-268		36-164	178	162	76	
Magnesium	0·9-1·3	8·4	1·5		24-163	7·8-24	32-38	22-93			17	79·5	24	47-128
Iron		2·0	2·9	8·3 / 3·5-20·5	1·4-4·9		2·8-4·9		7·7 / 1·1-20·7	2·65-8·42	2·8	26	6·3	169-389
Copper		0·01	0·04	8·7 / 4·5-18·0						0·25-0·83	0·29	0·5	0·6	0·6
Manganese		2·1	1·8	5·7 / 3·0-9·2				2·5-5·0		0·06-6·15	0·30	7·7	0·6	
Chlorine			4·1		170-820	180-190	120-210	81-267						

Country honey type	Germany blossom[1]	U.S.S.R. raspberry-fireweed[2]	U.S.S.R. buck-wheat[2]	Yugo-slavia[11]	Australia eucalyptus[4]	Sweden clover	Sweden heather	Austria[6]	Italy[12]	S. Africa floral	Hungary[13] acacia	Australia[14] eucalyptus	Australia[14] clover	India[17]
No. samples	—	1	1	43	1 each of 4 spp	2	3	4	—	17	12	1	1	—
		Percentage of ash							Parts per million parts honey					
Phosphorus	0·7-5·4	4·6	3·0		26-39	48-89	59-80	344-610	31·1 126-62·4	19·9-58·5	4·2	54	56	17·2-34·2
Sulphur			8·8		20-28									
Silica		52·7	1·4		52-280	24-8·5	70-115					23	136	

Aluminum U.S.S.R. —0·9% ash[3]
    Australia —2 samples—59, 5 ppm[14]
Iodine U.S.S.R. —0·020-0·026 ppm dry wt[7]
Boron Germany —100 samples—0·6-12·5 ppm dry wt[8]
Manganese Hungary —12 samples—3·5 ppm[13]
    Germany —25 samples—0·4-44 ppm[9]
    Australia —2 samples—7·7, 0·6 ppm[14]
    Poland —160 samples—4·2-17·7 ppm[15]
Titanium U.S.S.R. —1 sample—0·08% of ash[2]
Molybdenum U.S.S.R. —1 sample—0·02% of ash[2]
Cobalt Yugoslavia —43 samples—ave. 0·9 ppm, range 0·5-1·5 ppm[11]
    Australia —2 samples—5, 0·2 ppm[14]
Zinc Hungary —12 samples—5·1 ppm[13]
    Australia —2 samples—1·6, 2·5 ppm[14]
Lead Hungary —12 samples—0·05 ppm[13]
    Australia —2 samples—1·8, 0·9 ppm[14]
    Japan —29 samples—0·2-6·3 ppm[16]
Tin Hungary —12 samples—0·2 ppm[13]
Antimony Australia —2 samples— <2, <1 ppm[14]
Chromium Australia —2 samples— <0·3, <0·2 ppm[14]
Nickel Australia —2 samples— <0·05, <0·03 ppm[14]

1 Nottbohm (1928)
2 Makarochkin & Yudenich (1960)
3 Svoboda (1933)
4 Jewell (1931)
5 Sundberg & Lundgren (1930)
6 Elser (1928)
7 Karger, Zhukova & Radzivon (1944)
8 Büttner (1948)
9 Gottfried (1911)
10 Anderson & Perold (1964)
11 Čermagić et al. (1963)
12 Sacchi & Bosi (1964)
13 Varju (1970)
14 Petrov (1970)
15 Miśkiewicz & Krauze (1969)
16 Tatsuno et al. (1968)
17 Kalimi & Sohonie (1964a)

Table 5.5/3
Trace elements in honey*

	Chistov & Silitskaya (1952)	Makarochkin & Yudenich (1960)	Santos Ruiz et al. (1949)	Mladenov (1968)†
Chromium	+			+
Lithium	+		+	
Nickel	+	+	+	+
Lead	+	+	+	
Tin	+	+	+	+
Zinc	+			
Osmium	+			
Beryllium		+		+
Vanadium		+		+
Zirconium		+		+
Silver		+		+
Barium			+	+
Gallium		+		
Bismuth				+
Gold				+
Germanium				+
Strontium				+

* Elements listed in Table 5.5/1 not included
† Quantitative data are in original article

copious precipitate with buckwheat honey. Later development was by Langer (1909), Galli-Valerio & Bornand (1910), and Thöni (1911). Amounts of precipitate, or extent of dilutions showing precipitation, were associated with the degree of admixture of artificial honey. Thöni (1912) used an anti-bee serum for the same purpose. These tests were extensively used in the early part of this century, and Thöni's method was included in the Swiss *Lebensmittelbuch* (Kreis, 1915). By using antiserum techniques, Langer (1915) was able to show the error of Küstenmacher's (1911) contention that the invertase of honey arose from the pollen rather than the bee.

The older work on honey proteins included their assignment to the various classic solubility groups. Moreau (1911a) noted that the common tests for proteins (Millons test, the xanthoproteic reaction, heat coagulation) showed their presence in honey; he reported albumins, globulins, and proteoses. Stitz (1930) found peptones, albumins, and some globulin, but not protamines, alcohol-soluble albumin, histones, albumoses, and albuminoids. Nucleoprotein was present.

In their work on the colloidal materials of honey, Paine, Gertler & Lothrop (1934) noted that these materials, obtained by ultrafiltration of honey, were more than half protein. The isoelectric point of the colloidal material in the honeys examined was close to 4·3. They also noted that the ultrafiltered honey still retained over half of its original nitrogen content, implying that the common practice of estimating protein in honey by multiplying the nitrogen content by 6·25 is misleading.

Helvey (1953) studied the colloidal materials of buckwheat (*Fagopyrum esculentum*) honey by several physical methods. Ultracentrifugation and moving-boundary electrophoresis both indicated three components, two proteins of approximate molecular weight 146 000 and 73 000, and an (assumed) polysaccharide with a weight of about 9 000.

Crude protein preparations from several types of floral honey and from stores of sugar-fed bees were examined by White & Kushnir (1967a), using gel filtration, starch-gel electrophoresis, and ion-exchange chromatography. The number of constituents in the protein fraction varied among the 11 floral types examined from 4 to at least 7; 4 of these components appeared to originate in the bees. Molecular weights of 2 of the bee-imparted proteins were indicated by gel filtration to be about 40 000 and 240 000; other protein materials, of plant origin, showed values of about 98 000 and above 400 000. The portion of nitrogen-containing material passing through the dialysis membrane varied between 35 and 65%.

In the continuing quest for means to distinguish between natural and artificial honeys, minor constituents of honey were frequently examined. In the formol titration, the amino group of an amino acid is blocked by reaction with formaldehyde in neutral solution, which then permits the carboxyl group of the amino acid to be measured. This long-known test was applied to honey by Tillmans & Kiesgen (1927), who proposed that it be used for this purpose. Gottfried (1929), confirming their results, proposed that honeys with a value of 0·3–1·1 (ml 0·1-N NaOH per 20 g honey) be suspected, and assigned values of 0·6–4·0 to genuine honeys. Schuette & Templin (1930) in the United States then applied the test to 15 normal honeys and found an average of 0·41 for their samples (range 0·25–0·76). These low values, together with a lack of reproducibility, convinced Schuette and Templin that the method was of little value. They noted that only 11% of the nitrogen content found by analysis (0·004% N of 0·036%) could be accounted for in the titration. It seems possible that the higher formol titration values of European honeys might reflect a greater content of honeydew honey.

Shortly thereafter, Lothrop & Gertler (1933) proposed a method for determination of amino acids in honey, and found values for amino nitrogen ranging from 0·0024 to 0·0066% for 10 honeys from which protein had been removed by precipitation. It is interesting that their average value for amino nitrogen determined by a rather cumbersome procedure, 0·0033%, is similar to the average value calculated by Schuette & Templin from formol titration (0·004%). Later, Schuette & Baldwin (1944) studied the free amino acid content of 37 honey samples, harvested in 3 years and representing 20 floral types. Using the Lothrop-Paine procedure, they found an average amino acid content of 0·0034% for light-coloured honeys and 0·0058% for dark honeys (basis not given).

Chistov & Silitskaya (1952) found the nitrogen in honey to be distributed among amines, proteins, amides, amido acids, and small amounts of amino acids.

Later investigations on amino acids in honey gained from the development of paper chromatography. Vavruch (1952) was unable to demonstrate free amino acids by this procedure, and reported several amino acids after tryptic hydrolysis of honey proteins. By concentrating the free amino acids with ion-exchange treatment, Baumgarten & Möckesch (1956) detected a total of 17 amino acids in 15 honey types, the average for each sample being 11. Of the 15 samples, at least 5 contained honeydew honey, which is known to be relatively rich in amino acids (Auclair, 1963, page 468). They proposed that the free amino acids originate from the bee, arguing that the relatively uniform occurrence over a widely varying group of honey types makes it unlikely that they could originate from pollen or nectar. Support for this view is given by the data of Maslowski & Mostowska (1963) who found no qualitative differences between 5 honeys from various plants and one obtained by sugar feeding. Bergner and Körömi (1968) also reported that stores from sugar-feeding of bees contained the same 19 amino acids found in mixed samples of honey. Phadke (1962) noted that tyrosine and tryptophan were present in dark, but not light, *Apis cerana* honeys. A preponderance of proline (45% in Finnish honey and 80% in an imported honey) was found by Komamine (1960), who reported 16 known amino acids and 3 of unknown identity in the honeys, which unfortunately were not characterized as to type. The next most abundant amino acid in the Finnish honey was glutamic acid; in the other sample it was leucine. Komamine noted that, since proline was the principal amino acid found in several pollens (Virtanen & Kari, 1955), at least a part of the honey acids probably originated therefrom. Table 5.6/1 shows the concentration of the various amino acids found by Komamine, recalculated for comparison with other data. The results of Maeda *et al.* (1962) are also shown in the Table; their amino acid analyses for three honey types were determined by the Stein & Moore ion-exchange procedure, which is considered much more reliable

Table 5.6/1
Free amino acids in honey
(mg per 100 g honey)

Honey type	Komamine (1960)		Maeda et al. (1962)			Mizusawa & Matsumuro (1968)					Michelotti & Margheri (1969)		Biino (1971)	
	1	2	3a	4	5	6	3b	7	5	8	9	10	11	12
Lysine	0·6	0·4	38·2	8·1	36·7	2·50	2·71	1·85	1·91	1·31	2·07	1·46– 2·8		
Histidine			6·7	2·6	10·7	0·94	0·92	0·61	0·93	0·63	0·75	0·56– 1·2		
Arginine	0·6	0·0	5·4	5·1	5·8	0·63	0·42	0·33	0·56	+	0·46	0·35– 0·53		
Aspartic acid	0·4	0·5	12·3	7·9	17·0	1·81	0·90	0·84	0·86	3·97	0·17	0·06– 0·53		
Threonine	0·2	0·2	2·6	0·8	4·5	0·39	0·42	0·35	0·26	0·26	1·10	0·45– 1·9		
Serine	0·5	0·5	23·6	3·2	11·8	1·43	0·70	0·65	0·34	0·62	1·19	0·84– 1·57		
Glutamic acid	2·5	0·5	19·0	8·3	13·0	1·85	1·91	1·36	1·61	1·34	1·42	1·25– 1·80	1·18	1·18
Proline	6·2	19·0	297	134	249	28·71	20·20	22·19	21·06	16·91	14·6	12·5 –17·1	53	83·5
Glycine	0·2	0·2	5·9	2·2	3·6	0·31	0·23	0·14	0·12	0·13	0·46	0·33– 0·54	0·45	1·80
Alanine	0·6	0·4	10·5	4·6	8·5	0·46	0·53	0·32	0·41	0·31	1·3	0·60– 1·65	1·42	2·86
Cystine			6·1	5·5	0·0	–	0·35	0·44	+	+	+	+		
Valine	0·6	0·3	9·7	3·0	7·3	0·52	0·45	0·19	0·46	0·33	0·91	0·71– 1·05		
Methionine	0·3	0·0	2·7	1·2	0·8	–	0·05	0·04	0·17	–	+	+ – 0·19		

	1	2	3a	4	5	6	7	8	9	10	11	12
Isoleucine	0·7	0·9	4·6	2·3	3·6	0·28	0·34	0·12	0·16	0·19	0·77	0·44– 1·1	0·52	0·52
Leucine			5·3	1·4	4·9	0·30	0·34	0·12	0·25	0·15	0·58	0·32– 0·95	0·52	0·52
Tyrosine			6·9	3·3	6·2	0·49	0·47	0·27	0·26	0·18	2·59	1·3 – 3·9	0·72	1·45
Phenylalanine			9·6	10·5	11·4	0·93	1·62	0·58	0·54	0·28	16·6	5·0 –42·0	2·98	4·28
Tryptophan			0·0	0·0	0·1	+	+	+	–	+				

+ indicates traces; – indicates absent; blank indicates not reported.

Honey Types

1. Finnish honey
2. Honey imported into Finland
3a. Rape (*Brassica campestris*); 3b gives average of 3 samples
4. Common lime (*Tilia europaea*)
5. Buckeye (*Aesculus turbinata*)
6. Chinese milk vetch (*Astragalus sinicus*), average of 3 samples
7. Mandarin orange (*Citrus unshiu*), average of 3 samples
8. Acacia (*Robinia pseudoacacia*), average of 2 samples
9. Average for 9 unspecified honey types
10. Range for 9 unspecified types honey (individual values in original)
11. Acacia honey
12. Honeydew honey

than paper chromatography for quantitative work. Twelve Japanese honey samples representing 5 floral types were analysed by this method by Mizusawa & Matsumuro (1968). Averages for each honey type are given in Table 5.6/1, which also shows the average values and ranges found by Michelotti & Margheri (1969) who analysed 9 Italian honey samples of unspecified type by the Stein-Moore procedure. Their values have been recalculated from the 'milligram per kilogram' in the original to permit easy comparison of values. All of the values are reasonably consistent, except that the data of Maeda *et al.* (1962) are uniformly about tenfold higher. Proline is confirmed as the predominant acid; lysine, glutamic, and aspartic acids follow.

Another report of amino acids in honey determined by the automatic analyser is that of Biino (1971). He proposed analysis for proline to indicate the extent of admixture of honeydew with floral sources in honey. His data* (recalculated) for 8 amino acids are included in Table 5.6/1.

It is of interest here to examine the data of Hadorn & Zürcher (1963a), who view the formol number (*see above*) as a useful constant, and who have improved its estimation by combining it with an improved free acid and lactone titration (White, Petty & Hager, 1958). Their average values are in the range 0·3–0·9 meq (milliequivalents) per 100 g; assuming an average equivalent weight of 100 for the amino acids, this corresponds to 30–90 mg%, which does not support the higher values found by Maeda *et al.* (1962). It is evident that further examination of the levels of these compounds in honey would be useful. For comparison, the levels of amino acids in honeydew (as excreted) are of the order of 120 to 1 830 mg %, fresh weight; proline is a very minor constituent (Auclair, 1963).

5.7 ENZYMES

The enzymes are among the most interesting materials in honey, possibly have received the greatest amount of research attention over the years, and have supported the greatest burden of nonsense in the lay and even scientific press. The use of enzyme activity in some countries as a test for overheating of honey seems to support by implication the occasional supposition by food faddists that the enzymes of honey have a dietetic or nutritional significance of themselves.

An account of the earlier history of the enzymes of honey is included in published material based on Gothe's dissertation at Leipzig in 1913 (Gothe, 1914a). He noted that honey does not contain lactase, proteases,

---

* Prof. Biino has confirmed privately that the data in his original article should read 'micromoles', not 'millimols'. Data in Table 5.6/1 have been adjusted accordingly.

or lipases, and possibly not inulase. Catalase was found, and diastase and invertase were present. He reviewed earlier work of Marpmann (1903), who reported proteolytic and alcoholic fermenting enzymes, and whose tests were later shown to be reacting to sugars and acids in the honey. The primary interest in honey enzymes at that time was a possible means for distinguishing between natural and artificial honeys, since the methods of sugar analysis of the time were inadequate. This necessity to detect adulteration was the *raison d'être* for much of the European research on honey and continues even today, with the possible shift of emphasis from detection of adulteration to detection of overheating of honey, which—apart from its obviously detrimental effect on flavour—is supposed to destroy vital, though frequently vague, biological or nutritional qualities of natural honey. Thus, we still see in the literature many articles on the effect of heat on diastase and invertase, and reports of assay results for a given year in a given country, and suggested modifications in assay methods. Relatively little information has appeared on the enzymes themselves—their sources, purification, characteristics, kinetics, and so on. In this review only a sampling of the literature on the enzymes of honey is practicable.

5.71 Honey 'diastase'

The starch-digesting enzyme(s) of honey have been known for many years. As already noted, because of their sensitivity to heat they have been used as indication of honey quality.

The older concepts of amylase activity classified the enzymes into two groups. The α-amylase (amyloclastic) group splits the starch chain randomly, producing dextrins, with a slow loss of iodine-produced colour, and relatively little increase in reducing sugars. The β-amylase (saccharogenic) group splits the reducing sugar maltose from the ends of the starch chain, with rapid loss of staining with iodine. In honey analysis, both procedures have been used in diastase assay: loss of colour reaction with iodine, and increase in reducing power. Because the α-glucosidase of honey can further split maltose (produced by β-amylase) and double its effective reducing power, the second procedure can be misleading.

As previously noted, very little work has been done on honey amylases as such. Lampitt, Hughes & Rooke (1930) studied the effect of pH and temperature on both the α- and β-amylase of honey and found them somewhat interrelated. For α-amylase, optimum pH was found in the range 5·0–5·3, the lower value applying at 22–30°C (71–86°F) and the higher value at 45–50°C (113–122°F). The optimum pH for β-amylase was 5·3. Most reports place the optimum pH for honey 'diastase' at 5·3. The lack of adequate control of acidity in the diastase number assay as outlined by Gothe (1914*b*), and slightly modified by Fiehe & Kordatzki (1928), was shown by Lothrop & Paine (1931*b*) to produce erroneous results in some cases; this was confirmed by Schuette & Pauly (1933). Both studies showed adequate buffering at *p*H 5·3 to be necessary, as did Weishaar's (1933), whose modification was recommended by Kiermeier & Köberlein (1954) and is in general use, though in some laboratories it is being replaced by the more objective method of Schade *et al.* (1958). This method was adapted for a wider variety of instruments and further standardized against the Gothe method (White, 1959*a*) and tested collaboratively (White, 1964*a*). Hadorn (1961) also modified the method of Schade *et al.** Schepartz & Subers (1965) have described an abridged, simplified Schade procedure which would enable a beekeeper or small packer to determine an approximate number without expensive equipment.

No work on these amylases has come to the writer's attention beyond the above study of optimal temperature and *p*H conditions, and many papers relating to the effects of heating honey on its diastatic activity (e.g. Auzinger, 1910*a*, Moreau, 1911*c*, Gothe, 1914*b*, de Boer, 1931, Fiehe, 1931, von Fellenberg & Ruffy, 1933, Bergeret & de Castro, 1943*a*, Kiermeier & Köberlein, 1954, Schade, Marsh & Eckert, 1958, Duisberg & Gebelein, 1958, Hadorn & Kovacs, 1960, Curylo, 1961, Hadorn, Zürcher & Doevelaar, 1962, Hadorn & Zürcher, 1962*b*, White, Kushnir & Subers, 1964, Langridge, 1966, Chudakov, 1966). No definitive reports on the isolation or mode of action of the amylase(s) have yet appeared, though with modern chromatographic methods such research would be facilitated. Schepartz & Subers (1966*a*) have described their preliminary attempt to use such procedures in separating the amylases of honey. A 200-fold purification of α-amylase was attained, but attempts to characterize the enzyme were unsuccessful because of its instability.

Of many reports on heat inactivation of the enzyme in honey that have appeared, those of Schmidt-Nielsen & Engesland (1938), Schmidt-Nielsen & Årtun (1938), and White, Kushnir & Subers (1964) have contained kinetic data. The last-named report points out that the effects of heating and of longer-term storage at lower temperatures are so similar that the half-life of the enzyme can be calculated with a single equation over the temperature range of 10–80°C (50–176°F).

Figure 5.71/1 shows the effect of temperature on the rate of heat inactivation of the contained diastase when full-density honey is heated.

For this inactivation, $\log k = 22 \cdot 764 - \dfrac{35\,010}{2 \cdot 303RT}$. The half-life in days

---

* Results in the author's laboratory, and informal exchange of samples, indicate that the two modifications produce different results on the same sample (White, Kushnir & Subers, 1964, Kerkuliet & Putten, 1973).

of honey diastase over the temperature range is shown in Figure 5.71/2 and is approximated as follows:

$$log\ t_{\frac{1}{2}} = \left(\frac{1}{T} - 0\cdot003000\right)/0\cdot000130$$

where T is the temperature in degrees Kelvin between 283 (10°C, 50°F) and 353 (80°C, 176°F).

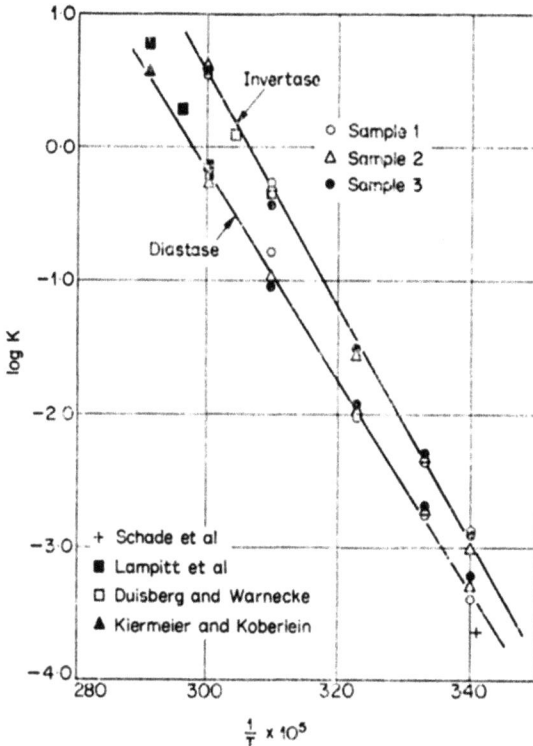

Figure 5.71/1 Effect of temperature on rate of heat inactivation of diastase and invertase in honey. ($\frac{1}{T}$ 0·00280 corresponds to 84°C, 183°F; 0·00300 to 60°C, 140°F; 0·00320 to 39°C, 102°F; 0·00340 to 21°C, 70°F.

(White, Kushnir & Subers, 1964)

Table 5.71/1 gives calculated half-lives for honey diastase when honey is stored at various constant temperatures. The necessity is obvious for adequate control of heat exposure for honey whose diastase must be preserved.

Gothe (1914*b*) concluded that the enzyme originated largely from the bees, a very minor portion possibly coming from pollen. Phillips, in his study of the ability of honeybees to survive on various carbohydrates (1927), found that the bees could not utilize raw or cooked starch or dextrins. He found no amylase on the washed lumen of the digestive tract, and though he recognized that the presence of glycogen in certain muscles of the ventriculus required the presence of the appropriate

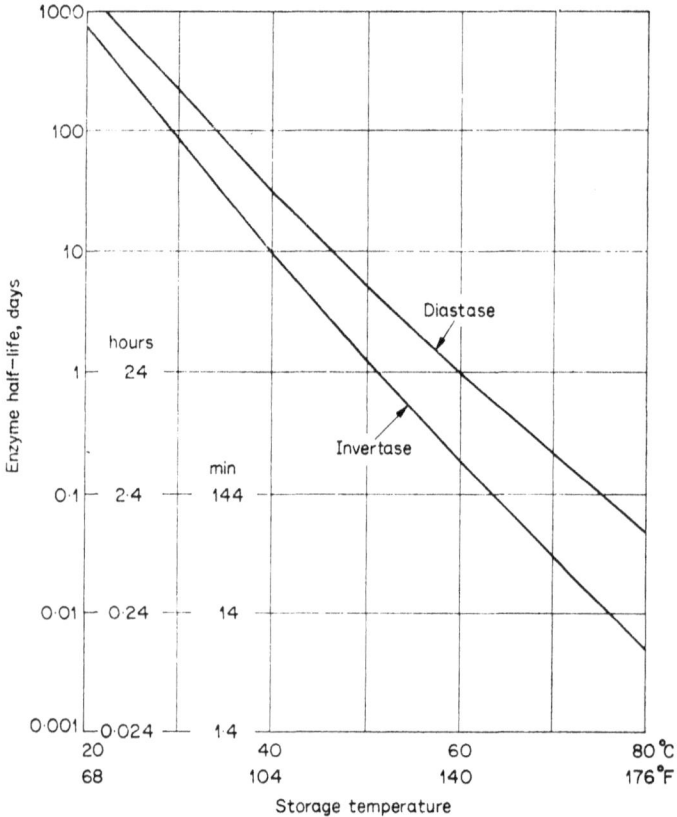

Figure 5.71/2 Approximate time required at a given temperature between 20°C (68°F) and 80°C (176°F) for the diastase and invertase activities of a honey sample to be reduced to one-half of the initial value.   (White, 1967*a*)

enzymes, he held that the origin of diastase was an open question; he appeared to favour pollen grains as the source. Later Vansell & Freeborn (1929) proposed that pollen was the source, on the basis of an apparent correlation between diastatic activity and pollen content, but they were unable to obtain high activities from pollen extracts. Their interest in the question arose from the low diastase values commonly reported for California honeys in Germany, and the low values found for unheated comb honeys. Lothrop & Paine (1931*b*) noted that the great variation in diastase value among honeys of different floral type supported Vansell's contention.

Table 5.71/1
Calculated half-lives of honey enzymes
(White, Kushnir & Subers, 1964)

Temperature		Half-life			
°C	°F	Diastase		Invertase	
10	50	12 600	days	9 600	days
20	68	1 480	,,	820	,,
25	77	540	,,	250	,,
30	86	200	,,	83	,,
32	90	126	,,	48	,,
35	95	78	,,	28	,,
40	104	31	,,	9·6	,,
50	122	5·38	,,	1·28	,,
60	140	1·05	,,	4·7	hr
63	145	16·2	hr	3·0	,,
70	158	5·3	,,	47	min
71	160	4·5	,,	39	,,
80	176	1·2	,,	8·6	,,

Fiehe (1932), in contrast to other workers, considered nectar to be the principal source of diastase, and the bees to be a minor source. Braunsdorf (1932) explained the considerable variability of diastase in 'sugar-fed' honey as the result of several factors—including race and colony strength of the bees, which affect the degree to which they 'work' their stores. He believed that diastase originates both from the nectar and the bees. Later Bartels & Fauth (1933) did not find any relationship between pollen level and diastase content, and suggested the possibility that high ambient temperatures in parts of California caused the low activities. Weishaar (1933) found only 1·5–2·5% of the diastase arising from nectar, 0·25–0·75% from pollen and the remainder from the bees; Gorbach (1942) compared bee and honey amylase(s). He noted an optimal pH for the former of 4·8–5·0, with a rapid diminution on both sides (limits 4·0–6·0). The fact that the optimum for honey was slightly different (5·1–5·2) he attributed to plant amylase. He further noted that bees digest starch with greater difficulty than maltose or sucrose, but more easily than melezitose. This is exactly in opposition to Phillips' results with starch and melezitose. Finally, Ammon (1949) on the basis of similarity between honey diastase and bee diastase—and its long-known presence in honey obtained from sugar-feeding—concluded that the bee is the source. Rinaudo *et al.* (1973), agreed with Ammon.

Lotmar (1935) has studied the apparent discrepancies in reports on the ability of honeybees to utilize starch and dextrins. She fed bees solutions of 8% starch dextrin with 8% sucrose, or with 8% or 16% sucrose solutions alone. The bees survived longer on the dextrin-sucrose solution than on the 8% sucrose. The time by which half of the bees had died (estimated from her chart) was 3·0, 2·5, and 3·5 days respectively. On water alone, half of the bees had died by 1·8 days. Sucrose was fed with dextrin because the latter is tasteless to bees. This procedure when applied to starch, of which only a 5% solution could be made, did not yield appreciable differences in death rates. The starch grains ($1·2-5\mu$ long and $1-2\mu$ wide) of various pollens were observed to be digested in the bee. The smallest non-pollen starch grains ($3·5-5\mu$ and $6·5-8\mu$) were also seen to be partly digested. Large grains, as of pea or wheat starch (about 45μ long), were not digested.

5.72   Honey invertase

The enzyme responsible for most of the chemical changes that take place when nectar is ripened to honey is the invertase (sucrase, saccharase). It has been known for many years that the bees add invertase to nectar, and it has been long recognized that invertase activity may continue in extracted honey. The subject is dealt with fully in 2.22.

The substrate for invertase is sucrose, which is hydrolysed (eventually) to glucose and fructose. Two general types of invertase are known, fructoinvertase and glucoinvertase. They differ in their mode of action, and earlier workers differentiated them by their behaviour with other sugars and by whether their action was inhibited by glucose or fructose (Neuberg & Roberts, 1946). Bacon and Edelman (1950) and Blanchard & Albon (1950) simultaneously demonstrated that the action of yeast invertase (fructoinvertase) on sucrose is actually a transfer of fructose molecules from sucrose to suitable acceptor molecules (which can be sucrose, water or certain other sugars). The suggestion (Nelson & Sottery, 1924, Papadakis, 1929) that honey invertase is a glucoinvertase was confirmed (White, 1952, White & Maher, 1953a) by demonstrating glucose transfer by the enzyme. It was found that separated honey invertase produced several oligosaccharides when acting upon sucrose. All these sugars are eventually hydrolysed to glucose and fructose by prolonged action of the enzyme. The principal of these sugars was identified as a new trisaccharide, α-maltosyl-β-D-fructoside (White & Maher, 1953b) (also known as fructomaltose, gluco-sucrose, erlose); it has since been found in honeydew of a scale insect (Wolf & Ewart, 1955a), in the digestive systems of many insects, and in blowfly excreta (Gray & Fraenkel, 1953, 1954), as a product of the action of hog intestinal invertase on sucrose (Dahlqvist & Borgström, 1959), and in honey (Siddiqui & Furgala, 1968a). It is likely that honey invertase is actually an α-glucosidase with activity against oligosaccharides having an exposed α-glucosyl group (White & Maher, 1953a).

It is of interest that Bacon & Dickinson (1955, 1957) subsequently established that melezitose, long known as a constituent of manna and honeydew, is also a product of transglucosylation by an aphid enzyme when hydrolysing sucrose. In this case glucose is transferred to the 3-position on the fructose in the sucrose molecule, whereas honey invertase transfers glucose to the 4-position of the glucose moiety of sucrose. The appearance of intermediates in the hydrolysis of sucrose by honey invertase complicates the determination of invertase activity. Erlose, the principal intermediate, is non-reducing and its $[a]_D^{25} = +121·8$ (White & Maher, 1953b), about twice that of sucrose. Appearing during the inversion in amounts up to at least 11% of the starting sucrose (when 25% remains) (White & Maher, 1953a), a considerable error can be introduced in invertase assay in honey if unspecific reducing-sugar methods are used (Moreau, 1911b, Gothe, 1914a, Gontarski, 1957, Rychlik & Fedorowska, 1960), or polarimetric (Duisberg & Gebelein, 1958, Hadorn & Zürcher, 1962a), or iodometric methods for glucose (Kiermeier & Köberlein, 1954). Even such a specific glucose method as glucose oxidase will not provide correct results for the honey invertase assay. Error could be greatly reduced by specifically determining free fructose, but even then a small amount of transfer of glucose to fructose occurs (Dahlqvist & Borgström, 1959). The methods noted above may be of use for routine invertase assay, but should not be expected to give results in agreement with one another.

The research of Nelson and his colleagues (Nelson & Cohn, 1924, Nelson & Sottery, 1924, Papadakis, 1929), mentioned above, is the most detailed study of honey invertase itself. These workers followed the reactions polarimetrically, under the long-accepted assumption that the enzymic hydrolysis of sucrose is a simple splitting of the molecule. Since we now know that the sucrose solution during hydrolysis by honey invertase can contain at least 11% of gluco-sucrose, among other saccharides, and that the specific rotation of the intermediate is nearly twice that of sucrose, about 22% of additional sucrose is simulated. This condition is present to greater or lesser extent during most of the hydrolysis. Rinaudo *et al.* (1973), report that bee-gland and honey invertases are inhibited by fructose, but not by glucose, and also (in contrast to other workers) that they have no action on maltose.

White & Kushnir (1967b) have reported preliminary efforts to characterize the α-glucosidase of honey, the so-called 'honey invertase'. Crude dialysed preparations from several bulk honeys, comb ('single-colony') honeys, and stores from sugar-feeding, were examined. Methods used were chromatography on ion-exchange cellulose, gel filtration with Sephadex G-200, and starch-gel electrophoresis using a high-resolution enzyme assay procedure developed for this work (White & Kushnir, 1966). Ion-exchange chromatography divided the crude α-glucosidase preparation into 3–9 closely spaced components. Material from sugar-feeding (with no plant constituent) was relatively unstable, and showed only one sharp peak. The α-glucosidase elution pattern from honey stored by a single free-flying colony resembled that from a sample of bulk honey derived from many colonies.

All honey α-glucosidase preparations yielded a single peak upon gel filtration with Sephadex G-200, and indicated a molecular weight of about 51 000. Starch gel electrophoresis showed the α-glucosidase complex to have 7–18 components (isozymes). The lowest number was for a sample of comb honey, the highest for a bulk honey clover. In one experiment, the isozymes showed differing degrees of heat sensitivity. Each of the 13 isozymes in one honey sample had the same ratio of activity upon sucrose to that upon maltose; this further confirms the α-glucosidase nature of honey invertase (White & Maher, 1953a), which was subsequently questioned by Siddiqui & Furgala (1968a).

The α-glucosidase complex from sugar-fed 'honey' was far less stable, and migrated more slowly than that from any of five floral honeys tested, during starch-gel electrophoresis. Taken together with the ion-exchange results, this implies an interaction between the bee and the plant nectar protein constituents.

As with diastase, a number of investigators have examined the effect of storage and heating upon invertase activity of honey (Moreau, 1911c, Achert, 1912, Kiermeier & Köberlein, 1954, Duisberg & Gebelein, 1958, Rychlik & Fedorowska, 1962b, Curylo, 1961, Gontarski, 1961, Hadorn, Zürcher & Doevelaar, 1962, White, Kushnir & Subers, 1964). Heat inactivation of honey invertase in buffer (pH 5·9 phosphate, 0·01M) proceeds 24 times as rapidly as in full-density honey (White, Kushnir & Subers, 1964).

Figure 5.71/1 shows the effect of temperature on the rate of heat inactivation of the invertase in natural honey. The enzyme is more susceptible to heat damage than is diastase. For the invertase, $log\ k = 26\cdot750 - \dfrac{39\ 730}{2\cdot303RT}$. An idea of the half-life of invertase in honey at constant temperature over the range 283°K (10°C, 50°F) to 353°K (80°C, 176°F) may be obtained from Figure 5.71/2. The equation is

$$log\ t_{\frac{1}{2}} = \left(\frac{1}{T} - 0.003083\right)\bigg/0.000113$$

in the temperature range given above.

Table 5.71/1 shows the half-life of invertase in honey under constant temperature conditions, as calculated by this equation at various temperatures. It can be seen that when invertase activity is to be preserved, attention must be given to proper storage and processing conditions for honey.

It should be noted that assays of either invertase or diastase of authentic unheated honeys give widely varying values. Since these enzymes originate largely in the bee, it would appear at first glance that enzyme levels should be at comparable levels in all genuine honeys. That this is not the case may be related to the degree of manipulation required by the bees to ripen the nectar (see 2.22). Schönfeld (1927) showed that as the sucrose content of bees' food increased, the nitrogen content decreased, as did the activity of invertase. Gontarski & Hoffmann (1963) also relate the more intensive working (required for thinner nectar) with higher enzyme activity. The apparent anomaly of high enzyme activity in the stores produced by feeding dry sugar is explained by bees' need to moisten and dilute the food before ingesting it and storing the product (Neprašova & Svoboda, 1956, Krupička, 1959, Simpson, 1964). Sipos (1964), concerned with the normally rather low enzyme content of Hungarian acacia (*Robinia*) honey, states that the enzyme content of stores is influenced by the quantity of daily flow and its sugar content; a rapid flow and high sugar content lead to lower enzyme content. The age of bees is also relevant; Hungarian acacia honey had the highest enzyme content when bees 25–30 days old predominated in the colony. It thus seems generally agreed that a rapid flow of concentrated nectar should result in stores of reduced invertase and diastase content.

While it is not the purpose of this review to examine the enzymes of the bee, mention should be made of the carbohydrases since the α-glucosidase in honey arises from the hypopharyngeal gland of the bee. Gontarski (1954), who reviewed the earlier literature, differentiated between the α-glucosidase of the gut with a pH-optimum of sucrose of 4·4–4·8, and the hypopharyngeal α-glucosidase with a corresponding value of 5·9–6·4. He noted that the gland enzyme was soluble, but the enzyme from the gut preparation was particulate; this is not unexpected considering the source, preparation, and function of the enzymes. In his Tables 6 and 7 and Figure 4 he shows the effect of adding glucose and fructose to the reaction of gland enzymes and sucrose. For fructose, inhibition is directly proportional to fructose concentration. For glucose, the figure shows a considerably greater and less proportional effect, but apparent errors in plotting the data from Table 6 and an apparent error on the blank value makes any conclusions, including his, impossible. It is interesting that he found no inhibition of the gut enzyme with glucose, whereas later, Maurizio (1962c) noted that transglucosylation (which would appear as inhibition) is much greater with the gut enzyme preparation than with gland enzyme preparation. On the basis of its nature as an α-glucosidase, Gontarski proposed that the 'so-called bee diastase' may be identical with the inverting enzyme. Differing stability to heat makes this highly unlikely; in fact this question appears to be settled by the approximate molecular weights reported by White & Kushnir (1967b) of 24 000 for honey amylase and 51 000 for honey α-glucosidase.

Maurizio has for some years studied the actions of the inverting enzymes of the pharyngeal glands and midgut of the bee in order better to understand the digestion and metabolism of carbohydrates by bees and the formation of honey. This work was described in a series of papers (Maurizio, 1957, 1959c, 1961, 1962a, 1962b). Since the subject is discussed in full in Chapter 2, no further comments are included here.

It is generally considered that honey invertase arises from the bee with contributions also from nectar enzymes.* Evidence for the latter is lacking, other than that some nectars may have invertase activity. It has been generalized that plant invertases are of the fructose-transferring type and animal invertases transfer glucose (Bealing, 1953, Gilmour, 1961, page 42). A prominent apparent exception was reported by Zimmermann (1954). In contrast to the demonstration using radio-active carbon that the nectaries of *Euphorbia pulcherima* secreted a transfructosidase with the nectar (Frey-Wyssling, Zimmermann & Maurizio, 1954), Zimmermann reported that the nectar of *Robinia pseudoacacia* contained an invertase which by transferring glucose instead of fructose 'im Gegensatz zuallen bisher untersuchen Invertasen mit einer saccharosespaltenden Transglukosidase zu tun haben' (Zimmermann, 1954). Such an enzyme had, however, been reported a year earlier in honey (White & Maher, 1953a). In her description of the experimental part, it is noted that over 100 fully unfolded flowers were used for obtaining the nectar. No mention is made of any precaution to exclude bees from the flowers, though Beutler notes (1953) the necessity (and difficulty) of excluding insects from flowers intended for nectar studies. Here again 'enzyme activity' was followed by papergramming nectars stored for at least 60 days. Very small amounts of bee-introduced invertase could provide extensive changes in much less time. Confirmation

* Rinaudo et al. (1973) conclude it is of bee origin.

of this report of the presence of a glucose-transferring invertase from a plant source with adequate controls would be desirable, since it complicates speculations (Siddiqui & Furgala, 1968*a*) on the biochemical origin of honey oligosaccharides.

5.73 Glucose oxidase

Early investigations found no evidence for oxidases in honey (Auzinger, 1910*a*, Moreau, 1911*b*). A glucose-oxidizing enzyme which formed an acid, probably gluconic, and peroxide, was demonstrated by Gauhe (1941) in the hypopharyngeal glands of the honeybee. She suggested that the organic acid formed acted as a preservative for honey. The presence of the enzyme in honey was not investigated. Gontarski (1948) described an enzyme in the hypopharyngeal glands that oxidizes vitamin C. He based his assumption on the disappearance of ascorbic acid from stores produced by feeding bees with sugar solutions containing ascorbic acid. The enzyme that oxidizes ascorbic acid was later shown to occur in honey.

A few years later Cocker (1951) reported that honey contained an enzyme which produced acid, based on the tendency of honey to become acid on standing after neutralization. White, Petty & Hager (1958) ascribed most of this pH drift after neutralization to lactone hydrolysis, but could not exclude enzymic action. The presence in honey of the glucose oxidase described by Gauhe was shown by White, Subers & Schepartz (1962, 1963), and its action on glucose was shown to produce hydrogen peroxide and gluconolactone. At the same time they showed the cause of the antibacterial effect known as inhibine to be hydrogen peroxide accumulating in the assay plates due to the action of this enzyme system. Adcock (1962) had independently shown inhibine to be destroyed by catalase. A chemical assay for hydrogen peroxide was developed (White & Subers, 1963) to replace the time-consuming biological assay for inhibine, and 90 samples were analysed, of which half were assayed for inhibine by the culture plate assay. The following relation was found between the empirical inhibine number I and peroxide accumulation P (measured in micrograms hydrogen peroxide accumulated per gram of honey in 1 hour under assay conditions): $I=0$, $P<3\cdot4$; $I=1$, $P=3\cdot4-8\cdot7$; $I=2$, $P=8\cdot8-20\cdot5$; $I=3$, $P=20\cdot6-54\cdot5$; $I=4$, $P=54\cdot6-174$; $I=5$, $P>174$. P for cotton honey was found to be especially high (225–360). Conditions affecting the peroxide value were discussed in this paper (White & Subers, 1963). Dustmann (1967*a*, 1967*b*) assayed 29 samples of European floral and honeydew honeys for peroxide accumulation and found honey from *Castanea sativa* to be consistently high, values ranging from 120 to 605.

Schepartz & Subers (1964) and Schepartz (1965*a*, 1965*b*) have studied the kinetics, inhibition, and specificity of the enzyme. It is unusual in having a high Michaelis constant (1·55M) and optimal substitute concentration (2·7M). It appears to be substantially inactive in full-density honey and becomes active upon dilution. Studies have been reported on the effect of heat (White & Subers, 1964a) and light (White & Subers, 1964b) on the accumulation of peroxide in diluted honey. It was found that the glucose oxidase activity in certain honeys was destroyed by visible radiation, the 425–525 nm region being the most effective. The presence in susceptible honeys of a heat- and light-stable non-volatile sensitizer was proposed, which facilitates oxidation of the enzyme, with the greatest effect at pH 3 and becoming negligible at about pH 7 and higher. Honeys were found which lost their peroxide accumulation activity unless weighed for the assay under reduced illumination, but the peroxide accumulation in others was substantially unaffected by direct sunlight. The importance is obvious of immediate adjustment of pH to about 7 before bacteriological assay of inhibine. In any event inhibine is not recommended as a quality factor for honey, because of wide variability in its sensitivity to light and heat and its distribution by floral source. The peroxide accumulation system in honey is at least as sensitive to heat as honey invertase and diastase.

The acid produced by this enzyme is the principal acid of honey (Stinson et al., 1960, Maeda et al., 1962). Since, during storage of full-density honey, acid is produced at a rate of only 0·002–0·012 μg/hr/g honey (White, Subers & Schepartz, 1963)—which is about 1/15 000 of the rate in diluted honey—the 70 mg of gluconic acid/100g honey reported by Maeda et al. (1962) would require over eight years' storage of full-density honey for its production. At a reasonable rate for diluted honey (100 μg/hr*) only seven hours would be required for the enzyme system to produce this amount of gluconic acid. Comparison of the above values for gluconic acid with the average lactone content shown in Table 5.2/1 (which represents only a fraction of the total gluconic acid) indicates that actual gluconic acid content of honey may be several-fold higher than the values of Maeda et al. Even so, the longer time that would be needed to produce larger amounts during ripening of honey is still in reasonable agreement with the time during which nectar is concentrated to full-density honey in the comb.

If the results of Gontarski (1948) on the 'vitamin C-oxidizing enzyme' are re-examined in the light of the glucose oxidase system, it is evident that his ascorbic acid oxidation could well be caused by the hydrogen peroxide produced by the system, rather than by a direct enzymic action with ascorbic acid as substrate. Unpublished observations in our laboratory with crude enzyme preparations from honey indicate that

---

* Measured as peroxide accumulation and not as acid production, and hence somewhat low.

ascorbic acid is oxidized by the system only in the presence of glucose, and not in its absence. Gontarski did note that the oxidation was initiated by an extract of hypopharyngeal glands, but he was uncertain if the enzyme was specific, or whether it was Gauhe's glucose oxidase. Schepartz (1966a) examined the effect of ascorbic acid on the glucose oxidase reaction; he found it to be a powerful accelerator, presumably acting by way of product removal and not on the enzyme itself.

5.74 Other enzymes

Honey has been examined for other enzymes from time to time. Marpmann (1903) claimed the presence of inverting, alcohol-forming, proteolytic, and peroxidase enzymes in honey. The first had previously been reported, the second may have been due to fermentation, and the last-named depended upon a colour reaction with peroxide and *p*-phenylenediamine which Auzinger (1910a, 1910b) later showed was due to a chemical rather than an enzymic reaction. Auzinger did agree that a catalase was present, using a 24-hour period for evolution of oxygen from peroxide. No evidence of oxidase or reductase activity was found. Moreau (1911a, 1911b) confirmed Auzinger's results and reported the catalase activity of twenty samples by the Auzinger procedure. Gothe (1914a) could not demonstrate lactase, protease or lipase activity, and questioned the presence of inulase. Gillette (1931) measured evolved oxygen from added peroxide over a two-hour period and suggested that the catalase arose from pollen or yeasts and not from the bee. Occasionally catalase activity is reported for honey samples. Fedotova (1957) reported a range of 0·68 to 9·2 mg peroxide for 33 samples of Uzbekistan honey, alfalfa (*Medicago*) honey showing the highest activity. Dzialoszyński & Kuik (1963) experienced some doubts about methods; they measured residual peroxide iodometrically, after 10 minutes' incubation. Results for 21 samples averaged 10·8 (range 0–36) mg peroxide per 100 g honey (per 10 minutes?). Gillette's value of 0·7 ml gas would correspond roughly to 7 mg peroxide in 10 minutes.

Schepartz (1966b) reviewed the literature on the occurrence of catalase in honey; he criticized all of the methods used as inappropriate for honey, stating that no unequivocal demonstration of catalase in honey had been accomplished. Using manometric and spectrophotometric procedures, Schepartz provided unequivocal evidence for its occurrence in honey. He determined the reaction with peroxide to be a first-order one, with the pH optimum 7–8·5, the substrate concentration optimum 0·018 M, and the Michaelis constant 0·0154 M.

In order to examine a postulate that the peroxide accumulation (inhibine) in honey should be inversely related to catalase content, Schepartz & Subers (1966b) determined catalase in 28 samples which had previously been assayed for diastase and peroxide accumulation. A direct correlation ($r = +0.76$) was shown between catalase and diastase activities, and an inverse correlation ($r = -0.71$) between catalase and peroxide accumulation; both were significant at the 1% level. Dustmann (1971) reported catalase and 'inhibine' (peroxide accumulation) values for eleven honey samples. The four samples with the highest inhibine values (380–662 μg peroxide per gram honey per hour) were devoid of catalase activity.

Giri (1938) showed that honey contains an acid phosphatase. Using glycerophosphate as substrate, 12 samples of Indian honey were analysed. The enzyme was effective over the pH range 3.5–6.5, showed greatest activity at 35°C (95°F), was activated by Mg^{++}. Dzialoszyński & Kuik (1963), using Giri's method slightly modified, assayed 25 Polish honeys and obtained an average value of 13.4 mg P/100 g honey/24 hr (range 2.2–52.7). Günther & Burckhart (1967) described a procedure for determining acid phosphatase content of honey using p-nitrophenyl phosphate as substrate, with a 3-hour incubation.

Zalewski (1965) examined over 400 samples of honey, pollen, pollen loads, nectar, and bees for acid phosphatase, using phenyl disodium orthophospate as substrate. Honey averaged 197.2 (range 29.8–2 140) micromols per 100 g dry matter. Converted to units used by Dzialoszyński & Kuik, this corresponds to an average of 5.07, range 0.76–54.8 (assuming a dry-matter content of 82%). A sample of stores from sugar-fed bees was found to contain about one-sixteenth of the average found in honey. High values in nectar and pollen indicate the activity to be largely of plant origin. Heating honey to 60°C or 140°F (10–60 min) reduced activity by about one-third, and to 75°C (167°F) by up to 80%. Storage (at unspecified temperature) destroyed 48–82% of the activity. Acid phosphatase activity was increased over 50% by magnesium ions.

5.8   MINOR CONSTITUENTS

Honey is a complex material; with newer investigations using increasingly acute research tools, the apparent complexity—in terms of numbers and kinds of constituents—appears to increase correspondingly. Identification of these materials is but the beginning; their source or means of entry into honey and their effects on physical, chemical, and biological properties must also be studied, as well as their occurrence among different honey types.

5.81   Honey colour

The substances responsible for the colour of honey are largely unknown. Schuette & Bott (1928) isolated carotene from a buckwheat honey. Later von Fellenberg & Rusiecki (1938) separated honey colours into two fractions, water-soluble and lipoid-soluble. In light-coloured honey, the water-soluble colour was less than the fat-soluble colour, and the opposite was true in dark-coloured honeys. These authors did not think that the carotenoids (fat-soluble) were identical with carotene.

Browne (1908) noted that, of the 92 honeys he analysed, 25 gave a positive test for polyphenolic compounds with ferric chloride, the 5 most intense reactions being from very dark honeys. Oxidation of these compounds can lead to coloured materials. Goodacre (1926) stated that formation of tannic acid in honey is favoured by exposure to air. The development of colour in honey in storage has been ascribed (Milum, 1939) to several factors: combination of tannates and other polyphenols with iron from containers and processing equipment; reaction of reducing sugars with substances containing amino nitrogen (amino acids, polypeptides, proteins); and instability of fructose in acid solution (caramelization). Phadke (1962) noted the absence of tyrosine and tryptophan in light honeys, as contrasted with its presence in dark honeys. Rychlik & Zborowski (1966) were unable to detect either rutin or quercetin in 50 honey samples tested by paper chromatography.

5.82 Aroma and flavour

A discussion of the factors contributing to the honey flavour and aroma will not be attempted here. Maeda *et al.* (1962) ascribed the 'taste' of honey to the sugars, gluconic acid and proline content. But it is evident to anyone who examines a variety of honeys that, although there seems to be a characteristic 'honey flavour', almost an infinite number of aroma and flavour variations can exist. We will here consider only the volatile materials of honey which contribute to the total effect; the field is scarcely developed sufficiently to allow generalizations. It is realized that the various sugars, amino and other acids, tannins and minor non-volatile substances, all contribute to honey flavour. Further, additional contribution to flavour effects of certain honeys may be due to glucoside or alkaloidal compounds specific to the plant source.

The literature on aromatic materials in honey before the introduction of gas-liquid chromatography (GLC) is scanty. Diacetyl (or other diketo-alkane) was identified by Schmalfuss & Barthmeyer (1929) in very small yield (less than 0·1 ppm) from a German heather honey. Methyl anthranilate was identified in orange honey by Nelson (1930); Lothrop (1932) suggested that its presence provides a specific test for orange honey. More recently, Deshusses & Gabbai (1962) used thin-layer chromatography for this purpose.

The development of gas chromatography has resulted in a great increase of interest in volatile materials from many sources. So far five reports on the application of this technique to honey analyses are at hand. Dörrscheidt & Friedrich (1962) examined an ether extract of the vacuum distillate from 3 000 kg of honey, but did not get satisfactory results. Best results were obtained by the direct analyses of a sample of vapour phase over honey, using a flame ionization detector; 1·5 cm³ of vapour permitted recording of 31 components, of which only 4 were common to the 6 honeys examined. It was suggested that such chromatograms might serve as 'fingerprints' for identification of honey types. None of the components was positively identified. Merz (1963) used GLC of ether extracts from honey samples as an objective means for assessing honey flavour. He found 5-hydroxymethylfurfuraldehyde (HMF) to be principal peak in organoleptically satisfactory honeys; in honeys of poor (bad, not deficient) flavour, other high-boiling components exceeded HMF in quantity. No other components were identified. Qualitatively, the chromatograms resemble those shown by Dörrscheidt & Friedrich for the ether extract. The volatile carbonyl components of honey were investigated by ten Hoopen (1963). These compounds were isolated as dinitrophenylhydrazones and separated by column and thin-layer chromatography and GLC, and identified by chromatographic behaviour, melting point, and ultraviolet spectrum. Rape and clover honeys yielded greater amounts of material than did thyme honey. Compounds identified were formaldehyde, acetaldehyde, acetone, *iso*butyraldehyde, and diacetyl.

Cremer & Riedmann (1964) have extended the work of Dörrscheidt & Friedrich, using the same instrument but with a capillary column instead of the packed column. They reported 50 components in the analysis of 10 honeys; 22 were identified, of which only 3 (formaldehyde, propionaldehyde, and acetone) were common to all samples. Their results are shown in Table 5.82/1. Aliphatic alcohols comprised over half of the material examined.

In a later report these investigators (Cremer & Riedmann, 1965) using a Golay column 100 m × 1 mm, have brought the total components separated to 120, more than half being identified. However, only phenylethyl alcohol, n-pentanol, benzyl alcohol and 3-methyl-1-butanol were added to the list of the preceding paper. They noted that a honey sample stored for a year showed an increase in n-pentanol, 2-methyl-1-butanol, 3-methyl-1-butanol, and n-propanol, and suggest that these compounds may originate by fermentation from the corresponding amino acids: norleucine, *iso*leucine, leucine, and α-aminobutyric acid. They reported that 16 of 22 honeys contained phenylethyl alcohol, and 14 of these also had benzyl alcohol; the former they related to phenylalanine as a precursor. The 6 honeys not containing the last two alcohols were also low in the four alcohols noted above, and were not organoleptically recognizable as honey. Oxidation of phenylethyl alcohol produces phenylacetic acid. It is of interest that most synthetic honey flavours contain large proportions of methyl, ethyl or propyl phenylacetate and also phenylethyl salicylate or phenylacetate (among many other components); according to Jacobs nearly all phenylacetic esters are characterized by a honey taste and odour (Jacobs, 1955).

Table 5.82/1
Aroma constituents of honey[1]

Carbonyls	Alcohols	Esters
Formaldehyde[2]	Isopropanol	Methyl formate
Acetaldehyde[2]	Ethanol	Ethyl formate
Propionaldehyde	2-Butanol	
Isobutyraldehyde[2]	n-Propanol	*Other*
Butyraldehyde	3-Pentanol n-Pentanol[3]	Diethyl ether
Isovaleraldehyde	Isobutanol	
Methacrolein	3-Methyl-2-butanol 3-Methyl-1-butanol[3]	
Acetone[2]	n-Butanol	
Methyl ethyl ketone	β-Methallyl alcohol	
	2-Methyl-1-butanol Phenylethyl alcohol[3] Benzyl alcohol[3]	

[1] Cremer & Riedmann (1964)
[2] Also identified by ten Hoopen (1963)
[3] Cremer & Riedmann (1965)

It is also of passing interest that a patent for preparing a synthetic honey flavour specifies heating a solution of a monosaccharide with phenylalanine and some derivatives (Morton & Sharples, 1959); thus perhaps a chemical reaction in storage is involved.

The fantastic sensitivity of the newer methods of analysis provides an embarrassment of riches—or what may appear to be riches—in such investigations as these. It is probable that only a few of the compounds that are or will be identified in honey in this manner will actually contribute toward the unravelling of the nature of honey aroma, though the technique may be of value for comparison and 'identification' of floral types of honey on a more or less empirical basis. Merz (1963), using the nose as a detector in GLC, reported only 4 of 20 distinct odour fractions as honey-like.

Methyl anthranilate (MA) is a minor constituent of citrus honey and contributes to its distinctive aroma. Lothrop (1932) described a colour test for this compound in honey which required 1 kg of honey. White (1966) adapted a procedure requiring 10 g of honey and analysed 21 citrus honeys (average 2·87 μg/g, range 0·84–4·37) and 12 non-citrus honeys (average 0·07, range 0·00–0·28). He proposed that content of MA might be a useful quality measure for citrus honey if further studies were confirmatory. Knapp (1967), after examining 59 Florida citrus honeys (average 3·29, range 1·6–4·9) and 14 other samples, supported the proposal, but noted that additional work with samples of known source was needed.

Hadorn (1964) noted that orange and lavender honeys, both containing MA, were occasionally deficient in diastase and invertase. He determined that inhibition of enzyme activity by endogenous MA was not responsible for the deficiency.

5.83 Vitamins

As the importance of vitamins in nutrition became known in the 1920s and 1930s, the occurrence of vitamins in honey was studied. Apart from the report of Dutcher (1918), who found negligible amounts of 'water-soluble vitamine' in honey, succeeding workers (Faber, 1920, Hawk, Smith & Bergeim, 1921, Scheunert, Schieblich & Schwanebeck, 1923, Taylor & Nelson, 1929, Hoyle, 1929, Kifer & Munsell, 1929, Trautmann & Kirchhof, 1932) uniformly found no evidence of vitamins A, B_1, B_2, C, D, and E in honey. Markuze (1935) reported that two samples of Polish honey showed presence of vitamin B_2, but no others. All of these investigators used biological assay methods.

By the early 1940s, microbiological and microchemical tests for the vitamins had been improved so much that the very low levels of the various vitamins in honey could be measured. Haydak et al. (1942)* and Kitzes, Schuette & Elvehjem (1943) assayed considerable numbers of honeys for thiamine (B_1), riboflavin (B_2), ascorbic acid (C), pyridoxine (B_6), niacin, and pantothenic acid, and a few for biotin and folic acid.

* Values in their Table 2, last column, should be multiplied by 0·01

Table 5.83/1 summarizes the results of these investigations. Both groups of investigators remarked upon the wide variability, both among different samples of similar honeys and among different honey types. The lack of agreement between the two groups for pyridoxine and niacin values was ascribed (Kitzes *et al.*, 1943, Haydak, 1955) to the use

Table 5.83/1
Vitamin content of honeys

Samples	Ribo-flavin	Panto-thenic acid	Niacin	Thia-mine	Pyri-doxine	Ascorbic acid
	(m i c r o g r a m s / 1 0 0		g	h o n e y)		
29 Minnesota[1]	61	105	360[3]	5·5	299	2 400
38 U.S. and Foreign[1]	63	96	320[3]	6·0	320	2 200
21 U.S. 3–7 yr. old[2]	22	20	124	3·5	7·6	—
19 U.S. 1–2 yr. old[2]	26	54	108	4·4	10·0	—
4 India[4]	12–54	—	442–978	8–22	—	2 000–3 400

[1] Haydak, Palmer, Tanquary & Vivino (1942)
[2] Kitzes, Schuette & Elvehjem (1943)
[3] Corrected from original data in publication as later shown (Haydak *et al.*, 1943)
[4] Kalimi & Sohonie (1965)

of chemical methods by Haydak *et al.* and of microbiological methods by the Wisconsin group. Later Haydak *et al.* (1943) found that commercial filtration of honey reduced the vitamin content (except vitamin K) by amounts from 8 to 45%. This was ascribed to the more complete removal of pollen (which is known to contain all the vitamins reported in honey) (Kitzes *et al.*, 1943). The Wisconsin group also found (Kitzes *et al.*, 1943) for a few samples of honey, 0·066 and 3·0 μg per 100 g of biotin and folic acid respectively. Using a chick assay, Vivino *et al.* (1943) found a vitamin K activity for honey, equivalent to about 25 μg of menadione per 100 g honey.

These levels of the various vitamins are so low that they have no real nutritional significance, particularly in view of the amount of honey normally eaten.*

All of the nine early investigators cited at the beginning of this section used bioassays in examining honey for the various vitamins. Seven of them tested for ascorbic acid (vitamin C) and found none (Faber, 1920, Hawk, Smith & Bergeim, 1921, Scheunert, Schieblich & Schwanebeck,

* It can be calculated that honey supplies about 3% of the thiamine and 6% of the niacin required for the metabolism of the contained sugars.

1923, Hoyle, 1929, Trautmann & Kirchhof, 1932, Markuze, 1935). As noted in Table 5.6/2, Haydak *et al.* (1942) reported an average value for 67 honeys of 2·3 mg per 100 g honey. They used a chemical method, and reviewed chemical determinations of ascorbic acid in honey prior to their paper: Griebel (1938) reported 160–280 mg/100 g for mint honey and 7–22 for others. The mint honey was bioassayed, and protected guinea pigs at 1 g per day, which corresponds to about 100–200 mg ascorbic acid/100 g (Rosenberg, 1942, page 337). Kask (1938) found an average of 4·9 mg/100 g (range 0–20) for Estonian honey, noting that the vitamin was relatively stable in honey. Werder & Antener (1938) found a range of 1·1 to 14·6 mg/100 g for 19 samples; Becker & Kardos (1939) found 31–89 mg/100 g for honey from *Castanea sativa* by chemical means, but bioassay did not demonstrate any activity. Kardos stated later (1941) that the indophenol-reducing substances of acacia, fruit-bloom, and *Stachys* honeys were not ascorbic acid, but protein decomposition products. Nevertheless, Griebel & Hess (1939) isolated the strong reducing substance from thyme and mint honeys as a derivative whose melting point agreed with that from the ascorbic acid derivative. Their chemical analyses for buckwheat, mint, and thyme honeys gave 19, 103, and 391 mg/100 g respectively. An analysis of mint and thyme nectars (Griebel & Hess, 1940) showed 288 and 222 mg/100 g, giving calculated values 394 and 888 for the corresponding honeys if no loss were sustained in ripening. Hansson (1949) reviewed the status at that time. As noted in 5.73, Gontarski (1948) described an ascorbic acid oxidizing enzyme in honey; the oxidation however is probably by hydrogen peroxide produced by glucose oxidase. This enzyme system operates well in diluted honey and only very slowly in full-density honey. Since the amount in honey is quite variable (White & Subers, 1963), rate of destruction of natural or added ascorbic acid would vary widely. Lüttge (1962c), using the Roe method, found 162 mg/100 g (dry wt) for a mint honey, present only as ascorbic acid. None of the other honeys tested contained any. The nectar of *Mentha aquatica* was found to have 10 mg ascorbic acid per gram of sugar. Kalimi & Sohonie (1965), using a microchemical method, found 2–3·4 mg/100 g in four Indian honeys, with one-fifth lost after storage at 28–30°C (83–86°F) for one year.

Since ascorbic acid is important only because of its biochemical significance, and honey is a very complex mixture, it would seem reasonable to defer to biological assay values for the determination of this vitamin in honey.

The bioassay value deduced from Griebel's (1938) statement (1 g honey/day protecting guinea pigs from scurvy) agrees reasonably with their chemical value for mint honey. Rahmanian *et al.* (1970) have found high ascorbic acid values (118–240 mg/100 g) for three samples of honey of unknown source from the mountainous Damavand area in Iran. The analysis was done by several procedures, including a specific chemical method (thin-layer chromatography of the dinitrophenyl-osazone); the results were about 90% as high as those found by the dichlorophenol-indophenol titration. The sample containing 118 mg/100 g was assayed with guinea pigs. Weight gains of animals fed daily 5 mg ascorbic acid or 4 g honey did not differ significantly. Rahmanian *et al.* reported an ascorbic acid requirement of 3–6 mg/day for the animals they used, and concluded that the honey actually contained 75–150 mg/100 g of vitamin C. They suggested the possibility of encouraging the use of honey from this region as a means of helping to relieve the marginal vitamin C deficiency often found in Iran.

Schepartz (1966*a*), reporting that ascorbic acid is a powerful activator of the glucose oxidase system by way of product removal, stated that because ascorbic acid is oxidized by peroxide from this system, it is not likely that ascorbic acid in the reduced form would be found in a honey containing significant glucose oxidase activity.

When considering the possible interactions between ascorbic acid, glucose oxidase and catalase in honey, it must be remembered that conditions in full-density honey depart so greatly from optimal assay conditions that little or no interaction is likely to occur. During ripening of the nectar to honey, conditions are much more favourable to enzyme action.

5.84 Hydroxymethylfurfuraldehyde (HMF)

This compound, which may be formed by the decomposition of fructose in the presence of acid, was originally thought not to be a constituent of honey in the hive. Modern sensitive quantitative measurements have shown, however, that even fresh honey contains small amounts of HMF (0·06–0·2 mg/100g; White, Kushnir & Subers, 1964, Hadorn & Kovacs, 1960, Winkler, 1955). HMF in honey has received considerable attention over the years: at least a hundred papers devoted to the subject are available. In the early days its presence in honey, as revealed by such qualitative tests as those of Fiehe (1908*b*) and Feder (1911), was taken to be evidence of the addition of invert sugar; but as soon as the Fiehe test was described it was severely criticized (Bremer & Sponnagel, 1909, von Raumer, 1908, Drawe, 1908, Hertkorn, 1909, Klassert, 1909, von Raumer, 1909), because heated, though authentic, honey gave a result interpreted as positive. Many modifications are recorded, but the use of a qualitative test for the purpose carries a number of disadvantages. Use of quantitative procedures would be preferable; numerical values for permitted levels could then be used (Hadorn & Kovacs, 1960). Several studies of quantitative methods have been made. For specific information the reader is referred to the original papers. Optical methods based on ultraviolet absorption by HMF are desribed (Schou & Abildgaard, 1931, Lampitt & Bilham, 1936, Winkler, 1955, Franzke & Iwainsky, 1956, Gautier, Renault & Julia-Alvarez, 1961c, Romann & Staub, 1961); paper chromatography (Franzke & Iwainsky, 1956, Gautier, Renault & Julia-Alvarez, 1961b); a 'quantitated' Fiehe reaction (Schade, Marsh & Eckert, 1958, Gautier, Renault & Julia-Alvarez, 1961a, Gonnet, 1963); and, probably the best, the chemical procedure of Winkler (1955) which does not require extraction.

When quantitative methods are used, it can be seen that the early confusion about the validity of the Fiehe test was caused by the differing amounts of HMF present from various causes (heating, storage, added with invert sugar) and the sensitivity of the tests in various hands. Hadorn & Zürcher (1962b) examined the effect of processing-heating on HMF. Gonnet (1963) has pointed out that room-temperature storage led to an increase in HMF, whereas cool storage retarded it. A review of the effect of storage and heating of honey on HMF content is in the paper of White, Kushnir & Subers (1964). The effect of storage and heating of three honeys is shown in Figure 5.84/1, where it can be seen that the logarithm of 'time needed to reach a given HMF level' is linearly related to temperature over a range of (at least) 20–75°C (68–167°F). The considerable differences in rate even among the few honeys examined precludes a mathematical expression of the relationship.

Duisberg & Hadorn (1966) used the Winkler method to examine over 1 600 honeys for HMF; they reported most honeys to contain less than 10 ppm. On the other hand Hallermayer (1969), in discussing quality evaluation of honey, noted a mean value of 33 ppm in 1 500 samples of commercial honey.

5.85   Toxic substances

Since tremendous numbers of organic compounds are synthesized by various plants, many with marked pharmacological activity, it is not surprising that occasionally such materials are found in honey. Possibly many more compounds are present in nectar, but are not detected in the honey because the physiological response (taste, toxicity) is not sufficiently apparent, or in some cases because their action on the bees prevents storage of the nectar. The compounds mentioned in this category are not honey constituents in the sense that they are common to all or most honeys; they have been reviewed by White (1973).

Instances of toxic reactions from ingestion of certain kinds of honey have been reported since antiquity. By far the largest number concern honey from the Ericaceae (*Rhododendron, Azalea, Andromeda, Kalmia* spp.). Various articles (Kebler, 1896, Fühner, 1926, Howes, 1949, Palmer-Jones, 1947a, Gruch, 1957, Carey *et al.*, 1959) review poisoning by honey from these sources. Recent instances of human poisoning by these honeys are described by Ungan (1940), Jones (1947), Povchenko (1950) and Scott *et al.* (1971). A toxic compound was isolated by Plugge from honey (Carey *et al.*, 1959) in 1883, and by others since that time. Pulewka (1949)

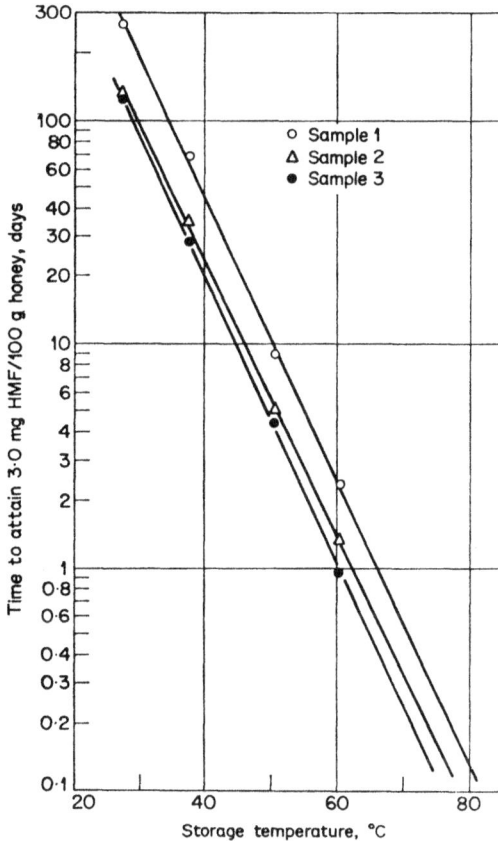

Figure 5.84/1 Effect of storage (processing) temperature on the accumulation of HMF in three honey samples.
(White, Kushnir & Subers, 1964)

and Popova *et al.* (1960) described biological tests for the toxic material in honey. White & Riethof (1959) described a chemical test for this toxic principle in honey, acetylandromedol. Scott *et al.* (1971), investigating a toxic British Columbia honey, isolated three toxic substances in the sample, which was of unknown origin. No acetylandromedol was detected, but three other related compounds were detected by thin-layer chromatography: andromedol, anhydroandromedol, and desacetyl pieristoxin B.

Hazslinsky (1956) described a toxic effect of a Hungarian honey which he ascribed to belladonna (*Atropa*) alkaloids from the nightshade, but Örösi-Pál (1956) claimed that the source was Egyptian henbane (*Datura metel*) and that the poisoning was by scopolamine, not atropine. Lehrner (1955) and Sviderskaya (1959) have described honey poisoning ascribed to atropine; in the latter article the source was said to be *Datura stramonium* and *Hyoscyamus niger*.

Wiley (1892, page 750) described an occurrence of poisoning by honey from the yellow jessamine (*Gelsemium sempervirens*) in which three persons died, of twenty affected. Analysis of the honey showed a large amount of gelsemine present.

Juritz (1925) discussed so-called 'Noors' honey from several species of *Euphorbia* in South Africa which produces a strong burning sensation in the throat. Sanna (1931) reported that a Sardinian honey with a sour, bitter taste was found to contain the glucoside arbutin, derived from *Arbutus unedo*.

Bitter honeys are not uncommon; Joachim & Kandiah (1940) attributed bitterness in a Ceylon honey to alkaloids from pollen of the Ceará rubber plant. Very little chemical work has been done with bitter honeys; likewise, little information is available on the materials responsible for the strong, sometimes nauseating odours and flavours of certain honeys, such as that from *Melaleuca* (Taylor, 1956), *Agave* (Pellett, 1947, page 97) and privet (*Ligustrum*).

Palmer-Jones (1947*b*) described an outbreak of poisoning in New Zealand, ascribed to eating honey. A new compound, mellitoxin, was isolated from the honey (Sutherland & Palmer-Jones, 1947*a*). It was shown to be related to tutin, a picrotoxin previously isolated from *Coriaria* species in New Zealand. In a search for the source of contamination in the field, it was noted that honey gathered during the blooming period of the suspected plant, *Coriaria arborea*, was not toxic. The source was then found to be honeydew on the leaves of the plant (Paterson, 1947). Subsequently Hodges & White (1966) isolated tutin and hyenanchin (mellitoxin) from a toxic honey, and Turner & Clinch (1968) tested 150 New Zealand honeys for toxin by both a biological assay and thin-layer chromatography. Samples found toxic by the mouse test (Clinch, 1966) showed spots corresponding to hyenanchin alone, or hyenanchin and tutin. Some indication of other similar substances was obtained.

5.86 Lipids

A qualitative investigation of the ether-extractable lipids in a cotton honey was described by M. R. Smith (1963, Smith & McCaughey, 1966). He found glycerides, sterols, and possibly phospholipids. Using thin-layer and gas chromatography, he identified the acids as palmitic acid 27%, oleic acid 60%, with small amounts of lauric, myristoleic, stearic, and linoleic acids. Twelve compounds were separated by thin-layer chromatography; ten were unsaturated, one was acid, with three gave a positive reaction to the antimony trichloride test for carotenoids.

Traces of beeswax may be microdispersed in extracted honey; they are probably introduced from the uncapping procedure. This subject is not considered here; such wax should have the general composition of beeswax (Callow, 1963).

5.87 Miscellaneous materials with biological activity

It is probable that materials additional to those described are present, and will be reported, in honey as isolative and analytical procedures improve. For example, (+)-2-hydroxy-3-phenylpropionic acid was isolated from a toxic honey sample (Hodges & White, 1966). Two materials not yet discussed are choline and acetyl choline. Neumann & Habermann (1950–51) showed that honey contained a material that caused contraction of isolated muscle preparations, and reported that the activity was equivalent to a content in honey (and in stored syrup) of 0·2–2·5 μg acetyl choline per gram of honey. By pharmacological methods Marquardt & Vogg (1952) showed that this cholinergic* factor of honey is probably acetyl choline. About thirty times as much choline was also probably present. Assay of 156 honeys showed acetyl choline levels of 0·06–5 mg/100 g (Marquardt, Aring & Vogg, 1953). Goldschmidt et al. (1952) also believed that choline was the cholinergic substance, and reported honey to contain 6 mg/100 g of choline. Later (Goldschmidt & Burkert, 1955b) they presented further evidence that the factor was acetyl and not propionyl or formyl choline. Schuler (1957) agreed. Watanabe (1955a) found no acetyl choline in a floral nectar (Japanese camellia) or in six pollens. The pollen of Alnus sieboldiana did contain acetyl choline, and all pollens contained choline. Confined bees were fed (Watanabe, 1955b) invert sugar both without and with (10 μg/ml) acetyl choline. The corresponding stored 'honeys' were found to have a biological activity corresponding to about 0·2 μg/ml of acetyl choline, stores from the food containing acetyl choline showing slightly more. Watanabe concluded that acetyl choline in honey is not derived from pollen or nectar, but is formed by biosynthesis and secreted into the honey. A review on the subject is in the literature (Marquardt & Spitznagel, 1956).

Smith et al. (1969) assayed eight honey samples of different floral sources from various regions for several previously alleged biological activities. Yeast growth stimulation greater than that from 1 μg of biotin was found; no real effect was seen on wrist-joint stiffness in guinea pigs.

---

* A material acting as a chemical transmitter of nerve impulses from parasympathetic nerve ends to the effector organ.

Some samples increased the numbers of roots on plant cuttings, but no oestrogenic activity for rats was found. Rats gained more weight on diets in which honey replaced sucrose or cornstarch.

We have seen that at least 181 substances are known to be present in honey. This total will increase with time, but our increasing knowledge of its complexity must not diminish the pleasure of eating honey, the only sweetener that need not be processed before we enjoy it.

CHAPTER 6

PHYSICAL CHARACTERISTICS OF HONEY

by Dr. Jonathan W. White, Jr.

CHIEF, PLANT PRODUCTS LABORATORY,
UNITED STATES DEPARTMENT OF AGRICULTURE

Thence there flows
Nectar of clearest amber, redolent
Of every flowery scent
That the warm wind upgathers as he goes.
MARTIN ARMSTRONG,
COLLECTED POEMS (1931)

6.1 INTRODUCTION

From the physical viewpoint, honey as extracted from the comb is an aqueous dispersion of material covering a wide range of particle size—from inorganic ions and saccharides and other organic materials in true solution, colloidally dispersed macro-molecules of protein and polysaccharide, to spores of yeasts and moulds, and the largest particles, pollen grains.

Since the sugars are by far the most important constituents, the gross physical attributes of honey are largely determined by the kinds and concentrations of the carbohydrates. That these properties are expressed in ranges rather than by constants, reflects the variability in honey composition (largely in solids content) that was so evident in Chapter 5. Other properties, much less studied, are ordinarily reported as constants, though they probably would be found to vary over a range if enough samples were examined.

Though honey is superficially a syrup and an average of 84% of its solids consists of glucose (dextrose) and fructose (laevulose), its properties (viscosity, refractive index, density) differ somewhat from those of an invert-sugar solution of the same water content. These properties vary in a regular manner with the moisture content (or solids content) of honey, but some uncertainty as to actual values is caused by a lack of accuracy in methods for determining water content and by the possible effects of differences in ratios of the various sugars and in amounts of the more important minor components. Even so, each of these properties has been used as a means of measuring the moisture content of honey—a value of great importance to the honey producer, packer, and merchant, since it bears a direct relation to the likelihood of undesired fermentation.

6.11 Moisture determination

Because a knowledge of procedures for direct determination of moisture content of honey is important in comparing the results of various investigators on the physical properties, a brief review follows.

These procedures may be considered in three categories: evaporation with measurement of weight loss, or with measurement of volume of water removed, and chemical determination. The first is most used; because of the sensitivity of honey sugars to heat, drying at a reduced temperature under reduced pressure is required. Generally an inert drying aid is added to increase bulk and porosity of the mass. Water may be added to the weighed sample to facilitate handling. The high hygroscopicity of dry honey requires the greatest care in manipulation.

As long ago as 1903 (Shutt & Charron, 1903), it was recognized that even in a vacuum at 70°C (158°F), fructose decomposition prevented the attainment of a constant weight; a temperature of 60–70°C (140–158°F) was recommended. Bryan (1908), however, questioned whether 70°C was sufficiently high to remove all water, but his reasons were based on a misconception; Fabris (1911) reported that 100° drying *in vacuo* gave results 0·3–0·5% higher than three other procedures, all of which gave agreeing values: drying at 60°C; a distillation method; drying in a dry air stream. Auerbach & Borries (1924) developed a method in which a 1 ml of a 50% solution is mixed with broken clay plate in a drying boat and dried at 60° in a current of dry air. Röttinger (1926) deposited a centigram sample on a roll of dry filter paper and heated at 100° *in vacuo*. Other procedures for increasing the surface area and preventing surface hardening during drying have been proposed, without bettering drying on sand below 70°C *in vacuo* (Marvin & Wilson, 1931, Rice & Boleracki, 1933, Schuette, 1935, Terrier, 1953, Kottász, 1958b). The Association of Official Analytical Chemists (Horwitz, 1965) uses a 1-g sample, mixes with sand, and dries at less than 70°C at a pressure not over 50 mm Hg until the weight is constant within 2 mg (which corresponds to about 0·2% water). Fulmer *et al.* (1934) proposed that the sample weight be increased to 5 g, claiming greater accuracy, but their data do not support this.

Distillation with turpentine, with measurement of recovered water, was one of the procedures tested by Fabris (1911) for honey moisture determination; for three samples results agreed with those of vacuum drying at 60°C. Abramson (1953) found that the Karl Fischer chemical titration for water gave values 0·29% moisture higher than vacuum drying at 70°C, with lower experimental error (0·14 against 0·33). Hadorn

(1956) confirmed that vacuum drying at 100°C (212°F) was unsatisfactory, as was Terrier's procedure. Further study of this application of the Karl Fischer procedure might be useful.

The direct drying procedure in any of its modifications is at best slow and cumbersome. Indirect methods have been studied, including refractometry at 20 or 40°C (68 or 104°F), density by pycnometer, relative density by spindle, and viscosity. Wedmore (1955) has written an excellent critical review of moisture determination in honey. He correlated the results of various investigators and proposed equations for the instrumental methods. Unfortunately, before his death he had completed only Part 1 of a projected six-part study of the subject, but it included his general conclusions. Pícha (1965) compared 12 procedures for determining moisture, using 15 honey samples. There were 3 refractometric, 5 drying, and 3 relative density methods, and a wet oxidation procedure. No significant differences were found among the methods; drying methods were recommended if the time required is not a factor. Refractometric methods, which are simple but of lower sensitivity, were sufficiently accurate to be recommended for processing control.

6.2 REFRACTIVE INDEX

As noted above, the primary interest in this property of honey is to provide a rapid, accurate, and simple measure of its moisture content. Early workers (Utz, 1908b, Bryan, 1908) noted that the moisture values obtained when converting refractometric readings by means of sucrose tables were higher by 1–2% moisture than those from vacuum drying. This was interpreted (Bryan, 1908) to mean that the latter method might not remove all water. Not until Auerbach & Borries (1924) studied the procedure was the necessity of special calibration for honey recognized. They calibrated the refractometer at 40°C (104°F) against a vacuum-drying procedure, using 23 samples of which only 10 were, however, fresh floral honeys. Auerbach & Borries provide the following relationship between dry substance and refractive index at 40°:

$$\text{dry matter } (T) = 78 + 390 \cdot 7 \ (n_{40} - 1 \cdot 4768)$$

This may be solved for n_{40}, to give $n_{40} = 0 \cdot 002559T + 1 \cdot 2772$ where T must be 78 or more.

In spite of this work, Marvin & Wilson (1931, 1932), Schenk (1934), Marvin (1933), and Snyder (1933) used sucrose tables for refractometric determination of moisture in honey. However, Chataway (1932) provided the definitive study of the relationship, calibrating the refractometer at 25°C (77°F) with vacuum-oven determinations for 60 honey samples and providing temperature correction factors which have been

corroborated (Snyder, 1933). Her values agreed quite well with those of
Auerbach & Borries when the latter were converted to 25°C. Fulmer *et al.*
(1934) felt that since the data from Auerbach & Borries, and Chataway,
and the vacuum drying methods, gave values lower by 1·7% moisture
than those obtained from sugar tables, it was advisable to modify the
vacuum drying method to give higher water values! The modification
they used was an increase of the honey sample to 5 g. Their refractometer
calibration equation was then:

$$\text{percentage moisture} = 400 \ (1 \cdot 5380 - n_{20})$$

These values are about 1% moisture below those in the Schönrock table,
and 0·7% moisture higher than Chataway's. After several papers
(Marvin & Wilson, 1931, 1932, Marvin, 1933) in which honey refractometric values were converted to Brix (% sucrose), Marvin (1934) finally
published a table relating water content and refractive index, without
attribution, which agreed with the Chataway data to within 0·0001–
0·0002 units. Experimental data were not published. Eckert & Allinger
(1939), in their analytical study of California honeys, determined moisture
by drying (A.O.A.C.) and also by refractometer. They stated that they
used the methods and tables of Marvin; however, the papers they cited
contained only the Schönrock sucrose conversion. Study of the Eckert &
Allinger values for moisture determination by refractometer shows that
they actually used either the Chataway table or the 1934 table of
Marvin noted above, and certainly not a sucrose table. Their data, therefore, may not provide an independent confirmation of Chataway's
results, as thought by Wedmore (1955). Torrent (1949) did confirm the
Chataway table. Wedmore gives the following as the best relationship
obtainable from the data of Chataway, Eckert & Allinger and Torrent:

$$\text{water content} = \frac{\bar{1} \cdot 73190 - \log \ (n_{20} - 1)}{0 \cdot 002243}$$

Table 6.2/1 shows the refractive index of honey at moisture contents
from 13·0 to 22·0%, as calculated by Wedmore (1955) from this relationship. The Table includes corresponding values for 40°C (104°F), calculated
from the Auerbach & Borries (1924) equation.

Several subsequent workers compared the water content determined
by refractometer and by other procedures, direct and indirect. In general
the deviation between water content by two instrumental methods
(refractive index, density, viscosity) was considerably less than between
the drying procedure and any other, indicating the relative imprecision
of the drying procedure (Abramson, 1953, Hadorn, 1956). In some cases
it is possible to infer n_{20} values from published equivalent values (Brix),
convert them to moisture values by the Chataway table, and compare

Table 6.2/1
Refractive index of honeys of different water contents[1]

Water content (%)	Refractive index (20°C)[2]	Refractive index (60°F)[3]	Refractive index (40°C)	Water content (%)	Refractive index (20°C)	Refractive index (60°F)	Refractive index (40°C)
13·0	1·5044	1·5053	1·4998	18·0	1·4915	1·4925	1·4870
13·2	1·5038	1·5048	1·4993	18·2	1·4910	1·4920	1·4865
13·4	1·5033	1·5043	1·4988	18·4	1·4905	1·4915	1·4860
13·6	1·5028	1·5038	1·4983	18·6	1·4900	1·4910	1·4855
13·8	1·5023	1·5033	1·4978	18·8	1·4895	1·4905	1·4850
14·0	1·5018	1·5027	1·4973	19·0	1·4890	1·4900	1·4845
14·2	1·5012	1·5022	1·4968	19·2	1·4885	1·4895	1·4840
14·4	1·5007	1·5017	1·4962	19·4	1·4880	1·4890	1·4835
14·6	1·5002	1·5012	1·4957	19·6	1·4875	1·4885	1·4829
14·8	1·4997	1·5007	1·4952	19·8	1·4870	1·4880	1·4824
15·0	1·4992	1·5002	1·4947	20·0	1·4865	1·4875	1·4819
15·2	1·4987	1·4997	1·4942	20·2	1·4860	1·4870	1·4814
15·4	1·4982	1·4992	1·4937	20·4	1·4855	1·4865	1·4809
15·6	1·4976	1·4986	1·4932	20·6	1·4850	1·4860	1·4804
15·8	1·4971	1·4981	1·4927	20·8	1·4845	1·4855	1·4799
16·0	1·4966	1·4976	1·4922	21·0	1·4840	1·4850	1·4794
16·2	1·4961	1·4971	1·4916	21·2	1·4835	1·4845	1·4788
16·4	1·4956	1·4966	1·4911	21·4	1·4830	1·4840	1·4783
16·6	1·4951	1·4961	1·4906	21·6	1·4825	1·4835	1·4778
16·8	1·4946	1·4956	1·4901	21·8	1·4820	1·4830	1·4773
17·0	1·4940	1·4951	1·4896	22·0	1·4815	1·4825	1·4768
17·2	1·4935	1·4946	1·4891				
17·4	1·4930	1·4940	1·4886				
17·6	1·4925	1·4935	1·4881				
17·8	1·4920	1·4930	1·4876				

[1] The values for 20°C and 60°F are Wedmore's (Wedmore, 1955) calculations. The 40°C values are calculated from Auerbach & Borries' equation (Auerbach & Borries, 1924).

[2] If the R.I. is measured at a temperature above 20°C, add 0·00023 per °C above 20°C before using the Table.

[3] If it is measured at a temperature above 60°F, add 0·00013 per °F above 60°F before using the Table.

these values with those reported in the publication by vacuum drying. Table 6.2/2 includes a comparison of the average deviations between the two procedures so calculated. Since refractometric values are relatively more precise, the high average deviations reflect the uncertainty in the determination of moisture by drying. The superiority of Chaatawy's data is evident in the small value obtained for average deviation from her data.

Zalewski (1962) compared pycnometry (20% solution) with refractive index at 40° and 20°C (68°F). For the last, the A.O.A.C. book is cited (which contains Chataway's table), but since Zalewski's values for solids at 20°C were on average 2·1% lower than those by the other two methods, it seems possible that the sucrose table therein was used. No specific citation was given. Abramson (1953) lists S (between duplicates) for 50 samples by 70° vacuum drying as 0·33; for refractive index with Chataway conversion (50 samples) as 0·06, and for Karl Fischer titration (148 samples) as 0·14.

Table 6.2/2
Average deviation between moisture content of honey determined by direct drying and by refractometry

Investigator	No. samples	$\bar{d}$
Bryan (1908)[1]	22	0·47
Auerbach & Borries (1924)	10[2]	0·51
Auerbach & Borries (1924)	17[3]	0·47
Chataway (1932)	60	0·12
Marvin & Wilson (1932)	21[4]	0·76
Fulmer et al. (1934)	25	0·20
Eckert & Allinger (1939)	99[5]	0·28
Torrent (1949)	30	0·12
Sacchi (1955)	72	0·30[6]

[1] Dry substance converted to n_{28} by Geerling's table as given, converted to moisture by Chataway table, compared with vacuum-drying values.

[2] Fresh floral honeys only.

[3] All floral honey samples.

[4] First 21 samples in publication: n obtained from Schönrock table, converted to moisture by Chataway table, compared with vacuum-drying values.

[5] Laevorotatory samples only.

[6] After correction of errors in Sacchi's Table 2 (see text).

Sacchi (1955) has published a rather extensive study of moisture determination by refractometry for Umbrian (Italian) honey. Unfortunately, she chose the Fulmer et al. (1934) conversion table. She did find a better fit with her drying data if the equation of Fulmer et al. was modified by subtracting 0·32. Examination of her Table 2 shows 13 errors which, when corrected, revise the equation given by Sacchi to

$$\text{percentage water} = 400 \, (1.5380 - n_{20}) - 0.35.$$

This reduces the 0·7 difference between the higher values of the Fulmer *et al.* (1934) conversion table and the Chataway values to half that amount.

It is impossible to separate the discussions of refractive index of honey and of its water content. The limiting factor in improving the accuracy of the Wedmore-derived relation is the independent direct method for moisture determination. Since the Karl Fischer method may have a lower error than oven drying (Abramson, 1953), and since a higher correlation coefficient was found (0·894) between Fischer and refractive index than between drying and refractometer (0·856), calibration of the refractive index method against moisture by Fischer titration might be considered. Abramson could not determine from his data which procedure should be the reference. From a practical viewpoint it is debatable whether further accuracy in moisture determination by refractometer would be significant, in view of the variations in honey composition. The calibrations given in Table 6.2/1 are at present more accurate than necessary for the hand refractometers that are in considerable use by honey producers and packers. Pearce & Jegard (1949) have calibrated such a refractometer against A.O.A.C. drying, and report a standard error of ±0·4% for the calibration. A standard error of ±0·5% was found for drying, and of ±0·4% for the refractometer. Thus the hand refractometer is much more convenient than the A.O.A.C. vacuum oven method, but not appreciably more accurate.

6.3   DENSITY AND RELATIVE DENSITY

The density of a substance is its mass per unit volume. In some countries the density of honey is expressed in pounds per gallon (U.S. or Imperial). The relative density (specific gravity) is the ratio of the mass of a given volume of a substance (at a stated temperature) to the mass of the same volume of water (at a stated temperature). Since water has a density of 1·0000g per ml at 4°C (39°F), relative density of a substance at any temperature (referred to water at 4°) is equal to the density at that temperature. The relative density of a liquid is determined by direct weighing of a known volume; it may also be determined by use of a calibrated hydrometer floating partially immersed in the liquid, or in other ways. There are numerous arbitrary calibrations of hydrometers for various purposes; some of those encountered in sugar analyses are Brix, Balling, Twaddel, and Baumé. In general, the use of hydrometers is potentially much easier and less expensive than pycnometry, but the nature of honey introduces such difficulty and uncertainty to the former that the two procedures are comparable.

6.31   Direct weighing methods (pycnometry)
Tables relating relative density to dry substance of sucrose solutions have

long been available, and have been much used in honey analysis. Fiehe & Stegmüller (1912), in comparing vacuum drying with density determination by pycnometer, noted differences in dry matter up to 1·5% with solutions of apparently equal density. They gave the relation $T = (d_4^{15} - 0·99913)/0·000771$ relating dry matter (T) with density (d). Auerbach & Borries (1924) determined d_4^{20} for 20% (w/v) honey solutions, using a 50-ml pycnometer, and also dry matter by direct drying using the same samples. For ten fresh floral honeys, the following relationships was obtained by the method of least squares:

$$T = (d_4^{20} - 0·99823)/0·00076763.$$

This was simplified to

$$T = 1\ 302·7\ (d_4^{20} - 0·99823),$$

the density value being that of a solution of 20·000 g honey in 100 ml water. In comparing values for water content calculated from this relation with those found by direct drying, for 17 samples, the average deviation obtained was 0·42% water. For comparison, a similar value for their refractometric procedure (drying vs. n) was 0·47% water.

Snyder (1933) compared density (in pounds per gallon) for 18 honey samples as determined: (a) by direct weighing of $\frac{1}{4}$ or $\frac{1}{2}$ pint; (b) by a pycnometer, using undiluted honey and converting the resultant d_{20} to weight per gallon from a sucrose table; (c) by refractometer converted to Brix and thence to weight per gallon by sucrose tables. The average values for the 18 samples by these procedures were 11·867, 11·867, and 11·859 lb/gal respectively. The average difference between (a) and (b) was 0·011; (b) and (c) 0·009; (a) and (c) 0·012. These differences are equivalent to 0·19, 0·16, and 0·21% water in the sucrose tables used. No relationship between moisture content of the honeys and density was determined or reported in this study.

Marvin (1933) described three procedures for determining the density of honey; one was the weighing of a standard pint or gill measure; another was the conversion of refractive index by sucrose tables to weight per gallon, both as described by Snyder (1933). Average values for 37 floral honeys were 11·838 and 11·845 lb/gal, respectively; the average difference was 0·015, equivalent to 0·26% moisture. Again, no independent determination of moisture content was made. This small difference is in contrast with the difference in moisture content between the sugar and honey calibrations of the refractometer in terms of solids (or water) content. Apparently, the refractive indices of honey and of a sucrose solution of equivalent density differ only slightly, the average difference being about 0·0006 in refractive index. By contrast, sucrose solutions and honeys of equivalent moisture differ by about 0·0040 in refractive index, or about 1·6% moisture. When Marvin (1934) published a revised table relating refractive index, weight per gallon, and water content, the refractive index and weight per gallon relationship was not changed, though the water values in the revised table correspond to the Chataway equivalents.

Hadorn (1956) found an average difference between the Auerbach & Borries refractive index calculation and pycnometric determination of dry matter of 0·17% solids. The averages for the ten honeys were only 0·01% apart.

6.32 Hydrometry

Use of hydrometers for relative density determination in honey came many years after the development of these instruments for technical and research measurements in the sugar industry. Pique (1914) described a hydrometer for honey musts which had three graduations: relative density, weight of honey per hectolitre, and percentage of alcohol which should result from proper fermentation. Some use of hydrometers in honey processing was noted by Chataway (1932). In considering the use of the hydrometer for undiluted honey she noted that two such instruments were then in use in Canada, and examined both. One, designed for small honey samples, showed very poor reproducibility (over 2% moisture); the other (larger) was somewhat better. Later (Chataway, 1933) she designed a large, sensitive Baumé hydrometer for honey, and tested 38 honey samples whose moisture content was also determined by refractive index. In this work, earlier erratic results were eliminated by placing a layer of water on the honey surface after the hydrometer was in place. Readings were made at about 120°F and corrected to 68°F (20°C); they were also corrected for the presence of the water layer. Average moisture content for the 38 samples by refractometer was 17·42%, and by a calibration curve constructed from the hydrometer values, 17·43%. The average deviation between values by the two methods was 0·15% moisture.

Marvin (1933) described the use of a hydrometer for determining the weight per gallon of honey. This density measure was used because recently issued U.S. Department of Agriculture grades had specified a minimum density of 11·75 lb/gal at 68°F. Two procedures were described: use of a Brix hydrometer in warm full-density honey, and the Brix dilution method using a Brix hydrometer in a 1:1 dilution, then doubling the reading. Conversion to weight per gallon from Brix was made from standard sugar tables. Results from this latter method were compared with those from direct weighing and averaged (for 37 honeys) 11·915 lb/gal against 11·838 by weighing. The difference is equivalent to 1·35%

moisture. This value is close to the -1.3 correction which must be applied to Brix values of molasses when determined by the double dilution procedure, and is needed in that case to correct for the excess volume contraction of molasses over sucrose when diluted (Browne & Zerban, 1941, page 29). Marvin noted the higher values but ascribed no cause. Some of the physical difficulties of hydrometry in a heavy viscous liquid such as honey may be overcome by enclosing the sample in the instrument and suspending it in water. White (1967b) has made a preliminary evaluation of this type of hydrometer, the Eichhorn type, for moisture determination in honey. Sources of possible error were noted, and he concluded that the accuracy of his model was at least as good as that of the hand refractometer, perhaps better.

Wedmore (1955), while admiring Chataway's work on refractive index of honeys, felt that her work on relative density of honey 'though not yet superseded, is not in the same class'. He discussed two calibration charts (percentage of water vs. degrees Baumé) of Chataway: that in the original 1933 publication and one published later (Chataway, 1935). When converted to the same temperature basis they differ somewhat, particularly in the lower moisture ranges. Wedmore thought that this resulted from (a) using too few samples with a low moisture content (less than 15.5%), and (b) the straight-line relationship used for the later conversion table; in the 1933 paper a curve was shown to be necessary, and this is also true for other sugar solutions.

In Wedmore's Table 6, column 6 is entitled 'Author's new determination' and lists relative density values at 20°/20°. Careful reading of the paper leads one to believe that this refers not to independent experimental work but to his fitting of a new line to the original Chataway data which he obtained from her Figure 2, from which Wedmore 'reproduced the original experimental results by the use of a reading microscope' (Wedmore, 1955). The relative density values in Wedmore's Table 6, column 6, differ from the 1935 Chataway table, and Wedmore noted that the relative density figures in the Chataway 1935 table suffered not only from the use of the linear relationship, 'but also from some error made during conversion to S.G.; it seems impossible now to trace this to its source, by its magnitude or otherwise . . . her published S.G. figures tend to give too low a water content, the difference in S.G. representing a difference in water content of about 0.2%'. The source of this difference now appears clear. In a letter written in 1937 to a U.S. Department of Agriculture official which has recently come to hand, Dr. Chataway commented on a Baumé-Brix conversion table in a 1933 Department honey-grading circular, pointing out that it did not agree with her table because two different Baumé scales were involved. The U.S. scale was the U.S. Bureau of Standards Bates and Bearce modification (Browne & Zerban, 1941, page 81) established in 1918 and relating Baumé to relative density at 20°/20°C. Dr. Chataway used the older 'American Standard' Baumé which related to relative density at 60°/60°F. Since the differences between relative density values 20°/20° calculated from Baumé are about 0·0012–0·0016 in relative density in the proper direction, it is evident that Wedmore assumed that Chataway used the modern Baumé (20°/20°) when in fact she used 60°/60°, so her lower values resulted from her proper correction of the relative density 60°/60° values obtained from the Baumé equation* to relative density 20°/20° values, which Wedmore did not do. An example will perhaps clarify this explanation. Wedmore notes that Chataway's values (in Baumé degrees) in the middle of the range are practically identical with his newly calculated figures. Her Table 2 (Chataway, 1933) gives for 17·4% moisture a Baumé value at 68°F of 42·89. Converting by her value of 0·024 per degree F, we obtain 43·08 Bé at 60°F (her 1935 table gives 43·09). Since the Baumé scale Chataway used was the older American scale, the relative density at 20°/20° is obtained as follows:

$$\text{R.D. } 60°/60°F = 145/(145 - 43·08)$$
$$= 1·42268$$

Using Wedmore's conversion factors to convert 60°/60°F to 20°/20°C, we have R.D. 20°/20°C = (1·42268 × 1·00081) − 0·0027 = 1·42113, which rounds to 1·4211. The corresponding value in the 1935 Chataway table is 1·4212. If we assume (as apparently Wedmore did) that the 'new' Baumé scale was used, we get

$$\text{R.D. } 20°/20° = 145 - (145 - 43·08)$$
$$= 1·42268$$

which rounds to 1·4227. The value given by Wedmore in his Table 6, column 6, is 1·4226 for his 'new' determination.

We must therefore conclude that Wedmore's new curve was obtained from Chataway's experimental Baumé values, but erroneously converted to relative density. We cannot then accept his Table 5, 'Proposed figures for the specific gravity [relative density] of honeys of different water contents' because the values he labels relative density 20°/20°C are in fact relative density 60°/60°F and must be converted (as indicated above) to obtain the 20°/20°C table.

Table 6.32/1 shows the Wedmore revision of Chataway's data, converted correctly to relative density. Departure from the 1935 Chataway table is primarily at the lower-moisture end, and the two tables are coincident between 17·2 and 19·2% moisture.

* Degrees Baumé = $145 - 145/(\text{R.D. } 60°/60°)$; Bur. Stds. Bé = $145 - 145/(\text{R.D. } 20°/20°)$.

The specification 'modulus 145' used by Chataway (Chataway, 1935) is not sufficient to identify the Baumé scale she used. It is true that other scales use different moduli, but the Bates-Bearce scale differed

Table 6.32/1
True specific gravity of honeys of different water contents*

Water content %	Specific gravity 20/20°C	Specific gravity 60/60°F	Water content %	Specific gravity 20/20°C	Specific gravity 60/60°F
13·0	1·4457	1·4472	17·0	1·4237	1·4252
13·2	1·4446	1·4461	17·2	1·4224	1·4239
13·4	1·4435	1·4450	17·4	1·4211	1·4226
13·6	1·4425	1·4440	17·6	1·4198	1·4213
13·8	1·4414	1·4429	17·8	1·4185	1·4200
14·0	1·4404	1·4419	18·0	1·4171	1·4187
14·2	1·4393	1·4408	18·2	1·4157	1·4173
14·4	1·4382	1·4397	18·4	1·4143	1·4159
14·6	1·4372	1·4387	18·6	1·4129	1·4145
14·8	1·4361	1·4376	18·8	1·4115	1·4131
15·0	1·4350	1·4365	19·0	1·4101	1·4117
15·2	1·4339	1·4354	19·2	1·4087	1·4103
15·4	1·4328	1·4343	19·4	1·4072	1·4088
15·6	1·4317	1·4332	19·6	1·4057	1·4073
15·8	1·4306	1·4321	19·8	1·4042	1·4058
16·0	1·4295	1·4310	20·0	1·4027	1·4043
16·2	1·4284	1·4299	20·2	1·4012	1·4028
16·4	1·4272	1·4287	20·4	1·3996	1·4012
16·6	1·4260	1·4275	20·6	1·3981	1·3997
16·8	1·4249	1·4264	20·8	1·3966	1·3982
			21·0	1·3950	1·3966

* Wedmore's (Wedmore, 1955) revision of Chataway's (Chataway, 1933) data as corrected (*see* text). By definition, values for S. G. 20°/20° calculated from Baumé are 'true' specific gravity, i.e. they correspond to weight *in vacuo*. To obtain 'apparent' specific gravity, i.e. corresponding to weight in air with brass weights, the correction to be added to the true value varies from 0·00047 at 21% moisture to 0·00055 at 13% moisture. An average correction of +0·0005 is satisfactory. The term 'relative density' is now preferred to 'specific gravity'.

from the older American standard only in the use of R.D. 20°/20°C rather than R.D. 60°/60°F, and it uses the same modulus. The newer scale appears to be in general use in the U.S.A. (Browne & Zerban, 1941, page 82; Bates et al., 1942, page 249). Chataway anticipated confusion, for in the letter noted above she remarked that the Bates–Bearce scale 'is still recognized, apparently, as it appears in the fourth, 1936, edition of the *American Official Agricultural Handbook* * . . . but it can hardly be considered correct. In effect it establishes a second American Baumé scale with no convenient title to distinguish it from the one more generally recognized.'

The relatively large range of honey densities requires that particular care be taken to ensure thorough mixing when honeys are blended. Layering of different honeys in a tank can be quite pronounced; in fact, Fix & Palmer-Jones (1949) state that the reason why the top layer of honey in settling tanks is the most dilute is density difference rather than absorption of moisture from the air. Heating with mixing is recommended to avoid such layering.

6.33 Other methods

A very approximate measurement of the 'consistency' of honey is described by Hansson (1966). A cone is supported point down, touching the honey surface, and is then released. Consistency (on a scale 1 to 5) is evaluated from the rate of sinking and the final depth. A field test for the maximum water content of honey that is inexpensive, easy, and rapid, is described by Aganin (1965). To determine whether honey contains more or less than 22% water by volume, a drop is added to a water solution of calcium perchlorate having the same relative density; if it rises to the surface it is less dense than the solution, and hence is presumed to contain more than 22% water.

6.4 VISCOSITY AND THIXOTROPY

This subject has been reviewed by Pryce-Jones (1953) in his article 'The rheology of honey' in Scott-Blair's book. No significant fundamental contribution to the field has come to hand since that time, so this discussion will be restricted to a brief review.

6.41 Viscosity for determining moisture

Von Fellenberg (1911) attempted to use viscosity measurement to

---

* This table appears in the 1965 Tenth edition. It gives 'true specific gravity', corresponding to weights *in vacuo*, calculated directly from the formula. Table 3 in Browne & Zerban (1941) also gives the Bates-Bearce scale, but is calculated to give 'apparent relative density' at 20°/20°C, which corresponds to weighing in air.

detect the addition of glucose syrup to honey. Considerable variation in
the viscosity of different honeys reduced the value of the method. In
the study of the effect of moisture content on various physical attributes
of honey already discussed, Chataway (1932) included viscosity measurement. Using a falling-ball viscometer, she reported a nearly straight-line
relationship between *log* viscosity and *log* moisture content. Using this
curve, moisture contents for the 60 honeys were calculated and compared
with those obtained by the A.O.A.C. vacuum drying procedure. The
average difference for all samples was 0·20% moisture; elimination of
5 buckwheat samples which did not fit the curve reduces this to 0·14%
moisture. Chataway noted that a difference of 0·1% in moisture gives
viscosity differences of 4–6%. The viscosity of honey is highly temper-
ature-sensitive; Chataway constructed a correction chart by which the
times of fall at any temperature between 15·0 and 30·0°C (59–86°F)
to 0·1°C could be corrected to 25·0° before conversion to moisture values.

Oppen & Schuette (1939) found a very low correlation between
refractive index and moisture content of honey by the A.O.A.C. drying
method (no data are given), so they investigated the use of viscosity
for determining moisture content. They criticized Chataway's apparatus
as permitting errors up to 8% because of wall effects, due to the use of a
too-narrow tube. Using an apparatus with a more favourable ratio of
ball diameter to tube diameter, they determined viscosities of 30 samples
at 40°C (104°F), and of 15 at four other temperatures. An equation
relating viscosity, moisture content, and temperature was developed, and
a graph was presented from which moisture content could be obtained
from the time of fall of the ball in their apparatus (Figure 6.41/1). The
time is determined at a known temperature between 30° and 50°C
(86 and 122°F) for a steel ball of diameter 0·16 cm (0·06 or $\frac{1}{16}$ in) to fall
20 cm through the centre of a 25-mm standard-wall pyrex tube containing
the honey, after gravity acceleration in the honey for 8 cm. This time-
temperature point is located on the graph, and a line parallel to the
nearest diagonal line intersects the W scale at the water content of the
honey.

The average difference between moisture values found by Oppen &
Schuette from the chart and from the A.O.A.C. method is 0·20%. Since
they claimed their procedure to be more accurate than Chataway's, her
lower average deviation may arise from a better technique for A.O.A.C.
moisture determination.

6.42 Absolute values for viscosity

Neither Oppen & Schuette nor Chataway provided absolute values for
viscosity. Lothrop (1939) did so in a study of the effect of composition
of honey on viscosity. Even when adjusted to equal moisture content,

the viscosity (at 40°C, 104°F) varied from 3·10 poise for alfalfa honey (*Medicago sativa*) to 4·11 for sumac honey (*Rhus*). Only four honeys were within the range 3·10–3·14 poise. (Two honey samples, one from honey-dew in Hawaii, and one from tarweed (*Hemizonia*) with an anomalous

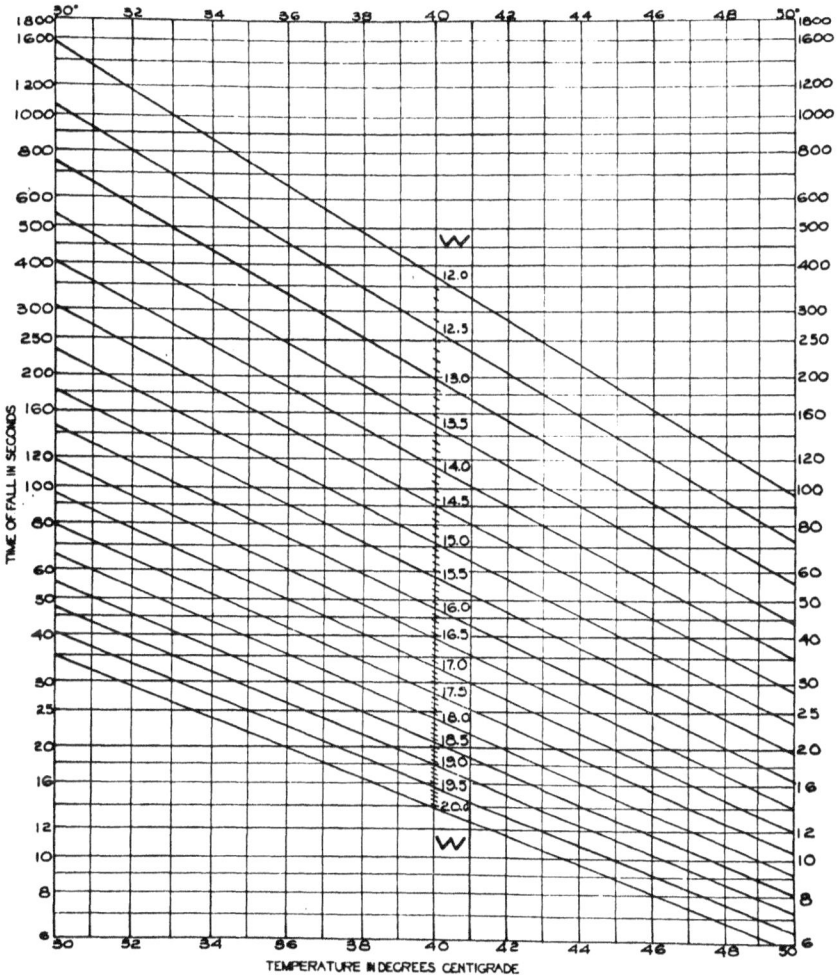

Figure 6.41/1 Relation between (*a*) time of fall of a ball through honey in the Oppen & Schuette (1939) viscosity apparatus at given temperatures and (*b*) moisture content (*W*) of honey (*see* text for details).

composition, are not included in this discussion.) Lothrop believed that viscosity variations in honey are due to non-sugar materials, particularly 'dextrins', and that colloidal materials also help to determine the viscosity.

The work of Munro (1943) appears to be the most extensive. Using a MacMichael viscometer, he determined viscosity for sweet clover honey (*Melilotus*) at 6 moisture contents, sage honey (*Eriogonum*) at 3, and white clover honey (*Trifolium repens*) at 9. He included several further samples, some with colloids removed. The viscosity of each was measured at a number of temperatures, about 3°C apart, over the range 5–80°C

Table 6.42/1
Viscosity of honey

Type	Moisture content (%)	Temperature (°C)	Viscosity (poise)
Sweet clover[1]	16·1	13·7	600·0
(*Melilotus*)		20·6	189·6
		29·0	68·4
		39·4	21·4
		48·1	10·7
		71·1	2·6
Sage[1]	18·6	11·7	729·6
(*Erigonum*)		20·2	184·8
		30·7	55·2
		40·9	19·2
		50·7	9·5
White clover[2]	13·7	25·0	420
(*Trifolium repens*)	14·2		269
	15·5		138
	17·1		69·0
	18·2		48·1
	19·1		34·9
	20·2		20·4
	21·5		13·6
Sage[2]	16·5	25	115
Sweet clover[2]	16·5	25	87·5
White clover[2]	16·5	25	94·0

[1] Data of Munro (1943)
[2] Interpolated from Munro's data

(41–176°F). Whereas Oppen & Schuette obtained straight lines for *log* viscosity plotted against *1/T* over their range of 30–50°C (86–122°F), Munro's data over the much wider range show a slight curve, characteristic of a highly associated liquid. Table 6.42/1 gives typical viscosity values observed by Munro. To facilitate comparison, the Table includes a number of values obtained from the Munro data by graphical interpolation.

Munro (1943) noted that the viscosity of honey changes most rapidly as the temperature rises to room temperature, and stated that the rate of viscosity decrease drops markedly from room temperature to about 30°C, and thereafter shows relatively little change. These observations were based on his plotting of the data on a direct basis. The Pryce-Jones plot (Pryce-Jones, 1953) of Munro's values as *log* viscosity against *1/T* shows that the rate of change is relatively constant. Munro stated that heating honey above about 30°C (except for low-moisture types) produces such slight viscosity decreases as to be without practical significance. He noted that 1% moisture is equivalent to about 3·5°C in its effect on viscosity. However, MacDonald (1963) has examined the effect of temperature on honey pumping rates and on flow through pipes of various sizes. Table 6.42/2 shows his data for the flow of honey at constant

Table 6.42/2
Relative flow of honey in pipes*

Pipe diameter (inside)	*Temperature*		
	82°F (28°C)	102°F (39°C)	122°F (50°C)
¾ in (19 mm)	149	400	1 125
1 in (25 mm)	367	973	2 353
1¼ in (31 mm)	729	1 895	5 000
1½ in (38 mm)	1 263	2 609	6 792

* Rate of flow (in pounds per hour) through 4-in. (10-cm) length of pipe with 4-in. head. Data of MacDonald (1963).

pressure head of honey (4 in., 10 cm, to top of outflow) through a 4-in. length of horizontal pipe at three temperatures. The difference between flow rates at 102°F (39°C, well above Munro's 30°C) and 122°F (50°C) is quite significant; the relative increase is equal to that from 82°F (28°C) to 102°F (39°C), as expected from viscosity data. From the point of view of maintaining quality of honey, however, it is better to increase capacity in honey pumping and handling by increasing pipe and pump size than by increasing temperatures excessively. However, contrary to Munro's view, temperatures over 30°C (86°F) can be significant in facilitating honey handling.

6.43 Non-Newtonian properties
In addition to the behaviour of honeys as Newtonian liquids, described above, certain non-Newtonian rheological phenomena have been reported. Thixotropy, an isothermal reversible gel-sol-gel transformation induced by shearing and subsequent rest, is quite pronounced in honey from heather (*Calluna vulgaris*) and a few other sources. Without agitation, this honey will not flow sufficiently for extraction in a centrifugal extractor. Pryce-Jones (1953) has extensively reviewed the rheology of heather honey. He notes that manuka (*Leptospermum scoparium*) honey of New Zealand is also markedly thixotropic. The property is ascribed for both honeys to their relatively high content of certain proteins; the heather honey protein can render clover honey thixotropic. Deodikar *et al.* (1957) report that karvi honey (*Carvia callosa*) from India is also markedly thixotropic.

Another non-Newtonian response is 'dilatancy'; this is increased viscosity with increased rate of shear. Pryce-Jones (1952) notes that several honeys—*Opuntia engelmanni* from Nigeria and several *Eucalyptus* species—possess this property to a rather high degree; he ascribes it to their content of the polysaccharide dextran of molecular weight in the 1 250 000 range. Also known as *Spinnbarkeit* or 'stringiness', it is easily noted by the formation of long 'strings' on honey when a rod is dipped into it and rapidly moved away.

6.44 Diffusivity
The apparent diffusivity of water in honey was measured by Fan & Tseng (1967), using a laser interferometric microdiffusion cell. It was highly dependent on concentration and was of the same order of magnitude as that of a glucose solution in water.

6.5 OPTICAL PROPERTIES

Relative little attention has been given to the various optical properties of honey, with the exception of optical rotation.

6.51 Optical rotation
Among many other materials of natural origin, honey has the property of rotating the polarization plane of polarized light. This is one further

property that depends largely on the sugars of honey—their types and relative proportions. Since each sugar has a specific and consistent effect, and the total optical rotation is dependent on concentration, analysts early used optical rotation under various specified conditions as a means of sugar analysis. It was quite precise and accurate in the analyses encountered in the sugar industry and, with the relatively simple view of the sugars of honey then current, the use of the method was extended for sugar analysis of honey. The limitations of the method (White, Ricciuti & Maher, 1952) are such that modern honey analysts have largely abandoned it. One generalization that appears to remain valid is that floral honeys are laevorotatory and honeydew (or adulterated) honeys are usually dextrorotatory. This is a consequence of the normal preponderance in floral honey of fructose, which has a negative specific rotation, $([a]_D^{20} = -92\cdot4°)$ over glucose $([a]_D^{20} = +52\cdot7°)$. Honeydew types are usually somewhat lower in fructose content and contain melezitose $[a]_D^{20} = +88\cdot2°)$ or erlose $([a]_D^{25} = +121\cdot8°)$ (White & Maher, 1953b) which, together with glucose, usually give a positive net optical rotation.

6.52 Mutarotation

Many sugars are capable of existing in solution in several physical forms, which may have different optical rotations. Usually, a sugar exists in one form in the crystal. On dissolving, an equilibrium is reached between the several forms; during the equilibration process the optical rotation of the solution changes. This is mutarotation (multirotation), and the extent is quite characteristic of specific sugars.

Honey, although completely liquid, will exhibit a slow change in optical rotation after being diluted. In Browne's (1908) analyses, 92 samples of laevorotatory honeys showed a change of $-3\cdot5°V$ after standing for 20 hr. The change is not brought about by the difference in specific rotation of the sugars due to the concentration change; this would be in the opposite direction and largely due to fructose. A pronounced mutarotation would of course be expected when the honey being dissolved contains glucose crystals. Even when the honey is entirely liquid, the net change in optical rotation on solution is in the direction of glucose mutarotation, not that of fructose; furthermore, fructose mutarotation is about 12 times as rapid as that of glucose. It seems likely that glucose is involved in honey mutarotation. No studies of the cause or mechanism of honey mutarotation are at hand.

6.53 Colour of honey

Much of the literature on honey colour is simply descriptive, relating colour to floral source and processing of honey. As noted in Chapter 5, relatively little is known of the compounds responsible.

Colours of honey form a continuous range from very pale yellow through ambers to a dark reddish amber to nearly black. Greenish casts are fairly common. Colour ranges are generally characteristic for floral type. Brice *et al.* (1956) made a rather extensive physical examination of honey colour in establishing the current U.S. Department of Agriculture honey colour classes, recording spectrophotometric data for several characteristic honeys in all colour ranges, and calculating CIE colorimetric data. They concluded that the principal colorants of honey, maple syrup, caramel solutions and other sugar products are similar, basing this view on strong similarity in the plotted values of *log* A *v.* wavelength for the various products. Honey shows more minor departures from linearity than do other products.

Honey appears to be lighter in colour after granulation than when liquid. The crystal size affects the degree of lightening, the finest crystals imparting the lightest shade. Two explanations may be advanced for this: (*a*) the opacity of granulated honey greatly reduces the thickness of the honey layer actually being observed; and (*b*) increasing whiteness is imparted to a material by reducing its particle size. The increase in the proportion of surface-reflected light in the total may account for this type of lightening.

Colour of honey is important in its marketing. This is dealt with in Chapters 12 and 13. References to colour grading in various countries include systems used in Hungary (Kottász, 1958*a*), U.S.S.R. (Kottász, 1958*a*), Britain (British Standards Institution, 1952), France (Barbier & Valin, 1957), U.S.A. (United States Department of Agriculture, 1951), Canada (Canada Department of Agriculture, 1952, Townsend, 1969*a*), and Australia (1964).

It is well known that honey darkens in colour during storage. One of the most extensive studies of the effect of storage on honey colour is that of Milum (1948). He concluded that discoloration during storage is in part dependent upon the amount of previous discoloration, and that discoloration during processing tends to lessen the subsequent rate of discoloration. Table 6.53/1 shows a summary of Milum's data obtained by replotting his values and interpolating. These values are approximations and useful to indicate order of magnitude only. Great variation in darkening rate is found among different honeys, depending upon their composition (acidity, nitrogen and fructose contents). F. G. Smith (1967) examined the darkening effect of storage at several temperatures between 43° and 80°C upon several Australian honeys. The variability in response by different honeys was confirmed; one honey (from *Dryandra sessilis*) darkened twice as rapidly as any other. Smith noted a correlation between the time required at a given temperature to effect a 10-mm (Pfund) increase in colour and the time required to produce 3 mg

Table 6.53/1
Approximate rate of honey darkening in storage*

Temperature of storage		Darkening in mm Pfund/month		
°F	°C	Original colour < 34 mm	Original colour 34–50 mm	Original colour > 50 mm
50	10·0	0·024	0·024	0·024
60	15·6	0·08	0·125	0·10
70	21·1	0·27	0·70	0·40
80	26·7	0·90	4·0	1·50
90	32·2	3·0	7·7	5·0
100	37·8	10·0	14·0	11·0

* Calculated from data of Milum (1948)

hydroxymethylfurfuraldehyde per 100 g honey at the same temperature (White, Kushnir & Subers, 1964).

Irradiation by ultraviolet excites fluorescence in honeys, the emitted colours differing among various honey types. The work of Orbán & Stitz (1928) appears to be the only study of the phenomenon.

6.6   THERMAL PROPERTIES

The physical properties of honey with regard to heat have received only minimal attention, though the chemical and biological effects of heat have been studied extensively. Presumably, the design of the minor amount of processing equipment originally intended for honey has largely been made by cut-and-try methods or by extrapolation from sugar data.

6.61   Specific heat

Helvey (1954) has determined several heat-related properties of honey. Using conventional methods, he reported that the specific heat of honey containing 17·4% moisture is 0·54 at 20°C (68°F), and the temperature coefficient 0·02 cal/°C.

MacNaughton, according to Townsend (1954), has also determined the specific heat of honey. He used a sample about seven times as large as Helvey's, and a temperature range of 29–48°C (85–119°F). He obtained somewhat higher results, shown in Table 6.61/1; they are believed to be accurate to ±0·02. Both investigators ascribed small variations to the effect of honey composition.

Table 6.61/1
Specific heat of honey*

Moisture content (%)	Specific heat
20·4	0·60
19·8	0·62
18·8	0·64
17·6	0·62
15·8	0·60
14·5	0·56
coarsely granulated	0·64
finely granulated	0·73

* Data of MacNaughton (Townsend, 1954)

6.62 Thermal conductivity

Helvey (1954) determined the thermal conductivity of honey solutions, over the range 0% to 90% water, at various temperatures; he presented the results in a three-dimensional figure. The data in Table 6.62/1 were obtained from Helvey's Figure 5 by tracing upon graph paper. He also reported that finely crystallized honey at 20°C (68°F) has a thermal conductivity of 129×10^{-5} cal/cm sec°C. Detroy (1966) determined the surface conductance, or film coefficient, for honey within the processing temperature range, using a concentric-tube counter-flow heat-exchanger. Values were obtained at honey flow rates between 700 and 900 lb/hr (317–442 kg/hr) and speeds from 0·17 to 0·24 ft/sec (5·2–7·3 cm/sec). Values ranged from 34·1–40·1 BTU/sq ft/hr/°F (preheater water circuit) to 57·7–77·0 BTU/sq ft/hr/°F (flash-heater water circuit).

6.63 Freezing point of solutions

Full-density liquid honey becomes increasingly hard as the temperature is reduced, but water does not appear to crystallize from it.

Stitz & Szigvárt (1931a) studied the freezing point of honey. Because of its physical nature they were unable to obtain values for honey solutions more concentrated than 68%, for which they found a freezing-point depression of 12·01°C (21·6°F). Agreement was excellent between the freezing points found for 15% honey solutions and values calculated from the concentrations of glucose, fructose, and sucrose: −1·44 and −1·438°C (29·41°F), −1·49 and −1·49°C (29·32°F).

For ten honeys, the freezing point of 15% solutions ranged from −1·42 to −1·53°C (29·44 to 29·25°F).

Table 6.62/1
Thermal conductivity of honey*

Moisture content (%)	Temperature (°C)	Thermal conductivity (cal/cm sec °C)
21	2	118×10^{-5}
	21	125
	49	132
	71	138
19	2	120
	21	126
	49	134
	71	140
17	2	121
	21	128
	49	136
	71	142
15	2	123
	21	129
	49	137
	71	143

* Interpolated from graph of Helvey (1954)

6.64 Calorific value

Calculations by the Consumer and Food Economics Research Division of the U.S. Department of Agriculture (Watt & Merrill, 1963), using the Atwater system as reviewed by the F.A.O./U.N., give 1 380 Cal/lb (304 Cal/100 g) for the energy value of an average sample of honey.

6.7 CRYSTALLIZATION

Glucose monohydrate spontaneously crystallizes from many honeys, which are supersaturated solutions under ordinary storage conditions. Whether they are also supersaturated under hive conditions at higher temperatures is not known, partly because the carbohydrate composition of honey is more complex than the model systems examined to date.

We are here concerned with general aspects of honey crystallization. Discussion of controlled granulation of honey, and of delaying granulation in liquid honey, are to be found in Chapters 9 and 10.

6.71   Model systems and honey composition

A possible route to understanding honey granulation lies in a study of phase relationships in model systems of sugars. An early attempt was that of Jackson & Silsbee (1924), who studied several systems at 30°, and discussed saturation relations in honey in the light of their data on the glucose-fructose-water system. They found that glucose solubility decreased with increasing laevulose concentration. With glucose hydrate as the solid phase, they recorded solubility at 54·64% without fructose, dropping to 32·55% at 39·4% laevulose.

Basing their calculations on the honey analyses of Browne (1908), Jackson & Silsbee (1924) concluded that all honey is supersaturated at 23°C (73°F); calculated supersaturation coefficients were 2·86 for alfalfa (*Medicago*) honey, and 1·66 for tupelo (*Nyssa*) honey which is known never to granulate. This apparent conflict was ascribed to the 'sluggishness with which dextrose crystallizes from solutions of high laevulose content'. The discrepancy is not due simply to the too-high dextrose values generally obtained by the older methods of analyses, since calculations based on recent analyses also imply supersaturation. Lothrop (1943), in an unpublished thesis, pointed out that Jackson & Silsbee's measurements did not extend to the laevulose concentrations often encountered in honey. Because of the past unavailability of this work, it is discussed here in some detail. Lothrop did not accept the explanation given by Jackson & Silsbee, noting that some honeys do not granulate even after many years, and even after seeding with dextrose hydrate. Lothrop studied the solubility of dextrose in laevulose solutions at concentrations extending to those found in honey. He found an abrupt (so sharp as to be transitional) increase in dextrose solubility at a laevulose concentration of about 150 g in 100 g water. In the area of lower solubility (85–90 g dextrose per 100 g water) the solid phase was dextrose monohydrate; beyond the higher solubility region (125–128 g per 100 g water) the solid phase was anhydrous dextrose. The equilibrium conditions were approached from both the undersaturated and oversaturated sides. Solubility curves were determined at 20°, 25°, 30°, and 52°C (68°, 77°, 81°, and 128°F) with both forms of dextrose, singly, as the initial solid phase. Solubility of sucrose (which does not hydrate) did not show the abrupt increase with increasing laevulose concentration. The curve for dextrose at 52°C (above the 50°C transition point for dextrose to dextrose hydrate) also did not show the break, and resembled the sucrose curve. Identification of the solid phase was by microscopic examination, the hydrate being stated to appear in hexagonal plate-like monoclinic crystals, with the anhydrous form as rhombic needle-like forms.

Lothrop believed that the change in solubility of dextrose was not related to the α-β-equilibrium, but rather to the degree of hydration of the dextrose in solution; anhydrous dextrose is known to be more soluble than the hydrate. He lists six arguments for this hypothesis, based on his data. Later Kelly (1954) published the complete diagram for the system at 30°C (81°F)—without knowledge of Lothrop's work. Kelly also noted an area in which anhydrous dextrose is in the solid phase, with an invariant point at which both forms of dextrose are in equilibrium. He proposed that the presence of fructose had the effect of reducing the transition temperature of the monohydrate from above 50°C (122°F) to something less than 30°C (86°F) for solutions saturated with fructose. He noted that published analyses of honey relate to the area in which anhydrous glucose is the solid phase at 30°C. Since dextrose does not normally granulate from honey until the temperature is appreciably below 30°C, the crystallization seems to occur below the transition temperature so that it would appear as the monohydrate.

However, Villumstad (1952) has described the simultaneous occurrence in granulated honey of both needles and plates of dextrose, though he does not speculate on the reasons for the different forms. He reported that examination of the chemical and physical composition of the different crystals was in progress. Dean (1974) has described a new metastable β-glucose hydrate which may explain some of these anomalies.

6.72   Prediction of tendency to granulate

In view of the variations in honey composition, a means of predicting the granulation behaviour of a particular batch of honey would be of considerable practical interest. Rational selection of honey for liquid-honey packs, and blending for packs of finely crystallized honey of desired hardness characteristics, could be done on a routine basis.

All attempts to accomplish this have been empirical, using various proposed indices to fit the granulation behaviour observed after storage. Unfortunately, data at hand from model systems are not helpful; Jackson & Silsbee's diagrams do not extend to areas of interest; Lothrop's cover a wider range but are also not complete; Kelly's data are valid only at 30°C, at which no honeys granulate. White, Riethof, Subers & Kushnir (1962) have discussed the relation of the granulating tendency of honey to its composition, based on their observations and analyses of nearly 500 honey samples. Using statistical procedures, they showed highly signi-ficant correlations between granulating tendency and several previously proposed indices, the unadjusted dextrose/water ratio of Austin (1958) giving the highest value. The laevulose/dextrose ratio, much used in the past, was the lowest-ranking index. When applied to individual samples rather than to group averages, a slightly higher score was found for

Jackson & Silsbee's $(D-W) \div L$ value. The difference was quite small, and since Austin's factor does not require determination of laevulose, it is preferred. The average D/W ratios for 477 honeys placed in ten granulation classes (White *et al.*, 1962) are shown in Table 6.72/1.

Table 6.72/1
Average dextrose-water ratios for honeys
classified by granulation characteristics[1]

Extent of granulation[2]	No. samples	D/W
none	96	1·58
few scattered crystals	114	1·76
1·5–3 mm layer of crystals	67	1·79
6–12 mm layer of crystals	68	1·83
few clumps of crystals	19	1·86
$\frac{1}{4}$ of depth granulated	32	1·99
$\frac{1}{2}$ of depth granulated	19	1·98
$\frac{3}{4}$ of depth granulated	16	2·06
complete soft granulation	18	2·16
complete hard granulation	28	2·24

[1] Data of White, Riethof, Subers & Kushnir (1962)
[2] Granulation observed in heated honey after 6 months undisturbed storage at 23–28°C; honey in $\frac{1}{2}$ lb or 1 lb jars (0·23, 0·45 kg).

Codounis (1962) has also studied honey crystallization in relation to composition. In his view, the index (Brix minus dextrose) ÷ dextrose is more useful than other indices, including D/W. Examination of Codounis' Table 4 shows that by 'Brix' he means total solids, or 100 minus water content. Thus, either the Codounis or Austin index can be calculated from the other,* and they are thus of equal value in predicting granulation, in spite of Codounis's statement to the contrary. It should be noted that the values in Table 6.72/1 are calculated from true dextrose values and not from those obtained by the nonspecific hypoiodite or other method without prior removal of interfering sugars. Codounis has noted, and we concur, that as a rough rule it can be taken that honeys with less than 30% dextrose rarely if ever granulate. Siddiqui (1970) was unable to relate granulation (as determined by the method of White *et al.*) with any of the ratios L/D, $(D-W)/L$, or D/W for 95 samples of Canadian honey in which sugars had been determined by paper chromatography. No data have been published; possibly the relative inaccuracy of the paper–chromatographic method obscured the relationship. Siddiqui further

* Codounis index $= (100/D)-(1/\text{Austin index})-1$

states that 'such predictions were, of course, not possible because the factor actually involved is the presence or absence of appropriate crystal nuclei'. Apparently Siddiqui failed to appreciate that nuclei are eliminated in the method of White *et al.*, as a result of the requirement that samples are heated to clarity (as indicated by the polariscope) before an undisturbed six months' storage. Presence or absence of nuclei certainly influences the onset of crystallization, but its extent and speed are dependent on the D/W ratio.

Now that a rapid and accurate photometric method is available for determination of true glucose in honey (White, 1964b), the use of D/W ratios for prediction of granulating tendency should become practical.

6.8 HYGROSCOPICITY

Because of its nature as a highly concentrated sugar solution, honey is remarkably hygroscopic for a natural product. Interest in this property arises for two reasons. First, honey absorbs moisture from the air under certain conditions, and thus becomes diluted, and more liable to fermentation. Secondly, honey can impart the desirable property of softness or non-drying to food products in which it is incorporated.

When honey is exposed to air, a gain or loss in its moisture content will take place, depending upon the temperature, the moisture content of the air, and the vapour pressure of water in the air, which is usually expressed as relative humidity.

For each honey a relative humidity exists at which no gain or loss of moisture takes place; this is the equilibrium relative humidity. It will vary with moisture content of the honey and with the gross composition, the latter having only a minor influence. Because of the high viscosity of honey, moisture absorbed at the surface can diffuse only very slowly throughout the mass, so there may be a relatively rapid dilution at the surface. The great density differences between honeys of different moisture contents also favour the maintenance of a dilute layer at the surface, rather than dispersion of absorbed moisture throughout the mass. For example, Martin (1958) has shown that within 7 days a honey sample at 22·5% moisture exposed at 86% R.H. had 26% moisture at the surface, whereas 2 cm below the surface no change was found; after 24 days the moisture content at the surface was 29·6%; at 2-cm depth it was 23·0%; at 6-cm depth no change was evident even after 95 days. Containers were 5·5 cm in diameter.

When honey is exposed to a relative humidity lower than its equilibrium value, drying will take place. Martin (1958) noted that moisture loss was more rapid at intermediate values (20–40% R.H.) than at 0% R.H. He ascribed this to the formation of a dry film on the surface which retards further evaporation. Dyce (1931*a*), Nico l(1937), Hansson (1942), and Villumstad (1951) have examined the ability or inability of certain types of screw-cap containers to prevent passage of moisture into honey contained in them.

Table 6.8/1 shows the relation between equilibrium relative humidity of a clover honey and its water content.

Table 6.8/1
Approximate equilibrium between relative humidity of air and the water content of a clover honey*

Relative humidity (%)	Water content (%)
50	15·9
55	16·8
60	18·3
65	20·9
70	24·2
75	28·3
80	33·1

* Interpolated from the data of Martin (1958).

Hansson (1942) noted that although the water vapour pressure of honey between 10° and 40°C (50° and 104°F) corresponded in general to 60° R.H., at about 30°C it appeared to correspond to about 75% R.H. He concluded that the decrease in water content of nectar in the hive to the values normally found for honey is purely a physiochemical phenomenon. Contrary to Hansson, Bartlett (1962) found that honey containing 18% water, in a closed system, maintained a R.H. of $59 \pm 4\%$ regardless of temperature, between 4–43°C (40–100°F). Advantage of this was taken in rearing and shipping certain insect parasites and predators that can be fed on honey.

6.9 COLLOIDAL PROPERTIES AND SURFACE TENSION

6.91 Honey colloids

As previously mentioned, honey contains small amounts of colloidally dispersed material. Probably the most extensive work with honey colloids is that of Lothrop and his colleagues (Lothrop & Paine, 1931*a*; Paine, Gertler & Lothrop, 1934). They reported that the colloidally dispersed material in honey showed an isoelectric point of 4·3, being positively charged at more acid pH values and negative in less acid honeys. Flocculation by colloids of opposite charge was demonstrated (Lothrop & Paine, 1931a), and the colloid content of various honey types was determined by ultrafiltration (Paine et al., 1934). The nitrogen content of the material indicated about 55–65% protein; in the samples of lower colloid content, 15–25% was found to be fat-soluble and to resemble beeswax.

Turbidity of honey becomes more pronounced upon dilution, as the peptizing effect of the sugars on the colloidal material is reduced. If dilution is carried out near the isoelectric point, pronounced flocculation may occur. Reconcentration of honey which had been diluted to less than 10% solids does not entirely reverse the aggregation process, but when the solid content exceeds 20% the turbidity begins to decrease (Paine et al., 1934). Removal of the colloidal material, by flocculation with bentonite and filtration, produces a clear honey that shows minimal Tyndall effect.

Roughly half of the nitrogen content of honey is removed by ultra-filtration, and the colour is reduced somewhat. Viscosity is decreased by only a minor amount (Paine et al., 1934). 'Dextrin' content, deter-mined by alcoholic precipitation, is not affected (Paine et al., 1934).

6.92 Surface tension

Surface tension of honey is an important property; in processing of honey, a low value may lead to excessive foaming and scum formation. Paine et al. (1934) examined 25% solutions of 7 floral honeys and 1 honeydew honey. They found that at 20° ultrafiltration produced an average change in the surface tension from 47·0 to 60·2 dyne/cm; they noted an accompanying decrease in foaming and retention of air bubbles.

It has been noted that the thixotropic properties of heather (*Calluna vulgaris*) honey are due to gel-sol-gel transformation of a protein con-tained therein. Removal of the protein produced a true Newtonian liquid; addition to a clover honey made it thixotropic (Pryce-Jones, 1953, page 160). Mitchell and his colleagues (Mitchell, Donald & Kelso, 1954, Mitchell, Irvine & Scoular, 1955, Kirkwood, Mitchell & Smith, 1960, Kirkwood, Mitchell & Ross, 1961) have done extensive analyses of heather honey and proposed an analytical index to distinguish between heather honey and honeydew honey.

6.10 ELECTRICAL CONDUCTIVITY

Very little research on this property of honey has been recorded. Vorwohl (1964) cites Elser's (1924) interest. Stitz & Szigvárt (1931b) measured the conductivity of several honeys at 50% solution at 20·5C (68·9°F) and found values ranging from 0·868 to 3·645 × 10⁻⁴/ohm cm. In general, values increased with ash content. The effect of temperature and concentration was examined; a maximum value was generally found at between 30 and 35% solids. Vorwohl (1964) found maximum values at 20–25% solids, with values for undiluted honey around 10⁻⁶ to 10⁻⁷/ohm cm, approaching the values for distilled water. He measured the conductivity of 40 single-species* samples of honey. His values for a 20% solution ranged from 0·85 to 8·47 × 10⁻⁴/ohm cm. Other samples ranged from 0·6 to 1·46 for floral honeys (heather 7·7) and from 6·3 to 16·41 for honeydew honey. He proposed the use of conductivity with pollen analysis for identifying honey sources and for determining the proportion of honeydew honey. Kaart (1961) proposed the measurement of electrical conductivity as more rapid than chemical analysis for determining suitability of honey for winter stores for bees.

Conductivity values will depend on the concentrations of mineral salts, organic acids, proteins, and possibly complexing materials such as sugars and polyols.

6.11 FERMENTATION

The discussion here is limited to undesired or spoilage fermentation of honey. Section 10.3 discusses methods for preventing undesired fermentation, and Chapter 16 describes controlled fermentation to produce alcohol.

Much of our present knowledge of undesirable honey fermentation dates from the late 1920s and early 1930s, when several groups of investigators in the northern U.S. and Canada (Fabian & Quinet, 1928, Marvin, 1928, Lochhead & Heron, 1929, Wilson & Marvin, 1929, Lochhead & Farrell, 1930a, 1930b, 1931a, 1931b, Marvin, 1930, Marvin et al., 1931, Wilson & Marvin, 1931, Dyce, 1931, Wilson & Marvin, 1932, Lochhead, 1933) studied the problem. Loss and damage to commercial honey was extensive in these areas; an understanding of the factors involved soon led to adequate control methods.

6.11.1 Yeasts and moisture content

It is generally agreed that all honeys contain osmophilic (sugar-tolerant) yeasts in greater or lesser amount, and will ferment if the moisture content is high enough, and the storage temperature suitable, if granulation occurs, if the yeast count is high enough in relation to the moisture content, and if ash and nitrogen contents are favourable.

Numerous strains of these sugar-tolerant yeasts have been isolated

* The samples were obtained experimentally from small colonies caged upon various plantings (Z. Demianowicz, 1964).

from fermenting honey. Table 6.11.1/1 lists those most frequently encountered. The sources of the yeasts have received some attention. Fabian & Quinet (1928) reviewed the earlier literature (from 1884) on the micro-organisms, including yeasts, in nectars. Lochhead & Heron (1929) isolated numerous osmophilic yeasts from nectar. These yeasts also may enter honey from the body of the bee (Klöcker, cited by Fabian & Quinet, 1928) apiary soil (Lochhead & Farrell, 1930a), and honey-house air and equipment (Lochhead & Farrell, 1931b).

Table 6.11.1/1
Yeasts isolated from honey

Type	Reference
Nematospora ashbya gossypii	Aoyagi & Oryu, 1968
Saccharomyces bisporus	,, ,, ,, ,,
,, torulosus	,, ,, ,, ,,
Schizosaccharomyces octosporus	Lochhead & Farrell, 1931b
Schwanniomyces occidentilis	Aoyagi & Oryu, 1968
Torula mellis	Fabian & Quinet, 1928
Zygosaccharomyces spp. (2)	Nussbaumer, 1910
,, barkeri	Lochhead & Heron, 1929
,, japonicus	Aoyagi & Oryu, 1968
,, mellis	Fabian & Quinet, 1928
,, mellis acidi	Richter, 1912
,, nussbaumeri	Lochhead & Heron, 1929
,, priorianus	Fabian & Quinet, 1928
,, richteri	Lochhead & Heron, 1929

It has frequently been recorded that fermentation—when it does occur —almost invariably follows granulation of honey. The removal of dextrose hydrate from solution leaves a higher-moisture liquid phase in which moisture may not be uniformly distributed. In the moisture range found in honey, a small increase in moisture can produce a considerably increased liability to fermentation. In partly granulated fermenting honey, the higher moisture content in surface layers may also result from absorption of moisture from the air above.

In his study of the hygroscopy of honey, Martin (1958) made yeast counts at different depths of the exposed samples. In honey originally containing about 700 000 cells/g, exposed to 66% R.H., where the surface moisture did not exceed 21·5% only very limited aerobic (and no anaerobic) growth took place. The same honey exposed to 86% R.H. developed heavy surface growth within 17–30 days (at which time surface moisture values were 28–29%), though anaerobic growth was quite limited lower down, where moisture content remained at 22·5%. The increase of pollen content at the surface of honey when it absorbs moisture would also favour yeast growth. Krumbholz (1936) has reviewed fermentations of high-density sugar solutions (including honey) and believes that there is no valid upper limit to the sugar concentration that the osmophilic yeasts can tolerate. Lochhead (1933) has summarized their investigations on the relation of moisture content and fermentation as follows, based on 319 honey samples:

Moisture content	Liability to fermentation
below 17·1%	safe regardless of yeast count
17·1–18·0%	safe if yeast count $<$ 1000/g
18·1–19·0%	,, ,, ,, ,, $<$ 10/g
19·1 – 20·0%	,, ,, ,, ,, $<$ 1/g
above 20·0%	always in danger.

Stephen, however (1946), reported in a study of over 700 Canadian honey samples that the incidence of fermentation was greatest in samples containing 17–18% moisture; above 19%, the lowered granulation tendency indirectly retards yeast growth.

6.11.2 Storage temperature
Wilson & Marvin (1931, 1932) recommended that honey be stored either below 52°F (11°C) or above 70°F (21°C), thus defining the range favourable to fermentation. The lower half of this range is that most favourable for honey granulation. In well ripened honey, fermentation is said not to occur at temperatures above about 80°F (26·7°C) (Wilson & Marvin, 1932). Other factors must, however, be considered when thinking in terms of higher temperature storage. Honey in unheated winter storage is more liable to fermentation in the spring, since temperatures then become more favourable, and the honey will usually have granulated during the winter.

6.11.3 Products of fermentation
Fermentation in full-density honey is quite slow, extending over periods of six months to a year. Krumbholz (1936) characterizes osmophilic yeasts as having slight fermenting power (50–70 g alcohol per litre) and a high tolerance to sugar concentration—the opposite of the non-osmophilic yeasts. Marvin et al. (1931) found that in 50% honey solution five honey yeasts completed fermentation in about 40 days and produced 4·0–5·6 g carbon dioxide, 3·8–5·0 g alcohol, and 1·5–3·1 milliequivalents of acidity, over 90% non-volatile. Borries (1934), studying natural honey fermentation, found small amounts of ethanol and CO_2 produced. Only

part of the latter is evolved; the remainder was recovered under diminished pressure. Less than 1% of the sugar was fermented, and no increase in acids could be detected.

It has since been found (Spencer & Sallans, 1956, Spencer & Shu, 1957, Peterson et al., 1958, Hajny et al., 1960) that various osmophilic yeasts isolated from honey can, under suitable conditions, convert 60% of a 10–20% glucose solution to polyols, such as glycerol, D-arabitol, erythritol and mannitol. In fermenting 20% glucose solution, aeration and low phosphate content favour the production of the polyols, whereas in the absence of aeration the production of ethanol is increased and of polyols is greatly reduced. It appears unlikely that polyols would be produced during natural fermentation of honey.

CHAPTER 7

MICROSCOPY OF HONEY

by Dr. Anna Maurizio

FORMERLY RESEARCH WORKER AT FEDERAL BEE RESEARCH DEPARTMENT,
LIEBEFELD-BERN, SWITZERLAND

Bright honey drops from flowers bee–distilled
TIMOTHEUS OF MILETUS (DIED 357 B.C.),
PERSAE

7.1 INTRODUCTION

In many European countries the food laws demand a declaration of the origin of food offered for sale, and this legislation covers honey. It can be implemented only if a reliable method exists for determining the geographical origin; and, for honey, the microscopical investigation of its constituents provides such a method. So honey microscopy has been included in the directives for carrying out food control orders e.g. in Germany (*Erläuterungen zur Verordnung über Honig* 1930, Evenius & Kaeser, 1970), Switzerland (*Schweiz. Lebensmittelbuch*, 1967).

The beginnings of honey microscopy go back to the turn of the century, when Pfister (1895) first suggested the possibility of determining the geographical origin of honey by the grains of pollen contained in it. By that time investigations had already been carried out on the structure and morphology of pollen (Guillemin, 1825, Fritzsche, 1832, Mohl, 1834, Fischer, 1890), and these were now given a new sphere of application.

About ten years later, research work followed on North American honeys by Young (1908) and on Swiss honeys by Fehlmann (1911). Fehlmann's work was especially important as the first indication of a microscopical distinction between flower honey and honeydew honey.

At the start of the 1930s there was an upsurge of interest in the microscopical investigation of honey in Germany, and details were worked out by Armbruster and his collaborators (1929, 1934/35), Griebel (1930/31) and, above all, by Zander (1935, 1937a, 1941, 1949, 1951), whose comprehensive work laid the foundation of current methods for determining the origin of honey samples. The years that followed brought an extension of honey microscopy in various European countries. Honeys from different localities, and typical combinations of pollen grains, were

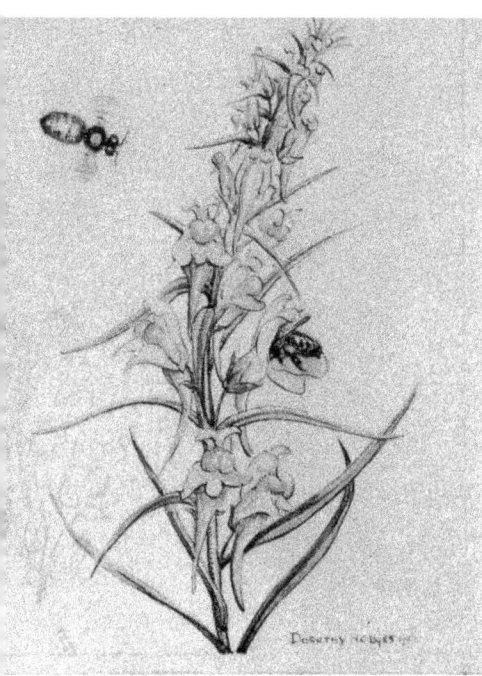

Figure 1.12/4 (see page 10) A honeybee collecting nectar from the long-spurred corolla of toadflax, *Linaria vulgaris* (Scrophulariaceae). As the bee pushes her way into the flower to reach the nectar at its base, her thorax presses hard against the anthers; this compresses the pollen into a yellow 'patch' which the bee carries away on her thorax. A second bee approaching the flower already carries such a 'patch' from a previous visit to another plant of the same species.

Figure 1.12/5 (see page 10) A honeybee collecting nectar from *Helenium* (Compositae). She stands entirely on the central disc florets.

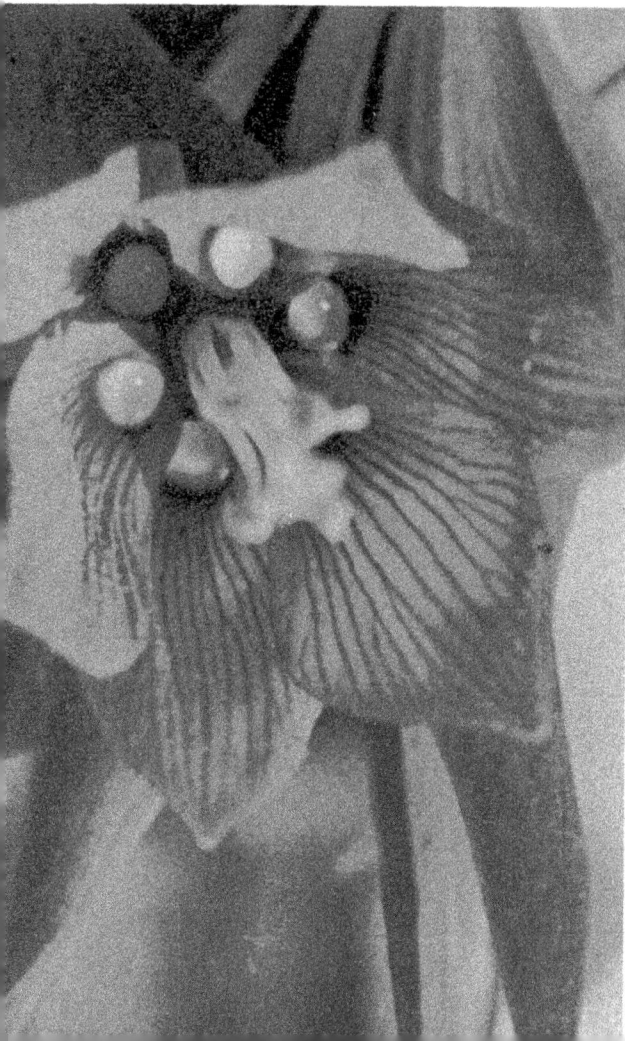

Figure 2.121/2 (see page 79) Floral nectaries of *Fritillaria imperialis*, showing large drops of nectar.
(after Schremmer)

PLATE I

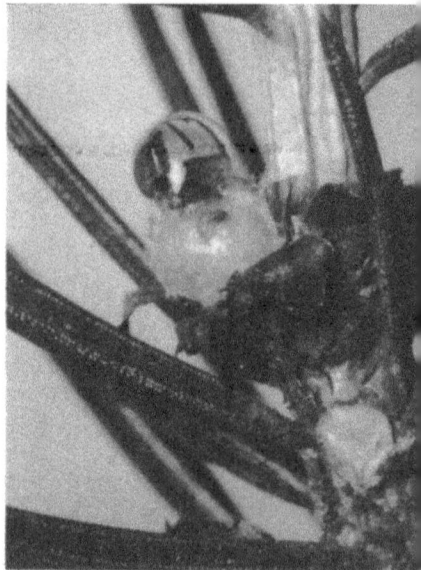

Figure 2.131/2 (see page 94) (above) Buchneria pectinatae, with a drop of honeydew, on Abies alba.
(right) Physokermes hemicryphus, with a drop of honeydew, on Picea abies.　　　　(photo: Hättenschwiler)

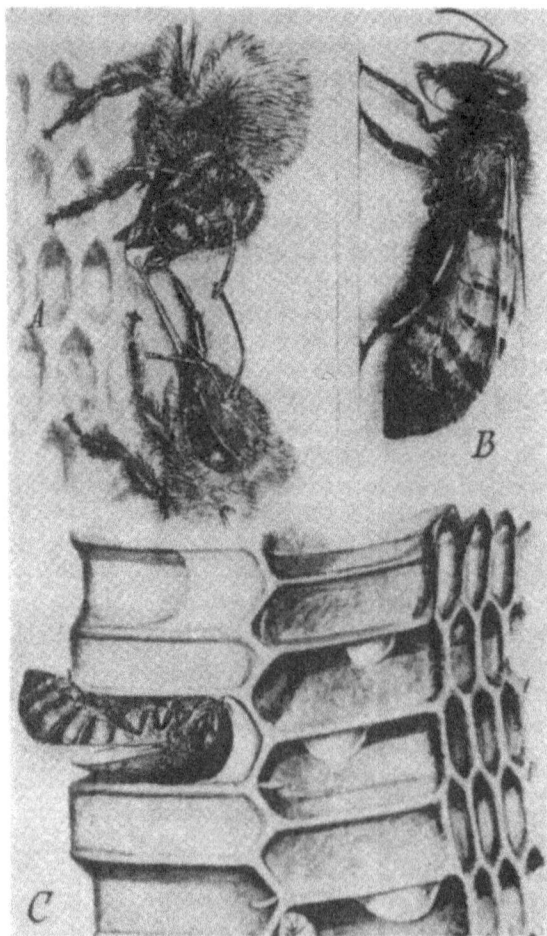

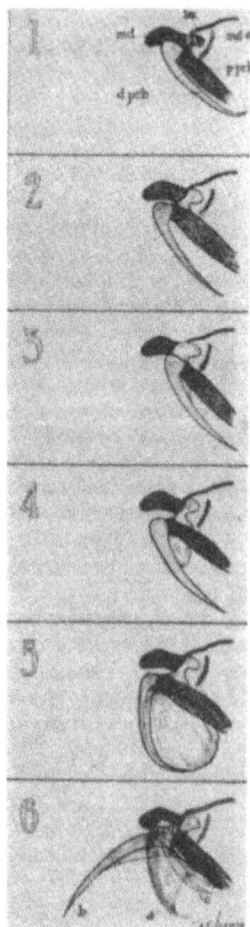

Figure 2.222/1 (see page 103) Honey-ripening process
  A A returning forager passing a drop of nectar from its glossal groove to a house bee.
  B A bee elaborating honey; the nectar drop is mixed with saliva and concentrated.
  C The half-ripened honey is transferred to a cell in the comb.　　(after Park, 1925, 1949)

Figure 2.222/2 (see page 103) Details of the honey-ripening process. A drop of nectar is pumped out rhythmically from the honey sac, spread out thinly, and sucked up again.
(after Park, 1925, 1949)

PLATE 2

Figure 3.13/1 (see page 107)   Beekeeper inspecting a hive in spring; note the hive tool in his right hand.
(photo: F. G. Smith)

gure 3.13/2 (see page 108)   Apiary in jarrah/wandoo woodland, W. Australia.   (photo: F. G. Smith)

PLATE 3

Figure 3.34/1 (see page 111) Migratory apiary in Western Australia.
From left to right: Beekeeper's living caravan; truck with boom loader, settling tank and bulk (44-gallon) containers; honey-extracting caravan with steam boiler at end; stack of empty hive boxes (supers). (photo: F. G. Smith)

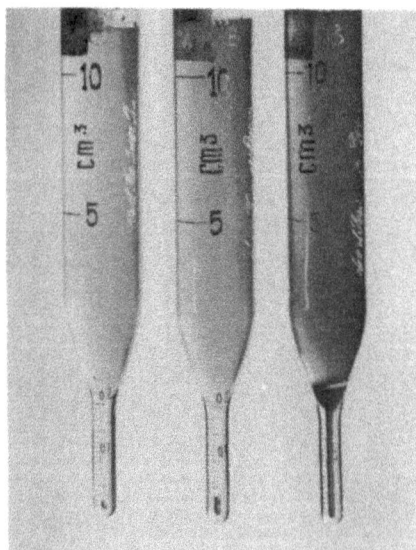

Figure 7.3/1 (see page 243) Determining the amount of sediment in honey.

Trommsdorff leucocyte tubes, each containing the sediment from 10 g honey: left and centre, extracted honeys; right, pressed honey. (photo: Liebefeld)

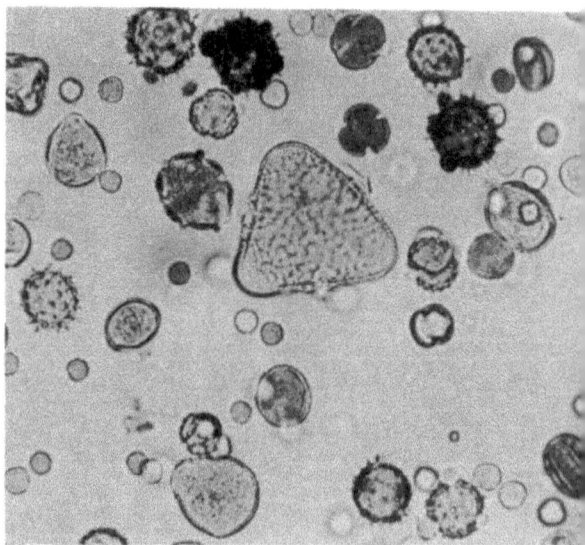

Figure 7.48/1 (see page 251) Pollen spectrum of a hone from South America (pollen grains of *Mimosa pudica Bombax* sp., Compositae H and Labiatae M). × 30 (photo: Liebefeld

PLATE 4

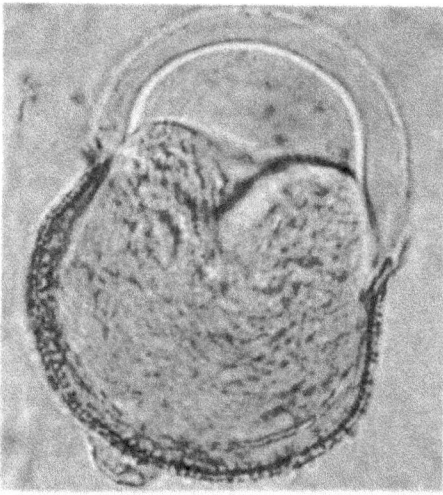

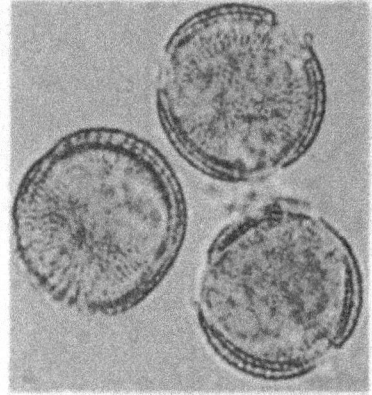

(a)

(b)

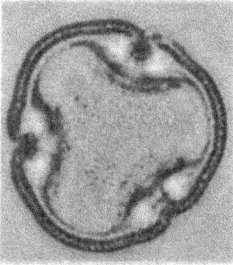

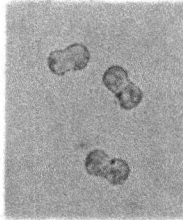

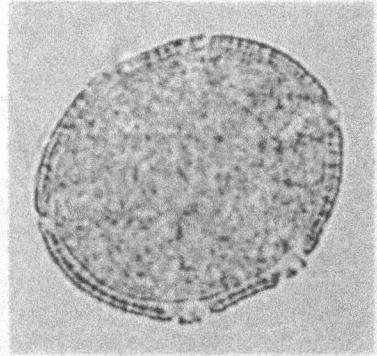

(c)

(d)

(e)

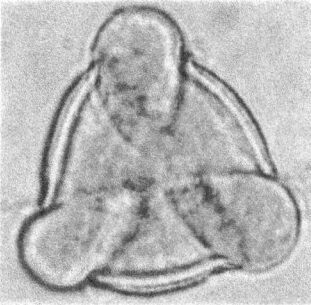

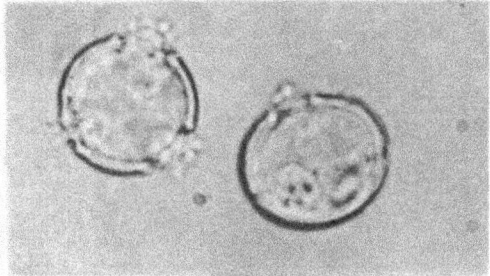

(f)

(g)

Figure 7.42/2 (see page 247) Pollen grains of various shapes (× 1 000).
  (a) *Iris sibirica* (side view, 1 furrow, Liliaceae type)
  (b) *Brassica napus* (top and side views, 3 furrows, exine with reticulate structure;
      Cruciferae type)
  (c) *Tilia* sp. (top view, 3 germinal pores)
  (d) *Myosotis* sp. (side views, 3 furrows situated in the 'waist')
  (e) *Salvia splendens* (top view, 6 germinal furrows, Labiatae S type)
  (f) *Acer pseudoplatanus* (top view, 3 germinal furrows)
  (g) *Aesculus hippocastanum* (3 germinal pores, situated in furrows and covered
      with incrustations)   (photo: Liebefeld)

PLATE 5

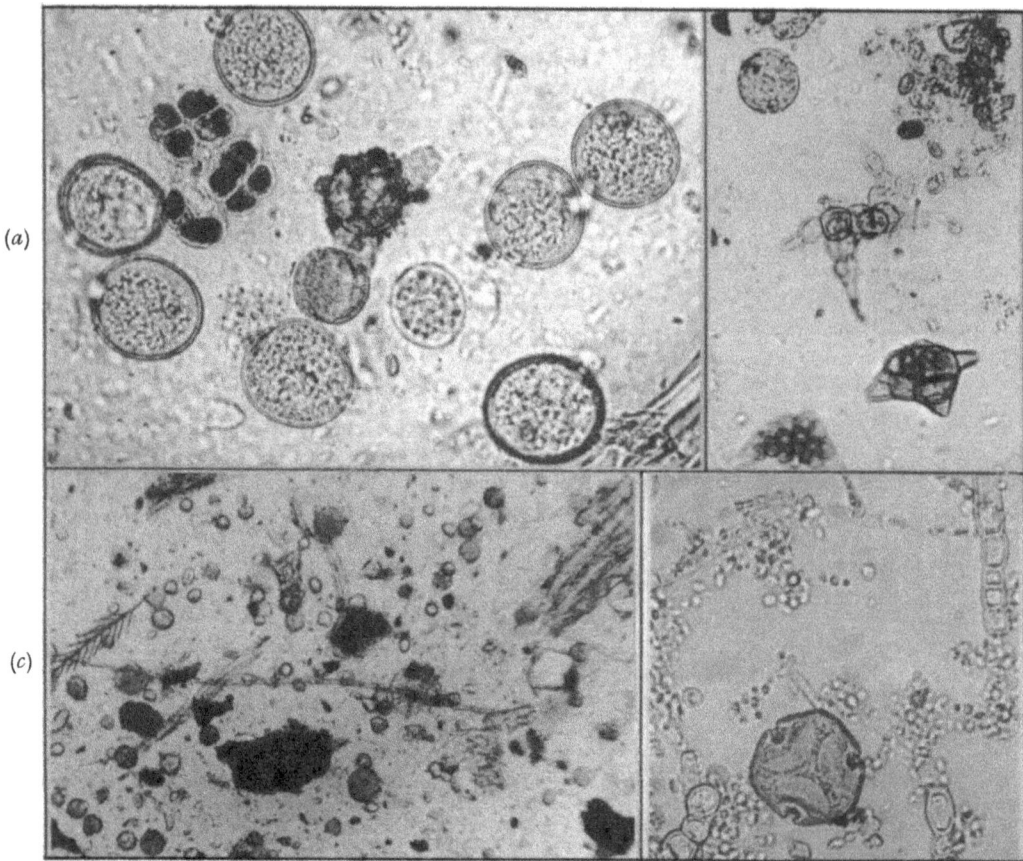

Figure 7.43/1 (see pages 247, 248) Photomicrographs of honeydew honey, and contaminated fermented honeys: (*a*), (*b*) honeydew honeys (algal cells, variously shaped spores of sooty moulds, pollen grains from anemophilous plants) × 500; (*c*) badly contaminated honey (soot particles, bee hairs, fragments of wood, as well as pollen grains) × 111, (*d*) fermented honey (yeast cells, chlamydospores of *Pericystis alvei*) × 333.   (photo: Liebefeld)

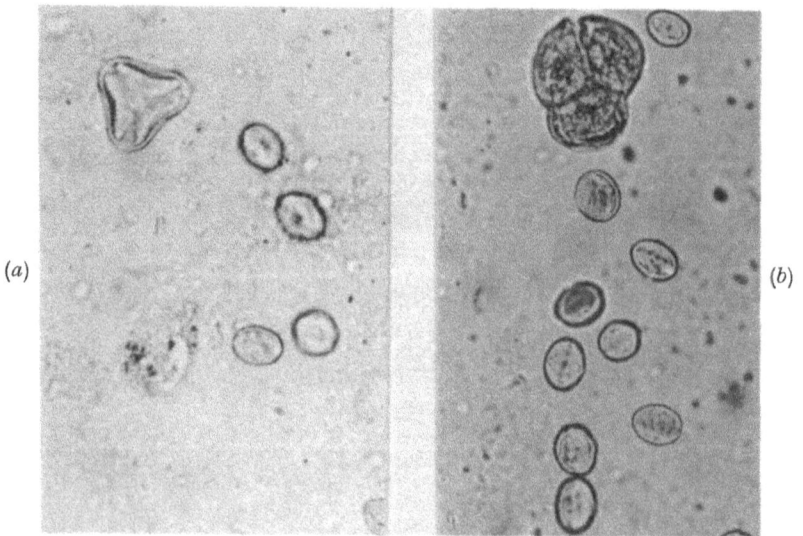

Figure 7.48/2 (see page 252)   Pollen spectrum of honeys from sweet chestnut: (*a*) Honey from Yugoslavia (pollen grains of *Castanea sativa* and *Loranthus europaeus*) × 370; (*b*) Honey from southern Switzerland (pollen grains of *Castanea sativa* and *Calluna vulgaris*) × 370.

PLATE 6   (photo: Liebefeld)

Figure 9.23/1 (see page 272) Uncapping/extracting outfit for beekeeper with up to 500 colonies. *Left to right:* steam-heated uncapping knife fixed above tray, with pipe to baffle-tank; revolving stand for uncapped frames; 30-frame radial extractor into which uncapped frames are loaded; supers of extracted frames. (photo: Ontario Dept. of Agriculture & Food)

Figure 9.23/2 (see page 272) Automatic uncapper: the comb passes slowly between two vibrating knives; A. R. Mallory, Stirling, Ontario, Canada. (photo: University of Guelph)

PLATE 7

Figure 9.24/1 (see page 274)   Radial extractor in operation.

Figure 9.33/1 (see page 276)   Tubular heat-exchangers and pressure strainer at Honingcentralen, Oslo, Norway.

PLATE 8

described; moreover, critical assessments were made of the methods themselves, their possibilities and limitations. At the same time closer contact was established with the other fields of pollen research which, under the designation 'palynology', had formed an independent scientific discipline since 1945. Honey microscopy is now linked with this as a sub-section 'melissopalynology'. As time went on, pollen analysis was established as a useful auxiliary technique in other spheres of bee research: the investigation of nectar and pollen sources of colonies of bees in different regions; the estimation of the value of the nectar and pollen yields from certain species of plants; diagnosis of bee poisoning, and so on.

The microscopical determination of the origin of a honey is based on the fact that every natural honey contains microscopic particles. Some of these constituents get into the raw plant materials of honey (nectar and honeydew), and some are contributed later—during the ripening processes in the colony, and during the harvesting of the honey by the beekeeper. This statement does not apply to honeys from which all solid components have been eliminated by excessive filtration during processing. In order to avoid misleading statements on the origin of honey, recent Swiss food regulations, for instance, stipulate the minimum mesh size for honey strainers: 0·2 mm (Verordnung über den Verkehr mit Lebensmitteln, Art. 217/3, 1971).

When a bee visits a flower, she comes into more or less close contact with the anthers, according to the structure of the flower. Some of the ripe pollen falls into the nectar, is sucked up with it into the bee's honey sac and thus gets into the ripening honey in the cells of the comb, and can still be found in the extracted honey. And just as the nectar in honey is characterized by the pollen grains from the flowers, so honeydew as a source of honey is characterized by minute green algal cells, and spores of sooty moulds, from the surface flora of the plants.

Apart from this direct and—from the point of view of determining the origin—most important way in which pollen grains and other solid constituents can get into the raw materials of honey, and thence into the honey itself, there are other indirect ways. For instance, many pollen grains stick to the hairs on the bee's body during visits to flowers, and these grains can enter directly into the unripe honey lying in open cells in the hive (Louveaux, 1959, Maurizio, 1952). It is also possible for pollen grains suspended in the air to fall on to the exposed surface of nectar, or on to sticky honeydew lying on leaves, and needles of conifers. It is chiefly the pollen grains of wind-pollinated plants that are introduced in this way, and their presence thus characterizes certain European honeydew honeys. Finally, pollen can get directly into honey during its processing by the beekeeper. This occurs extensively during the ex-

H—9

traction of honey from the combs by primitive pressing methods, or when honey is extracted from brood combs which contain much pollen. Sometimes cells of pollen occur in combs in the honey chamber, and this pollen can get into the honey during extraction. This is especially true for heather honey if combs are treated with a loosener before extraction (Evenius, 1933, 1958, Louveaux, 1966a, Lunder, 1955, Zander, 1932).

The problems of honey microscopy can be divided into the following sections, which are now treated in detail, according to the information at present available:

Determination of the sediment content of honey;

Determination of the geographical origin of honey (qualitative pollen analysis);

Determination of the botanical origin of honey (quantitative pollen analysis).

7.2 APPARATUS AND REAGENTS USED IN HONEY MICROSCOPY

The following are needed:

laboratory centrifuge 2 500–3 000 rpm

centrifuge tubes of 10–50 ml capacity (for quantitative pollen analysis, tubes calibrated at 100 ml and 10 ml)

test-tubes calibrated at 20 ml

leucocyte tubes (Trommsdorff type)

platinum loop or Pasteur pipettes (Breed pipettes for quantitative pollen analysis)

water-operated or other vacuum pump for quantitative pollen analysis

Pyrex micro-filtration apparatus of the Millipore Filter Corporation XX 10 025 00, or filter holder SX 00 025 00 and 50-ml glass syringe, filters 25 mm in diameter with pore-width 5, 3, and 1μm; Millipore immersion oil, refractive index 1·515 (for quantitative pollen analysis)

slides and glass covers (20 × 20 mm or larger)

microscope with mechanical stage, 320–450 × and 800–1 000 × water-bath

thermostatically controlled oven or electric hot plate (40–45°C, 104–113°F)

ethyl ether

Kaiser's glycerine gelatine, which is made as follows: 7 g gelatine are cut up and soaked for 2 hr in 42 ml distilled water; then 50 g glycerine (glycerinum conc. d = 1·26) and 0·5 g crystallized phenol are added, stirring all the time. The mixture is heated (not boiled) for 15 minutes and then filtered through moistened glass wool.

Canada balsam (diluted with xylol) or a suitable sealing lacquer.

7.3 DETERMINING THE CONTENTS OF THE SEDIMENT IN HONEY

The amount of solid constituents is less characteristic of the *origin* of the honey than of the *method of extraction*. It is determined as follows (Figure 7.3/1—*see* Plate 4):

Weigh out 10 g of well mixed honey in a test-tube with a calibration mark at 20 ml. Fill up to the mark with distilled water, and dissolve the honey, using a water-bath. Centrifuge the solution in a Trommsdorff leucocyte tube for two periods of 5 min. The sediment from 10 g honey collects in the lower calibrated part of the tube.

In Germany and Switzerland, the average amount of sediment of carefully harvested and extracted honey lies between 1·4 and 2·0 μl per 10 g (Evenius, 1933, 1958, Evenius & Focke, 1967, Maurizio, 1939, Zander, 1932). Certain honeys derived mostly from one species, e.g. cow parsnip (*Heracleum*), heather (*Calluna*), and certain honeydew honeys, often contain rather more than the average amount of sediment for flower honeys. As a rule the solid constituents of extracted honeys do not exceed 10 μl per 10 g. Carelessly produced pressed honey, honey extracted by heating the combs, and honey 'dripped' from a cloth bag containing loose combs, usually contain substantially more sediment; in extreme examples this may be more than 1 ml/10 g (Gassparian & Vorwohl, 1974). A high content of sediment is also characteristic of fermented honeys; here the sediment consists largely of yeast cells.

7.4 DETERMINING THE GEOGRAPHICAL ORIGIN OF HONEY (QUALITATIVE POLLEN ANALYSIS)

The pollen grains and other microscopic constituents are distributed homogeneously throughout the honey, and change position only very slowly (Goillot & Louveaux, 1955). As a result of the low water content and high sugar content of honey, pollen grains in it present a swollen appearance. The technique for producing honey-sediment preparations and comparative standard preparations (reference slides) must take this into account. The solid constituents must be removed from the honey and concentrated into a suitable preparation; on the other hand the pollen taken directly from the anthers must be freed from fat and brought to the same condition as that in the honey preparation. There is, however, no need for preparatory fossilization (as is usual in other branches of palynology) when dealing with honey and its comparative standard preparations.

7.41   Methods and techniques

These have been worked out by a number of specialists and published by the International Commission for Bee Botany of the I.U.B.S. (1962, 1970).

Honey-sediment preparations

Ten grams of well mixed honey placed in a test-tube or other glass vessel in a water-bath (at about 40°C, 104°F) are dissolved in 20 ml cold distilled water. The solution is centrifuged for 5 minutes and the supernatant liquid poured off, leaving only a few drops with the sediment. The sediment is stirred with a heated platinum loop or a Pasteur pipette and a drop of it placed on a slide and spread out with the loop (or blown out of the pipette) to a surface of approximately 20 × 20 mm. The smear is left to dry in an oven or on a hot-plate (not above 40°C) and mounted permanently with a cover-slip spread with a drop of liquid glycerine jelly. After a few weeks the preparation is sealed with liquid Canada balsam or a suitable lacquer.

Reference slides

Fresh pollen from ripe anthers is placed on a slide or a watch-glass, and freed from fat with a drop of ether; for small flowers whole anthers can be used. The remains of the anthers are removed after evaporation. The degreased pollen is embedded in glycerine jelly to make a permanent slide—a drop of glycerine jelly melted in a water-bath is put on a cover-slip, and this is placed over the layer of pollen. The finished preparation is sealed with a dilute solution of Canada balsam (diluted with xylol) or a suitable lacquer.

With the passage of time this basic method has been modified and supplemented to some extent. For honeys with a high colloid content, for instance, dilute sulphuric acid (5 g concentrated acid to 1 litre water) can be used instead of water for diluting the honey. If so, the sediment must be twice washed with water after centrifuging, and then centrifuged again. In some countries it is usual to stain the honey sediment and the reference slides, either directly or by the use of stained glycerine jelly. The dyes basic or acid fuchsin, gentian violet, methyl green, saffron, etc. are used. Instead of Kaiser's glycerine jelly, other glycerine-jelly products or other mounting media, such as euparal or diaphan, are occasionally used (Brown, 1960). In subtropical and tropical areas with high temperatures and humidities (in which glycerine jelly preparations easily melt and become mouldy), it has been found useful to seal the pollen in glycerine or liquid jelly with a ring of paraffin between the slide and cover-slip (Erdtman, 1966, 1969).

Louveaux (1961) has described a filter method for making honey-sediment preparations by using Millipore filters. The same author has devised a method for the microscopical determination of honey in foods such as gingerbread, Louveaux, 1954, *see also* Sturm & Hanssen, 1961).

The pollen grains and other microscopic components of the honey preparation are identified and counted. A microscope fitted with a mechanical stage, and having a magnification of about 320–450, is best suited for routine work, but for scientific work more powerful oil-immersion object glasses should be used, giving a magnification 800–1 000 ×.

7.42 Pollens

The identification of the pollen grains is based on their size and shape. In honey their swollen condition is an important factor, since many details are visible in the swollen grain which cannot be detected in dry pollen.

Details of pollen morphology are not within the scope of this book, so only the general principles are given here. Details are available in the literature (Beug, 1961, van Campo, 1954, 1967, Erdtman, 1966, 1969, 1970, Erdtman & Vishnu-Mittre, 1958, Evenius & Focke, 1967, Faegri & Iversen, 1964, Faegri & Ottestad, 1949, Hodges, 1952, Hyde & Adams, 1958, Ikuse, 1956, Iversen & Troels-Smith, 1950, Kremp, 1968, Louveaux, 1968a, 1970, Maurizio & Louveaux, 1965, 1967, Shimakura, 1973, Wodehouse, 1935, Zander, 1935–1951).

Pollen grains, the male reproductive cells of higher plants, as a rule appear as single cells, but they are often found in groups. On the pollen grain (Figure 7.42/1) two perpendicular axes are recognized, the polar axis (P) and the equatorial axis (E). The contents of the pollen grain are mostly enclosed within a double wall. The inner wall—the intine, consisting of cellulose—lies immediately next to the plasma body. The outer wall—the exine—generally consists of two layers; the inner layer is the nexine or endexine, which is fairly homogenous, and the multiple outer layer, sexine or ektexine, which is composed of a complex of rods (columellae, bacula).

In a few plant species the exine forms a closed layer round the contents of the pollen grain, but most pollen grains possess openings in (or thin parts of) the exine, through which the pollen tubes generally emerge. They can be in the form of furrows (colpus, length/breadth ratio greater than 2) or pores (porus, length/breadth ratio less than 2). The germinal apertures also allow distension of the grain, leading to swelling, when it absorbs water. Number, position, formation, and size of the germinal apertures constitute characteristics for determining the pollen.

The surface of the exine is sculptured in characteristic ways. Distinction is made between knobs (verrucae), grains (gemmae), rods (bacula), clubs (clavae), spines (echini), borders (valla), network (reticulum), etc.

The size of a pollen grain is greatly dependent on the mounting medium, and in particular on its water content, so comparative measurement should be made only on pollen grains in the same condition of swelling, i.e. by using the same mounting medium. In bipolar pollen grains the polar and equatorial axes are measured, i.e. the distance between the two poles (P), and the distance between two planes lying parallel to the polar axis touching the grain equatorially (E). Very often two mutually

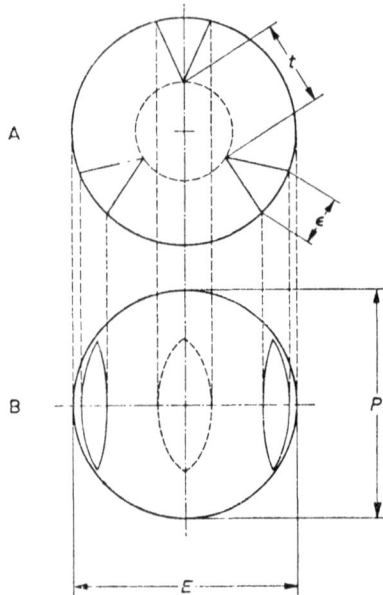

Figure 7.42/1 Drawing of a pollen grain with three furrows. (A) from above; (B) from the side.

$P=$ polar axis, $E=$ equatorial axis, $\epsilon=$ diameter of the furrow in the equatorial plane, $t=$ distance between the furrows
(drawing: Louveaux (from Maurizio & Louveaux, 1965)

perpendicular equatorial axes are measured (E_1, E_2). With pollen grains of other shapes the greatest diameter and the diameter perpendicular to it are specified. The relative magnitudes are important, especially the pollen-form index; in bipolar grains this is the ratio between the polar axis and the greatest equatorial axis $(P:E)$. The following classes are distinguished according to the value of the pollen-form index: pollina perprolata $(P:E>2 \cdot 0)$, prolata $(P:E = 2 \cdot 0 - 1 \cdot 33)$, subsphaeroidea $(P:E =$

1·33–0·75, with the subclasses subprolata, sphaeroidea, and suboblata), oblata (*P:E* 0·75–0·50), peroblata (*P:E*<0·5) (Erdtman, 1966, 1969, 1970, Faegri & Iversen, 1964, Iversen & Troels-Smith, 1950, Kremp, 1968).

The characteristics of a pollen grain include also the dimensions of the germinal apertures (length and width for furrows, diameter for pores); thickness of the intine and exine; height and width of the exine structures (columellae), and the sculpturing of the exine. Illustrations of pollen grains should comprise a number of photographs taken from different optical sections (foc.0–foc.5, Iversen & Troels-Smith, 1950, Kremp, 1968). *See* Figure 7.42/2—Plate 5.

7.43 Honeydew constituents

Among the other microscopic components of the honey sediment, algae and sooty moulds are especially important as 'honeydew indicators' (*see* Figure 7.43/1—Plate 6). The algal cells come chiefly from the bark algae from the surface flora of living leaves and conifer needles. The most numerous are clumps of *Pleurococcus*; less common are cell-aggregates of *Chlorococcus* and single cells of the *Cystococcus* type (Maurizio, 1959*a*). Among the honeydew algae there are sugar-tolerant species as well as species sensitive to sugar, so there may be an increase of algae in the honeydew layer. The appearance of green algae in honey sediment seems to depend on climatic factors. In central and northern Europe green algae are regularly present in honeydew honeys from conifers and deciduous trees, but they are often absent in honeydew honey from dry areas, such as *Cedrus* and *Quercus* species. Brazilian honeydew honeys are poor in algae and fungus spores (Barth, 1971*b*). Honeydew honey from larch (*Larix decidua*) is often poor in algae; on the other hand those from spruce and fir (*Picea abies* and *Abies alba*) in the Alpine regions are particularly rich in algae.

Occasionally other algae are found in honey sediment, for instance Chlorophyceae, Cyanophyceae, Diatomaceae, and Desmidiaceae, which probably come from the water and have no significance as to the origin of the honey. Honey from some countries, that has been filtered under pressure, often contains large numbers of sea silico-flagellates and Diatomaceae introduced from the diatomaceous-earth filters.

Among the fungus spores appearing in honey sediment, several groups can be differentiated. Only sooty moulds (*fumago*) are true 'honeydew indicators'. According to Neger (1918) 'sooty moulds' are fungi which have dark mycelia; they are purely epiphytes on living leaves and twigs, and do not penetrate into the plant tissues either by haustoria or by the mycelium. These fungi belong chiefly to the group of *fungi imperfecti*, of whose systematic relationship little is known. A mass development of sooty moulds occurs only in the presence of honeydew. The high sugar content of honeydew brings about selection among the epiphytic fungi, and only the osmophilic species, which can live and multiply in highly concentrated sugar solutions, count as true sooty moulds.

Neger (1918), Reynolds (1971), and Zander (1949) described important kinds of sooty moulds which occur in honeydew and honeydew honeys. Only a few of them can be identified with certainty. According to Zander *Coniothecium* species are dominant in the honeydew of deciduous trees, and species of *Atichia, Hormiscium*, and *Triposporium* types in the honeydew of conifers. The sediment of honeydew also contains many other types, of whose systematics nothing is known—they were designated by Zander according to their shape: crescent, horned, club, many-tipped, etc. [*See* Note on page 257.]

Besides sooty moulds, one occasionally finds in honey sediments hyphae and spores of fungi associated with bees (e.g. *Pericystis alvei*, Betts = *Ascosphaera alvei* = *Bettsia alvei*, pollen mould, Skou, 1972), and of plant-pathogenic fungi: Uredinaceae (rusts), Ustilaginaceae (smuts), Peronsporaceae, etc. (Maurizio, 1959*a*). Such components have little significance in determining the sources of the honey.

7.44 Other microscopic constituents

Other microscopic constituents of plant or animal origin which may be found in honey sediment are calcium oxalate crystals, soot, yeasts, small wax elements of honeydew-producing insects, plant and insect hairs, butterfly scales, particles of chitin, and of tissue from bees, mites, etc.

Some of the above particles originate in the raw materials of the honey, and are characteristic of its origin. For instance one often finds a thick sediment of calcium oxalate crystals in pure *Tilia, Castanea* and *Mentha* honeys; these crystals are in the form of macles or envelopes, and can also be identified in the nectar of these plants (Z. Demianowicz, 1963, Maurizio, 1963). The sediment of *Heracleum, Calluna*, and *Castanea* honeys is characterized by the presence of a 'finely granulated substance' which cannot be defined more closely and may possibly consist of bacteria (Evenius, 1933, Zander, 1935, 1937*b*). Honeydew honeys from the vicinity of large towns and industrial centres can occasionally contain a thick sediment of soot particles. A direct indication of this environment as the origin of the honey is the presence of the small wax elements of *Physokermes hemicryphus* in the honey sediment (Gontarski, 1951). Isolated starch grains of the potato or cereal type often occur in honey; a large number of aleuron grains indicate that the honeybee colonies have been fed pollen substitutes (soya-bean flour) at an improper time (Fossel, 1968, Vorwohl, 1966).

Other particles can lead to inferences as to the condition and method of harvesting the honey. For instance, fermenting honey contains many germinating yeasts, occasionally also hyphae and spores of the pollen mould (*Pericystis alvei* Betts, *see* page 248). Bacterial fermentation, which is less usual, is demonstrated by an abundance of bacteria. The sediment of honey produced by primitive pressing procedures often contains not only a large amount of pollen, but also particles of chitin and of tissue from adult bees and brood (tracheae, fat-body cells, etc.) and of wood, fibres, etc.

7.45 Counts made on the honey preparation

The counting will vary according to its purpose, and we distinguish between 'orientation analysis' and 'complete analysis'. An orientation analysis consists of the identification of the predominating elements and a search for certain characteristic constituents. Complete analysis includes the determination of all the constituents present in the sediment, and establishing their relative proportions by counting. For this purpose 100–300 pollen grains are counted, together with the corresponding honeydew constituents. Grains of anemophilous and entomophilous nectarless plants are recorded separately, and the honeydew constituents (sooty moulds and algae) separately from the pollen grains. The percentages of individual constituents are then worked out. The variability of results, in relation to the number of samples and the number of pollen grains counted, has been statistically investigated by Faegri & Ottestad (1949) for soil samples, and by Vergeron (1964) for honey. According to Vergeron, frequency percentages are statistically significant only if 1 200 pollen grains are counted. With honeydew honeys, which contain little pollen, it is sufficient to count 100–150 pollen grains and the corresponding honeydew constituents. After the count the different components are separated into frequency classes: predominant pollens, more than 45%; secondary pollens, 16–45%; minor pollens 1–15%. With honeys in which one form predominates (e.g. *Castanea*, *Myosotis*) a second count excluding the predominant pollen is recommended. Further details of counting are given by the International Commission for Bee Botany (1970).

7.46 Evaluation of the microscopic analysis of honey

Identifying and counting the microscopic constituents of honey sediment allow conclusions to be drawn as to the plant habitat in which a honey has been produced by the bees, and therefore as to the geographical origin of the honey. The most reliable clues are the pollen grains of entomophilous nectar-yielding plants, as they supply direct indications of the sources of the raw material. But also pollen grains of entomophilous plants that yield little or no nectar (e.g. *Papaver*, *Helianthemum*, *Cistus* species, *Olea europaea*, *Filipendula ulmaria*) and pollen grains of anemophilous plants (e.g. Gramineae, *Rumex* and *Plantago* species, *Cannabis*, *Humulus*, *Morus*, *Betula*, *Corylus*, conifers) can help to identify the geographical origin of the honey (Maurizio, 1959a, Zander, 1937b).

It is not only the most numerous pollen grains in a honey sample (the predominant and secondary species) that are important, but also minor pollens and, above all, the combination and proportion of all grains— the *honey type*—are characteristic. Corresponding to the climatic, geographical, botanical, and agricultural conditions, several types of honey are harvested in one region, often in a small area, and each of these types has a microscopic picture which may be sufficiently characteristic to permit the identification of the geographical location of a honey.

7.47   Honey types

A thorough study of the honey types in different countries constitutes the basis for determining the geographical origin of samples of unknown honeys. The number of honey types which are harvested in any one area is relatively limited. The description of European honey types is already well developed; honeys from outside Europe have been studied to some extent in European institutes as imported honeys have become available. More recently a start has been made with similar studies in countries of North Africa, North and South America, Asia, and Australia (Arai *et al.*, 1960, Austin, 1958, Barbier, 1958, 1963, Barbier & Pangaud, 1961, Barth, 1969, 1970a, b, c, 1971a, b, de Boer, 1933, Bozchilova & Anchev, 1967, Chaubal & Deodikar, 1965, Deans, 1957, Deodikar & Thakar, 1953, Drimtzias, 1968, Evenius, 1932, 1953, 1958, 1960, Evenius & Focke, 1967, Focke, 1968, Fossel, 1955, 1956, 1958a, 1958b, 1960, 1962, 1966, 1970, Génier, 1966, Griebel, 1930/31, Hammer *et al.*, 1948, Hazslinsky, 1938, 1943, 1952, Kropáčová, 1969, Louveaux, 1956b, 1966a, 1966b, 1968a, 1968b, 1970, Lunder, 1945, 1955, Martens *et al.*, 1964, Martimo, 1945, Maurizio, 1936a, 1936b, 1940, 1941, 1946, 1947, 1949a, 1951, 1955, 1958a, 1960a, 1960b, 1966, 1971, Maurizio & Louveaux, 1965, Nair, 1964, Panelatti, 1961, Patterson & Bach, 1968, Pelimon, 1960, Poszwiński & Warakomska, 1969, Pritsch, 1957a, 1957b, 1958, Ruttner, 1956, 1961, 1964, Santos, 1961, Sabatini, 1973, Serwatka, 1958, Tone & Coteanu, 1969, Vergeron & Louveaux, 1964, Vieitez, 1948, Vishnu-Mittre, 1958, Vorwohl, 1968b, 1970, Wolthers, 1955, 1956, Woźna, 1966, Zander, 1935–1949).

7.48   Determination of origin

The knowledge of honey types, and of the pollen combinations characteristic of them, now permits a certain distinction between honeys of European origin and honeys from other continents. More than this, it enables us to define honeys from different areas of Europe, and to detect mixtures of honeys from different locations. The analysis of commercial honey blends made up of honeys from several localities requires a thorough knowledge of the honey-flow conditions of the whole country, because one is not dealing with a single honey type but with a combination of several honeys produced under different conditions. Honeys from which the pollen and other plant constituents have been removed by pressure-filtering through diatomaceous earth or similar material are an exception, because they contain no definable trace of their origin.

Distinguishing between honey imported from overseas and Central European honey is one of the easiest tasks for the pollen analyst. 'Infallible indicators' of oversea honey (*see also* Griebel, 1931, Zander, 1937*a*, 1941) are the presence of Magnoliaceae, *Mimosa pudica*, and Compositae of the H and S types as predominant or secondary pollens. These are especially characteristic of honeys of Central and South America (Figure 7.48/1—*see* Plate 4). *Eucalyptus* species (and other Myrtaceae) are chiefly found as predominant pollens in honeys from South America, Australia, and Africa, but also in honeys from the Mediterranean region. Honeys from Florida are characterized by the combination of *Ilex-Nyssa-Citrus-Serenoa* pollens. Characteristic minor pollens in honey from overseas are the following: *Acacia* (also from the Mediterranean area), *Bombax, Bursera, Canna, Commelina, Bravaisia, Coffea, Manihot, Croton, Gossypium, Musa, Eriogonum* (California, Mexico) Convolvulaceae, Caesalpiniaceae (Africa), Polemoniaceae, *Ephedra* (also Mediterranean), and others.

It is more difficult to identify the provenance of honeys from different parts of Europe. Only a well-grounded knowledge of the honey types of one's own country makes a correct diagnosis of these honeys possible, as the following example shows. Some years ago large quantities of so-called forest honey were exported from Yugoslavia to countries of Central and Western Europe. Results of a microscopical examination suggested that it was a honey harvested from sweet chestnut and honeydew, and therefore difficult to distinguish from certain honey types from other areas of Europe. Sweet chestnut honeys are harvested in Europe: in France, Spain, Italy, south and west Switzerland, south Germany, Austria, Yugoslavia, Rumania, and occasionally also in the Netherlands, Belgium, Denmark, and England. In all these honeys pollen of *Castanea sativa* predominates, forming even as much as 99% of the total. The secondary pollens make it possible to distinguish several geographical types of this *Castanea* honey.

Thus, for instance, *Castanea* honey from upper Italy and southern Switzerland is characterized by the pollen combination *Castanea-Trifolium repens-Rubus-Centaurea jacea-Heracleum-Luzula*, often accompanied by *Robinia, Tilia, Chamaerops*, and *Vitis*. In the southern valleys of the Alps there are additions of pollens from mountain plants such as *Rhododendron, Lotus (corniculatus), Campanula, Helianthemum, Polygonum bistorta,* and *Myosotis.* Honey from Styria shows the combination *Castanea-Centaurea jacea-*Leguminosae (*Trifolium repens, T. pratense, Onobrychis, Lotus*)-Rosaceae-*Cucurbita-Aesculus-Myosotis;* in honeys from western Switzerland and the French border area, *Castanea* and Leguminosae pollens are accompanied by those from Cruciferae, Rosaceae, Umbelliferae (*Heracleum, Daucus, Pastinaca, Astrantia), Cornus sanguinea, Echium,* and *Centaurea jacea. Castanea* honeys from the Pyrenees and the Alpes Maritimes contain, besides *Castanea,* a distinctly Mediterranean addition, with *Erica* sp., *Olea,* Labiatae, *Cistus,* often *Calluna* and *Eucalyptus.* South-eastern European imported honeys are recognized by the pollen combination *Castanea-*honeydew constituents-*Robinia-Onobrychis-Trifolium incarnatum-Centaurea cyanus-C. jacea-Fagopyrum-Cannabis-Verbascum-Mentha pulegium.* Occasionally these honeys contain pollen grains of *Loranthus europeaus;* this is a mistletoe native to south-eastern Europe, but absent in western and central Europe (Maurizio, 1960b, Ruttner, 1961, 1964). So the presence of these pollen grains in the sediment of an unknown honey sample is adequate evidence that it originated in south-eastern Europe. (*See* Figure 7.48/2—Plate 6.)

In the same way it is possible to distinguish between *Calluna* heather honeys from different areas of Europe (north Germany, Poland, Lithuania, Denmark, Netherlands, Scandinavia, British Isles, and Atlantic coastal areas of France, Spain, and Portugal). Orange-blossom honeys of different origins (Florida, California, Spain, southern Italy, Israel, North Africa) can similarly be distinguished. Many more examples could be quoted in which microscopical examination can localize the geographical origin of related honeys.

7.5 DETERMINING THE BOTANICAL ORIGIN OF HONEY (QUANTITATIVE POLLEN ANALYSIS)

Even today, it is essentially easier to define the geographical origin of a honey sample than to determine its botanical origin, i.e. the proportions contributed by individual plant sources.

The difficulties in allocating the plant proportions arise particularly from the fact that the number of pollen grains that get into the unit volume of nectar and honey depends on the abundance of anthers and pollen, and on the structure and floral biology of different plant species. Some flowers yield much pollen, and their nectar is exposed (like *Castanea*). Others are very small (like *Myosotis* and other Boraginaceae). The nectar of these flowers receives a much greater dusting with pollen than the nectar of flowers with a wide calyx and which yield much nectar (like *Robinia*), or flowers whose nectar is separated from the anthers by special devices (*Salvia* and other Labiatae). Certain cultivated plants (lavandin, some *Citrus* species, etc.) present a special case in that their anthers yield little pollen or are completely sterile.

The water content of the nectar, the distance of the hive from the forage source, and beekeeping practices, can all influence the number of pollen grains in the unit volume of honey. If bees collect very dilute nectar (from which much water must be evaporated), the honey produced will be richer in pollen than that from more concentrated nectar (in which the pollen content per unit volume is little changed). The action of the proventriculus also affects the amount of pollen in the honey sac, since it can withdraw solid components from the honey sac, while hardly reducing the amount of fluid (Bailey, 1952, 1954, Maurizio, 1949*b*, Todd & Vansell, 1942). Accordingly, the further the source of forage from the hive, i.e. the longer the bee takes to fly home with her full honey sac, the poorer in pollen the honey will become.

It has been appreciated for a long time that the percentages of different pollens in the honey sediment do not correspond exactly to the proportions of nectar contributed to the resultant honey by the individual plant species. This lack of correspondence is not very important when determining the geographical origin of honey, but it is crucial when assessing the contributions of individual nectar plants, and must be corrected as far as possible. The influence of beekeeping methods, the distance from the hive to the forage sources, and the differences in nectar concentration of these sources, can hardly be judged objectively. On the other hand, the differences in flower biology, which determine how much pollen gets into the bee's nectar load, can be understood and are currently being closely studied.

The quantitative pollen analysis of pure unifloral honeys, and of experimentally produced single-species honeys, involves the determination of the absolute content of pollen grains and other plant ingredients in a standard volume of the honey, and this has provided a competent method for dealing with the problem. It allows an estimate to be made of the amount of pollen which falls on to the unit volume of nectar of an individual plant species, and so makes it possible to apply corrections to the percentages of the pollen counts in the honey sediment (Andrásfalvy, 1969, Barbier, 1958, Z. Demianowicz, 1960, 1961, 1962*a*, 1962*b*, 1964, 1968*b*; Demianowicz *et al.*, 1955, 1956, 1957, 1959, Hammer *et al.*, 1948, Kropáčová, 1969, Louveaux, 1961, Maurizio, 1939, 1949*a*, 1955, 1958*b*, Okada *et al.*, 1968, Pritsch, 1957*a*, 1957*b*, Vorwohl, 1970, 1973, Kropáčová & Nedbalová, 1971, *see also* International Commission for Bee Botany, 1970).

7.51   Methods and techniques
So far three methods for preparing quantitative preparations of honey sediment have been referred to. They are briefly described here.

Method according to Maurizio (1939, 1949a)
Two measures of 50 g honey are weighed in separate small Erlenmeyer flasks previously marked at 100 ml, and filled to the mark with distilled water. (With limited material, or honey very rich in pollen, smaller amounts can be used, keeping to the same proportions.) The honey is dissolved, using a water-bath, and the solution centrifuged in tubes holding 100 ml for 5 minutes. The supernatant liquid is poured off, leaving 1–2 ml of sediment; this is stirred with a heated platinum loop and transferred quantitatively to a calibrated centrifuge tube. After another centrifuging the remaining liquid is poured off, the sediment stirred and dispersed in a volume of liquid appropriate to the quantity of sediment. Experience shows that 0·5 ml is likely to give the most suitable dilution for the sediment from 50 g of many cleanly produced extracted honeys. For extracted honeys that are rich in pollen, or pressed honey, the sediment must be diluted to 1 ml or more.

The stored sediment is drawn up into a Breed pipette calibrated to 0·01 ml, blown out on to a slide and spread out over an area of 1 cm². For each honey, duplicate preparations are made, and two separate smears. The smears are allowed to dry in the air, protected from dust. On drying, the honey solution forms a smooth, glazed surface which can be microscopically examined without a cover-glass. Alternatively, a suitable haemacytometer can be used to determine the absolute content of the particulate constituents of honey.

Method according to Demianowicz (1955, 1961)
(a) This method applies to experimentally produced single-species honeys. A drop of honey is placed undiluted on a weighed cover-glass which is then reweighed. The glass is placed upside down on a slide. The quantity of honey must be chosen so that when the cover-glass has been placed on the slide no honey oozes out at the edge.
(b) For unifloral honeys Maurizio's method is followed in principle. The drawing of squares (1 cm²) in Indian ink on the slide is recommended as it prevents the sediment dispersion flowing out of the prescribed area.

Method according to Louveaux (1961)
Honey (10 g, or 20 or 30 g if it contains little pollen) is weighed and dissolved in distilled water to which sulphuric acid has been added

(5 g acid to 1 litre water). The solution is filtered through a nylon strainer (0·1 mm mesh), then centrifuged for at least 5 minutes. The supernatant liquid is carefully drawn off with the vacuum pump until only 4–5 mm is left above the sediment. The tube is refilled with distilled water and centrifuged, this process being repeated several times until the liquid is free from acid. Finally a few drops of detergent are added to the water. The solution containing the sediment is then poured into the cylindrical upper part of a filtration apparatus and drawn through a Millipore filter by a vacuum pump. The cylinder is refilled with water several times, and this water also drawn through the filter. (Instead of the vacuum filter apparatus, a Millipore filter holder and a glass syringe of 50-ml capacity may be used. The honey solution is put into the syringe and forced through the filter.) In either case the filter is then removed and dried on a hot-plate at about 60°C. Then 4 drops of immersion oil are placed on a slide and the dry filter laid on top; 2 more drops of oil are put on the cover-glass, and the filter, which by then is transparent, is covered. Details are given by Louveaux (1961) and International Commission for Bee Botany (1970); for apparatus, *see* Section 7.2.

7.52 Counts made on the quantitative preparation

The same principle is used in all three methods. With the aid of a mechanical stage and a graticule the plant constituents (pollen grains, algae, and fungal spores) are counted in a number of fields of view, then the mean value is calculated for one field. (Maurizio counts 100 fields in each of the 4 parallel smears, i.e. 400 fields of view in all; Demianowicz counts a number of strips across the preparation.) The number of pollen grains, algae, and fungal spores in 1 g or 10 g of honey is calculated from the following data: mean number of plant elements; known area of the field of view; original quantity of honey; (in Maurizio's method) the known quantity of sediment.

7.53 Evaluation of quantitative pollen analysis

The quantitative investigation of a large number of honeys, from different countries and from different plant sources, has shown that five groups can be distinguished according to the absolute content of plant constituents (Maurizio, 1949a, 1955, 1958b). The majority of all cleanly produced extracted honeys belong to an average group (II) with 20 000–100 000 plant constituents per 10 g. The lowest group (I) consists of honeys extremely poor in pollen—less than 20 000 plant constituents per 10 g; such honeys are from acacia (*Robinia pseudoacacia*), lavandin (*Lavandula vera* × *L. latifolia*), lime (*Tilia* sp.), Salvia (*Salvia* sp.), varieties of red clover (*Trifolium pratense*) so far studied, orange (*Citrus* sp.), and lucerne (*Medicago sativa*). Group III contains honeys rich in pollen (100 000–500 000 plant constituents per 10 g), such as sweet chestnut (*Castanea sativa*), forget-me-not (*Myosotis* sp.), hound's tongue (*Cynoglossum officinale*), and most honeydew honeys. Group IV (500 000–1 000 000) and Group V (over 1 000 000) contain centrifugally extracted honeys extremely rich in pollen, and honeys extracted by pressing the combs.

The results of the quantitative analysis are usually expressed in a conventional formula. For example 32/38/0·7-II means that the honey concerned contains 32 000 pollen grains + 38 000 fungus spores + 700 algae cells in 10 g; the total content of plant constituents per 10 g is 71 000, so the honey belongs to Group II.

These investigations have been confirmed and supplemented by the work of Demianowicz and her colleagues in Poland, who determined the absolute pollen content of honeys produced experimentally, by using colonies of bees caged with a single plant species (Demianowicz *et al.*, 1955, 1956, 1957, 1959, 1960, 1961, 1962a, 1962b, 1964). So far honeys from 46 species and 22 families have been studied in this way. The next stage is to set up coefficients, using the above results, by which the percentage pollen count obtained can be corrected to show the true contribution of each individual plant species to the honey concerned; among the honeys poor in pollen (with a pollen content less than 20 000 per 10 g and a conversion coefficient under 1 000) are included: *Chamaenerion angustifolium, Cucumis sativus, Robinia pseudoacacia, Anchusa officinalis, Borago officinalis, Centaurea jacea, C. cyanus, Helianthus* sp., *Lamium album, Salvia nemorosa, S. officinalis, Hyssopus officinalis, Tilia cordata.* Among the honeys extremely rich in pollen content, over 5 million per 10 g, and a coefficient of 400 000 to 14 000 000, are *Cynoglossum officinale* and *Myosotis sylvatica* (Demianowicz, 1964).

It is not yet known with certainty how accurately the percentage counts from the pollen determination, corrected by these conversion coefficients, express the contribution of the individual plant sources to a honey sample. The pollen content of the honey is dependent on other factors besides the floral biology of the plant (cf. Demianowicz, 1968b).

Based on the quantitative analyses made so far, the International Commission for Bee Botany (1970) suggests the following principles for the practical determination of the botanical origin of specific honeys:

Over-represented pollens: Honeys in which the predominant pollen is *Castanea sativa* or *Myosotis* spp. or *Cynoglossum officinale* can be regarded as unifloral only if the predominant pollen represents more than 90% of the total. *Mimosa pudica* is another source with over-represented pollen.

Under-represented pollens. Pollens listed below are considered to be under-represented. Honey may be judged to be largely of the origin stated if the pollen frequency reaches the following percentage:

Lavandula spica × *L. latifolia* (lavandin)	10–20%
Citrus (in part)	10–20%
Salvia (European)	20–30%
Tilia	20–30%
Robinia	30%
Medicago	30%

Other under-represented pollens are *Chamaenerion*, *Rosmarinus*, Cucurbitaceae, and *Banksia* spp.

All other pollens so far studied may be regarded as of standard value. Hence a honey may be considered unifloral, from the source whose pollen is predominant, if that pollen represents more than 45% of the total pollen count.

NOTE

Additional information

Use of pure cultures on artificial nutrients has made it possible to isolate and identify further fungi. Especially characteristic of honeydew and honey from fir are *Capnophialophora pinophila* and *Tripspermum pinophilum* (Borowska, 1971, 1973, Borowska & Demianowicz, 1972, Demianowicz *et al.*, 1972).

Other publications on statistical significance are by Faegri & Iversen, 1964, Maher, 1972, Mosimann, 1965.

Other descriptions of honey types are by Battaglini & Ricciardelli (1971, 1972, 1973), Kropáčová & Nedbalová (1971), Lieux (1972), and Vorwohl (1972, 1973).

CHAPTER 8

BIOLOGICAL PROPERTIES OF HONEY

by a group of authors

And bring withall pure honey from the hive
To heal the wound of my unhappy hand.
WILLIAM LILY, THE WOMAN IN THE MOON (1597)

8.1 INTRODUCTION

This short chapter has presented more difficulties than any other in its preparation. Professor Haydak wrote the first draft shortly before he died in 1970. This was considerably longer than what is now published, and described in some detail experiments carried out during and since the 1930s. The text presented here could not be discussed with Professor Haydak. It has resulted from extensive discussions between the Editor and Dr. H. Duisberg (Germany), Dr. T. A. Gochnauer (Canada), Dr. R. A. Morse and Dr. J. W. White (U.S.A.), Dr. P. Wix (England), and many others. All concerned believe that the time has come to make a realistic appraisal of the position, and to clear away some of the misconceptions that are published from time to time.

Honeys vary according to their plant origin and the conditions under which they were produced. Nevertheless, it is possible to establish mean values, ranges, and standard deviations for the physical properties and chemical constituents of honey. These are presented in Chapters 5 and 6 respectively. Physical and chemical characteristics may be changed by processing and storage of the honey, but in ways that are in general reproducible and predictable. Biological properties of honey are subject to all the types of variation associated with living organisms, and are immeasurably more difficult to establish with certainty.

From a scientific viewpoint—and also now from a practical viewpoint, because of the proliferation of legal prohibitions of unsubstantiated claims for goods marketed—it is important to distinguish between: (a) what is assumed to be true because it 'has been known (i.e. stated) from earliest times'; (b) what a large body of experience suggests is likely to be true; (c) what has been proved by experiment to be true (under the conditions of the experiment); (d) what is true under certain conditions but irrelevant if taken out of its proper context.

At least 2 000 papers and articles have been published in scientific and medical journals, and elsewhere, describing various biological effects of honey. In addition there are a number of books on the subject. Many of these have an introductory section on uses of honey in antiquity and in later historical times, and then move on to recent applications, in a way which implies that *use* for a purpose proves *effectiveness* for that purpose. Many statements have been made, and quoted and requoted, which are not based on fact, but whose very reiteration seems to give them a semblance of fact.

It is impossible in this book to assess the validity of methods used in the many experiments and observations on biological effects of honey. A number of those carried out in or before the 1930s used techniques which have since been superseded, and lacked statistical treatment with adequate controls, such as would be demanded today. Many examples of this work are referred to and described in books published in the past few decades. Attention may be drawn especially to those cited below; all except the first have appeared since the end of the Second World War.

Beck's *Honey and health* (1938) gives an interesting but uncritical discussion of the subject; the bibliography of this work is omitted from later editions by Beck & Smedley (1944, 1971). Spöttel's book on honey and dried milk was published in Germany (D.D.R.) in 1950; it adopts a more scientific approach, and has many references especially to European work. Another book on honey and milk was published by Simonis in 1965.

In the late 1950s Lavie in France undertook an extensive study of antibiotics in honey and other materials in the honeybee colony. This was published as a thesis in 1960 and is partially reprinted, with other work, in Volume 3 of Chauvin's *Traité de biologie de l'abeille* (1968); see especially pages 116–154.

Steyn in South Africa wrote an updated review in 1970, and in the same year Herold's book *Heilwerte aus dem Bienenvolk* was published in Germany (B.R.D.). This last book includes an extensive account of reported effects of honey as a 'healing substance'. There is also a host of popular books which make no pretence at providing scientific evidence. Duisberg (1975) has published a careful account of effects of honey on human beings so far reported. He finishes with the recommendation that those who take honey on health grounds should use it as the bees produce it, i.e. unprocessed.

We will now look in a little more detail at some specific aspects of possible biological effects of honey.

8.2 ANTIBACTERIAL EFFECTS

There is a vast amount of literature reporting antibacterial effects of honey. Work on the subject falls into three periods: before 1937, when the concept of *inhibines* was introduced; from 1937 to 1962, when the inhibine effect was shown to be due to small amounts of hydrogen peroxide in dynamic equilibrium in honey solutions; and since 1963.

In 1919 Sackett had shown that certain bacteria quickly perished in heat-sterilized honey, diluted honeys being more effective than undiluted. Dold, Du & Dziao (1937) were the first to examine the antibacterial effect of honey in any detail and they ascribed it to a material they called 'inhibine'. This was thermolabile and possibly light-sensitive, but when protected from light and heat it seemed to be quite stable.

Subsequent investigations showed that natural unheated honeys of various origins were effective against both gram-positive and gram-negative bacteria (Prica, 1938, Khristov & Mladenov, 1961, Lindner, 1962). The bactericidal action was due neither to the normal acidity of honey, nor to its high sugar content, enzymes, nitrogenous or other compounds, but to a special bactericidal substance which was thermolabile, destroyed by direct sunlight, susceptible to a low pH, and influenced by many other factors. It was classed with the inhibine of Dold (Prica, 1938, Plachy, 1944, Stomfay-Stitz & Kominos, 1960, Gonnet & Lavie, 1960, Khristov & Mladenov, 1961). An inhibitory effect was reported on *in vitro* growth of the tubercle bacillus (Lennartz, 1947, Pothmann, 1950) and against *Salmonella* (Bahr, 1939). Dold & Witzenhausen (1955) published a method for assaying inhibine by its effect on the growth of organisms such as *Staphylococcus aureus* on nutrient agar plates containing graded amounts of honey.

In 1944 Plachy reported that samples of honey produced at altitudes over 1 000 m had at least twice the bactericidal activity of samples from lower areas. However, most of the honeys in high areas were all or partly from honeydew, whereas those in the valleys were from nectar. Other workers have found that honeydew honeys have more bactericidal activity than floral honeys. In addition to aqueous solutions of honey, extracts prepared with alcohol, ether or acetone were found to have a strong inhibitory action (Vergé, 1951). In 1969 M. R. Smith and his colleagues reported on investigations carried out in 1962, in which they confirmed the activity of honey against *Micrococcus flavus*, *Sarcina lutea*, *Bacillus cereus*, and *B. subtilis*.

We now move on to the next phase of the story. White and his colleagues in the United States demonstrated that the antibacterial effects shown in the inhibine assay of honey result from the accumulation of hydrogen peroxide, which is produced by a natural glucose oxidase system in honey (see 5.73). This work was published in a preliminary note in 1962 and in a series of papers 'Studies on honey inhibine' (1963a, 1963b, 1964a, 1964b). Working independently, Adcock in England (1962) had found that both inhibine and peroxide values of honeys could be destroyed by catalase; he therefore suggested a possible connection between the two, but could not account for it.

The inhibine value, which is empirical, is determined by factors governing the speed of the formation and destruction of hydrogen peroxide, details of which are discussed by White (5.73 and references quoted above). It is related to the peroxide accumulation as follows; 0–5 are inhibine values (White & Subers, 1963):

0	1	2	3	4	5
	3·4	8·7	20·5	54·5	174

The lower figures are the accumulation of hydrogen peroxide (μg) per gram honey in 1 hour under assay conditions.

Whereas Dold's inhibine is enzyme-produced hydrogen peroxide and is heat-sensitive, Lavie (1960, 1963) found another group of antibacterial factors in honey, which are light-sensitive but relatively heat-stable. These factors were destroyed by heating at 80°C (176°F) for 30 minutes, were extractable with hot alcohol, acetone, and cold ether, and were volatile at 95°C (203°F). They could be preserved in a refrigerator for over 2 years.

Lavie concluded that his antibacterial factors are introduced by the bees during the elaboration of nectar into honey; indeed he found that the bees themselves and most of the materials in a honeybee colony contain antibiotic materials. (Pavan had published his first paper on antibiotics of animal origin in 1948.) On the other hand, honey seems to contain no anti-fungal material, and prevents the development of fungi merely by its high sugar concentration (Lavie, 1960).

In an investigation of the glucose oxidase activity of honeys from food-storing social species of the Apidae, Burgett (1973) found that 'inhibine' was present in the honey of *Apis* (all species) and of the subfamilies Meliponini and Bombinae. Honey from two species of *Myrmecocystus*, replete-forming ants, showed no inhibine activity, so this may well be confined to the Apidae. There were large variations within the various species of Apidae tested, which are probably attributable to differences in the geographical and hence floral sources of the honeys. For all species the rate of production of hydrogen peroxide (μg per g honey) was equal to or greater than that reported for *Apis mellifera*.

8.3 GENERAL PHARMACOLOGICAL EFFECTS

Honey has been used in medicine throughout historical times, and probably earlier (see 19.3). In the past fifty years or so there have been many reports of *in vitro* experiments showing effects of honey on animal tissues and organs; these cannot necessarily be translated into terms of human physiology, however.

It is difficult in clinical observations, or even experiments, to obtain proper controls. In early trials, especially, a certain treatment may have been followed by success even though the reason for this success was not correctly established at the time. The most useful results in future research will probably come from more detailed biochemical studies rather than from clinical trials. Advances in biochemical techniques in the last decade or so enable separations of complex organic substances to be carried out which were impossible earlier. Also, new instruments allow measurement of much smaller quantities of micro-constituents than formerly.

The benefit of combined chemical and biological approaches has been shown by recent research on many biologically active substances. The two disciplines together can provide measurements of the presence and activity of compounds in a natural mixture that constitutes, say, a food substance or a plant or insect secretion. This combined approach may yield information additional to that obtained by earlier methods of separating and purifying biologically active compounds, determining their structure, and testing the activity of the pure compounds. Although such investigations may never discover a universal panacea in honey, whether natural or processed, they may well serve to define any and every beneficial activity much better than before, and they might provide an improved basis for any claims made—for which present food and drug authorities will require convincing proof.

Some of the applications tried and reported to be successful within the past few decades are referred to below. The topical application of honey to burns and wounds has been described by a number of clinicians. The viscosity of honey makes it a good barrier compound, its water solubility allows easy removal, and its mild non-corrosive properties prevent any additional harm to either damaged or healthy tissue. A few of the many reports on the successful use of honey for wounds are by Temnov (1944), Gubin (1945), Bulman (1955); for infected wounds *see* Gundel & Blattner (1934) and Zaiss (1934); for burns Phillips (1933) and Voigtlander (1937).

One recent assessment (Manjo, 1970) is that the best thing to do to a wound is to leave it without any application or dressing—unless infection necessitates antibiotic treatment—but that if treatment is considered desirable, honey is likely to be safer than most others. Remarkable successes from its use appear from time to time, one being by Cavanagh *et al.* (1970), who poured honey twice daily into extensive wounds, after their breakdown following operations for carcinoma of the vulva. *In vitro* studies of bacteria, cultured from the wounds of all twelve patients, showed that undiluted honey killed them. The honey was easily applied at home after (early) discharge from hospital. Blomfield (1973) has reported on ulcers.

The relatively high fructose content of honey has led to its use to speed up alcohol metabolism in sobering drunken patients (e.g. Martensen-Larsen, 1954, Molnár et al., 1966; see also Brown et al., 1972). Balogh et al. (1964) found that honey had a greater effect than fructose, and this was attributed to its enzyme content, especially catalase. The relevance of fructose metabolism to diabetes was discussed at the International Symposium on the Clinical and Metabolic Aspects of Laevulose (1971).

The acetyl choline content of honey is discussed in 5.87.

Chauvin (1968b) reports on the uses of honey in respiratory infections, various digestive diseases, and malfunctions of the heart. The other books cited in 8.1 quote many other applications of honey: topical, oral, or in a preparation suitable for injection.

In general the use of honey is less likely to harm a patient than most other preparations, and on many occasions it has proved beneficial. For specific uses see Chapter 15, and e.g. Gennaro et al. (1959), Duisberg (1975).

8.4 NUTRITIVE VALUE

Honey has been a valuable survival food for primitive peoples. By reason of its sweetness, it is also a very attractive food, worth harvesting in spite of the pain of bee stings received in the process.

In a classical experiment published in 1936, Professor M. H. Haydak set out to subsist for a prolonged period on milk and honey. This was done in order to obtain information useful for patients who must have a liquid diet. Haydak lived for three months on a diet of cow's milk and honey, 100 g of honey per quart (litre) of milk. His ability to work was normal, and he did not feel sluggish or tired. Clinical observations were limited, but showed maintenance of weight, normal bowel movement, absence of protein or sugar in the urine, and a slight rise in haemoglobin content of the blood. Towards the end of the experiment signs suggesting vitamin C deficiency were noticed; these were cured promptly by adding orange juice to the diet. Later (Haydak et al., 1944) five adults whose age ranged between 22 and 44 years were put on the same milk-honey diet, supplemented with vitamins B and C, and iodine. The doses of vitamin C had to be increased during the experiment for some subjects, because signs of scurvy appeared. There were two test periods, and two control periods during which the subjects ate their customary diet. The test and control periods were alternated, and lasted for about four weeks. The subjects maintained their normal activities. At the end of the experiment they were in normal health, and no untoward effects were reported. These experiments showed that the milk and honey mixture, properly supplemented with vitamins, can be used as the sole food for adult men for some months.

A healthy man or woman receiving an adequate diet will normally

Table 8.4/1
Nutrients in honey in relation to human requirements

1 Nutrient	2 Unit	3 Average amount in 100 g honey	4 Table used for Column 3	5 Recommended daily intake U.S.A.	6 Recommended daily intake U.K.*
Energy equivalent	kcal	304	6.64 (text)	2 800	2 600–3 600
Vitamins:					
A	i.u.			5 000	2 500
B₁ (Thiamin)	mg	0·004–0·006	5.83/1	1·5	1·1–1·4
B₂ complex:					
Riboflavin	mg	0·02–0·06	5.83/1	1·7	1·7
Nicotinic acid (niacin)	mg equiv.	0·11–0·36	5.83/1	20	18
B₆ (Pyridoxine)	mg	0·008–0·32	5.83/1	2·0	(1–2)
Pantothenic acid	mg	0·02–0·11	5.83/1	10	(10–20)
Folic acid	mg			0·4	0·05–0·1
B₁₂	μg			6.0	3–4
C (Ascorbic) acid	mg	2·2–2·4	5.83/1	60	30
D	i.u.			400	100
E	i.u.			30	(10 mg)
H (Biotin)	mg			0·3	
Minerals:					
Calcium	g	0·004–0·03	5.5/2	1·0	0·5
Chlorine	mg	0·002–0·02	5.5/1		(5–9)
Copper	mg	0·01–0·1	5.5/1	2·0	(2·0–2·5)
Iodine	mg			0·15	0·15
Iron	mg	0·1–3·4	5.5/1	18	10
Magnesium	mg	0·7–13	5.5/1	400	(150–450)
Manganese	mg	0·02–10	5.5/1		(5–10)
Phosphorus	g	0·002–0·06	5.5/2	1·0	(1·2–2·0)
Potassium	g	0·01–0·47	5.5/1		(2–4)
Sodium	g	0·0006–0·04	5.5/1		(3–6)
Zinc	mg	0·2–0·5	5.5/2	15	(10–15)

* Figures in brackets indicate actual daily intakes, *not* recommended daily intakes.

enjoy eating honey, although it is not essential to his well being. The nutritive value of honey is determined by its components, which are set out in Table 8.4/1; entries are derived from data in Chapters 5 and 6. The amount of honey represented (100 g, about 4 oz) is chosen as being the most that a normal person is likely to eat in a day.

Amounts of various nutrients in 100 g honey (Column 3), derived from the Tables cited in Column 4, may be compared with the minimum daily requirements published by the U.S. Food and Drug Administration in 1973 (Column 5) and the daily intakes recommended by the Department of Health and Social Security, U.K., in 1969 (Column 6).

Figures in Columns 5 and 6 relate to an adult man; those for an adult woman, or for a person under 20, are in general the same or lower.

The Table shows that honey contains many substances that are valuable nutrients. The *amounts* of the various constituents (Table 8.4/1) are, however, small, so it is unlikely that eating honey will correct deficiencies of trace elements or other substances in the diet. A similar table could be compiled from data in Table 5.6/1, giving the amounts of 18 amino acids, but the total weight of amino acids in 100 g honey is only a few mg (cf. page 177), and the weights of each one are of scientific interest rather than nutritional significance.

For infants, children, old people, and invalids, honey can be a more easily digested and more palatable carbohydrate food than, say, sucrose; the conclusion of Vignec & Julia (1954) is typical of many: 'It would seem from these observations that honey has a definite place in infant feeding'. There have been a very large number of reports (*see* 8.1) that honey has proved beneficial, based on clinical trials (in heart patients, after strokes and operations, and so on), and from tests on man and various other mammals. There is a suggestion (Albanese *et al.*, 1952) that the rapid assimilation of fructose may be associated with increased nitrogen retention, and also that the presence of invertase in honey is helpful for old and sick people. If the benefits are due to a unique mystical component of honey, as some would claim, then this substance has not yet been identified, although 181 substances are known to be present in honey to date.

As a food, honey is a readily acceptable and easily digestible source of carbohydrates. Honey can be obtained in a form as near to a 'natural' food as almost any other, and many consumers are at present greatly attracted by this.

8.5 Conclusion

'Eat honey, my son, for it is good' was advice given by Solomon, King of Israel, around 1000 BC (Proverbs, 24, 13). All the knowledge and scientific research presented in this book endorses the 'goodness' of honey as a food for man. Its main sugars, fructose and glucose, are absorbed directly into the blood, and provide a rapid source of energy without need of digestion. On the other hand, non-sugar components of honey—vitamins, amino acids, and minerals—are present in such minute amounts compared with those in a normal diet (Table 8.4/1) that honey is not in general a useful source of them.

This need not surprise us, for honey is primarily a food for bees, not man: by inverting the sugar collected from plants, converting sucrose into fructose and glucose, the bees are enabled to create a high-energy food which, sealed in the cells of the comb, will be preserved in the nest unspoiled by fermentation, until it is needed in a period of dearth (Section 5.1).

Section 8.2 shows that honey is effective in killing bacteria, and 8.3 refers to some of the many successful uses of honey in treating wounds, sores, and burns. In this sense, and since honey is widely accessible and non-injurious, we may perhaps be permitted to end this chapter with a comment from Sura 16 of the *Koran*, which refers to honey 'wherein is healing for mankind'.

SECTION 3

PREPARATION OF HONEY FOR MARKET

CHAPTER 9

PROCESSING AND STORING LIQUID HONEY

by Professor Gordon F. Townsend

PROFESSOR OF APICULTURE, UNIVERSITY OF GUELPH, CANADA

I have known honey from the Syrian hills
Stored in cool jars

V. SACKVILLE-WEST,
THE LAND (1926)

9.1 INTRODUCTION

Honey is a natural food product. This is one of its main attributes, and every effort must be made to maintain the delicate flavour and aroma associated with freshly extracted honey, or—for those who have the opportunity to sample it—with honey in the comb fresh from the hive. It is natural for many honeys to granulate (crystallize) upon storage. The onset of granulation, and the rate at which it takes place, depend on the plant source of the honey and the manner in which it is handled. During the course of natural granulation, honey may appear at one stage as half liquid and half granulated, but honey should not be sold in this condition; for best consumer appeal, it should be either a clear liquid or smoothly granulated.

The main problems encountered in packing honey are the presence of excessive moisture, air, pollen, or beeswax particles. Most of these problems can be overcome, provided that proper care is exercised during the production as well as during the extraction of the honey crop.

In recent years there has been considerable stress on quality in food products. As a result, much greater attention is now being placed on the preparation of honey for market. This chapter attempts to bring together in concise form the principles of the operations that lead up to the packing of honey in liquid form. Heating for 'pasteurization' is discussed in 9.33, and pressure-filtering in 9.7. Methods that involve neither of these processes are summarized in 9.34. Chapter 10 deals with granulated honey.

9.2 PROCESSING UP TO AND INCLUDING EXTRACTION

9.21 Production

The preparation of good-quality honey starts in the apiary or bee yard, and Chapter 3 gives the broad outlines of procedure there. Even under the stress of commercial conditions, the effect of colony operations on the end product (honey) must be constantly kept in mind. For instance honey should be produced in separate honey supers,* and not in combs used for rearing brood. Use of brood combs will increase the pollen content of the honey (7.53), leading to special problems when the honey is filtered. It also darkens the honey; in one experiment (Townsend, 1969b) brood combs and super combs were alternated in the same super, and in every sample the honey stored in the brood comb was darker. Of eleven pairs of combs, the average colour difference was 124 mm on the Pfund grader (Sechrist, 1925) (6.53) with a range 10–470 mm. Much of this 'increase' in colour was due to honey being moved from the brood chamber to the darker combs in the super, but the experiment shows clearly how honey may be darkened by using brood combs in the supers. Some of the samples of honey from brood combs contained so much pollen that they had a cloudy appearance (Townsend, 1966).

Recent developments in commercial beekeeping may well overcome this problem: a number of operators are working towards the production of honey that can be uncapped and extracted by the super, instead of by the comb. This will probably necessitate using a shallower super than the standard Langstroth brood box, to facilitate lifting, getting the bees out of it (3.35), and uncapping; it may well involve supers with permanently fixed combs. If the super depth is of necessity less than that of the brood chamber and non-removable combs are used, supers will more easily be kept for honey production only. Some types of hive already have shallow supers (e.g. Dadant, British National), but for different reasons.

9.22   Moisture control

Moisture content is the major factor which determines the keeping quality of honey. Honeys show a marked variation in water content (6.11), depending on the atmospheric humidity both before and after the honey is removed from the hives, and on other factors. Honey, being hygroscopic (6.8), may absorb a considerable amount of moisture if it is permitted to remain too long above a bee-escape (3.35), or if it is stored in a cool damp honey house. The lower the moisture content the less chance there is that fermentation will take place (6.11.1). The optimum relative humidity for maintaining a 17.8% moisture content in honey is about 60% (Martin, 1939).

The processes by which bees reduce the high water content of nectar, to say 20% in the ripe honey, are described in detail in 2.22. Bees fanning

---

* 'Super' is the name given to a hive box superimposed on the brood chamber, specifically for honey storage, hence also 'super combs'. The depth of a super may be the same as that of a brood box, or less (*see* page 108).

with their wings force warm air from within the hive over the comb surfaces. The beekeeper can help the bees to make their action more effective: combs of honey that would normally take 21 days to ripen can ripen in 6 days if the supers are moved slightly forward on the hive, to allow cross-ventilation (Rheinhardt, 1939). The beekeeper can profitably use the same principle in the honey house, to reduce the moisture content of honey before extracting. Air at 38°C (100°F), passed over combs of honey at the rate of 250 m/min for 12 hours, removed as much as 0·7% moisture from almost fully capped combs, and 3·1% from uncapped combs (Stephen, 1941). The relative humidity of the air used in these experiments was 33%, although the optimum relative humidity for drying honey is apparently somewhere between 0 and 20% (Martin, 1939). Recirculation of the air in the room should be possible (Stephen, 1941), since the amount of moisture removed from the combs of honey is not large compared with the capacity of the volume of air passing over them. Moisture in honey can be reduced from 20% to 18·4% in 24 hours in such a drying room (Fix & Palmer-Jones, 1947). These rooms (also known as warming rooms or hot-rooms) are now found in 'honey houses' in many of the major honey-producing areas of the world. There are various types and designs of these rooms, but they all follow the principle of passing a large volume of warm, relatively dry, air through the stacked supers of honey combs. The temperature in the room must be thermostatically controlled so that it does not rise above 38°C; the usual operating temperature is 32–35°C (90–95°F). Care must be taken that the air is not forced to one part of the room in such a way that it overheats one stack of supers and causes the combs in them to break down (Townsend, 1965). The same principle may be applied on a small scale by adapting an electric fan and heating unit so that warm air is blown up through a single stack of supers (Line, 1955). Moisture removal may be more necessary in some climates or in some seasons than in others. In very dry areas, it may be necessary to reverse the process, and *increase* the moisture content of the honey, to facilitate its extraction from the combs. This is perhaps best done by using radiant-heat units in cement floors, with circulating air, and sprinkling water over the floor.

Warming honey at 35°C (95°F) for up to two days before extracting does little harm; it has been found that longer periods at this temperature considerably increase the hydroxymethylfurfural content of the honey, although they have little effect on its colour (Townsend, 1969b).

Drying rooms have also been used for removal of moisture from honey already extracted. The honey with the greatest moisture content rises readily to the surface of the tank, and by passing air over the surface this honey is dried somewhat; its density increases thereby and it moves to a lower part of the tank, letting further high-moisture honey come to the surface. In a commercial plant so established, the reduction in moisture content reported for the top layer of honey was 1·2% in 30 hours. The optimum temperature for drying moist honey on the surface is 32–35°C, 90–95°F (Fix & Palmer-Jones, 1947).

9.23 Uncapping

'Uncapping' is the process of removing the thin wax covering with which each cell of ripe honey is sealed by the bees. It is the first stage of extracting the honey from the combs.

Uncapping equipment may vary from a hand knife dipped in hot water, to various types of automatic units. The most commonly used knife is hand-operated and steam-heated, but electrically-heated knives are becoming popular. Figure 9.23/1 (see Plate 7) shows the general set-up of one of the many types of automatic uncapper that have been developed; the most common is perhaps a power-driven steam-heated knife mounted horizontally—or, more recently, vertically. A few units have a pair of power-driven steam-heated knives, and a track to carry the comb between them, so that both sides are uncapped together (Figure 9.23/2—see Plate 7).

The Sioux Bee Automatic Uncapper (formerly Bogenschutz) has revolving knives, and will uncap nine frames a minute (Grout, 1963, p. 315). Several of the automatic types of uncapper tear off the wax cappings in very small particles, all of which must be subsequently removed from the honey. A number of the machines have now been improved to overcome part of this difficulty, but any beekeeper who introduces machines of this type must be prepared also to introduce methods by which the wax can be removed from the honey. This can be accomplished (see 9.4) by the use of baffle tanks, or by warming the honey and allowing time for gravity settling. If the small particles of wax are left in the honey when it goes to the packer, it may be heated in the packing plant to a temperature beyond the melting point of wax before the wax is removed, thus damaging the flavour of the honey. It is up to the beekeeper to see that these small wax particles are not left in the honey.

A considerable proportion of the honey will be incorporated with the cappings, no matter what uncapping method is used. Many ways have been devised for separating this honey from the cappings. The simplest is gravity straining, but this leaves so much honey behind that it is often followed by pressing. In some outfits the cappings are allowed to fall into a centrifuge-type cappings drier while it is running. These driers are quite popular in Canada and remove most of the honey in good condition. Whether a press or centrifuge is used, the compressed cappings will still contain 50% honey by weight (Grout, 1963, p. 313).

Figure 9.33/2 (see page 276) Plate-type heat-exchanger at the Manitoba Honey Producers Co-operative, Winnipeg, Canada.

Figure 9.42/1 (see page 278) Small-scale extracting equipment, including provision for warming the honey with electric soil-heating cable before straining it.
(photo: Ontario Department of Agriculture and Food)

PLATE 9

Figure 9.5/1 (*see page 281*)   Liquefying drums of granulated honey in
the 'melting room'; R. W. Maguire,
Minesing, Ontario, Canada.

(photo: E. J. Dyce)

Figure 9.91/1 (*see page 286*)   Central
honey storage tank holding 1 000
tons, with tank holding 55 tons for
bulk transport by rail; Ralph Stone,
Billings, Montana, U.S.A.

PLATE 10

Figure 10.62/1 (see page 301) A large Votator at the Manitoba Honey Producers Co-operative Ltd., Winnipeg, Canada, which cools the pasteurized honey to room temperature. During the cooling process the starter is metered into the Votators and thoroughly mixed with the honey.

Figure 10.62/2 (see page 301) Two small Votators (seen in the background in Figure 10.62/1). The honey passes from one Votator to another in the process of cooling it to 14°C (57°F).

PLATE II

Figure 10.62/3 (see page 302) A standard dairy horizontal coil vat which is used at the Saskatchewan Honey Producers' Co-operative to cool honey and introduce the starter.

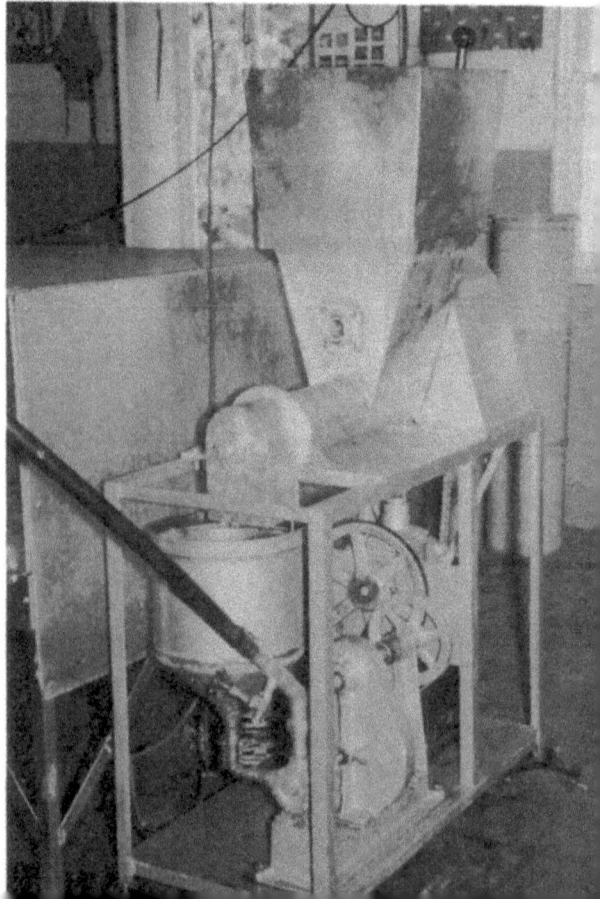

Figure 10.63/1 (see page 302) Home-made breaker used to grind starter by William Hamilton at Nipawin, Saskatchewan, Canada. The starter accumulates in a small hopper at the front of the machine, whence it is metered into the cooled liquid honey by a small honey pump.

PLATE 12

Figure 11.4/1 (see page 310) Section honey, with a device for packing each section in a polythene bag. (photos above and below: C. E. Killion)

Figure 11.5/1 (see page 311) Thin foundation being fitted into a special shallow frame. Note the space at the bottom, and the protruding foundation at the top, which is 'mashed down' with the heated tool in the foreground.

PLATE 13

Figure 11.5/2 (see page 311) Miniature centrifugal extractor for drying extraneous
honey from cut comb. (photo: C. E. Killion)

PLATE 14

Figure 12.1/1 (see pages 314, 315) Illustration at the head of a honey publicity leaflet (1788).

PLATE 15

Figure 15.22/1 (see page 380) Ornamental relief on the front of the bakery shop in Budapest devoted to products made with honey. It is based on a gingerbread mould in the shape of a mounted hussar.

(photo: Eva Crane, 1963)

Figure 15.22/2 (see page 380) Gingerbread from Torún in Poland, from a nineteenth-century mould of a similar type to Figure 15.22/3. (B.R.A. Collection B70/79)

Figure 15.22/4 (see page 380) A gingerbread 'heart' from Ljubljana in Slovenia (1956) with superimposed decoration showing bunches of grapes and a man drinking. (B.R.A. Collection B69/22)

Figure 15.22/3 (see page 380) Dutch gingerbread mould carved out of wood (nineteenth century) in the shape of a bishop with his mitre. (B.R.A. Collection B69/21)

PLATE 16

Figure 15.22/5 (see page 380) A gingerbread house, complete with occupants and livestock in the garden (1963), from Prague, Czechoslovakia. (B.R.A. Collection B70/80)

Cappings melters are quite common. They will completely separate the honey and wax in one operation, but considerable and constant care must be taken if heat damage to the honey is to be avoided. In most circumstances honey from a cappings melter should not be mixed with the rest of the extracted honey.

Many types of equipment are available and used for handling cappings during the extracting process—whirl-drys, draining boxes, etc. One of the most common is the Brand-type wax melter. Most of the Brand-type melters have a steam-heated grid on the surface, usually made of copper. It is almost impossible to operate this unit effectively without injuring both the honey and the beeswax. In the first place, heat from steam should never come into direct contact with honey at any time. In the second place, copper should never come into contact with liquid beeswax at any time, since it adds colour to the beeswax which is difficult to remove even by bleaching. Thirdly, the overheating of the honey and the beeswax together near the copper surface will always damage and darken the honey. This damaged honey could lower the grade of a whole batch of honey if it is mixed in with other honey as it goes through the line. Surveys on the operation of these steam-heated units have shown that the honey was *always* darkened (Townsend, 1969*b*). In recent years at the University of Guelph we have developed, and have been using, an all-electric Brand-type melter. The complete bottom of the unit is jacketed with water thermostatically controlled to about 52°C (126°F). A hood is placed over the top, fitted with radiant electric heat units; the hood is adjusted to such a height (according to the wattage of the heat units) that the heat melts only the wax on the surface; this wax runs off in liquid form without being overheated. Tests show that this unit does less damage to either the wax or the honey than any other Brand-type melter (Townsend, 1969*b*).

9.24 Extracting

Extracting honey from the combs is almost always done by using a centrifuge, which spins the honey out of the combs against the cylindrical wall of the extractor. (The exception is heather honey, which is thixotropic (6.43); sometimes this is pressed from the combs; if a centrifuge is to be used the gel must first be broken by using a 'loosener' on the combs.) The small-scale beekeeper often uses a 2- or 4-frame basket extractor, which may be either a reversible or a non-reversible type. The framed combs are placed vertically in the 'baskets' which support them, and they are *tangential* to the (vertical) spinning axis. One man can extract 450 kg per working day with a 4-frame reversible extractor; in the non-reversible type the combs must be reversed by hand to extract honey from the other side. The *radial* extractor is used for extracting the bulk of the world's honey crop. Here the combs are supported (without baskets) radially like spokes of a wheel, and a greater number can be accommodated—commonly 30, 45, or 50 frames (Figure 9.24/1—*see* Plate 8). The largest size will extract up to 3 tons a day. The normal shaft speed of a radial extractor is 300 rpm, and provision is made for a slow increase of speed. Many of the larger honey houses use three radial extractors. During centrifugal extracting the honey is inevitably cooled through evaporation in the whirling machine, so a 'warm-room' (9.22) does not ensure a supply of warmed honey for settling or straining.

It is usually convenient to have the honey flow into a sump tank when it leaves the extractor. In this tank the coarse material from the cappings may be removed quite readily by a series of baffles. The baffle openings are alternately at the bottom and top, forcing the honey to flow first under then over a 2·5-cm opening across the tank. It is advisable to use five or six baffle plates, to give two or three skimming areas. In a large-scale operation, a series of large baffle tanks at various levels in a warm-room is a very satisfactory arrangement. Large-scale handling of honey through baffle tanks, in which it can settle for one or more days before being put into containers, is quite often sufficient for bulk shipments of honey, without using a strainer (9.4). The use of a sump tank is essential for most systems, so that the load may be taken off the strainers, or some other way devised for ensuring that the bulk of the coarse material is removed with ease. In many plants the sump tank is jacketed with heated water, so that the honey can be warmed before it passes through a strainer.

9.25 Honey pumps

The incorporation of air in honey leads to one of the greatest problems the packer has to contend with, and it is commonly caused by the faulty operation of honey pumps. The pump should be of a type suitable for handling viscous materials. It should be operated no faster than is necessary to keep the way clear for oncoming honey, and if possible it should be set so that the pump runs fairly steadily. In most circumstances the speed should not exceed 40 revolutions per minute. When the honey does not flow into the pump fast enough, a partial vacuum is created, with the result that air is sucked into the pump through bearings and connections; this air becomes incorporated into the honey in tiny bubbles. So it is always better to use a larger honey pump than is necessary, and to run it as slowly as is consistent with keeping the tanks clear. A small accurate vacuum gauge may be inserted in the upper part of a T in the pipe leading from the honey supply to the pump.

If a slight vacuum shows on the gauge, the pump speed should be reduced (Dyce, 1953). It must be emphasized that minimal friction, and a sufficient supply of honey, can be guaranteed only if the pipes leading to and from the honey pump have a diameter at least as large as—and preferably larger than—the openings in the pump. The honey should always be gravity-fed to the pump and never sucked from the tank.

9.3 HEATING HONEY, AND THE USE OF MINIMAL HEATING

9.31 Effects of heat on honey

Warmth is a great ally in the handling and packing of honey, the most extreme heat being usually applied to prevent fermentation or to delay granulation. But although heating is essential at various stages of honey handling, every care must be taken during these processes to control the period during which the heat is applied, as well as the temperature itself. The presence in honey of an excessive amount of hydroxymethylfurfural-dehyde—HMF—(5.84) has been considered in many countries as evidence of overheating, and it implies loss of freshness of the honey; darker honeys tend to be injured by heat faster than lighter honeys (White, Kushnir & Subers, 1964). The effect of heat is cumulative, so the effects of processing and storage of honey must be considered as one (White, 1964). Keeping a 300-kg drum of honey for 5 days in a warm-room at 48°C (118°F) (to liquefy it) doubled its HMF content; at 43°C (109°F) the increase was 25%. Even the commonly recommended heat treatment of 63°C (145°F) for 30 minutes will increase the HMF content (White, Kushnir & Subers, 1964). Therefore as little heat as possible should be applied to honey, and for as short a time as possible.

9.32 In the honey house

The first heating of honey usually takes place in the honey house prior to uncapping; this is often carried out in a warm-room (9.22) where the honey is stored for about a day at 32–35°C (90–95°F) so that it can be extracted cleanly from the combs.

In some systems, the honey is warmed again in the extractor. As honey runs down the walls of the extractor it may be passed over coils which are heated with warm water, or the bottom of the extractor may be heated with warm water. This second warming overcomes the cooling which takes place during extraction (9.24) and speeds up subsequent settling and straining. The next point where heating may take place is in the sump tank. By jacketing the sump tank and surrounding it with warm water, its temperature may be kept around 46°C (115°F). If sufficient heat is not obtainable here to get the honey warm enough for straining, then the pipe which carries the honey to the strainer can be jacketed and surrounded with warm water.

9.33 In the packing plant

A number of methods are available for heating to temperatures high enough* to kill yeasts (and thus to control fermentation) or to delay granulation. Methods and equipment which will heat and then cool the honey *very rapidly* are the best.

The simplest system for heating and cooling honey is the double-jacketed tank with a stirring device to mix the honey. The double jacket carries either hot or cold water. Honey can be heated quite rapidly in a tank of this type, but it takes a long time to cool, and wall scrapers are necessary where heating and cooling are done in the same tank.

The dairy coil-type pasteurizer (Figure 10.62/3) follows almost the same procedure as the double-jacketed tank, except that the hot water and the cold water pass through a revolving coil. This system works much faster than the double-jacketed tank, and the revolving coil itself serves as the mixer. The foregoing two methods are often used by bee-keeper packers (*see also* 10.62).

A tubular heat-exchanger, through which the honey passes in a layer not more than 7 mm thick, is used in the smaller packing plants and by beekeeper packers (Figure 9.33/1—*see* Plate 8). The hot (or cold) water passes on each side of the honey, flowing in the opposite direction to the honey. The cost of this equipment is much lower than that of equipment used in the larger packing plants, but there is a limitation on volume, owing to the pressures involved (Townsend & Barrington, 1953). The tubes used for heating the honey should be manufactured in such a manner that they can be taken apart for cleaning (Townsend, 1970).

The plate-type heat-exchanger (Figure 9.33/2—*see* Plate 9) is the most common in larger packing plants. It can be used for both heating and cooling, and is very successful for liquid honey: the honey is brought to quite a high temperature in one section, then passed through a filter press, and on through the cooling section for quick cooling down to the packing temperature. The same type of equipment is used for cooling milk, but for honey it is necessary to use extra-heavy gaskets to withstand the high pressures (E. Braun, 1954, Palleson, 1952). The equipment is made of stainless steel, is quite compact, can be adjusted to any size of operation, and is easily cleaned. The only drawbacks are cost, and the limitation on volume due to increased pressures if the honey must be cooled below 38°C (100°F).

---

* This is commonly referred to as pasteurization, but the purpose of the process is quite different with honey, which contains no micro-organisms infectious to man such as the tubercle bacillus.

9.34 Packing honey with minimal heating

It is not *necessary* to use any heat whatsoever when packing honey in either liquid or processed form. At least a minimal application of heat is however very useful, and most packers use heat at various stages, sometimes to quite a high level (9.33).

The reasons for heating of honey are several: firstly, to warm it sufficiently to facilitate both straining and fast handling; secondly, to destroy yeasts that may be present so that the keeping quality of the honey is assured, particularly if the moisture content is above 17%. Other reasons for heating honey are to destroy all of the large granules that may be present, to assist in filtering (9.7), and so that it will stay liquid for a long period, even though it is of a type that would normally granulate fairly soon.

Much of the honey that is sold directly from the beekeeper to the consumer is not heated, and in recent years there has been an increasing demand in many countries for honey which has not been heated. It is usually advisable to warm the honey to 38°C (100°F), to facilitate straining and pumping. However, even this is not essential if the packer is prepared to handle the honey at a slower rate. Reasonably large volumes of honey can be strained at room temperature in the O.A.C.-type strainer (9.43) and bottled directly from the strainer, so all that is required is a sump tank to remove the bulk of the coarse material and a pump to transfer the honey directly to the O.A.C. strainer and into a settling tank to remove any air bubbles. For a liquid pack, the honey is filled directly into the containers. This same honey can be processed quite readily (Chapter 10) by adding 10% of finely granulated honey directly to the liquid honey and mixing thoroughly at room temperature, taking great care not to incorporate air. The container should then be stored in a cool place, preferably at 14°C (57°F) for at least ten days. It does not need any expensive equipment to pack honey in this form if heat is not used, but one cannot pack the same volume each day; continuous-flow processing would not be possible, so batch handling is necessary.

Several precautions should be taken. The honey should not contain more than 18% moisture, and preferably less, as any honey with a higher water content is likely to ferment during prolonged storage. Only moisture-proof containers should be used, and they must have tightly fitting lids. For a liquid pack, the honey should not contain too many air bubbles or pollen grains; these could give it a cloudy appearance, and also act as nuclei for the formation of crystals, leading to early and perhaps uneven granulation. For liquid packs it is wise to select honeys which granulate slowly, such as acacia (black locust, *Robinia pseudoacacia*). Few honeys produced in the Northern Hemisphere will stay liquid very long, and some will granulate within weeks of packing. This explains the preference for processed or granulated packs.

9.4 STRAINING

9.41 Principles

The method and the equipment to be used for straining honey will vary with the size of the operation, and according to whether or not straining is being carried out in the honey house or in the packing plant. No matter where it is done, or what the size of the operation, the following basic factors must be considered. The bulk of the refuse material should be removed by settling or baffling, prior to straining. The honey should be warmed to about 35°C (95°F) to reduce the viscosity to a value where straining is rapid. The straining should always be done below the surface of the honey, and the straining area should be as large as possible. Provision should be made for rapid cleaning. If a screen of some type is used for the final straining, then it should conform to the requirements of the country concerned. For example, Canada Number 1 grade demands straining through standard bolting cloth with 35 meshes per cm. Standard bolting cloth of this mesh has apertures about 0·175 mm across. Therefore, if a metal screen is used for the final straining, it should have openings no larger than 0·175 mm.

9.42 Small-scale straining

On a small scale, the honey may be transferred by pail from the extractor into a can holding say 27 kg, by pouring it through a screen with about 3 meshes per cm. The honey then flows through a pipe (3 cm in diameter) wrapped with soil-heating cable and outer insulation (Figure 9.42/1— see Plate 9). By controlling the rate of flow of the honey out of this pipe, it is possible to raise the temperature of the honey to 35°C (95°F) when it passes on into the strainer. Twenty metres of soil cable (total 600 watt), if properly insulated, will raise the temperature of 45 kg honey by 10°C (18°F) in 1 hour.

The honey then passes through a straining cloth, in the form of a bag, which is supported by a screen with about 1·6 meshes per cm, suspended in a container of sufficient height to allow a baffle on one side to take the honey from the bottom and remove it near the top. This method of straining helps to prevent the formation of air bubbles, and also keeps most of the refuse material out of contact with the straining cloth.

The strainer is raised sufficiently for the honey from the baffle outlet to flow directly into a storage tank. The strainer and storage tank may be made from standard galvanized containers (Townsend, 1965).

9.43   In the honey house

One of the most widely adopted strainers for the honey house is the O.A.C. strainer (Figure 9.43/1), named after the Ontario Agricultural College where it was developed. It consists of a series of four cylindrical screens of different meshes, one inside the other. (To facilitate drainage, the bottom of each screen is fitted with a gate which can be opened from the top.) The honey enters at the centre and passes outwards from one screen to the next, each being finer than the last. After passing through the finest (outermost) screen, the honey is drawn off by a baffle near the top of the tank. If a sump tank is used first, and if no granulation has taken place in the honey, this strainer can be operated very satis-factorily at room temperature, provided the straining area is large enough; at 35°C (95°F), it will handle a very large volume (le Maistre, 1936).

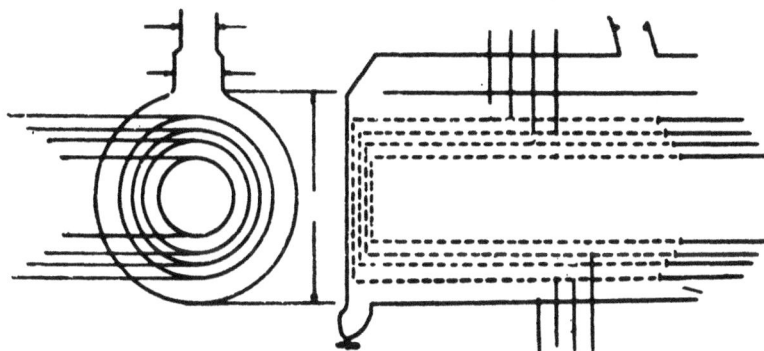

Figure 9.43/1   O.A.C. honey strainer, diagrams to show elevation and plan.

If cloth or nylon is used for straining instead, it should be below the surface of the honey, and supported by a metal screen. In any cloth strainer, it is advantageous if there are a few folds; when the cloth becomes clogged, draining can be facilitated considerably by pulling out the folds and freeing a clean straining area.

A very quick and neat arrangement, which will handle a large volume of honey in the honey house, incorporates a baffled sump tank, and a tubular heat-exchanger (heated with warm water) which raises the tem-perature of the honey to 35°C (95°F) (Figure 9.33/1); from this cylinder the honey is discharged into an O.A.C. type strainer. The pump handling the honey should be set to operate at the speed of the extracting opera-tion, so that the honey is passing continuously through the heating unit and through the strainer.

9.44   Other methods of 'straining'

With the ever-increasing demand for faster honey extraction, and to satisfy the larger commercial operations which are developing, new types of 'strainers' have come into use. One recently put on the market is the 'Spinfloat Honey-Wax Separator'; this operates on the same principle as the cream separator, and will handle more than 1 360 kg of honey an hour if the honey and wax mixture is at a temperature of 40°C (104°F). This process has been developed and patented by Cook and Beals Inc. of Loup City, Nebraska. It works well with the Bogenschutz type of uncapping machine, if care is taken when setting it up to avoid incorporation of air in the honey.

9.45   In the packing plant

Honey arriving at the packing plant should be clean and relatively free from small particles of wax and other materials, but unfortunately this is not usually the case. Nevertheless many packers make no provision for removing refuse material from the honey prior to the final straining —which is quite often carried out at high temperatures. In order to avoid the damage to honey resulting from heating it in the presence of beeswax, all packers should at least give honey a preliminary straining at a temperature not over 45°C (113°F), prior to putting it through the final strainers or filter-press. This can be accomplished by modifying the O.A.C. strainer (le Maistre, 1936) to connect it to an in-line type, or a standard in-line pressure strainer. Several standard types of pressure strainer are available. Whatever is used, it should be available in duplicate, so that—when the pressure builds up—the line can be switched over to the second strainer while the first is cleared. Prior to straining, it is essential to operate some form of baffling, settling or skimming, to remove the bulk of the coarse material and so to take the load off the strainers. This is quite often accomplished by letting the honey flow from the warm-room into a double-jacketed tank, where gentle mixing and hot-water heating brings the honey to 40°C (104°F); the honey is then skimmed. The use of the filter-press is dealt with in 9.7.

9.5   MELTING HONEY FOR REPACKING

Honey is usually stored in bulk containers and liquefied at some later date for repacking. The application of heat is the only satisfactory method for bringing honey from the granulated to the liquid state. Where not many containers are to be dealt with at one time, heating them in a water-bath is much more efficient than heating them in an air-bath (Büdel & Grziwa, 1959). It is, however, not practical to use water-baths for liquefying honey for packing purposes on a commercial scale, and therefore most liquefying rooms (called 'melting rooms') are heated by hot air. A number of factors govern the design of an efficient 'melting room'. Air in motion will transfer heat to the wall of the honey containers more than twice as rapidly as still air (Büdel & Grziwa, 1959). In order to speed up liquefaction of the honey, provision must be made for the honey to escape from the 'melting room' as soon as it is movable. The most suitable procedure is to allow the semi-liquid honey to run into a double-jacketed tank outside the heating room (where further heating is possible). Alternatively it can run into a catch-tank under the heating racks, where further heating can be carried out from the bottom, to dissolve crystals as they descend to the lower part of the tank. To facilitate the liquefying process, many packers have found it advisable to allow a small amount of steam to escape into the melting room. This is particularly beneficial if the moisture content of the honey is low. The optimum temperature for operating the room is 60–70°C (140–158°F) (Louveaux & Trubert, 1958). If the honey is allowed to escape as it becomes liquid enough to flow, the temperature of the room will not rise much above 57°C (135°F) during the melting process (Townsend, 1961). In most circumstances the honey will liquefy overnight.

When the honey to be liquefied is in drums, these are usually placed on their side on a sloping rack, with a screen support to prevent the core of honey leaving the drum as a solid mass (Figure 9.5/1—see Plate 10).

Barrels are coming into common use for bulk transport of honey. Only barrels with removable tops should be used. The most satisfactory method of emptying them is to invert the barrels of granulated honey over a 'grid' of hot-water heated pipes; a suitable temperature for the water is 60°C (140°F), although some packers use 80°C (176°F) to speed the operation. The pipes should not be more than 6 mm apart, and the room should be heated as usual.

Liquefying coarsely granulated honey can be troublesome if the liquid honey is allowed to flow away from the granulated part too rapidly. Coarse granulation generally occurs in honey that has been heated; it can be avoided by deliberate seeding of the honey after cooling (White, 1958); see Chapter 10.

9.6 BLENDING HONEY

Almost all packers blend honeys to meet certain requirements of the trade, yet there is very little published information on this subject. Stephen (1934) prepared a series of blending charts based on the Pfund colour grader. He drew attention to the rapidity with which the colour increases when blending takes place. This was particularly noticeable when the original honey was lighter than 70 mm on the Pfund grader. A practical system for calculating the blend of honey by using optical density or Pfund readings has now been developed (Townsend, 1969a, 1971).

A most important factor in blending honey is adequate mixing. When honeys of different moisture contents are mixed, the honey with the higher moisture content will layer out on the surface, particularly at temperatures around 32°C (90°F). Honeys of different moisture content will not separate if they have been heated, and mixed at a temperature above 71°C (160°F) (Fix & Palmer-Jones, 1947). Blending at low temperatures may be accomplished by using an in-line mixer developed for continuous-flow regranulation (Townsend, 1971).

9.7 FILTERING HONEY

Filtering honey means filtering under pressure, and this is not the same operation as straining. The filtering removes from honey the pollen and colloids, and some of the small air bubbles which normally are not removed by straining. Filtering can also be used to remove all of the colour from honey, but this is very rarely done—only when a badly coloured or flavoured honey, which has no market value, can be reduced to an invert sugar solution and sold as such. Processing honey without any filtering is discussed in 9.34.

Many beekeepers, packers, and honey analysts object to the filtering of honey (e.g. 7.48), on the grounds that honey from which the above materials have been removed is no longer truly honey. The filtering is done in North America to give a clear brilliant product, which the public in general seem to want. A filtered honey will remain liquid for a much longer period than unfiltered honey, and for this reason also most liquid honey sold in North America is now packed in a filtered form. The presence of fine particles of suspended material (pollen, etc.), colloids, and air bubbles, can produce a distinct cloudiness that detracts from the general appearance of honey, to a public accustomed to clear syrups and jellies. The substances removed are primary factors in the formation of surface foam and scum; the suspended particles and fine air bubbles also play a part in hastening granulation.

The method of processing honey by pressure filtering was first described by Lothrop & Paine (1934). Procedures and equipment have varied from packer to packer in recent years, but the general process has not changed to any great extent. The method consists of mixing a small proportion of inert filter-aid (diatomaceous earth) with the honey, heating it, and then passing it through a filter-press; this consists of plates on which are placed sheets of filter paper to remove the filter-aid with the materials adhering to it.

The filter-aid material is prepared ahead of time, to allow for removal of both air and moisture. The dry filter-aid is prepared in the form of a soft cake, by taking the requisite weight of diatomaceous earth, suspending it in water and bringing it to the boil to drive out the air which is adsorbed on the surface. The excess water is then removed by a vacuum filter, taking care that suction stops before any air is drawn into the cake. A honey solution (honey:water = 4:1 by weight) is immediately poured over the cake, and suction is continued until the layer of honey solution just disappears into the surface. The soft cake is then added to the honey in the mixing tank, and dispersed by means of an agitator running at such a speed that no air is incorporated; about 2 kg of filter-aid is needed for 450 kg of honey. The honey is then heated to 60–71°C (140–160°F), and pumped through the filter-press to remove the filter-aid and adhering materials.

In a recent variation of the process, the filter-press is first coated with the filter-aid by circulating a honey and filter-aid mixture through the filter without previously mixing the filter-aid with the whole of the honey. Many packers now cool the honey down to a 'filling' temperature (i.e. suitable for bottling) immediately after it leaves the filter press. With the introduction of fast cooling, some of the packers have moved the filtering temperature up to 77°C (171°F), which helps to delay granulation (Austin, 1953). In this process it is important to use a rapid-filter grade of diatomaceous earth; those most commonly used are High-flow Super Cell, Dicalite Special Speed Flow, and Johns-Manville Cellite 512.

Lothrop & Paine (1931a) described a process for removing a considerable proportion of colloids, colour, and flavour from certain honeys. The process essentially consisted of diluting the honey 50% with water, warming it to 66°C (151°F), and suspending bentonite in the honey at a concentration of 5%. After 20 minutes of agitation, the mixture was filtered under pressure with 2–2½% of a suitable filter-aid on a pre-coated surface, and then evaporated under vacuum until the density was that of honey. Lothrop & Paine found that there was better precipitation of the colloids at the isoelectric point; therefore in most of the work with bentonite the honey was brought to pH 4·3 by using a slurry of hydrated lime. White & Walton (1950) described a similar procedure, but used activated carbons such as Darko×51 and Neucar C190N in place of bentonite; diatomaceous filter-aids are also essential in the process. By using this procedure, it was possible to end up with a product which was very similar in composition to the original honey used (except for slightly lower nitrogen and slightly higher ash contents), but which had lost practically all its colour and flavour.

Kranz (1967) described a patented method for clarifying honey by distilling under vacuum, and replacing the distillate; it is claimed that the method de-aerates and clears the honey without damaging it or removing the aromatic components.

Wilson Honey Company (1963) described a somewhat similar process, using vacuum-plate heat-exchangers; the heat removes excess moisture and air during the process, which is capable of handling 1 300 kg of honey an hour.

9.8 GRANULATION OF HONEY

Granulation (crystallization) is one of the greatest problems the packer has to contend with in the preparation of a liquid pack of honey. A number of factors influence granulation, of which the most important is perhaps the fact that honey is a supersaturated solution of naturally granulating sugars. Honeys from different plant sources differ greatly as to their granulating tendency; this subject is dealt with in 6.72.

Many attempts have been made to delay the granulation of honey. Kaloyereas & Oertel (1958) have described the use of ultrasonic waves (9 kilocycles per second) for 15–30 minutes, which was reported to destroy all yeast cells and to delay granulation for 15 months. The same paper describes the use of two additives, reporting that 0·3% of *iso*butyric acid or of sorbic acid delayed granulation in honey, and moreover that feeding either of these materials to the bees during the nectar flow (so that it came mixed with the honey) also delayed granulation. No confirmation of these findings is known to the author. Moreover such additives would be unacceptable in most countries. Walsh (1960) recommends increasing the moisture content of the honey to 20%. Austin (1953) recommends heating the honey to 77°C (171°F) for 5 minutes, cooling rapidly to room temperature, bottling, and keeping the bottled honey at 0°C (32°F) until shipped to the wholesaler.

The most common practices among packers at the present time include the following. The honey is heated to 77°C (171°F), and filtered through a filter-press to remove all particles such as pollen grains that might act as nuclei for crystallization, the honey being held at 77°C (171°F) for at least 5 minutes; it is then cooled rapidly to 57°C (135°F), at which temperature the jars are filled and immediately capped. Some packers take further precautions, such as vacuum-cleaning the jars prior to filling, and cooling them after they have been filled. These practices, together with selection of honeys that tend to remain liquid for a fair length of time, have overcome most of the problems of unwanted granulation during normal shelf life.

9.9 STORAGE OF HONEY

It was pointed out at the beginning of this chapter that the main attractions of honey are its delicate flavour and aroma. Apart from improper processing, nothing else will injure honey so much as long storage under unsuitable conditions. Quite a number of the honey-importing countries have recognized this, and are using chemical analyses to determine whether honey offered for shipment has been abused; accumulation of hydroxymethylfurfuraldehyde, HMF (5.84), and reduction of the enzymes diastase and invertase (5.7) are two danger signals. In spite of this, many large shippers and packers in honey-exporting countries pay little attention to honey storage conditions. White, Kushnir & Subers (1964) suggest that, because of the nature of HMF formation, increase in HMF content is likely to be accompanied by darkening and loss of flavour. Storage at 20°C (68°F) will not prevent some HMF production; this varies according to the honey, and depends upon moisture content (Schade *et al.*, 1958). It therefore follows that honey should be stored at a temperature as low as is feasible, and marketed within the crop year if possible.

9.91   Storage of bulk honey

Two major dangers must be taken into consideration in the storage of raw (i.e. unprocessed) honey. The first—which will cause the greatest loss over the shortest time—is fermentation. Honey with less than 17·1% moisture is safe irrespective of yeast content; between 17·1% and 18%, the honey is safe provided the yeast count does not exceed 1 000 per gram; between 18·1% and 19%, it is safe if the yeast count does not exceed 10/g; and between 19·1% and 20%, it is safe if the yeast count does not exceed 1/g. All honeys with more than 20% moisture are in danger of fermentation (Lochhead & McMaster, 1931). This subject is also discussed in 6.11.

Honey in storage should always be held in tightly closed moisture-resistant containers. Considerable loss has been experienced in packing plants because moisture has been absorbed by the honey under conditions of high humidity.

Various methods have been suggested for preventing fermentation during storage. Addition of sodium benzoate at the rate of 0·025% will control fermentation (Lochhead & Farrell, 1930b); while this would possibly be permitted under most food laws, it is not desired by the beekeeping industry, and it has therefore never been practised. Storing honey below 11°C (52°F) will prevent fermentation (Marvin, 1928); if this cannot be done, and if the moisture content of the honey is in the range where fermentation is likely to set in, then the yeasts must be destroyed before the honey is stored for any length of time. This is done by rapid heating and cooling of the honey: yeasts are destroyed by heating at 63°C for 7½ minutes, 69° for 1 minute, or 71° instantaneously (145, 156, 160°F respectively) (Townsend, 1939).

The second danger is damage by unsuitable storage temperature (*see also* 6.11.2), especially if stocks of honey are held for any period of time. A reduction of 5–9°C (9–16°F) in the storage temperature of honey will reduce the rate of HMF production to one-third, enzyme loss to one-fifth, and darkening rate to one-sixth (White, Kushnir & Subers, 1964). 'White' honeys stored at 21°C (70°F) will darken at the rate of 1 mm Pfund every four months, light amber honeys at almost 3 mm in the same period and at the same temperature. The rate increases very rapidly as the temperature rises above 10°C (50°F) (Milum, 1948). At 71°C (160°F), half the invertase is destroyed in 40 minutes and half of the diastase in 4½ hours; in 5 hours sufficient HMF is produced for the honey to be rejected on some export markets (White, Kushnir & Subers, 1964).

Ralph Stone of Montana (Stone, 1967) is now providing beekeepers with storage tanks holding 25 and 50 tons. The honey is picked up from this storage in trucks with a holding capacity of 20 tons or tank cars holding up to 55 tons. The honey is moved directly to the packer, or is stored at headquarters in 1 000-ton storage tanks (Figure 9.91/1—*see* Plate 10). All the storage and transport equipment is provided with equipment for heating and pumping the honey.

9.92 Storage of packed liquid honey
Honey can be preserved in a liquid state over a prolonged period if it has first been stored at 0°C (32°F) for at least 5 weeks. Honey samples stored for 5 weeks at 0°C, and then at 14°C (57°F), showed no granulation for 2 years, whereas samples of the same honey stored at 14°C without the previous storage at 0°C had crystallized within 5 weeks (Austin, 1953).

'Stack heat', resulting from the high thermal capacity of large stacks of bottled honey, can be a serious problem, as it is often the practice to fill the jars at 57°C (135°F). The importance of avoiding stack heat is shown by the following example: honey heated to 49°C (120°F) for bottling was piled in stacks in a room held at −12°C (+10°F). Honey in the centre of the stack cooled by only 17°C (31°F) in 11 hours (Austin, 1953). For optimum conditions in the storage of liquid honey, the cartons should be spaced out with forced circulation of air around them, and the honey should be held at a temperature of 0°C (32°F) for at least 5 weeks; the stocks could then be removed to a higher temperature for further storage. Precooling of the bottles prior to placing them in cartons is also strongly recommended.

9.10 QUALITY CONTROL
This aspect of honey processing is discussed briefly here, with particular reference to North American practice. Chapter 12 deals with the subject at greater length, and especially in relation to the European Codex

(1969) and the 1974 E.E.C. Directive. Both treatments are self-contained, and any differences in recommendations reflect differences between practices in the two continents.

Quality control is now referred to as the hallmark of the food industry, the keynote of competition. The slightest deviation—in flavour, colour, aroma or consistency—from the usual quality associated with a brand will mark the product for rejection. Future sales and business in most food industries are built up on quality control. Because of the varying sources of supply, and the ease with which honey may be injured, the honey industry is one that needs quality control. All incoming shipments should be checked for colour, moisture, flavour, aroma, source if possible, and sediment. From this information it is possible to determine whether the shipment should be designated for bottled liquid or processed granulated honey, whether it should be sold immediately or can be stored for some time, and how it can best be used in blending. Most packers routinely retain samples of their final packs, to determine the shelf life and to have a record of their performance.

Certain market areas become used to one type of flavour, and a major deviation from this will lose customers. All honey shipments should be checked for flavour and aroma, and any with a noticeable deviation should be excluded from the domestic market. Flavours will vary from year to year, and should be watched very carefully in areas where the honey crops have been low, and where new nectar plants have been grown. Examples of honeys that may be accepted well on one market, but refused on another, are honeydew, rape, buckwheat, and heather.

9.10.1 Moisture content

Moisture content is very important in both the liquid and the processed granulated pack. A variation of 1% in liquid honey can influence the rate of granulation, ease of serving, and even flavour. Moisture influences the texture of granulation in regranulated honey. It may even be wise to vary the moisture content of honey offered for sale according to market and season. The glucose/water ratio (see 5.31) is possibly the main factor controlling the granulation of liquid honey. The higher the moisture content the less chance of granulation, so it is wise to consider packing liquid honey containing 18% rather than 17% moisture. In no circumstances would it be considered wise in North America to pack liquid honey with more than 19% moisture; the product would be too thin. For any moisture content above 17%, however, honey packs should be handled in such a way that fermentation will be avoided.

Moisture content must also be taken into account when bulk honey is provided to the baking trade. Variations in moisture content, as well as in flavour, will introduce problems in duplicating certain recipes.

Methods for determining the moisture content of honey are covered in 6.11. The most useful and the most reasonably priced of the field instruments for the purpose is a hand refractometer manufactured by the Atago Optical Works, Tokyo, Japan. This instrument is specially designed for honey, with necessary corrections and adjustments built in, and with a field of view that is easy to read.

9.10.2 Colour

There is no standard method throughout the world for the classifying of honey by colour; each country has its own approach. When blending is practised, as is usual with most packers, a fine colour classification of honey is essential. Most packers find the most useful instrument for this purpose is the Pfund colour grader (Sechrist, 1925), manufactured by the Koehler Manufacturing Company in the United States. Although there is rather a wide margin of error in reading this instrument, it has been used as a basis for developing colour standards for honey. Perhaps the most useful field type of grading instrument is the set of permanent glass colour standards for extracted honey developed in the United States Department of Agriculture; it is based on a simple comparator, with permanent glass standards, which was developed for maple syrup. A description of this equipment is available (Brice et al., 1956), with a complete documentation of light transmittance values, etc. A similar instrument, with slight changes in the White and Extra White classes, has been adopted in Canada. Optical density can now be used for direct determination of the quantities of different honeys needed for blending to a specific colour grade (Townsend, 1969a).

9.10.3 Pollen and sediment

All honey samples should be checked for excess quantities of small wax particles and pollen. Excess wax should be removed at low temperatures; if there is excess pollen, the honey should be set aside for regranulation. If a liquid pack is to be prepared by the use of a filter-press at high temperature, the use of honeys with a high pollen content is practically eliminated, and the removal of wax prior to heating is mandatory; otherwise the honey will inevitably be darkened and the flavour impaired. Pollen will concentrate on the filter pads, and the continued application of heat will injure the flavour of the honey as this passes over the pollen. The wax and pollen content of honey can be measured quite readily by diluting a sample of honey with 50% water, warming, and passing through a sediment-measuring unit by suction (Townsend, 1965).

9.10.4 Detection of granulation

It is often advisable to follow the onset and development of crystallization in liquid packs of honey. For this purpose polarized light may be used to advantage, since it enables crystals to be detected which are not visible to the naked eye. The honey container is placed between two polaroid screens with a light behind, the screens being correctly oriented in relation to each other. With this piece of equipment small crystals can be detected quite readily (White & Maher, 1951).

9.11 HONEY CONTAINERS

Honey containers vary from small tubes for samples to large drums holding 300 kg, used for bulk shipping. Formerly most of the honey for retail sale was packed in tinned metal containers, waxed cardboard or glass. In recent years, with the introduction of new materials, the trend is toward plastics. Availability, cost or weight may determine the preference in some circumstances, but attractiveness and usefulness are important considerations. There are several basic factors. A honey container must be odourless; there must be no exposed metal which will react with honey; the container must be moisture-proof; the honey must be readily removable from it.

Steel or galvanized containers are usually protected with either tinning or a baked-on lacquer. The lacquer must be durable enough (it must not scratch easily), and the tinning must be of a type that does not have any exposed pock marks. If a seam in the container is necessary, it should be covered well with a good lacquer.

Moisture-proofing of containers is most important. Some work carried out in 1942–1943, when there was a shortage of certain materials, gave the results shown in Table 9.11/1 (Townsend, 1943). Tinned metal containers with press-on lids and glass jars with screw-on lids were almost

Table 9.11/1
Moisture absorption by honey stored for 30 days: percentage gain in weight

Type of container	Relative humidity of atmosphere	
	90–95%	75%
Glass	0·07%	0·00%
Metal (tinned)	0·05%	0·00%
Fibre board (waxed)	0·50%	0·00%
Paper (waxed)	0·90%	0·00%
Fibre board (not waxed)	8·50%	1·00%

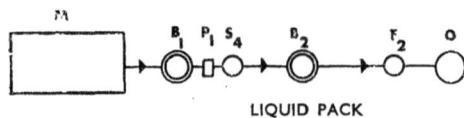

LIQUID PACK

Figure 9.12/1 Layout of processing/packing plant, handling up to 200 kg per hour, suitable for a beekeeper operating in his honey house.

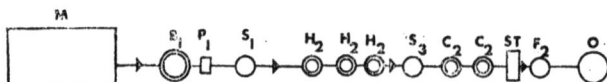

Figure 9.12/2 Layout of processing/packing plant, handling up to 200 kg per hour on a continuous-flow basis, suitable for a beekeeper-packer or small-scale packer.

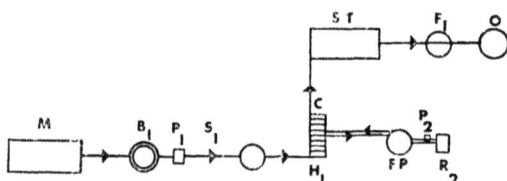

Figure 9.12/3 Layout of processing/packing plant, handling 200–750 kg per hour.

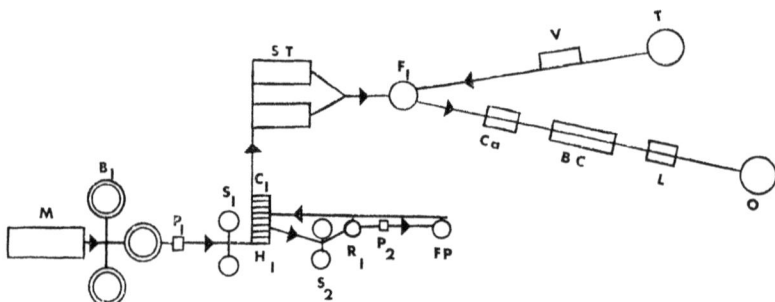

Figure 9.12/4 Layout of processing/packing plant, handling 750 kg per hour or more.

Key
B_1 jacketed tanks (blend, skim, heat to 40°C (104°F)
B_2 jacketed tanks for heating and cooling
BC bottle cooler (water spray, tap temperature)
C_1 plate-type cooling exchanger (cool to 57°C (135°F))
C_2 tube-type cooling exchanger (cool to 57°C, *see* diagram)

Ca bottle capper
F_1 automatic bottle filler
F_2 filler, tap-type
FP filter-press
H_1 plate-type stainless steel heater-exchanger (heat to 77°C (171°F))
H_2 tube-type stainless steel heat-exchanger (heat to 77°C, *see* diagram)
L bottle labeller
M 'melting' room ($=$liquefying-, warm-, hot-room)
O full-bottle collector
P_1 pump
P_2 pump, stainless steel, suitable for filter-aid
R_1 tank for pressure relief and coating filter-press with filter-aid
R_2 tank for pressure relief and coating tank for filter-press
S_1 Strainer (in-line) 35 mesh per cm
S_2 strainer (in-line) 60 mesh per cm; tank size to hold honey 3–4 minutes
S_3 strainer (in-line) 60 mesh per cm; small, with layer of sand over screen
S_4 strainer (O.A.C. gravity type), 35 mesh per cm in the final screen
ST storage tank above filler
T empty bottle starter
V bottle vacuum cleaner

air-tight. Waxed paper or waxed fibre containers picked up some moisture at high humidity.

Unwaxed board containers were definitely unsatisfactory (Villumstad, 1951).

9.11.1 Bulk containers
In the past, most bulk honey was packed in 30-kg tins or drums. In recent years the trend has been towards 300-kg lacquered drums with completely removable lids, or 30-kg corrugated-board cartons and 300-kg drums lined with a pliofilm (plastic) bag. The corrugated-board carton cannot be shipped safely until the honey has granulated. Gasoline-type barrels with small openings are not suitable, as the granulated honey cannot be readily removed without injuring it.

9.11.2 Consumer containers
Formerly most consumer containers were of metal, glass or paper-board. The trend at the present time is to opaque and clear plastic, and glass. Some regranulated honey, as well as liquid honey, is finding its way into glass and clear plastic containers. Clear plastic may find increased use in the near future for export shipments or prepacked liquid honey, as it is much lighter in weight than glass and it is proving popular for novelty non-drip containers for table use. Unfortunately most plastic materials at present available are brittle. On a few occasions plastic has been blamed for giving an unpleasant odour to honey, but this has always been traced to the carbolic in the ink used for stamping the name and

address on the lid. If these stamped lids are allowed to air-dry before they are stacked on top of each other, this trouble can be avoided (Townsend, 1965a—unpublished).

Honey may be quite satisfactorily packed in aluminium containers (Hugony, 1953), but it readily picks up zinc from zinc-plated containers unless these are heavily waxed (Kalinowska, 1957).

9.12 PROCESSING AND PACKING PLANT LAYOUTS

The type of equipment used must depend upon the circumstances, which may vary widely. Four layouts for processing honey, finishing with a liquid pack, are described in Figures 9.12/1 to 9.12/4, using what the author believes to be the most suitable equipment available at the present time for the size of the operation. Various advantageous optional steps may be included (such as vacuum-cleaning of containers, fast cooling of the liquid package, and cold storage). The large-scale commercial method should include all these options. A very good description and comparison of commercial methods of packing honey in Canada and U.S.A. has been published in France (Borneck *et al.*, 1964).

When designing a plant, or planning to use any new equipment, it is most important to deal with firms which have had experience with honey, unless the operator is prepared to do some experimenting himself and this can prove rather costly.

CHAPTER 10

PRODUCING FINELY GRANULATED OR CREAMED HONEY

by Dr. E. J. Dyce

EMERITUS PROFESSOR OF APICULTURE, CORNELL UNIVERSITY, U.S.A.

Brown-gleaming comb wherein sleeps crystallized
All the hot perfume of the heathery slope.
MARTIN ARMSTRONG,
COLLECTED POEMS (1931)

10.1 INTRODUCTION

Of all the physical phenomena observed in honey, that of crystal formation is possibly the most interesting, and certainly one of the most important to the honey industry. Crystallization or granulation must be prevented in honey which is to be offered to the trade in liquid form; equally important, it must be controlled in honey which is to be sold in granulated form. Most honeys tend to granulate within a few days or weeks after they are extracted from the combs, although a few, such as sage (*Salvia*) and tupelo (*Nyssa*) may remain liquid for months or even years (6.71).

Most normal honeys are supersaturated with respect to glucose (dextrose) at normal room temperatures, and part of the glucose tends to separate out in the form of crystals, giving the honey the appearance of solidity (6.7). As the dextrose crystals form, they build up an internal network of crystals which holds all the other constituents of the honey in an immobile state. Since dextrose crystals are pure white, honeys become lighter in colour as the granulation progresses. Light honeys may become white when completely granulated, but the darker honeys still retain a brownish cast.

Over the years most countries throughout the world have sold, and still sell, the bulk of their honey in the granulated form. Among the main exceptions to this are the United States and Australia, where liquid honey is more customary. Yet the neighbouring countries, Canada and

New Zealand, have continued to sell most of their crops as granulated honey, on both the home and export market.

Honey is potentially a perishable product, and loss from fermentation has presented serious problems. Heating honey to about 71°C (160°F) for a few minutes, and subsequently selling it in the liquid form, is the simplest and most practical method of controlling fermentation (9.3). It is therefore not surprising that beekeepers in some countries offered their crop for sale mainly as liquid honey. But in following this practice, much honey has been reduced in quality, owing to the almost universal practice of overheating the honey in an effort to delay regranulation as long as possible. Chapter 9 describes the ways to minimize the danger of this damage through overheating.

10.2 HISTORY OF THE FERMENTATION AND CRYSTALLIZATION PROBLEM

During and immediately after World War I, sweet clover (*Melilotus*) was introduced to various countries as a forage crop. Its use spread rapidly, especially in Canada and the United States, and this greatly increased the amount of honey produced. By 1920–1922 there was not enough demand for the honey, prices rapidly declined, and it soon became necessary to hold substantial portions of each season's crop from one year to the next. Much loss was experienced from fermentation, and bee-keepers were in serious financial difficulties. In Canada the Ontario Honey Producers' Co-operative was founded in 1923 to deal with these problems; they estimated their annual loss from fermentation as at least $10 000.

An effort was made by the Ontario Co-operative to develop an export market, especially in Great Britain, but fermentation and coarse granulation of honey retarded this development also. In England, for example, many buyers were willing to pay up to 20% extra for good quality, finely granulated honey. Many small packers and shopkeepers followed the practice of softening honey which had granulated in the bulk containers just enough to permit it to be poured into retail glass jars. The honey remained soft and was pleasant to eat, and it did not form coarse crystals or return again to its former hard state.

Honey with fine crystals is more palatable and satisfying to the person who eats it than honey with coarse 'gritty' crystals. Strong-flavoured honeys are usually mellowed and greatly improved in flavour when finely granulated, and persons who dislike either liquid or coarsely granulated buckwheat honey often enjoy it in the finely granulated form. For successful marketing, therefore, granulated honey should contain no coarse crystals.

10.3 EFFORTS TO CONTROL FERMENTATION IN GRANULATED HONEY

It became abundantly clear soon after the end of World War I that honey producers needed much help. Many beekeepers believed that if fermentation and granulation of honey could be controlled in a practical way, it would do more than anything else to bring back prosperity to the beekeeping industry. The situation became especially acute in Canada and the United States, with the result that investigators there started working on the problem.

Honey packers keep a close watch for bulged metal tins and other signs of fermentation in their bulk honey containers. It was well known that granulated honeys ferment more readily than liquid honeys, whether or not the honeys have been heated to keep them liquid (6.11). This is especially true in late winter, after granulation is complete, and when the weather commences to turn warm in the spring. The reason why granulated honeys ferment more readily is that, when dextrose crystals are formed in honey, the liquid phase has a higher water content than the entire honey had when it was all liquid (6.7).

When dextrose forms crystals in a water solution, the composition of the crystals is that of dextrose hydrate, and the water of crystallization comprises only 9·1% of the solid material. Since the water content of honey is around 18%, the liquid phase remaining when crystals form has an increased water content, and provides a favourable medium for the growth of yeasts. Only a slight increase over the original water content of the honey is required for fermentation to proceed. Sugar-tolerant yeasts are normally present in all honeys, and are able to multiply if and when conditions prove suitable. Whether or not the honey is granulated, it can sometimes happen that the moisture content is slightly higher near the surface than elsewhere, but the *significant* change in moisture content in honey is that at the onset of granulation, when dextrose hydrate crystallizes out.

Marvin (1928) suggested that since yeasts are incapable of growth at temperatures below 11°C (52°F), fermentation might be controlled by keeping the honey in cold storage. Some honey has been held in cold storage, but the practice has never been used to any great extent, mainly because of the expense involved. However, to reduce the possibility of fermentation, honey should be stored in dry, cool buildings (9.9); honey is hygroscopic and readily absorbs water (6.8).

Wilson & Marvin (1932) found that the optimum temperatures for honey fermentation are between 13° and 21°C (55–70°F), and that above 27°C (80°F) fermentation develops slowly in unripe honeys, but probably not at all in well ripened honeys. Honey deteriorates in quality roughly according to the temperature and the period of storage at that temperature. In other words, honey held at 30–32°C (85–90°F) for periods of days or weeks is likely to be damaged by heat just as much as honey held at 71°C (160°F) for several hours (White, Kushnir & Subers, 1964). Holding honeys at high temperatures does not provide a practical method for preventing fermentation.

Lochhead & Farrell (1930b) tested several preservatives as a means of controlling fermentation. They selected chemicals of low toxicity, which most laws permit in foods, for the prevention of spoilage. Of the materials tested, sodium benzoate, sodium sulphite, and sodium bisulphite were considered the most promising. Ever since the days of honey adulteration and the passage of the Pure Food Laws in the U.S.A. in 1906, beekeepers have been opposed to the use of any additives for honey which might be construed as adulteration. Possibly for this reason, the use of chemicals to control the fermentation of honey never became popular (see 9.91).

10.4 EFFORTS TO PRODUCE FINE GRANULATION IN HONEY

The quality of granulated honey is influenced by the size of the crystals, so control of this size is of great importance. Most foods which compete with honey in the shops are highly refined and standardized, and as a result many consumers of honey desire a uniform product. There is, however, now a growing demand for non-standardized honey, i.e. packs of different varieties from specific areas or plant sources. But all consumers want good quality and well prepared honey; if inferior honey is purchased, the consumer's faith in the product is lost and other foods are chosen instead. It was therefore highly desirable that a process be devised for preparing for market a finely and uniformly crystallized honey which would not ferment.

Well before 1930, a few beekeepers in Canada, New Zealand, and a few other countries had added naturally granulated honey as a 'starter' to freshly extracted honey, in the hope that the new batch of honey would form fine crystals. The method did not become popular, because the honey often fermented, and there was no control over the size of the crystals that developed. Fine crystals developed in some batches of honey, and coarse crystals in others.

Some beekeepers attempted to bring about a condition of fine granulation by grinding naturally granulated honey in various types of mill. The grinding process invariably incorporated many small air bubbles in the honey. The flavour of honey was improved by this process, but the honey developed an objectionable appearance and often fermented prematurely. This came about because the incorporated air bubbles gradually rose to the surface of the honey, forming a thick scum, while the larger crystals settled to the bottom. For this reason the grinding of honey to produce fine granulation never became popular in any country.

Heating honey to control fermentation seemed to be the best method, but the crystals which appeared later were so large and coarse that the honey was entirely unsuitable for the market. Nevertheless, the only course left for preventing fermentation was *first* to kill the yeasts and *then* to control the crystal size in the honey. The control of crystallization would not only solve the fermentation problem; it would also give a standard texture to the produce and thus improve its quality.

10.5 RESEARCH ON THE GRANULATION PROBLEM

In 1928 Dyce (1931*a*) commenced a detailed study of the fermentation and crystallization of honey. His experiments brought to light much new information, of which the following are some of the most significant points.

10.51 Influence of temperature on crystallization

It was found that honeys of average consistency crystallized most rapidly at a constant temperature of 14°C (57°F), and not at fluctuating and low temperatures—as commonly stated. Honey with a low moisture content crystallized faster at 15°C (58–59°F), and those with more moisture granulated better at slightly lower temperatures. Few crystallized satisfactorily at temperatures above 16°C (60°F), the crystals which developed being too large and gritty. Below 10°C (50°F) crystallization was greatly retarded, and below 4½°C (40°F) there was almost no crystal growth. Samples with starter added and left at −1°C (30°F) for nearly two years had no visible signs of crystallization.

The rapidity with which dextrose crystals form in honey is affected by changes in viscosity which occur with changes in temperature: honey is 'thinner' when warm and 'thicker' when cool. Dextrose crystals are unable to grow unless there is molecular movement within the honey. When the temperature is reduced, supersaturation of the dextrose in honey is increased—favouring crystal growth—but the lower temperature also increases viscosity, and this reduces the diffusion of dextrose and so retards the formation of crystals. There is a critical temperature for each honey at which the effects of supersaturation and of viscosity balance out, and at this temperature crystallization progresses most rapidly. For the honeys tested, this critical granulation temperature was, on average, about 14°C (57°F). This was the most valuable discovery in the entire work.

10.52   Nuclei and their effect on crystallization

Various materials stimulate crystal formation in honeys. Pure dextrose hydrate crystals are highly effective, as is crystallized honey. Honeys containing air bubbles, considerable amounts of pollen, or other particulate elements, crystallize much more quickly than honeys which are free of such materials. When the lid of any one jar in a batch of cool liquid honey is removed, even for a second, the honey in that jar will granulate more quickly than the rest. Particles floating in the air enter the honey and act as nuclei for crystal formation. High 'pasteurizing' temperatures (9.33) will usually eliminate nuclei which consist of crystals or air bubbles.

Some beekeepers prefer to get combs well cleaned of honey by the bees before storing them for winter, whereas others like to store combs without such cleaning. The two methods are often referred to as dry and wet storage. Samples of honey from combs that had been stored wet during the previous winter developed smaller crystals than the honey from combs that had been stored dry. It would appear that minute dextrose crystals developed in the film of honey left in the wet combs, and when the honey crop was extracted in the following year, these crystals became well mixed in the honey, and led to the formation of crystals which were much smaller than normal.

It is well known that crystals do not form as quickly in comb honey as in extracted honey from the same apiary. In the process of extracting, the honey is subjected to much agitation, and this affects the rate of subsequent granulation. Extracted honeys that were run through a honey pump granulated faster and developed finer crystals than others, presumably because some crystals present in the honey were fractured, and the consequent increase in the number of crystal nuclei in turn hastened crystallization.

10.53   Growth of dextrose crystals in honey

It is interesting to observe the growth of dextrose crystals in honey. A drop of honey containing a few small dextrose crystal fragments is placed on a glass slide and observed under a microscope every day. The slide is kept at 14°C (57°F) to ensure maximum growth of crystals. At first needles are formed; these gradually broaden into flat plates, with an area at the blunt or lower end from which other crystals grow in all directions. The upper end of the crystal plate is pointed, but the point is not exactly in the centre of the crystal.

When rapid crystallization occurs under ideal conditions, the crystals seldom grow beyond the stage of flat plates. Such plates are characteristic of fine granulation and feel smooth to the tongue. When crystallization is retarded, many crystals gradually grow from the base of each crystal, forming large star-shaped masses, and these give honey a 'gritty' texture. Such large crystal masses are often seen at the bottom of jars of liquid honey in grocery stores.

10.54   Amount of starter required

The number of crystal nuclei (or the amount of granulated honey), used as the 'starter' for 'seeding' liquid honey, directly influences the size of crystals that are formed. If many starter crystals are present, and they are evenly distributed throughout the honey, the dextrose in excess of saturation is rapidly drawn to these points of growth. Each crystal grows at approximately the same rate, so the surplus dextrose is depleted before the crystals become too large. If on the other hand only a few nuclei are present, most of the dextrose must diffuse or travel a long way before reaching a nucleus. The rate of crystal growth is therefore very slow, and the resulting crystals are small in number and large in size. So it is important to add sufficient nuclei to the liquid honey to obtain the results desired. The starter is usually in the form of previously processed honey, and a minimum of 5% and maximum of 15% by weight was found to give good results; a practical working average is 10%.

When 5% of finely granulated honey was thoroughly stirred into cool liquid honey and the resulting mixture was left undisturbed at room temperature, the crystals which eventually developed were always very coarse. When the same honey mixture (also held at room temperature) was stirred every day for several days, the resulting crystals were small, and the rate of granulation was greatly increased. This was presumably due to the fracturing of crystals by the daily agitation of the honey, which provided a greatly increased number of crystal nuclei. In commercial practice the time and labour required by such a method of producing finely granulated honey would be too costly. In some countries beekeepers stir their seeded honey each day for two or three days, at relatively low temperatures. This procedure, also a costly one, occasionally produces a good finely granulated honey.

10.55   All normal honey may be finely granulated

Honeys from various sources were secured from different parts of the United States and Canada (Dyce, 1931a). All of them reacted to the process and developed fine crystals. The honeys low in dextrose did not granulate quite as rapidly or as firmly as those high in dextrose, but the final product was mostly quite satisfactory. Large batches of honey were processed, and they reacted to the process in the same way as in the small-scale experiments.

After three years of research in Canada and the United States and the accumulation of much new information about the physical properties of honey, Dyce (1931*a*) developed a process by which fermentation and crystallization in all normal honeys can be controlled.

10.56   Response of heated honeys to induced granulation

Honeys which had been heated to destroy the yeasts, and then cooled to room temperature, were found to respond to seeding and regranulation in about the same way as freshly extracted unheated honeys. Honeys that were heated even to excess to dissolve coarse crystals also responded to the process. This was also an important addition to our knowledge of granulation.

Most of the scientific facts leading up to the development of the process have now been briefly reviewed, and it is possible to discuss each step of the commercial process for preparing finely granulated honey for market. It is, however, not necessary to use large expensive equipment to make good granulated honey. If the basic principles are followed, excellent granulated honey can be made in a small pail, using a kitchen spoon to stir in the starter.

10.6   THE DYCE METHOD OF PROCESSING HONEY

10.61   Heating, blending, and straining

It should be assumed that every honey contains sugar-tolerant yeasts which may cause fermentation whenever conditions become favourable for their growth. It should also be assumed that most honeys, even those recently extracted, contain some dextrose crystals. For these reasons all honeys to be finely granulated should be heated sufficiently to destroy the yeasts and to dissolve any dextrose crystals present. Honeys high in water content should be blended with honeys low in water content, so that the honey processed will not contain more than $17\frac{1}{2}$ or 18% of water. This precaution will usually result in a spreadable product which is not too hard or too soft. After the honey has been heated, it should be thoroughly strained to remove particles of wax and any other foreign material which might be present. These processes are dealt with in Sections 9.3 and 9.4, and only a few points will be made here.

The use of two steps in the heating process is recommended. The honey is first heated to 49°C (120°F) and then strained to remove all foreign materials. It is essential to remove any wax particles before they liquefy. The honey should be heated to 66°C (150°F), preferably in a modern-type heat-exchanger, and then given a final straining. It is not necessary to heat the honey to 71°C (160°F) or to filter it (9.7), but straining should be sufficiently thorough to remove all visible particles. This may be accomplished by running the honey through a wire-mesh screen which has at least 40 meshes/cm (100/in.). All the steps in handling honey up to this point are discussed in Chapter 9.

10.62 Cooling

The honey should be cooled as rapidly as possible to room temperature— or below this if the equipment is sturdy enough to operate with the honey at lower temperatures. Different types of equipment are used to cool honey; these are discussed in Chapter 9, which also stresses the care that should be taken to avoid the incorporation of air. When too many air bubbles are present, they rise to the surface of the honey and form an objectionable scum. Honey pumps that are run too fast are one common source of the incorporation of air.

The more honey is cooled the more viscous it becomes. When the temperature of honey is reduced in double-jacketed tanks or in pipes, the honey lying next to the cold walls forms a thick insulating layer. Unless this layer of cold honey is constantly removed, it greatly retards the transfer of heat, and the speed at which the rest of the honey is cooled. Equipment is available to remove this viscous layer as it is formed; large cooling rolls such as are used in the lard industry are very efficient. The honey is picked up on one side of the roll as this turns, and is scraped off on the other side. It is important that the long knife which removes the cooled honey from the roll should be installed below the surface of honey, to minimize the incorporation of air.

Each cooling roll should be installed directly above (but in contact with) the honey in the tank. As the honey is cooled and removed from the roller it is slowly mixed back into the tank of warm honey. With this equipment the temperature of an entire batch of honey may gradually be reduced to 24°C (75°F), or lower, depending on the cooling agent or refrigerant. The tanks are normally equipped with agitators near the bottom; these mix the cool honey with the warm honey, and also mix in the starter after the honey is sufficiently cooled. The rolls should turn very slowly—again to avoid incorporation of air bubbles.

In the 1950s the Votator appeared on the market; this is a relatively small but highly efficient unit for cooling viscous materials. It is completely closed, and designed to fit in a pipe-line of a continuous-flow process. The first model that was tested in a honey plant was not satisfactory, but with some modifications the new models (Figures 10.62/1 and 10.62/2—*see* Plate 11) are capable of cooling the honey down to 14°C (57°F). By doing this—instead of adding the starter at room temperature—less starter is needed, and less time is required for the honey to crystallize. The Votator unit is made of stainless steel; it is quite expensive to purchase and rather costly to operate.

Double-jacketed cooling units built on the principle of an ice cream freezer are satisfactory, providing the scrapers are successful in removing the viscous honey from the walls, and are turned slowly enough to avoid the incorporation of air bubbles. A few of the coolers of this type have been carelessly built, and they are not satisfactory for honey.

Several honey plants use standard dairy vats, with coils or pipes that revolve about a horizontal axis (Figure 10.62/3—*see* Plate 12). Cold water or a refrigerant is passed through the revolving coil. These coil vats are economical to purchase, but are not as efficient as some other types of cooler. It may be necessary to raise the walls of the vat, so that the coil is always immersed at least 15 cm (7·8 in.) below the surface of the honey, which will prevent the incorporation of air bubbles. It is difficult to cool the honey below 24°C (75°F). As the coil turns in the honey, some of the cool viscous honey which accumulates on it is removed, but not enough is rubbed off for rapid cooling. However, this type of coil-vat does an excellent job of incorporating the starter. Dairy vats with coils rotating about a vertical axis are not satisfactory for cooling honey.

10.63 Introducing the starter
When the temperature of the honey is reduced to 24°C (75°F) or below, about 10% by weight of starter (fine, creamy, previously processed granulated honey) is thoroughly mixed into the honey. It is not necessary to add a larger amount of starter, but the crystals should be fine; large crystals added to honey will continue to grow, and the final product will be 'gritty'. In a closed honey-packing system the correct amount of seed may be metered into the cooled honey by a small honey pump operating at the required constant speed (Figure 10.63/1—*see* Plate 12).

The temperature of the honey at which the starter is added need not be as low as 14°C (57°F), but it must be low enough to prevent liquefaction of the tiny crystals. The consistency of the honey at about 24°C (75°F) is such that a thorough mixing of the starter with the liquid honey is easily accomplished. If the temperature of the honey is much lower, mixing is more difficult unless special heavy-duty equipment is used. With honeys of exceptionally low water content, it is desirable to introduce the starter at a slightly higher temperature; for honeys with a higher water content a temperature slightly below 24°C (75°F) is more favourable.

The honey to be used as starter is usually placed in rather large containers, such as 60-lb (27-kg) honey cans with their tops cut out. Flared containers, larger at the top than at the bottom, are preferred. The honey to which the starter has been added is placed in a cold room, and left to harden, like honey packed in smaller containers for the retail trade. When the honey is hard and completely crystallized it may be removed from the cold-room and allowed to soften slightly at ordinary room temperatures. It is then run through some type of meat or paint grinder which will break it up into a creamy consistency (Figure 10.63/1). A few packers use large meat grinders for this purpose, and some have built large special heavy-duty breakers which will handle cold blocks of crystallized honey.

10.64   Preparing the starter

One of the first steps in the commercial application of the process was to prepare yeast-free finely granulated honey for use as starter. This was accomplished by grinding a small amount of coarsely crystallized honey that had been previously heated. The ground honey was stirred into a few kilograms of pasteurized cool liquid honey and the mixture was placed in a cold-room at 14°C (57°F). After the honey was crystallized, in about six days, it was removed and again run through the grinder to break up the dextrose crystals. This ground honey was used as starter for a larger batch of pasteurized honey. In this way it was not difficult to build up an ample supply of yeast-free finely granulated starter, which is one of the most important requirements in the entire process. The use of unheated naturally granulated honey as a starter would defeat the whole purpose of the process.

Nowadays it is unnecessary to build yeast-free starter by the method just described, since any finely granulated honey that is prepared for market by a reputable honey packer who uses the Dyce process will be suitable. However, each time the starter is added to a new batch of honey, it should first be run through an efficient grinder. If this procedure is not followed, the resulting honey—and the next supply of starter— will become progressively coarser. Failure to carry out this requirement has caused packers more difficulties than any other single point in the process. It is impossible to pack high quality finely granulated honey year after year without regularly grinding the starter. A few packers consider this phase of the process so significant that they grind their starter in a special room not open to the public.

10.65   Filling the containers for market

After the starter is thoroughly mixed into the pasteurized and cooled honey, this is left to settle for an hour or two, skimmed, and then run into containers of the desired size for market. The settling period is necessary to permit the majority of the large air bubbles to rise to the surface of the honey, so that the final product does not appear frothy.

Small air bubbles which rise to the surface of processed granulated honey give a white appearance which is referred to as surface frosting. A slight frosting is difficult to avoid in most packing plants, but every effort should be made to avoid a thick scum, which has an objectionable appearance, and raises questions in the mind of the consumer as to the purity of the product.

Comparatively little granulated honey is packed in glass, which magnifies any slight imperfections. When for example the starter is not thoroughly mixed into the honey, little ribbons of starter (containing more air bubbles than the starter itself) are readily seen through the glass. And it is almost impossible to grind the starter without incorporating some small air bubbles. A small speck of foreign material, not removed by the straining process, may appear much magnified by the glass.

When honey is completely granulated and then exposed to freezing temperatures, for instance during transportation, the honey will shrink and pull away from the sides of the glass. Air then slips into this space, and when the jars are returned to room temperature, air bubbles are trapped between the honey and walls of the glass, giving the honey a streaked and frothy appearance which is also sometimes referred to as frosting. This is another reason why granulated honey is not usually packed in glass. Control of the problem is discussed in 10.67.

Completely granulated honey appears solid; yet only about 15% of the honey is actually in the solid state. When dextrose crystals form in honey, they grow in all directions, and mesh with each other at all manner of angles. In doing so they form a strong internal structure, which holds the parts of the honey that are still in the liquid phase between the crystals, in what appears to be a solid mass. If the internal crystalline structure of honey is disturbed by stirring or heating, the entire honey is softened. When this occurs after honey is completely granulated, the honey will never again return to its earlier solid state unless the honey is liquefied and processed all over again. For this reason, the granulation of honey to be sold in a finely granulated form should take place in the retail containers in which it will be marketed.

10.66   The constant-temperature cold-room

The honey must be placed in a well insulated, dry, cold room until it is completely granulated. The refrigeration equipment should be able both to remove surplus moisture from the room and to maintain the correct temperature. It is recommended that a constant temperature of 13°C (55°F) be maintained in the cold-room; this is sufficiently below the desired temperature of 14°C (57°F) to compensate for the slight rise in temperature when new batches of honey are placed in the cold room, so that the rate of crystallization and the resulting crystal size are not affected. The temperature of the cold-room should never be allowed to rise above 15°C (59°F) or go below 10°C (50°F). The honey is usually firm in three days and ready for market in about six days. If the temperature of the honey is much below 13°C (55°F), crystallization may take several days longer.

10.67   Conditioning honey after it is granulated

Finely granulated honeys that are properly processed will remain firm and creamy at normal room temperatures. But they will soften at high temperatures like naturally granulated honeys. If a honey is too hard for table use, it should be left at a temperature of about 30°C (85°F) until it is 'conditioned', or soft enough to spread. Once granulated honey becomes soft, it will not again return to its original hardness, even if it is stirred to break up the crystals. Conditioning is an important step in the processing of granulated honey that is acceptable to the customer.

Over the years, one difficulty experienced has been the 'pulling away' of finely granulated honey from the sides of retail glass jars, often referred to as frosting. This gives the honey an unattractive appearance. The problem sometimes, although not invariably, arises when the honey has been exposed to very low temperatures. McDonald (1964a) found that if the processed honey is removed from the refrigerator at the end of five days, and conditioned or softened a little at a temperature of about 32°C (90°F) for three days, this difficulty is not experienced.

10.68   Fermentation in processed honey

Since 1930, many thousands of tons of honey have been processed by the Dyce method, and as far as we know not a single sample has fermented. This was an unexpected bonus to the development of a process for producing finely granulated honey. No fermentation has in fact been found in the many damaged containers in grocery stores, whose contents were exposed to air for several weeks. Since the crystals of finely granulated honey are so small, and so evenly distributed in the liquid phase of the honey, it is possible that no portion of the honey can increase its water content sufficiently to permit the growth of sugar-tolerant yeasts.

Another unexpected advantage of the process is that if the honey had previously commenced to ferment, even to the extent where the flavour and odour of fermentation were readily detectable, after processing and regranulation the same honey has a flavour which is mellowed to a point where the fermentation is hardly detectable. The processing tends in general to reduce or mellow any pronounced flavours in honey. It will not, of course, remove strong objectionable flavours in badly fermented honeys.

10.69   Patents of the Dyce process

Shortly after the Dyce process was developed, it was patented (No. 1 978 893, 1931) by the inventor. The rights in the United States were given to Cornell University, and the Cornell Research Foundation administered the patent until it expired in 1952. There were two reasons for taking out a patent. The first was to prevent an individual or company dishonestly claiming the invention and securing a monopoly on the process. The second was to ensure that only finely crystallized honey of high quality would be prepared by the process and placed on the market.

Licences were granted to a few of the largest honey packers in the United States, who were willing to go to the expense of installing proper equipment and to follow instructions that would ensure a high-quality pack. Only a token royalty was charged, and the money derived from the patent has been gradually invested in research towards improved methods of preparing honey for market. The programme as planned is accomplishing this objective, and is continuing to be of definite benefit to beekeepers.

In Canada a similar patent was taken out (Dyce, 1933), and given to the Province of Ontario; no royalty was charged, and no effort was made to restrict the use of the process. As a result, much inferior crystallized honey appeared on the market. To correct this situation, the Canadian Government restricted the use of the process and granted licences only to packers who installed proper equipment. Over half of the honey produced in Canada is offered for sale in the granulated form, so the process is of great economic importance to beekeepers in Canada. This is true in other countries where most of the honey is sold granulated. No patents on the process have ever been taken out in other countries.

CHAPTER 11

PRODUCING VARIOUS FORMS OF COMB HONEY

by Carl E. Killion

LARGE-SCALE BEEKEEPER AND AUTHOR OF STANDARD BOOK
ON COMB HONEY PRODUCTION

This—worketh virgin waxen comb,
Six sided pipes to form her home,
The which, inspired by duty
She weaves and dovetails, all in fine
Neat handiwork, that doth combine
Security with beauty.
GREGORY OF NAZIANZUS (329–389)

11.1 GENERAL CONSIDERATIONS

There are a number of things that the beekeeper could do to improve the quality of honey sold in the comb, whether this is produced as sections, cut-comb honey or chunk honey. A 'section' is traditionally a miniature wooden frame filled with comb of capped honey; it usually weighs about $\frac{1}{2}$ kg or 1 lb (*see* Figure 11.4/1). 'Cut-comb' honey is, as the name implies, cut from a large comb. 'Chunk' honey is similar, but sold in a jar filled up with liquid honey. The present short chapter does not attempt to review the special equipment and management for producing a superior quality honey in general, but it points out some things that may help in getting the best results when working for comb honey. A degree of perfection is expected in comb honey that is not demanded in combs of honey for extraction (Chapter 9), and this must be created *by the bees* while the honey supers are still on the hive. After the combs have been taken off, the ways in which man can improve the quality of comb honey are very limited indeed.

A little extra care here and there, or special precautions at a critical time, may mean the difference between success and failure later on. In liquid honey it is largely the *weight* that governs the value of the crop; in comb honey we naturally want weight—although perhaps less of it— but *quality* above all.

Some regions are not suitable for producing comb honey, especially section comb; for instance areas with overlapping flows of light and dark honeys are not good. An area having an abundance of honey plants producing a light-coloured honey is usually preferred, although there are certain areas from which darker comb honey may be marketed in limited quantities.

The fact that honey is completely sealed, without open cells, does not necessarily mean that it is completely ripened (2.222). In some years of extremely high humidity, bees are unable to get the moisture content low enough to prevent fermentation (6.11). Many tests made during this type of season show a moisture content as high as 25%. At other times, when humidity is very low and the season hot and dry, some frames of unsealed honey may show a moisture content as low as 16–17%. Unsealed honey is *never* suitable for comb honey, because of packaging problems, but even a comb of sealed honey cannot be judged properly by just looking at it.

The author's book (Killion, 1951) discusses the production of comb honey in detail. Other publications (Bland, 1962, United States Department of Agriculture, 1957a, 1967) deal with special aspects.

11.2 BEEKEEPING FOR COMB HONEY

The writer has made some changes in the beekeeping equipment used for comb production, since these were described in his book (Killion, 1951). The changes have been made to get maximum brood rearing in the brood chambers, and to provide built-in ventilation and insulation in both the brood chambers and the honey supers. These conditions increase the likelihood of getting a better quality product, and add to the appearance of it.

Modern innovations have changed the packing of comb honey (*see* 11.5), and also its production. Until recently it was necessary to use queen excluders when producing shallow-frame comb honey, but within the past few years manufacturers have been preparing a foundation with cells larger than worker size, and so far the queens have refused to lay eggs in these larger cells.

There is not much difference in the management of colonies up to the time of the main flow (3.33), whether the beekeeper is working for extracted honey or for comb honey. Only the strongest colonies should be used for either sections or shallow-frame comb honey (cut-comb or chunk); weaker colonies are likely to produce light-weight comb, or none at all. Only the whitest and most tender wax should be used in the foundation (3.13) for honey to be sold in the comb in any form.

Many comb-honey producers use a double brood chamber right

through the year. The writer has preferred to use the double brood chamber for winter and up to the start of the honey flow, then reducing the brood nest to a single hive body. This crowding of the colony, necessary to get well filled sections or frames, will cause the bees to build burr comb between supers, and between the super and the hive body. This burr comb should be scraped away from both the top and bottom of the super at the time it is removed from the hive. There may be considerable dripping of honey when this is done; the supers should therefore be returned to the colony so that the bees dry up any open cells or smeared honey. In a short time these supers can be removed, completely free of extraneous honey. If the burr comb is left on the bottom of the frames (even though they are dry), there may be some bruising of the comb when supers are placed on a flat surface without proper clearance underneath.

Smaellie (1957) has published an account of methods used in New Zealand.

11.3 ATTENDING TO THE MOISTURE CONTENT

When comb supers are brought into the honey house, there are two important things to do: to check the moisture content of the honey, and to make plans for fumigating against any subsequent damage from wax moths. If the moisture content is too high (6.11), then removal of the excess moisture is recommended, because otherwise the honey may start to ferment during storage. It has already been stated that in some seasons of high humidity, sealed honey may contain as much as 25% moisture. Even if the water content is only slightly above 18·6%, the writer himself would plan to remove some of the moisture immediately. To delay doing so will allow more yeast cells to develop, with consequent damage to the flavour or even complete spoilage of the honey by fermentation.

The first requisite for moisture removal (see 9.22) is a room as airtight as possible and as small as will hold the supers and the necessary equipment. Supers should be placed on strips of wood or other material to allow free passage of air beneath the stacks. They should be criss-crossed in such a way that air can circulate through and around them, about 12 to 15 supers to each stack.

The equipment needed is a heating unit and a dehumidifier. The heat source should have a blower fan to circulate the air, and other circulating fans can be used to move warm air to all parts of the room. The dehumidifier should have a capacity a little larger than is needed, to speed up the drying operation.

The room temperature should be 27–29°C (80–85°F). Warm moist air will then condense quickly as it passes the cooling fins on the dehumidifier. The warm air within the room soon becomes very low in moisture content, and as this dry air passes over the comb surface, it absorbs moisture through the porous cappings.

Moisture should be checked twice daily, to show when the desired level of 17% has been reached. The dehydrating process is then discontinued; the room temperature may be lowered to 21–23°C (70–74°F) and all circulating fans turned off. Loss of moisture must not be allowed to continue to a level below 17%.

Wax moths can seriously damage comb honey if this is not fumigated properly (Roberts & Smaellie, 1958). For fumigation the supers are stacked tightly, leaving no cracks through which the fumigant can escape. One of the best fumigants is carbon disulphide, but this is highly inflammable, so extreme caution must be used; also, its use may not be approved by food legislation. When this material is used, about 30g are poured into a small pan placed on top of the supers; an empty super can conveniently be used to hold the pan, and a cover is placed on top. One well known fumigant, paradichlorobenzene (PDB), should never be used for comb honey, as the odour will permeate the cappings and spoil the flavour of the honey.

11.4 SECTION HONEY

Good comb honey is, to the beekeeper, a thing of beauty, and he must keep the beauty of the comb intact until the comb is packaged and ready for the buyer to see it: it is the appearance of comb honey that starts the sale. So, with section honey the propolis must be removed from the wooden sections; this is facilitated by painting the exposed wooden parts of the sections with hot paraffin wax before the supers are put on the hive. Sections should be well filled, and firmly attached to the wood. There is more breakage in light-weight poorly attached comb than in the heavier well filled sections. The cappings should be free of any stain, from propolis or of soot from the smoker; the sealing of the cells should be even—the pattern of the capping has always been a critical point in the desired appearance. Bruised cappings, finger smudges on the wood, or cells of pollen in the comb, all detract from the general appearance. It is best to sort the sections into weight groups before packaging.

For many years the single- and double-tier cases with a glass front were used entirely for packaging section comb honey. Then came the cellophane bag and wrapper, which permitted the packaging of comb honey in cardboard cartons (Figure 11.4/1—see Plate 13).

Today most of the chain stores refuse to handle section comb honey, because it is so fragile, and damage by breakage is too high. Beekeepers are finding that their best outlets for section comb honey are roadside markets and smaller grocery stores.

Descriptions of the production of sections are available from different countries (Roberts, 1958, Rope, 1962, Savage, 1971).

11.5 CUT-COMB HONEY

Some of the larger markets will handle cut-comb honey if it is properly packaged to prevent breakage and leakage of honey. It is now possible to get polythene bags (Kennedy, 1963) into which either sections or cut-comb honey may be placed, and the package heat-sealed. The bag with the honey is then placed in a window-front carton. This double package prevents damage by the two most common causes—squeezing by shoppers and leakage from careless handling.

The wooden sections (11.4) are thus rapidly being replaced by cut-comb honey. The honey is produced in shallow frames (of various depths), and the comb is then cut out of the frame, into pieces of any desired size, for packaging as cut-comb honey or (11.6) chunk honey.

Some beekeepers use melted beeswax for fastening the foundation in these shallow frames. The writer prefers to have a slotted top bar; the foundation is placed in the frame through this slot, and should extend a few millimetres above it, so that it can be mashed over with a heat sealer. The foundation should extend to 3–4 mm from the bottom bar (Figure 11.5/1—see Plate 13). The space left will prevent any buckling of the comb by heat or by weight of the bees.

To remove honey from these shallow frames, the usual procedure is to cut the whole comb out, either on a board or on hardware cloth resting above a tray. Knives for this purpose must be heated, so that the comb is cut easily and without tearing the cells. Some use a paring knife, others a multiple blade that cuts the entire comb out in one operation. Most of the multiple-blade knives are heated with hot water. Some bee-keepers use a heated wire and report excellent results.

When packaging either chunk honey or cut-comb honey, the liquid honey must be drained from the cut edges. The cut pieces may be drained on drying trays, or put in a special extractor (Figure 11.5/2—see Plate 14) and the loose honey removed by centrifugal force. If the liquid honey is not drained from the edges, granulation will be hastened, whether in the package of cut-comb honey or in the jar of chunk honey.

For cut-comb honey the bulk comb honey is cut into pieces varying in size from an individual serving (50–60 g, 2 oz) to larger ones weighing nearly $\frac{1}{2}$ kg (1 lb). The cut pieces of comb honey are drained in the way described, and either wrapped in cellophane or placed in polythene bags and heat-sealed, and then packaged in containers of various styles with suitable labels.

The latest and most popular container for cut-comb honey in the U.S. is a clear plastic box, about 4⅝ in. square and 1⅛ in. deep (11 × 11 × 3 cm). The cover is held in place with plastic tape, printed with the net weight of the product and the name and address of the packer. The tape is available in several colours, allowing for individual preference.

Comb packed in these clear containers is an invitation to many customers, since both sides of the comb and the cut edges are clearly visible. Although plastic boxes are more expensive than polythene bags and window-front cartons, the display they present is far more beautiful. These boxes can be re-used as storage containers.

Other descriptions of the production of cut-comb honey have been published (Meeks, 1968, Mraz, 1955, Rope, 1963).

11.6 CHUNK HONEY

When packing chunk honey, the pieces of comb are placed directly in a glass jar or tin pail of the size required by the trade. (Experience has proved that best results are obtained if the moisture content of the comb honey used for chunks in jars is reduced to not more than 17%, because of fermentation problems—*see* 6.11, 11.3.) The containers are then filled with extracted honey which has previously been heated to 66°C (150°F) and allowed to cool to 49°C (120°F) by the time of filling. The moisture content of the liquid honey may be as high as 18·6%. The extracted honey should be run down the inner side of the container to prevent incorporation of air bubbles. As soon as each container is filled, the lid is placed on tightly, and the container laid on its side so that the comb will not be crushed by its own buoyancy in the warm honey. After cooling thoroughly, the containers may be packed in shipping cases.

Each producer uses slightly different methods (Stephen, 1957, White, 1959*b*).

Chunk honey should not be stored for long after packaging, or granulation may set in, and this would spoil the sale value of the pack. The surface of the comb tends to hasten crystal formation in the liquid honey adjacent to it. So chunk honey should be delivered to the store freshly packed and in small lots which can be sold and consumed before granulation occurs. If the storekeeper has an oversupply and the honey granulates, it should be replaced immediately with freshly packed honey. Any such returned containers of honey may be placed in a warm oven at a temperature safely below the melting point of beeswax (say about 63°C, 145°F) until the honey is entirely liquefied; they can then be placed on the market again. Alternatively the containers can be placed in hot water, the comb melted, and the contents emptied into a tank where the beeswax is separated from the honey after cooling. This process is likely to darken the honey and to reduce its quality.

Consumers who like both comb honey and extracted honey will usually pay a higher price for this special chunk-honey pack, which has both kinds of honey in the same container. Also, some who are suspicious of honey 'purity' will buy honey which has a piece of comb visible in it. The glass package is usually preferred by discriminating buyers.

To maintain its reputation as a special pack, the chunk-honey package must look its best. The honey must be of good flavour, and the comb must be white, free from pollen, and cut neatly to a length that will extend from the top to the bottom of the container, and to a breadth that will permit it to slip readily through the mouth of the container. Because of this, glass containers with large openings are always used for chunk honey. The liquid honey surrounding the comb honey should be of good flavour and exceptionally clean and clear. This is a picture package—the comb is the picture, and the liquid honey with its attractively labelled glass container is the frame.

CHAPTER 12

HONEY QUALITY CONTROL

by P. E. W. Rodgers

MANAGER, PRESERVES SECTION,
RECKITT & COLMAN FOOD RESEARCH DEPARTMENT

Richd Hoy, At his Honey-Warehouse N°. 175, Piccadilly
Sells the choicest & purest Honey only, Fine Minorca
Honey, D°. Narbone Honey, D°. Breakfast Honey, D°. Honey Comb,
Box & Glass Bee-hives contrived so as Ladies may
have them on their dressing Tables without the least
danger of being stung.

[SEE FIGURE 12.1/1—PLATE 15] (1788)

12.1 GENERAL CONSIDERATIONS

Some aspects of quality control in honey have already been discussed in the chapters on honey processing (9 and 10). The present chapter reviews the subject on a wider basis, including the requirements of the European Codex standard for honey (1969), which is itself summarized in 13.2.

In order to set the highest possible standard for honey, certain chemical and physical properties of honey must be considered. Basically honey is a concentrated sugar solution, whose average composition is given in Chapter 5. Many honeys also contain dextrins—sugars of high molecular weight, non-sweetening, and rather gummy in character. Traces of other substances are also present, and it is these very small traces which give the honey its character and enable us to distinguish one honey from another by taste and colour. The major components of most honeys show a similar analytical picture, which does not explain why one honey is darker in colour than another, or why honeys have different flavours.

Speed and accuracy of examination of any characteristic are of great importance when large quantities of different types of honey are involved. It will be seen that the choice of method depends to some extent on the stage of its journey from the hive to the consumer the honey is at, when the analysis is done. For instance it is unlikely that many beekeepers will be able to invest in expensive equipment for the purpose of measuring the quality of their honey. At this early stage of production, very simple and inexpensive methods would be most appropriate.

The wholesaler or shipper has a different problem, in that he may be dealing with honey already blended, or may have to decide how best to mix honeys from several sources. Added to this, he has to inform potential customers of the quality of the honey, and probably send them small samples. One would therefore expect the shipper to have reasonable facilities for determining the moisture content of honey samples, as well as some of the other criteria referred to later.

A delightful illustration from a honey publicity leaflet of 1788 is shown in Figure 12.1/1—*see* Plate 15.

12.2 FACTORS THAT DETERMINE HONEY QUALITY (*see* note on page 325)

12.21 Importance and determination of moisture

The sugars of honey are responsible for its typical sweetness and wholesomeness (Chapter 5). They also prevent or suppress the onset of fermentation if they are present in sufficiently high concentration. The percentage of water decreases as that of the sugars increases, and honeys with $18\frac{1}{2}\%$ water or less are very unlikely to ferment in normal circumstances, particularly if granulation does not take place. The control of the water content is an important requirement of the proposed Codex Alimentarius standard for honey (1969), which sets an upper limit for moisture of 21% for honey in general. The relationship between sugar and water contents of honey can be measured in several different ways (5.3, 6.3). Firstly, the water may be driven off a weighed quantity of honey and the resulting dry matter re-weighed. There are several alternative procedures for doing this which yield reproducible results (A.O.A.C., 1970, Auerbach & Borries, 1924). The disadvantages of these procedures are that they are time-consuming and require some rather specialized equipment. Secondly, the density of the honey may be measured, and the corresponding percentage of water looked up in the appropriate table. Or the relative density may be measured with a hydrometer, but because of the high viscosity of honey, the instrument takes a long time to settle, and reading is difficult unless a Chataway (1933) type hydrometer is used. The Newkirk (1920) picnometer provides an accurate but slow method of density measurement, in which all the air is removed from the honey before this is weighed at constant temperature. Thirdly, Oppen & Schuette (1939) found a relationship for some honeys between their viscosity and water content. This method works quite well for pure sugar solutions, the solution becoming less viscous with increasing water content. However, honeys are not pure sugar solutions, and the protein materials present in very small quantities in many honeys can exert a marked influence on the viscosity (Pryce-Jones, 1953). Fourthly and finally, a refractometer provides the most rapid and convenient method, by using the relationship that exists between the refractive index and the concentration of sugar in solution. The relationship is not exactly the same for every sugar, and special tables have been developed for honey, which express directly the percentage of water in relation to the refractive index (6.2); *see also* Wedmore (1955). Alternatively, it is possible to use the sucrose scale fitted to many refractometers which is also known as the Brix scale; it must then be stated that the result is expressed in ° Brix. Again, this reading can readily be converted to refractive index and thence to percentage of water, using the table for honey. The most convenient refractometer for accurate work is the Abbé type; hand refractometers are small and convenient, and can give useful results in the field. The refractive index is sensitive to temperature, and this must be taken into account when making readings, which is generally done at a standard temperature of 20°C (68°F); otherwise the results must be adjusted to this temperature by using a correction factor given in the table.

Of the methods listed above, the beekeeper will probably limit his examination to the use of a hydrometer or density bead, whereas the shipper and large-scale supplier of bottled honey to the public would be expected to use more elaborate equipment, for the sake of both accuracy and speed.

12.22   Colour and flavour
Having established the water or soluble solids content of a given honey, one may turn to other factors which help in determining the overall quality.

It may be thought that colour should be regarded as an important criterion of quality, but honey is a natural product, and the colour of freshly extracted honey from the same plant source varies according to climatic factors during the honey flow, and the temperature at which the honey was ripened in the hive. Colour is not the decisive factor when judging the quality of fresh honey but colour *is* important when a large-scale packer is endeavouring to supply the public with a standardized product, and honey bought on the world market is commonly graded by colour, light colours fetching a higher price (Chapter 14).

A simple form of colour comparator is that designed by the United States Department of Agriculture, using glass colour standards for honey (Brice *et al.*, 1956). The liquid (non-crystalline) honey is viewed against the light in a standard square glass tablet-bottle, alongside rectangles of coloured glass, in a special holder. This comparator enables one to place a honey sample in one or other of the colour categories specified by the United States Department of Agriculture.

The Pfund colour grader is a more elaborate instrument, which is calibrated against the above comparator. It consists of a standard amber glass wedge, with which the liquid honey is compared visually, the honey being contained in a wedge-shaped cell. The lightness or darkness of the honey is expressed in millimetres distance along the amber wedge, whose scale is also sub-divided into honey colours, as set out in Chapter 13. It is noteworthy that the relationship between distance along the wedge and colour name is not the same for all countries.

Other ways of standardizing the colour of honey are using standard colour chips or cards (which give only approximate results), and the Lovibond comparator, which has not gained the same international use as the Pfund instrument, for honey.

Flavour is the next property which must be considered, and of all the factors contributing to honey quality it is the most difficult to describe. At the present time it is not possible to determine the flavour of honey by instrumental methods, although this may be possible in the future (12.3). Many of the professional honey tasters are expert in differentiating between shades of quality, and beekeepers become used to judging their own and similar honeys from season to season. Indeed, honey competitions rely on the expert opinion of such people.

The shipper and packer of honey are faced with a very different problem. Whereas the honey judge for a competition may be faced with several dozen samples over a period of a few days, the shipper—and particularly the packer—have to classify hundreds of samples every week. This work is particularly important with imported honeys, on which most of the large-scale packs in the United Kingdom are based. It is possible for sub-standard honey to be included accidentally, and if inadvertently mixed with 'good honey', the whole batch may be spoilt.

Honey for large-scale packing is generally imported in 300-kg steel drums, in quantities of several hundred tons at a time. The drums themselves can give trouble, as a few are not new, but reconditioned. If the washing process is not carried out properly the honey will be contaminated by the previous contents. The present answer to this problem is to examine every drum of honey and to isolate sub-standard material. It is known that the human palate can become de-sensitized through tasting too many honey samples at one time, and this must be allowed for. The following factors must be considered when tasting honey to assess its flavour:

(a) The honey should be free from foreign flavours or taints, such as oil, creosote, or insecticides.

(b) It should be free from acid or vinous flavours associated with fermentation.

(c) The flavour should be clean and wholesome, and free from the flavours associated with obnoxious plant sources, or with over-heating of the honey.

(d) If the floral source of the honey is named, the flavour should be reasonably true to that of normal honeys reputed to have come from that source (clover, orange blossom, eucalyptus, etc.). The only reliable way of doing this is to keep reference samples of the main types of honey and to replace them at regular intervals.

So far, the matters dealt with are of common interest to the beekeeper, shipper, and packer. Apart from a visual inspection of the honey for foreign matter, colour and general clarity, and a critical tasting of his product, it is not usual for the beekeeper to go any further. It is, of course, most important that all honey is inspected in this way; and poor-flavoured, dirty or cloudy honey must be down-graded, or prevented from reaching the consumer. Honey that is heavily contaminated with beeswax, due to bad management, or honey containing fragments of wood, etc., should be warmed gently and strained through muslin or fine stainless steel mesh in order to clarify it. On the other hand, there is little the beekeeper can do about opalescent honey.

12.23   Influence of sugar composition

The sugar spectrum of a honey has a decisive effect on its physical properties. The main sugars of honey are laevulose (fructose) and dextrose (glucose), and on average these together account for around 75% of the honey. The proposed Codex Alimentarius (1969) standard for honey requires a minimum reducing-sugar content of 65% for flower honeys, and 60% for honeydew honeys. Both laevulose and dextrose are defined as reducing sugars, and are usually determined by one of the copper reduction methods such as that of Lane & Eynon (1923) or Somogyi (1952). Other sugars present in significant amounts are the disaccharides sucrose (cane sugar) and maltose, and the trisaccharide melezitose. Of these only the first is significant as far as honey standards are concerned. A maximum sucrose content of 5% is required by the food regulations of many countries (13.2–13.5), and by standards such as the British Standard for grading of honey (1952), and the proposed Codex Alimentarius (1969).

Sucrose is defined as a non-reducing sugar, and as such does not react when the copper reduction methods are used. If, however, sucrose is hydrolysed either by mineral acid (Walker, 1917) or by the enzyme invertase, each molecule combines with a molecule of water, and laevulose and dextrose are formed in equal quantities. The difference between the results of reducing-sugar determinations before and after hydrolysis is therefore a measure of the content of non-reducing sugars (oligosaccharides)—including sucrose—and this is sometimes expressed as 'apparent sucrose'.

Due to the complex mixture of sugars present in honey, the simple methods of analysis referred to above are open to errors. In general the straightforward determination of reducing sugars will suffice for most purposes, as will the figure for sucrose, providing it is below the stated maximum. If the figure for sucrose exceeds the maximum, however, it is advisable to check it by using a more sophisticated method, if facilities are available, because other sugars that are non-reducing may be recorded as sucrose.

Of the many proposed methods for accurate determination of honey sugars, that of White & Maher (1954) is probably the most reliable. The method uses the selective adsorption of the sugars on a column of carbon powder, followed by elution and analysis of the fractions by conventional methods. The method is accurate, but because it is very time-consuming it is unsuitable as a regular control procedure. More recently gas-chromatographic methods have been developed (Pourtallier, 1967), and these are largely automatic in operation, although they require a considerable amount of elaborate equipment.

A high sucrose content (8% or above) is generally associated with heavy sugar feeding (Hadorn & Zürcher, 1963b), which results in a pale-coloured honey with a weak and nondescript flavour. The main purpose of setting a limit on the sucrose content is to discourage the production of 'honey' by feeding sugar to the bees, or the subsequent adulteration of honey by the direct addition of sucrose.

An important property of honey dependent upon its sugar composition is the tendency to crystallize or 'set' (see also Chapter 10). The factor responsible for crystallization is the dextrose component, which is deposited from the honey as the monohydrate. It has been established that the other sugars usually present in honey (together with the laevulose) can influence this process; evidence has been presented for sucrose by Jackson & Silsbee (1924), for maltose by Austin (1958), and for melezitose by Gontarski (1960). The first two sugars act in a similar manner to laevulose in reducing the solubility of dextrose. Melezitose is itself a rather insoluble substance, which can crystallize rapidly in honey; traces are found in most honeys, but it rarely occurs in sufficient quantities to crystallize out except in honeys containing a high proportion of honeydew, particularly from mountainous areas (Valin, 1956).

As a rule, the laevulose content of honey is greater than the dextrose content, although it occasionally happens that the two are equal or that the dextrose concentration is even slightly greater, particularly in honeys from the Chinese mainland. The importance of the ratio between laevulose and dextrose concentrations was demonstrated by Bosch *et al.* (1932). These authors calculated the ratio of laevulose to dextrose, the L/D ratio (a term introduced by Dyce, 1931*b*). They found that crystallization took place rapidly in honeys with L/D ratios of 1·0 to about 1·2, but that it was delayed when the ratio exceeded 1·3. Other figures have since been quoted (Milum, 1956), but the present author has confirmed the validity of the Bosch ratios over a number of years. More recently another ratio has been suggested, the ratio of dextrose to water (D/W); it was reasoned by Austin (1958) that other sugars with similar effects to laevulose should be disregarded. White, Riethof, Subers & Kushnir (1962) found the ratio D/W superior to all others for expressing the tendency of a honey to crystallize (6.72).

If the reducing-sugar content (R) of a honey has already been determined, the L/D ratio can be calculated if the dextrose content (D) is also known. A convenient method is by hypoiodate oxidation, as developed by Hinton & Macara and modified by Lothrop & Holmes (1931). $R\text{-}D$ gives a measure of the laevulose content L. Although this method is rapid and convenient, it is open to a considerable number of errors. First, the reducing-sugar determination gives a figure for all sugars that have some reducing power under the conditions of the determination, generally expressed as the equivalent amount of invert sugar prepared from pure sucrose under standard conditions. So maltose and other sugars with reducing groups will be included with dextrose and laevulose in the final figure. Secondly, the hypoiodate oxidation of dextrose is not very specific, and some oxidation of laevulose occurs at the same time; this over-oxidation can be corrected in the calculation if the laevulose content is known. More serious is the oxidation of maltose and all sugars with a free aldehyde group; this cannot be corrected for, unless the concentration of the individual sugars is known. The net result of these errors is that the value for reducing sugars may be only slightly too high, but that the dextrose figure may in certain circumstances be substantially increased, thereby reducing the figure calculated for laevulose.

It has been found that all direct methods of analysis such as those described above are subject to such errors; but, in spite of this, a very useful indication is obtained as to whether or not the honey is likely to granulate. When investigating alternative methods, White, Ricciuti & Maher (1952) found that the variance due to methods was as great as that of honeys from 14 different floral sources. This discovery led to his development of the selective adsorption method for determining the sugars of honey (White & Maher, 1954).

The procedure for obtaining a D/W ratio is somewhat simpler and less liable to error. The dextrose is determined by hypoiodate oxidation, and the water content is derived from the refractive index.

An improved method for dextrose determination has been developed, depending on the selective oxidation of dextrose by the enzyme glucose oxidase. This has been shown to be rapid and accurate, thereby lending itself to the precise determination of the D/W ratio (White, 1964b). Recent modifications have ensured stability of reagents over a long period.

The performance of honeys in relation to their L/D ratio has been referred to already. In general honeys with a high L/D ratio are selected for packs which are to be marketed as clear (liquid) honey. Honeys with a low L/D ratio may be selected for granulated honey; after seeding and conditioning they may be marketed without fear of reversion to the liquid state.

The corresponding D/W ratios are not quite so well defined, as less experience has been accumulated. Published results (White, Riethof, Subers & Kushnir, 1962) indicate that D/W ratios of 1·7 and below are associated with non-crystallizing honey, and ratios of 2·1 and above with those that crystallize rapidly.

Rates of crystallization of honey are affected by the presence or absence of crystal nuclei, and their presence invalidates the use of the above ratios as indicators of granulation (Siddiqui, 1970).

12.24 Indicators used to detect ageing, heating, and adulteration with invert sugar

The sugars, and also some of the minor constituents of honey (chiefly those contributing to the colour and flavour), are adversely affected by high temperatures. This temperature damage can be caused either by storage at temperatures in excess of 27°C (80°F) for long periods (months), or by the use of excessive temperatures (above 75°C, 167°F) when extracting, liquefying, and clarifying the product.

The effect of excessive temperatures or age on the honey sugars can be recognized by the production of 5-hydroxymethylfurfuraldehyde (HMF), which is a breakdown product of sugar solutions—particularly those containing dextrose and laevulose. The reaction is not specific for overheated honey; it occurs in many compounds and formulations containing sugars, particularly where acid hydrolysis of sucrose with the help of elevated temperatures is required to form invert sugar. Use is made of this fact to determine whether a honey has been adulterated with commercial invert sugar, addition of which causes a large increase in the HMF content.

A trace of HMF is always present naturally in honey, but this rarely exceeds 10 mg/kg in fresh material. Through adverse and very prolonged storage, or overheating, this quantity can exceed 30–40 mg/kg and may rise to 100 mg/kg or more. A content of more than 150 mg/kg is taken to indicate adulteration with invert sugar. In Germany, Renner & Duisberg (1968) have suggested that figures in excess of 40 mg/kg are an indication of 'heated honey', and the proposed Codex (1969) standard imposes a limit of 40 mg/kg for normal honeys.

Several methods are available for the determination of HMF, the best known being the Fiehe (1908a) test, which depends on the reaction of extracted HMF with acidified resorcinol; this has been adapted to give a quantitative procedure (e.g. Schade et al., 1958). Other methods available include the use of ultraviolet absorption, the Feder (1911) reaction, and Winkler's direct (1955) colorimetric method which has the advantage of simplicity and speed. It is more sensitive (but perhaps less specific) than the Fiehe test, which becomes significantly positive only at a level of 30 mg/kg of HMF (Hadorn & Kovacs, 1960). The Winkler method is sensitive to 2–3 mg HMF per kg.

Other constituents affected by excess heat are enzymes and the substances described as inhibines. All honeys contain small quantities of the enzymes diastase and invertase, both of which are destroyed if heated for too long at too high a temperature, and diminished by long-term storage. Of the two, diastase has the higher temperature resistance, and is the one more often used as an indicator of overheating. In some countries regulations demand that these enzymes should be intact, and the proposed Codex standard for honey requires a minimum diastase number of 8.* The diastase in honey can be measured by spectrophotometric or more simple means. The original method of Gothe (1914b) has been developed into an elegant procedure based on the degradation of starch, by Kiermeier & Köberlein (1954); the degraded starch is mixed with iodine, and the end-point taken when no blue colour remains when examined by eye.

The colorimetric method is based on the same starch reaction; it has been fully examined by Hadorn (1961), who has suggested constructive modifications for both the visual and spectrophotometric methods. These, and changes made by White (1959a), have been incorporated into the official Codex (1969) method.

Invertase is a sucrose-splitting enzyme, and this property is used in its determination. The two main methods in use are the polarimetric method of Duisberg & Gebelein (1958) and the titrimetric method of Kiermeier & Köberlein as modified by White (1964b).

---

* Diastase number expresses the diastase activity as the number of millilitres of 1% starch solution that are hydrolysed by the enzyme in 1 g honey in 1 hr at 40°C (104°F).

Inhibines are substances in honey which inhibit the growth of micro-organisms, particularly bacteria of certain kinds. This activity has been ascribed by Adcock (1962) and White, Subers & Schepartz (1963a) to traces of hydrogen peroxide formed naturally in honey as a result of the oxidation of glucose to gluconic acid by traces of the enzyme glucose oxidase. On this evidence it would appear that the determination of inhibine activity would not serve any useful purpose, since hydrogen peroxide could easily be added to honey.

In order to utilize the above tests, limits have been proposed for the two enzymes and HMF by Hadorn, Zürcher & Doevelaar (1962), and these figures have been considered when formulating honey legislation. Hadorn's minimum figures for acceptable honeys are 8 for diastase and 4 for invertase, with a maximum of 30 mg/kg for HMF.

12.25 Acid, ash, and insoluble solids

The determination of titratable acidity may give some information about the history of a honey sample, and it is demanded by several countries in their honey regulations. It is thought that a high acid figure may mean that the honey had fermented at some time, and that the resulting alcohol had been converted to acetic acid by bacterial action. In the past, honey acidity was often quoted as 'percentage formic acid', or as millilitres of normal alkali per 100 g honey.

It has been shown by Stinson et al. (1960) that the most important of the honey acids is gluconic, probably arising from the oxidation of glucose by the naturally occurring enzyme glucose oxidase.

The acidity of honey can vary widely, and existing honey regulations merely set an upper limit for titratable acidity. The limit quoted in the proposed Codex regulation is not more than 40 milliequivalents acid per 1 000 g honey as determined by direct titration. The formation of lactones makes it advisable to use a modified titration procedure such as that of Hadorn & Zürcher (1963a) if an accurate picture of the free acid and lactone content is required.

The formol number of honey is a measure of its amino acid content. It is also a quantity that may vary widely, and the usefulness of the test for it is doubtful. The determination is best carried out in conjunction with the titration of acids by Hadorn & Zürcher's (1963a) method.

The pH of a honey sample is not directly related to the free acidity because of the buffering action of the various acids and minerals present. The pH of honey is hardly affected by dilution with water, so solutions for this purpose may be easily prepared. The pH of honey generally varies between 3·5 and 5·5.

The ash content of honey depends on the material gathered by the bees during foraging. Nectar normally has a low ash content, but honey-dew is often high in ash. The proposed Codex standard (1969) requires an ash content not more than 0·6% for normal honey and 1·0% for honey-dew honey. Honeydew is a sweet sticky fluid excreted on plants by certain leaf-sucking insects (2.13). Honeydew honey may be dark in colour and unpleasant in flavour, so its identification by determination of ash content can be useful. Kirkwood et al. (1960) have derived an equation for discriminating between floral and honeydew honeys:

$$X = -8·3x_1 - 12·3x_2 + 1·4x_3$$

where $x_1 = p\mathrm{H}$, $x_2 = $ ash content (%), and $x_3 = $ content of reducing sugars (%). X is less than 73·1 for honeydew honeys, and greater than 73·1 for floral honeys. The ash is determined by standard combustion methods.

The water-insoluble solids of a honey sample consist of suspended wax particles, insect and vegetable debris, and pollen grains; a low content corresponds to high clarity. The proposed Codex (1969) standard places the upper limit at 0·1% in normal honeys and 0·5% in pressed honeys (heather and other thixotropic types). The amount of insoluble solids is determined by dissolving the honey in warm water and passing the solution through a tared fine filter and weighing the resulting deposit after drying.

12.3 THE FUTURE OF HONEY STANDARDS AND CONTROL

The primary aim of current international standards is to establish criteria by which honey may be positively identified, and protected from adulteration and substitution. Having differentiated honey from other sugar mixtures, the secondary aim is to establish a minimum quality grade below which the honey is not recognized as fit for consumption at the table. Various standards worked out by some countries and organiza-tions go further, and define a series of quality grades; these are often based upon colour and clarity and to some extent on flavour. It is possible that the present proposed European Codex standard for honey (1969) will eventually be extended, with some modifications, to become a world-wide standard; agreement with non-European countries must first be reached on the various quality parameters involved.

The existing international and national standards contain an assort-ment of criteria and different methods for assessing them (Chapter 13). It is expected that, following the lead given by the Codex, there will be a certain amount of rationalization and agreement on standard analytical methods. The Codex standard itself contains methods for determining all the compositional criteria demanded by the standard. The methods will doubtless be updated from time to time as knowledge of the subject increases.

In its final form the Codex will be regarded by signatory nations as the fundamental standard of quality for a genuine honey. This will not prevent the development of other standards by nations and organizations subscribing to the Codex, which could demand even stricter criteria of quality, based upon different natural honey components. The aims of these stricter standards would be to obtain a premium price for a type of honey probably produced in limited quantities. The reasoning behind these stricter criteria—except for flavour, colour, and appearance—will be difficult to justify on scientific grounds. This applies particularly to a demand for a high level of the enzymes invertase and diastase together with a low level of HMF (White, 1967a, Gebel, 1968, Hallermayer, 1969).

Non-signatory nations will be free to develop their own standards for internal purposes, but will still be affected in the export field by the Codex standard and the (very similar) E.E.C. (1974a) directive.

With regard to analytical procedures, it is quite possible that new criteria for quality will be suggested to replace those accepted today. Honey contains small amounts of a legion of compounds, including other enzymes such as acid phosphatase, and the amino acids which are significantly connected with flavour. It is hoped that any such new criteria will not serve merely as indicators, but will have a meaningful connection with quality.

Methods of analysis will certainly be improved, and where large numbers of honey samples have to be examined for quality, the new automatic analysing machines will be used, particularly by the larger suppliers of honey. With regard to tasting numbers of similar honey samples, some progress has been made with mechanical methods; Merz (1963) has separated some volatile honey flavour components by gas chromatography, and has used the patterns produced to differentiate between good and bad honeys. This is a promising method because—unlike the human palate—the machine does not become less sensitive with repeated use.

Note on Section 12.2

The complex nature of the factors that determine honey quality, and complications known to arise in the estimation of these factors, are discussed in Chapter 6. The account in 12.2 is intended to serve as a practical guide, and has been made as straightforward as possible.

SECTION 4

HONEY
AS A COMMERCIAL
PRODUCT

CHAPTER 13

HONEY STANDARDS LEGISLATION

by Arthur Fasler

MANAGER, FOOD LEGISLATION,
RECKITT & COLMAN FOOD RESEARCH DEPARTMENT

*The greatest care must be taken that whatever is prepared by hand—
bacon . . . mead, honey, flour, all should be prepared with the greatest
cleanliness.*

from DIRECTIONS TO STEWARDS IN CHARGE OF THE
VILLS (ROYAL ESTATES) OF CHARLEMAGNE,
EMPEROR 800–814

13.1 INTRODUCTION

In this chapter the food regulations of eighteen countries, and three
regional standards, are summarized as far as they apply to honey and
artificial honey. Regulations concerning beekeeping (which deal mainly
with bee diseases) or factory hygiene are not included.

Each country has documents with widely varying legal powers, and
the following is their approximate order of power:

1. *Laws* usually contain only basic requirements such as that foods
   must not be harmful, and that their presentation must not be
   deceptive; in addition laws often delegate powers for issuing more
   detailed regulations. In English-speaking countries, the basic law
   is usually called Food and Drug Act.
2. *Regulations, Orders, Decrees* are usually issued by the Government
   or one of its Ministries.
3. *Letters or circulars* are issued by the enforcing authorities to serve
   as guides for their administrative officers.
4. *Recommendations* may be agreed between manufacturers, possibly
   developed in co-operation with enforcing authorities. They are also
   called guide lines, codes of practice, standards.
5. *Voluntary grading schemes.*

Classes 1 and 2 have the force of law in the law courts. The other
classes are usually considered by the law courts as describing fair trade
practice or experts' opinions, and are taken into consideration as such
when deciding any specific case. All five classes have been abstracted in

the following Sections; 13.2–13.5 describe national regulations, and Section 13.6 compares a few points to show how the eighteen countries and three regional standards deal with them.

The summaries given here are fairly complete, but it is not considered useful to include every aspect; for instance labelling regulations may describe detailed requirements for letter sizes to be used, and the position of a declaration on the pack. The references quoted may be consulted for such additional details.

In writing these summaries it is assumed that the honey supplier (producer, processor, blender, packer, labeller, seller or his agent) tries to give the consumer value for money to the best of his ability. Matters such as grossly deceptive descriptions, or selling impure or strongly fermenting honey, have therefore not been included, although most countries mention them in their regulations. It may be assumed also that the addition of substances such as sugars or flavours to a product called honey is illegal in all countries, and this likewise has not been specifically stated.

The reasons for including certain specifications in food regulations are not set out in this chapter, but discussions of them can be found in Chapters 5, 6, 9, and 12, which also describe methods of analyses.

Acidity has usually been specified as 'percentage formic acid', but is better expressed as milliequivalents/kilogram of honey. The following conversions can be used:

$$1\% \text{ formic acid } = 208 \text{ meq/kg}$$
$$1\% \text{ gluconic acid} = 51 \text{ meq/kg}$$
$$1\% \text{ glucolactone } = 56 \text{ meq/kg.}$$

13.2   EUROPE

13.21   European Economic Community

A honey directive (E.E.C., 1974) has been issued which is very similar to the F.A.O. European Regional Standard (13.22) and requires that its provisions be incorporated into the honey legislations of the member countries. A few requirements are not harmonized and are left to national legislations, e.g. certain honeys may have up to 25% water, or the country of origin need not necessarily be declared.

13.22   F.A.O. European Regional Standard

GENERAL

A standard for honey has been established by a committee of representatives from European F.A.O. members. This standard has been sent to all F.A.O. members with the recommendation that it should be adopted and used as the basis for national legislation (Codex Alimentarius Commission, 1969).

DEFINITIONS

Honey is defined as the sweet substance produced by honeybees from the nectar of blossoms or from secretions on living plants, which the bees collect, transform, and store in honey combs.

The following designations are also defined:

Blossom honey
Honeydew honey
Comb honey
Extracted honey
Pressed honey.

STANDARDS OF COMPOSITION

Water	—heather honey	maximum	23%
	—other honeys	maximum	21%
Reducing sugar	—blossom honey	minimum	65%
	—honeydew honey	,,	60%
Sucrose	—honeydew, robinia, lavender, and banksia honeys	} maximum	10%
	—other honeys	,,	5%
Acidity		,,	40 meq/kg
Ash	—honeydew honey	,,	1·0%
	—other honeys	,,	0·6%
Water-insolubles	—pressed honey	,,	0·5%
	—other honeys	,,	0·1%
Diastase	—honey with low natural enzyme content and a maximum of 15 ppm HMF: Gothe scale	} minimum	3
	—other honeys: Gothe scale	,,	8
HMF		maximum	40 ppm

Methods of analysis are also described in this Standard.

Honey which does not comply with the requirement for diastase or for HMF may be sold as 'baking honey' or 'industrial honey'.

LABELLING

The label should give the name of the food, its amount, the name and address of the supplier, and the country of origin of the honey if it is imported.

13.23 Austria

GENERAL

In place of regulations, the Austrian Government prefer a Codex or collection of expert opinions (Austria, 1957). The standards given in this Codex are likely to be accepted by the courts even though they do not have the legal force of regulations.

DEFINITION

Honey is the sweet substance collected by honeybees from flower nectaries or exudations of other plant parts, which has been processed, enriched, and stored by the bees.

Honey can be characterized by the type of liquid collected by the bees (nectar or honeydew), by origin or plant source, and by the extraction process used.

STANDARDS OF COMPOSITION

Mature honey contains at the most 22% water. Sucrose content is usually below 10%. Honey should not have an acidity above 54 meq/kg.

Overheated or fermented honey may be used only for baking or mead preparation, etc., or for feeding bees.

GRADES

Voluntary beekeepers' standards exist (Büdel & Herold, 1960).

ARTIFICIAL HONEY

Artificial honey is a honey-like product based on inverted sucrose, to which up to a total of 20% of dextrose, or glucose syrup or both, plus flavouring and colouring substances have been added. There must not be more than 22% water, 20% sucrose, 0·4% ash and 40 meq/kg acidity.

13.24 Benelux (Belgium, Netherlands, and Luxembourg)

GENERAL

Recommendations for honey regulations to be introduced in Belgium, the Netherlands, and Luxembourg have been published (Benelux, 1963). All three countries have modified their regulations accordingly (Belgium, 1967, Luxembourg, 1966, Netherlands, 1965). Where national legislations differ from the E.E.C. Directive, they must be amended by 22.7.75.

DEFINITION

Honey is the sugary substance produced by bees exclusively from juices collected from plants.

STANDARDS OF COMPOSITION

Water—clover or briar honey [*Rubus fruti-cosus*] produced in Benelux	maximum	23%
—other honey	,,	22%
Sucrose	,,	5%
Acidity	,,	50 meq/kg
Ash in the solids	,,	0·5%
Water-insolubles in the solids	,,	1%
HMF	must be absent	

Belgian and Dutch regulations (Belgium, 1969, Netherlands, 1971) make it clear that 'absence' of HMF means 40 ppm or less.

LABELLING

For imported honey, the names of the countries of origin of the honey must be indicated, and a mixture of imported honeys must be declared as 'foreign honey'.

If the honey has been heated so that it has partially or totally lost its fermentative (enzymatic) properties, it must be called 'heated honey'.

The weight must be indicated on the label, and the name and address of the supplier is also required.

ARTIFICIAL HONEY

The use of this name is prohibited.

13.25 France

GENERAL

The various decrees, regulations, etc., dealing with honey have been conveniently collected together by Dehove (1970). Where present legislation differs from the E.E.C. Directive, it must be amended by 22.7.75.

DEFINITION

The word honey may be used only to describe the honey obtained by bees; it may not be used in connection with other products (e.g. artificial honey is not a permitted designation).

If the bees have been fed on sugar, etc., the honey obtained from them must be called sugar honey (*miel de sucre*).

If a honey is named by its origin or its source, it must conform to the following definition. The product must be derived exclusively from nectars and honeydews gathered by bees from living plants and stored by them in their combs; it must be extracted by a centrifuge, purified by straining, and matured; it must have a negative Fiehe reaction.

STANDARDS OF COMPOSITION
The maximum water content permitted is 25%. The honey must not have been caramelized by heat; for 'named' honeys the Fiehe reaction must be negative.

IMPORTED HONEY
For imported honeys the country of origin must be declared. Blending of imported honey is permitted on condition that the origins of the honeys used, and their percentages, are given.

LABELLING
The following are required: name of the product, name and address of the supplier, amount (for packs of more than 250 g), country of origin (if the honey is imported).

ARTIFICIAL HONEY
The use of this name is not permitted for any product.

13.26 West Germany (B.R.D.)

GENERAL
Germany probably has more detailed regulations concerning honey than any other country. The main body of the German law on honey is contained in the 1930 Honey Order (Germany 1930a) which is still in force. In addition there are voluntary grading schemes (Evenius & Kaeser, 1970) and a code of practice (Deutsche Lebensmittelbuch-Kommission, 1972). National legislation must be amended by 22.7.75 where it differs from the E.E.C. Directive.

DEFINITION
Honey is the sweet substance made by the bees by taking in nectar juices or other sweet juices found on living plants, by enrichment with substances from their body, by changes in their bodies, by storing in combs, and by maturation in the combs.
  The possibilities for permitted characterization of specific honeys are mentioned: by the bee forage (flower honey, honeydew honey); by the main plant species on which the bees fed (e.g. heather honey); by origin (e.g. Black Forest honey); by method of extraction (e.g. centrifuged honey); or by intended use (e.g. bakery honey).

STANDARDS OF COMPOSITION
The properties of normal honey are enumerated. Detailed lists describe when honey is considered to be adulterated or spoiled. The following limits apply:

Water —maximum 22% (except heather honey which may contain up to 25%)

Acidity—maximum 40 meq/kg.

Honey with certain defects may not be sold, or even given away (high acidity, presence of brood, honey from sugar-fed bees, etc.); honey with other defects may be used only in baking (fermenting or heated honey).

The addition of water, even indirectly as by feeding sugar syrup to bees, is forbidden (Evenius & Kaeser, 1970).

IMPORTED HONEY

There is no need to declare imported honey as such, but some importers do it voluntarily, and there have been attempts to make this compulsory (Büdel & Herold, 1960).

If the bees have fed *mainly* on non-German honey, the resulting honey must not be called German honey.

LABELLING

The Labelling Order of 1972 (Germany, 1972) applies to honey. The following information is required in German: name and address of supplier, nature of the product, amount.

Claims for health-giving properties of honey are permitted for honey in general, but not for any particular type or brand. 'X-honey is healthy' would be the subject of prosecution, but 'Honey is healthy' would not (Evenius & Kaeser, 1970). *See also* Deutsche Lebensmittelbuch-Kommission, 1972, Section 8, and Germany (1974) paragraphs 17, 18.

CONTAINERS

Honey in small containers must be packed in fixed amounts of $\frac{1}{8}$, $\frac{1}{4}$, $\frac{1}{2}$ or 1 kg (Germany, 1972).

There are detailed standards for plastics, which will have to be complied with if plastic containers are used; standards are being issued in the form of communications; for an English summary of the first twelve *see* B.I.B.R.A. (1963).

GRADES

Grade evaluation schemes are used, e.g. for assessing the entries in honey quality competitions (Evenius & Kaeser, 1970). Points are given for water content, diastase and invertase, cleanliness, appearance, packing, odour, and flavour, and prizes are awarded if a certain standard is reached.

ARTIFICIAL HONEY

The 1930 Order on artificial honey is still in force and gives standards for this product (Germany, 1930b). Artificial honey is defined as inverted sucrose which may contain other ingredients; it will always contain hydroxymethylfurfural. A list of acids permitted for the inversion is given. The product must contain no more than 22% water, at most 30% sucrose, a maximum of 0·4% ash, at most 20% glucose syrup, and its acidity must not be above 40 meq/kg.

CODE OF PRACTICE

The German Code of Practice states that, in addition to complying with the legal requirements, honey should also have a diastase number of at least 8 (Schade method) and that its HMF content should not exceed 40 ppm. For honeys with special claims, such as 'best', the HMF content should not exceed 20 ppm, and the invertase activity should be at least 7 (Hadorn number). Health claims should only be made if there is a sufficient scientific foundation for them.

13.27 Italy

An old law of 1890 is still valid; it merely states that honey must not be harmful, altered, or contain added substances (Puecher-Passavali, 1964). The general law (Italy, 1962) states that the natural composition of a food may not be changed. The label must contain the name of the product, the name and address of the supplier, and the amount. If a tin 'can' is used, the lead content of the tin in contact with the food must not be more than 1%. Curative claims are not permitted. National legislation must be amended by 22.7.75 to comply with the E.E.C. Directive on honey.

13.28 Spain

GENERAL

Spanish proposed legislation gives definitions and standards for honey (Spain, 1968). These proposals will come into operation at a date to be announced by governmental decree. So far (1974) no such decree has been issued. Apart from these proposals there is no honey legislation in Spain, but the requirements given below are a guide to possible future requirements.

DEFINITIONS

Honey is the natural sugary product made by bees (*Apis mellifera* and other species) from flower nectar and other plant exudates, without anything being added, and conforming to the specifications below.

The following specific designations are also defined:

Comb honey
Virgin honey
Crude honey
Centrifuged honey
Pressed honey
Gummy honey
Overheated honey
Beaten honey
Aromatic honey

A product called *meloja* is also defined; it is the syrup obtained by concentrating the liquid obtained by washing the combs.

STANDARDS OF COMPOSITION

Water	maximum	22·5%
Reducing sugars	minimum	70%
Sucrose	maximum	3%
Dextrins	,,	8%
Acidity	,,	5 meq/kg
Ash	minimum	0·1%
	maximum	0·6%
Water-insolubles	,,	1% in the total solids
HMF	,,	5 000 ppm
Diastase on the Gothe scale	minimum	8
	maximum	10

It is rather surprising to see a maximum stated for the diastase activity, especially such a low one. If this specification is really used by the Spanish authorities, most honeys examined would surely be found to be illegal in composition!

Other unusual features of these specifications are the inclusion of a minimum ash content, the very high maximum for HMF, and the very low maximum acidity.

LABELLING
The label statements must include the name of the product, the name and address of the supplier, and the amount.

ARTIFICIAL HONEY
The sale of artificial honey is expressly forbidden.

13.29 Sweden

REGULATIONS
Swedish legislation on honey is short (Sweden, 1972). Honey is defined as the product prepared by bees from flower nectar and other juices

from living parts of plants containing sugars. Any heat treatment above 45°C (113°F) must be declared, and the country of origin for imported honey must be indicated. The label must state the name of the product, the name and address of the supplier, and the amount (Sweden, 1971).

INFORMATIVE LABELLING SCHEME
An Institute of Informative Labelling (Sweden, 1964) issues codes in which requirements for a detailed description of products are laid down. For honey, the code gives standards and required declarations for colour, purity, enzymes, sucrose content, water content, consistency, calorific value, storage conditions, etc. This is a voluntary scheme.

13.2.10 Switzerland

GENERAL
Honey, like all other foods, is regulated by legal standards (Switzerland, 1936); the analytical methods to be used are published in an official methods book (Switzerland, 1967).

DEFINITION
Honey is bee-honey, i.e. the mature, sweet substance which bees obtain from flower nectar or other natural plant exudations, and which they process and store in their combs.

STANDARDS OF COMPOSITION
The water content must not be above 20%. Honey which has more water, or which is acid, fermenting, or not sufficiently clean, may be used for baking.

LABELLING
Requirements are: name of product, name of supplier (or a registered trade name), amount, and—for imported honey—origin or the description 'foreign'. Honey which has been heated, so that its enzymatic properties or its aromatic components have been lost, must be declared as 'overheated'.

Health claims can be made only if special permission has been obtained.

GRADES
Voluntary beekeepers' standards exist (Büdel & Herold, 1960).

ARTIFICIAL HONEY
Artificial honey is a sugar-containing product similar in appearance to honey. It must not contain more than 20% water. It may contain

natural flavouring substances, and up to 40 ppm sulphur dioxide as a preservative.

13.2.11 United Kingdom

GENERAL

No specific regulations for honey exist, except for the labelling of imported honey, but the general requirements of the Food and Drugs Act have to be met. The honey must be fit for human consumption and of the nature and quality requested by the customer; it must not be injurious to health, and it must be correctly described (O'Keefe, 1968).

A voluntary grading scheme exists.

National legislation must be amended by 22.7.75 to comply with the E.E.C. Directive.

IMPORTED HONEY

Imported honey must be described by its country of origin or, if it is a blend of honeys from more than one country, a conspicuous indication must be given that it originates from more than one country (United Kingdom, 1972).

LABELLING

The following minimum information must be given on the label: name and address of supplier, contents designation (including one of the indications mentioned above under imported honey, if applicable), and the minimum quantity in British units (United Kingdom, 1963, 1970).

A court decided that the use of imported honey should be declared on the label, even though it was only used to feed the bees during a dearth period (*British Food Journal*, 1965).

A label claim that 'honey strengthens the heart' was considered to be misleading by a Public Analyst. No proceedings were taken, however, since the label had gone out of print (*British Food Journal*, 1959).

CONTAINERS

Honey must be packed in the following fixed amounts: 1 oz, 2 oz, 4 oz, 8 oz, 12 oz, 1 lb, 1½ lb [28, 57, 114, 227, 340, 454, 680 g], and multiples of 1 lb [454 g] (United Kingdom, 1963).

GRADES

Quality standards exist which allow the use of the following grade designations: select, select crystallized, extra select heather. The scheme is voluntary, but the grade names may not be used unless the standards are met (British Standards Institution, 1952, Ministry of Agriculture, 1934). As an example, the following minimum standards must be fulfilled to allow a honey to be called 'select'. It must be made from nectar, have a relative density of at least 1·415 at 16°C (60°F) when clear, a minimum total solids of 80%, at most 5% sucrose, a maximum total acidity of 43 meq/kg. It must have been strained through a single thickness of a strainer equivalent to a standard bolting cloth of 54 meshes/in. The colour must be uniform and must be described as Select Light Colour, Select Medium Colour, or Select Dark Colour, as determined by standard colour glasses.

ARTIFICIAL HONEY
There are no specific regulations; the sale of artificial honey is legally acceptable.

13.3 NORTH AMERICA

13.31 Canada

GENERAL
There is a short section in the Food and Drug Regulations which deals with honey; there are detailed specifications in the Honey Regulations (Canada, 1954, 1967), and there is also a rather unhelpful honey standard (Canada, 1968).

DEFINITION
Honey shall be derived entirely from the nectar of flowers and other sweet exudations of plants by the work of bees.

STANDARDS OF COMPOSITION
Water	maximum 20%
Invert sugar	minimum 60%
Sucrose	maximum 8%
Ash	maximum 0·25%

LABELLING
Detailed requirements have been laid down. Labels must be approved and require the following information: name of product, name and address of supplier (including Province of origin), amount, lot number, class, and grade. After 1975 labelling will have to be in English and French.

CONTAINERS
Honey must be packed in the following fixed amounts (for consumer packs containing more than 5 oz of honey): 8 oz, 12 oz, 1 lb, 1½ lb, 2 lb, 4 lb, 8 lb [227, 340, 454, 680, 907, 1 814, 3 629 g].

GRADES

There are three honey grades, based on moisture content, defects such as turbidity or flavour taints, and freedom from foreign materials. For export honey the following colour names are defined:

Extra white	Pfund reading	up to	13 mm
White		,,	30 mm
Golden		,,	50 mm
Light amber		,,	85 mm
Dark amber		,,	114 mm
Dark		above	114 mm

The U.S.A. and Australian colour grades have different definitions (13.32, 13.51).

ARTIFICIAL HONEY

There are no specific regulations. Amongst the general regulations, those concerning the substances permitted in the sugar inversion process are probably the most important. Citric, phosphoric, tartaric acids, etc., may be used, but not, for example, sulphuric or hydrochloric acids. Sodium carbonate, sodium bicarbonate, etc., are permitted neutralizers.

13.32   United States of America

GENERAL

The enforcing authorities of the U.S.A. food laws use factory inspections, as well as analytical checks on the products, as control methods. Factory inspections seem to be of more importance than in other countries.

Official standards of composition for many classes of foods have been issued, but none exist for honey. There are voluntary grades, which can be used for guidance but do not have the force of law. The general provisions of the Food and Drug Act apply. Thus, no substance may be added to reduce the honey in strength; contamination of honey with filth (such as insect debris), or the presence in it of substances injurious to health, would make it illegal (United States, 1962).

DEFINITION AND STANDARDS OF COMPOSITION

The old Food and Drug Act of 1906 contained a definition of honey which, although it was not carried over into the present Act, may still be used as a guide on what is considered to be honey. Honey is the nectar and the sugary exudations of plants which have been collected, modified, and stored in combs by honeybees (*Apis mellifera, Apis dorsata*). Honey is laevorotatory and contains not more than 25% water, not more than 0·25% ash, and not more than 8% sucrose (White, Riethof, Subers & Kushnir, 1962, page 38).

LABELLING

The following information must be given on the innermost container holding the honey (United States, 1962): name and address of supplier, name of the product, amount.

GRADES

There are Federal as well as State grading schemes, all of which are voluntary (United States Department of Agriculture, 1933, 1951, 1957*a*, 1957*b*). The Federal scheme includes, in grading extracted honey: flavour, defects, clarity, water content. Points are awarded, and these determine the grade, which is one of four: U.S. grade A (or U.S. Fancy), U.S. grade B (or U.S. Choice), U.S. Grade C (or U.S. Standard), U.S. grade D (or U.S. Sub-Standard).

Colour is not regarded as a quality factor, but the designations have been standardized as follows:

Colour name	Pfund scale	Optical density
Water white	up to 8 mm	0·0945
Extra white	up to 17 mm	0·189
White	up to 34 mm	0·378
Extra light amber	up to 50 mm	0·595
Light amber	up to 85 mm	1·389
Amber	up to 114 mm	3·008
Dark amber	over 114 mm	

The Canadian and Australian colour grades have different definitions (13.31, 13.51). The optical density is measured at 560 nm in a layer of caramel-glycerine solution 3·15 cm thick, using glycerine as a blank.

For comb honey the factors used for grading are: appearance of cappings, attachment of the comb to the section, uniformity of honey, absence of pollen, granulation, honeydew, honey quality. There are four grades: U.S. Fancy, U.S. No. 1, U.S. No. 2, Unclassified.

ARTIFICIAL HONEY

There are no specific regulations for artificial honey. As long as there is no deception as to the type of product, artificial honey appears to be permitted. The label should contain the same information as for honey, but should also give a list of ingredients, and declare separately the use of any of the following substances: artificial colours, artificial flavours, preservatives.

13.4 LATIN AMERICA

13.41 Latin American Codex

GENERAL

A committee composed of representatives of twenty-one Latin American countries has agreed on a food codex, of which so far two editions have been published (Latin America, 1964). This is the first, and so far the only, set of internationally agreed standards for the whole range of foods.

It is difficult to establish how much the individual countries rely on this codex, but the following statements have been found:

Ecuador	Adheres to this Codex while preparing its own National Food Codex (Ecuador, 1963).
Panama	This Codex has been partly adopted (Panama, 1962,
Peru	Peru, 1963).
El Salvador	
Guatemala	This Codex is used (Averza, 1968).
Nicaragua	

DEFINITIONS

Honey is the natural product made by the domestic bees (*Apis mellifica*, *Apis ligustica*, etc.) with flower nectar and saccharine plant exudates, and stored by them in the combs.

Definitions for the following designations are also given:

Comb honey	
Virgin honey	(from unbrooded combs)
Crude honey	
Centrifuged honey	
Pressed honey	(obtained by pressure without heating)
Gummy honey	(obtained by pressure and heat)
Overheated honey	(above 70°C until enzymes are destroyed)
Beaten honey	(obtained by beating the combs)

STANDARDS OF COMPOSITION

Average values for the composition of honey are given, as well as the following limits:

Water	maximum 20%
Sucrose	,, 8%
Dextrins	,, 8%
Acidity	,, 54 meq/kg
Ash	,, 0·8%
Water-insolubles	,, 1% of the total solids
Fiehe reaction	must be negative
Lund reaction	minimum 0·6 ml precipitate

Limits for trace metals (required for all foods) are also given, e.g. 0·1 ppm arsenic, 10 ppm copper, 2 ppm lead.

LABELLING

The label should contain the name of the product, its amount, the name and address of the supplier, and the country of origin of the product if it is imported. It is expressly forbidden to mention medicinal or therapeutic properties on the label of any food.

CONTAINERS

There are regulations for food containers to ensure that toxic substances are not transferred to the food packed in them, etc.

ARTIFICIAL HONEY

A product to be called artificial honey should contain at the most 20% water, 1% ash, 100 meq/kg acid, and 50 ppm free sulphur dioxide. It may not contain artificial essences, artificial preservatives, or artificial sweeteners.

13.42 Argentina

GENERAL

The food regulations contain definitions and standards for honey and for artificial honey (Argentina, 1971).

DEFINITIONS

Honey is the sweet substance made by the domestic bees (*Apis mellifica, Apis ligustica*, etc.) from flower nectar or from exudations of or in living plant parts, which the bees collect, transform, combine with specific substances, and store in combs.

The following designations are legally permitted:

Comb honey
Virgin honey (from unbrooded combs)
Crude honey

Centrifuged honey
Pressed honey (obtained by pressure without heating)
Gummy honey (obtained by heat and pressure)
Overheated honey (above 70°C, 158°F, until enzymes are destroyed)
Beaten honey (obtained by beating the combs)

STANDARDS OF COMPOSITION

Water	maximum	18%
Sucrose	,,	8%
Dextrins	,,	8%
Acidity	,,	54 meq/kg
Ash	,,	0·4%
Water insolubles	,,	0·1%
		(0·5% in pressed honey)
Diastase Gothe scale	minimum	8 ⎱ immediately
HMF	maximum	40 ppm ⎰ after processing
Lund reaction	minimum	0·6 ml precipitate

LABELLING

The label, etc. has to be approved. It should give (in Spanish): name of the food, amount, name and address of supplier, number of the official certificate of approval, number of the manufacturing establishment. For home products the words 'Industria Argentina' are required. For imported products, the country of origin, and the name and address of the importer or distributor, are needed.

ARTIFICIAL HONEY

This is defined as a syrup based on sucrose, invert sugar, etc. It must not contain more than 20% water, 1% ash, and 100 meq/kg acid. Colours are permitted, and up to 60 ppm of sulphur dioxide, but the product must not contain any flavouring or preservatives.

13·43 Brazil

GENERAL

The food regulations contain definitions and standards for honey (Brazil, 1946, 1967).

DEFINITION

Honey is the natural sugary product made by bees exclusively from plant juices or exudates, extracted from completely sealed cells of the comb.
    The regulations also define 'centrifuged honey' (which must show a

strongly positive diastase reaction) and 'heated honey' (whose diastase reaction is negative or only slightly positive).

STANDARDS OF COMPOSITION

Water	maximum 20%
Sucrose	,, 8%
Dextrins	,, 5%
Acidity	,, 46 meq/kg
Ash	,, 0·35%
Fiehe reaction	must be negative
Lund reaction	minimum 0·6% precipate
	maximum 3% precipitate
Lugol reaction (a test for the presence of glucose syrup)	must be negative
Polarimeter	must be laevorotatory

LABELLING

The general regulations require that the name of the food, and the name and address of the producer, shall be given on the label.

ARTIFICIAL HONEY

The general regulations state that all substitute foods must be identifiable as being different from foods they are intended to replace. All manufactured foods must be registered by the Ministry of Health.

13.44 Mexico

The Mexican regulations have been codified (Mexico, 1965); there are no specific regulations for honey. All foods have to be registered by the Department of Health, and the label of a food must contain the following information: name of product, name and address of supplier, and registration number.

Quality standards have been issued (Mexico, 1958) which require that honey must be extracted at temperatures below 45°C (113°F). Honey is defined as the natural nutritious product, the major components of which are sugars, which is derived from the harvest obtained from plants by various insects, mainly bees (*Apis mellifica*) which store and mature it in their wax cells for the feeding of the bees and their brood. Analytical methods are also given, and the following limits are laid down:

Density	minimum	1·3966
Invert sugar	,,	63·9%
Sucrose	maximum	9%

Polarization $-21°$ to $-2°$
Acidity   minimum   8 meq/kg
                      maximum 52 meq/kg
Ash   minimum   0·01%
                      maximum 0·25%
Diastase
    For grade A honey   minimum 13·9 (Gothe)
    For grade B honey   no limit

13.5 AUSTRALASIA

13.51 Australia

GENERAL

Australia consists of six States and one Territory. Each State has food laws of its own. For honey they are very similar, and can be summarized together (New South Wales, 1908, Queensland, 1964, South Australia, 1964, Tasmania, 1960, Victoria, 1966, Western Australia, 1961). There are also regulations prescribing the grade descriptions for honey to be exported (Australia, 1964).

DEFINITION

Honey is the nectar and saccharine exudations of plants, gathered, modified, and stored by the honeybee.

STANDARDS OF COMPOSITION
    Water   maximum 20%
    Reducing sugars minimum 60% as invert sugar
    Ash   maximum 0·75%
Queensland and Victoria also specify a maximum of 5% sucrose. Maximum permitted amounts for trace metals are specified, honey being included under 'other foods* which may not contain more than 1 ppm arsenic, 2 ppm lead, 30 ppm copper, 1·5 ppm antimony, etc.

LABELLING

All six States require the following minimum information on the label: name of product, name and address of supplier. Western Australia, Queensland, and South Australia also require the amount to be stated, and in Tasmania the country of origin has to be stated.

EXPORT GRADES

Regulations for exported honey are valid for all States (Australia, 1964). They are similar to the State regulations quoted above, but have lower maxima for water (18·5%) and ash (0·5%). Colour grades are defined as follows:

	Pfund reading	
Extra white	up to	17 mm
White	,,	34 mm
Extra light amber	,,	50 mm
Light amber	,,	65 mm
Amber	,,	90 mm
Dark amber	,,	114 mm

The Canadian and the U.S.A. colour grades have different definitions (13.31, 13.32).

ARTIFICIAL HONEY

New South Wales, Tasmania, Victoria, and Western Australia state that words such as 'imitation' may be used only where specifically allowed by the regulations. Since there are no specific references to the use of such words in conjunction with the word 'honey', the sale of a product under the name 'artificial honey' would be illegal in these four States. The product itself, however, is not necessarily illegal, provided an acceptable designation is used.

13.52   New Zealand

GENERAL

Honey standards, label requirements, and limits for pesticides and for trace metal contents, are included in the regulations (New Zealand, 1973).

DEFINITION

Honey is the sugary product obtained from the comb of the honeybee.

STANDARDS OF COMPOSITION

Water	maximum	20%
Reducing sugars	minimum	60%
Ash	maximum	0·4%

LABELLING

The name of the product, the amount, and the name and address of supplier must be stated on the label. No claim may be made that the food value of honey is superior to that of sugar.

ARTIFICIAL HONEY

This is not mentioned in the regulations, and is assumed to be permitted provided the ingredients used (acids, etc.) are themselves permitted. The ingredients must be listed on the label.

13.6 COMPARISONS

13.61 General remarks

In addition to the legislation quoted in Sections 13.2–13.5, most countries have a body of experts' opinions which has to be taken into account. Thus, the Swiss regulations state that honey is not acceptable if heat has caused loss of its enzymic properties. No figures are given, but Swiss experts seem to agree that honey with a diastase value below 8 would be legally unacceptable (without reference to the diastase value of the honey before it was heated). (*See also* Chapter 12.)

Another point which cannot be judged from outside is the degree to which the legislation in any country is respected and adhered to. Where requirements are unrealistic, adherence is unlikely; for instance it is difficult to see how the Spanish honey trade can possibly adhere to the proposed Spanish legislation. White and his colleagues found that 21% of the U.S.A. honey samples they analysed had an ash content above the U.S.A. limit (White, Riethof, Subers & Kushnir, 1962). Further examples can be found by comparing the ranges of honey composition given in Chapter 5 with the legal requirements described in this chapter.

13.62 Standards of composition

The standards of composition of honey, as published in the legislation of the eighteen countries included in this chapter, vary widely both in the type of tests used and in the minimum and maximum values acceptable. Table 13.62/1 gives the permitted values for what would be legally acceptable as table honey in all eighteen countries. Very few honeys would comply with these figures. However, many tests are used in only a few countries, and some of the extreme values are used only by one or two countries. The final column in Table 13.62/1 gives values chosen so that each value by itself would be acceptable in at least fourteen of the eighteen countries. These are much more realistic, both in the number of tests required and in the limits specified.

13.63 Water content

The following are the maximum water contents legally acceptable for ordinary honey (in some countries there are other limits for special honeys):

18%   Argentina
20%   Australia, Brazil, Canada, Latin American Codex, New Zealand, Switzerland

21%   F.A.O. European Codex, E.E.C. Directive
22%   Austria, Benelux, West Germany
22·5% Spain
25%   France, U.S.A.

No limit is specified in Italy, Mexico, Sweden, and the United Kingdom.

Table 13.62/1
Characteristics required for table honey in all 18 countries considered, and in at least 14 of them

Component or test	Value acceptable in all 18 countries	Value acceptable in at least 14 of these countries
Water content	18% or less	20% or less
Sucrose content	3% or less	8% or less
Reducing sugars as invert sugar	70% or more	n.s.
Dextrins	5% or less	n.s.
Acidity as meq/kg	5 or less	50 or less
Ash	between 0·1 and 0·25%	0·4% or less
Water-insolubles in the solids	0·1% or less	1% or less
Diastase, Gothe value	between 8 and 10	n.s.
Fiehe reaction	negative	n.s.
HMF	40 ppm or less	n.s.
Lund reaction, precipitate	between 0·6 and 3%	n.s.
Lugol reaction	negative	n.s.
Trace metals	below certain limits	n.s.
Polarimetry	laevorotatory between $-21°$ and $-2°$	n.s.

n.s. = not specified

13.64 Enzymes

There is considerable controversy about the importance of the honey enzymes. Some maintain that they have beneficial physiological effects, and that it is therefore necessary to ensure that honey is rich in enzymes. Others regard the enzymes as useful indicators of careful processing and storage of honey. The validity of both these functions are, however, questioned by others (*see* Chapter 12).

In spite of these conflicting views, some countries do not specify limits for enzymes. An example is Spain, where the proposed diastase value of honey must be between 8 and 10, which will be attained by very few honeys. The European Codex, the E.E.C. Directive, and Argentina specify a minimum diastase value of 8 (or 3 for some special honeys). Grading schemes and a Code of Practice state values for enzymes (diastase, invertase) in Germany and Sweden. The preservation of the enzymes in table honey is a requirement in German legislation. In Switzerland, honey in which the enzyme activity has been destroyed has to be labelled 'overheated'.

13.65 Heating of honey

A slight amount of heat is often applied to honey during processing, to facilitate its transfer from one container to another, especially when the honey has started to crystallize. This is legally acceptable in all countries. However, if more heat is used, the enzymes of the honey are likely to be affected and, with still more heat, certain changes take place which interfere with the flavour of the honey.

The heating of honey is specifically mentioned in some legislations. In Austria and Germany, in the European Codex, and in the E.E.C. Directive, the use of overheated honey is restricted e.g. to use in baking. In France, caramelized honey may not be sold. In Benelux and Switzerland such honey must be labelled 'heated' or 'overheated'. There are definitions for 'overheated' honey in Argentina, Brazil, and Spain, and in the Latin American Codex, though the use of this designation does not appear to be compulsory.

The following tests are mentioned: HMF determination (Argentina, Benelux, Spain, European Codex, E.E.C. Directive); Fiehe test (Brazil, France, Latin American Codex); enzyme activity (*see* 13.64).

13.66 Health claims

Germany specifically permits health claims for honey, provided they are not restricted to a particular type or brand of honey, but these claims are restricted to those which can be scientifically supported. Italy and

the Latin American Codex forbid curative claims for foods. In Switzerland, health claims require special permission by the Federal Health
Authorities. In New Zealand, no claim may be made that the food value
of honey is superior to that of sugar. The position in the other countries
is not expressly stated in the regulations consulted.

13.67 Sugar feeding of bees

Honey derived from bees fed on sugar syrup, candy, etc., is not legally
honey in any of the countries here considered. France, Germany, Spain,
Switzerland, and the Latin American Codex expressly state that such
honey would be illegal; it would also contradict the definition in the
regulations for Argentina, Australia, Austria, Benelux, Brazil, Canada,
New Zealand, the European Codex, and the E.E.C. Directive. In other
countries, such a product would probably not be admitted, as not being
what honey is generally understood to be.

This does not exclude the use of sugar for feeding bees, but it does not
recognize as honey any product stored by such bees during the period of
sugar feeding. In Germany such a product may not even be called
artificial honey, but in France it may be sold as 'sugar honey'.

13.68 Honeydew honey

Honey obtained from bees collecting honeydew (*see* 1.2, 2.13) is included
in the definition for honey in Argentina, Australia, Austria, Benelux,
Brazil, Canada, France, Germany, New Zealand, Spain, Switzerland,
the European Codex, E.E.C. Directive, and Latin American Codex. In
countries where definitions are absent or insufficient (Italy, Mexico,
Sweden, United Kingdom, U.S.A.), honeydew honey is probably accepted
as honey. Honeydew honey can therefore legally be called honey in all
countries considered.

13.69 Artificial honey

The use of the term 'artificial honey', or a similar name, is expressly
forbidden in four of the six States of Australia, Benelux, France, and
Spain. There are regulations describing standards for a product to be
called artificial honey in Argentina, Austria, Germany, and Switzerland,
and in the Latin American Codex. For countries whose regulations do
not mention artificial honey, it can be assumed that this is a permitted
product, provided the ingredients used (e.g. the acid for the inversion)
are permitted, and the presentation is not deceptive. This is the case
for two out of six States of Australia, Brazil, Canada, Italy, Mexico, New
Zealand, Sweden, United Kingdom, and U.S.A.

13.7 SUGGESTIONS FOR IMPROVED LEGISLATION

For some countries, it is easy to find out what the legislators want; for others it is not.* One of the more difficult legislations, in the author's experience, is that of the United Kingdom: requirements are difficult to track down and difficult to understand. On the other hand, the food legislations for Argentina and Switzerland are written and published in such a way that information is easy to locate and to understand. It is suggested that the latter two rather than the former be taken as models for any new or modified legislation.

The following aspects should be considered for inclusion in honey legislations.

DEFINITION
There is good agreement between countries on the definition of honey. In some countries, the definition is extended by a statement of what should *not* be added to honey (e.g. colours, flavours, artificial sweeteners), but this is not really necessary.

SENSORY ASSESSMENT
This is very important, but unfortunately difficult to specify with any precision.

PHYSICAL PROPERTIES
These include viscosity, clarity, polarimetric behaviour, crystalline state, etc., and are not often specified. Their measurement is useful in the quality control of certain branded honeys, but not for legislative purposes.

CHEMICAL COMPOSITION
In the legislation of some countries only a few chemical properties are included, but in some a long list is specified. The main purpose is to make sure that honey is not adulterated with invert sugar, etc. However, the composition of honeys varies so much that it is difficult to specify limits which will include all pure honeys yet exclude adulterations. The question really is how much the legislators are willing to leave to the expert (public analyst, etc.), and how much they want to take out of his hands.

---

* In addition to those mentioned in the text, the following honey standards are known to have been published.

Bulgaria (1957a, 1957b), Egypt (1963), Hungary (1966a, 1966b), Iran (1966), Israel (1970), Poland (1967), Rumania (1969), Standards Association of Central Africa (1971).

If the limit for, say, ash is set at 0·25% (Canada, U.S.A.), then honey found to contain 0·4% will lead to a prosecution, however well the producer knows that his product is pure honey: there is no argument. In the absence of a specific maximum value for ash content, the producer can try to convince the authorities that his honey *is* honey.

The use of chemical tests is, however, very important for the quality control to be exercised by honey producers, shippers, and packers (Chapter 12).

LABELLING AND CLAIMS

Requirements on these are sometimes included in honey legislation. However, it would seem more appropriate that they should be covered by general regulations which state what minimum information is required for all foods (name of product, name and address of supplier, amount in container, etc.). Mis-descriptions, and claims not supported by facts, should also be specifically forbidden in such general regulations.

GUIDE LINES

In addition to legislation on honey, it is suggested that guide lines agreed between local honey producers and importers should be established with the authorities, based on local conditions, giving more detailed indications.

MODEL LEGISLATION

Based on the considerations outlined above, the following is suggested as sufficient text for the honey legislation of any one country:

Definition:	Honey is the sweet substance produced by honeybees from flower nectar or from other secretions on living plants.
Sensory assessment:	Honey must have its characteristic flavour, odour, colour, and appearance.
Chemical composition:	Guide lines will be established giving acceptable limits for various constituents of the main types of honey consumed in the country.

CHAPTER 14

WORLD TRADING IN HONEY

by R. B. Willson

CHAIRMAN, R. B. WILLSON INC., NEW YORK

By the goodness of God this Land doth yield great plentie of Honye
and Waxe, as not onlye dothe suffice the necessarye uses of the
Queenes Majestie and her Subjects to be spent within the Realme,
but also a great quantitie to be spared to be transported unto other
Realmes and Countreys beyonde the Seas by waye of Merchaundize, to
the great Benefite of her Majestie and the Realme.
FROM AN ACT PASSED BY QUEEN ELIZABETH IN LONDON (1580)

14.1 INTRODUCTION

The value placed upon honey by peoples other than those who produced
it is evident from earliest times. Taxes and tributes were often imposed
in the form of payments of honey and wax. The laws of Manu in Ancient
India, around 1000 BC, regulated the amount of honey a king might
claim from his subjects as a sixth of the production (Ransome, 1937).
Trade in honey must have existed then, because a Brahman—a member
of the highest caste—was forbidden to sell or trade in honey and wax.

As a 'voluntary' offering, honey has always been a favourite choice.
Jacob, in sending his sons to Egypt with gifts for Joseph, counselled
them: 'Take in your baggage, as a gift for the man, some of the produce
for which our country is famous: a little balsam, a little honey . . .'
(Genesis 43, 11); this is referred to again in 19.33.

Chapter 19 traces the history of honey in some detail, and the following
early records of trade in honey will suffice here. In the twelfth century
we read of a sales tax and customs duty on honey sold in Novgorod in
Russia, indicating that the obstacles to free trade existed early too. In
mediaeval Russian monasteries, honey (for mead making) took second
place only to wheat, and several purchases of a ton or more at a time are
cited in Dorothy Galton's book *Survey of a thousand years of beekeeping
in Russia* (1971).

By the tenth century the Maya Indians of Yucátan and parts of
Central America were trading in the honey of their stingless bees; cacao

355

was the drink of the ruling classes, and honey was the only sweetener they had to make it palatable.

The development of modern beekeeping, explained in Chapter 3, arose out of the invention of the movable-frame hive by the Rev. L. L. Langstroth in Philadelphia, U.S.A., in 1851; the story has been told many times (see e.g. Crane, 1963b, Johansson & Johansson, 1967). By 1900 honey was being moved into cities from the surrounding country in wagonload lots. After the end of World War I it was being stocked as a regular item in shops, and regularly crossing borders into international trade. More and more men had found they could make a vocation of beekeeping; production was soaring, and markets were being eagerly sought. Between the two World Wars important refinements were made in apiary management, and agriculture was moving into vast new lands where beekeeping would follow. Production still increased, but production costs decreased, and so honey became cheaper, and attractively competitive with jams and jellies for table use.

The production explosion in Mexico, Argentina, and Australia (14.22, 14.21, 14.23), in the two decades following World War II, remain for all time a classic example of how eagerly and with what great success human beings will pursue a way of making a living, even such a difficult one as beekeeping. It helps that there is—as there always has been—a ready market for honey, and for the most attractive of all commodities: cash.

Only a beginning has yet been made in honey production in Latin America. Road building programmes now in the blueprint stage will double the accessibility of much of this continent, and may double its honey production by the mid-1980s. No more than a guess can be made for Africa, and aggressiveness of the African honeybees (17.21) will not necessarily be a significant limiting factor to production, since they are already being used effectively (e.g. Guy, 1972a).

14.11 Honey as a world commodity

At the present stage in the development of world honey production, and of world trading in honey, countries fall into two general categories: those that are exporters and those that are importers. The most important exporters are Argentina, Mexico, People's Republic of China, Australia, and the United States. Less important but consistent exporters are New Zealand, Canada, Spain, Chile, Hungary, Rumania, Yugoslavia, Cuba, Guatemala, El Salvador, and Costa Rica.

The big importers of honey are West Germany (the Federal Republic, B.R.D.), Switzerland, Great Britain, France, Italy, Belgium, Netherlands—and now Japan. Small but consistent importers are Austria, Ireland, the Scandinavian countries, and Algeria. The United States is a fairly large importer as well as an exporter, and this is true to a some-what smaller extent for Canada.

The Soviet Union produces as much honey as the U.S.A. (4.2), but its honey trade is largely internal, and has not yet affected the world market much, although it has recently been exporting to Japan (Table 4.4.10/1), and on a barter basis to Western Germany, exchanging honey for greatly needed German products. This business is handicapped by the Russian stipulation that their containers (of heavy metal, like a milk can or churn) must be returned. The high cost of returning empty containers reduces the possibility that Russia could sell honey to far off places like the United States. Apart from the exceptions mentioned in 14.26, the same is true of other countries of eastern Europe.

The honey of world commerce varies greatly in quality, and its quality is assessed largely on the basis of a light colour and a mild flavour. Thus the mild white honeys of Argentina, the U.S.A., Canada, Spain, and the Balkan countries rate highest. (There are, however, exceptions; honey packers in Switzerland prefer the darker aromatic honey of Yucatán to mild light honeys.) Light amber and amber honeys from Australia, Chile, and China, with rather more flavour, rate somewhat lower; a few honeys are unpleasant in taste, or even rank and disagreeable, and these may be almost unsaleable.

Most countries have some good honey, and most countries have some bad honey. Any good honey may be spoiled by bad beekeeping (Chapter 3), or by bad handling, processing or storage (Chapters 9 and 10). Argentina produces sunflower honey (*Helianthus annuus*) that is dark, strong, and unpalatable. The U.S.A. has its share of miserable honey: cajeput in Florida, horsemint in Texas, safflower in Arizona and California, and an occasional big crop from heartsease (*Polygonum*), which is beautiful honey to look at—white and brilliant—but which, to use an an old Yankee expression, 'smells to high heaven'.

Whilst travelling in Europe, the author has tasted fine-flavoured honey of unknown floral source from China, and has eaten mild, pleasant-flavoured honey from certain eucalyptus species in Australia. Chile grades its honey Pile I, Pile II, and Pile III; some of the Pile I honey is good too, although not all.

The price of honey in the world market varies for the same reasons as with other commodities: supply and demand. Demand for honey is often increased after small fruit crops, because jams and jellies are scarce. Demand is also influenced by the price of honey itself.

A classic example to illustrate how the world honey market fluctuates can be quoted from the two years 1963 and 1964. In the closing months of 1962, prices for honey were around normal.* There was going to be

* A 'normal' price is the average for, say, the previous ten years.

a comfortable amount of carry-over from 1962 production. But as early as the end of January 1963, crop conditions and prospects were looking bad for the first time in *all three* of the principal honey-exporting countries of the world (Argentina, Mexico, Australia). By the end of February doubts were confirmed, and prices were rising every day. Exporters were wildly outbidding each other for the producers' honey, because the European importers (14.4) were practically begging for honey. The market soared, and kept soaring until late June and early July 1963, when honey that had sold for U.S. $200 a ton in 1962 was selling for $400 a ton. Honey, the world commodity, was in short supply.

Chaos developed. Many short sales* by exporters were not honoured. Importers at the buying end of the unshipped contracts were furious but frustrated, because *force majeur* was being invoked by exporters. The worst was yet to come. Some packers of honey in Europe, especially in Germany, bought the high-priced honey in 1963, expecting to sell as much as they had sold the year before at half the price. But the public had different ideas. They cut their purchases of honey so drastically that, at the end of the high-cost year, the packers found themselves with extremely heavy inventories.

Then came the crusher—abundant honey crops all over the world in 1964. Prices to producers naturally started out high, near to where they were at the close of 1963. The reverse of the 1963 pattern set in: a daily downward spiral that was to bankrupt some important people in the honey business, exporters and importers as well as packers. In two years the price of honey had run full circle.

The year 1971 developed like 1963, with short crops in Australia, Argentina, and Mexico. By 1 July honey supplies were non-existent, and new crops were being contracted at soaring prices. U.S.A. honey that had been sold at $330 per metric ton by the producer in 1970 now fetched $660, and by midwinter 1971/72 $770, with drums returned. Yucatán honey that sold in 1970 for $285 a ton F.O.B. Progreso was $450 in the fall of 1971, and $470 in early 1972. In spring 1972 one official sale of 1 000 tons was made to Japan at $495 a ton. Argentina beekeepers benefited similarly, their honey costing importers $680 per ton on the docks of Europe and the U.S.A.

It is remarkable that this development of the world honey market has *not* been accompanied by a slackening of demand, which the high prices would seem likely to induce. On the contrary the demand for honey has never been so great. There are two main reasons for this sensational change. First, Japan—which up to a decade ago was an exporting country for honey—is now a large importer, as the eating habits of the Japanese are being westernized. Secondly, in many parts of the world

* Short sales are those made, by sellers, of goods they have not yet bought.

young people (and others) have become champions of natural foods, and honey is one of them. An editorial in *Bee World* (1974) sums up the current position.

14.2 HONEY-EXPORTING COUNTRIES

14.21 Argentina

The largest honey exporter, as these data are compiled, is probably Argentina. In the main, Argentine honey goes for export to Europe, the United States, and now in substantial quantities to Japan. Formerly shipped in barrels made of algaroba wood from Brazil, it now travels in home-manufactured open-head steel drums with good liners. The drum heads are bolted on, so samples cannot readily be taken, and this presents difficulties with U.S.A. honey-handling procedures.

It is estimated that only 3 000–5 000 tons of Argentina's honey production is consumed at home. The balance is exported at top prices, for most of the honey is white and mild, of good body and delicious. It is produced largely from alfalfa, white clover and thistles (*Medicago sativa, Trifolium repens, Cirsium vulgare,* and *Cynara cardunculus*), and is widely valued for bottling, since it improves colour and flavour when blended with darker and cheaper honeys.

Table 14.21/1
Honey exports from Argentina, 1940-1973, in metric tons

Year	Amount	Year	Amount	Year	Amount
1940	152	1951	5 504	1962	18 542
1941	134	1952	5 281	1963	10 894
1942	768	1953	3 758	1964	19 616
1943	611	1954	12 469	1965	27 040
1944	2 465	1955	3 727	1966	15 640
1945	4 198	1956	5 042	1967	25 756
1946	5 572	1957	3 058	1968	10 158
1947	2 977	1958	2 956	1969	16 282
1948	1 470	1959	12 825	1970	21 640
1949	3 025	1960	15 904	1971	13 788
1950	3 744	1961	13 313	1972	20 138
				1973	17 856

Table 14.21/1 shows how consistently the total Argentine export trade in honey has grown since 1940. The proportions sold to different countries are set out in Table 14.21/2. In the years 1961–1965, from 40 to 70%

of Argentina's exports went to West Germany, with important quantities also to Britain and the U.S.A. In some years substantial quantities were sold to the Netherlands, France, and Italy—but the surprise among the buyers has been Japan (see 14.4.10).

Table 14.21/2
Honey exports from Argentina to different countries during 1962, 1965, and 1970, in metric tons

		1962	1965	1970
Europe	Germany (B.R.D.)	13 713	16 144	8 277
	Britain	1 070	1 818	2 927
	Netherlands	598	1 114	441
	France	907	207	—
	Italy	604	623	733
	Denmark	26	307	394
	Belgium	40	168	163
	Ireland	38	—	484
	Finland	46	—	—
North America	U.S.A.	1 368	2 548	313
	Canada	—	849	—
Asia	Japan	—	1 567	7 735
	Lebanon	1	—	—
Africa	Libya	10	—	—
Oceania	Australia	27	—	—
Miscellaneous		—	1 374	225
		18 448	26 719	21 692

Argentine honey is exported by as many as twenty-five strongly competitive free-enterprise companies. Two of these, Honex, S.A., headed by Joaquin Guckenheimer, and Tancia, S.A., headed by Federico Tannhauser, have recently done over 85% of the business between them.

14.22 Mexico

Fast closing in on Argentina, if not yet ahead on any recent five-year average, and destined long before the end of the 1970s to achieve the top place in honey exporting, is Mexico. As recently as the early 1940s beekeeping was unimportant in Mexico, but then two events occurred which, although not related in any way, together led to the most remarkable production development in world beekeeping history. They were the creation of the firm Miel Carlota, S.A., in Cuernavaca in the central highlands, and the discovery that the peninsula of Yucatán, in southeast Mexico, is sensationally good honey country.

Arthur and Ana Wulfrath migrated to Mexico from Germany after World War I, during which he had served as a first lieutenant of artillery in the German Army. When Mexico declared war against Germany in 1941, Wulfrath's import-export business was confiscated. He and his family were detained in their own custody and obliged to fend for themselves. It was here that Ana acted. She was hungry for honey such as she had always enjoyed as a girl on her parents' farm in Germany, so she bought her husband a bee hive for a Christmas present. At about this time Hans Speck, a ship's doctor from the seized *S.S. Columbus*, was also detained and placed in custody with Sr. Wulfrath. Getting bees for Ana's hive generated interest in bees by both these men, and from this small co-operative venture grew the world's largest beekeeping operation (Willson, 1955) which within twenty years had 50 000 colonies in more than 1 000 apiaries, producing over 5 500 tons of honey annually!

Miel Carlota's principal honey crop is from *Viguiera grammatoglossa* (*acahual* in Spanish), a member of the Compositae with myriads of yellow flowers some 5 cm across. It blooms in October and November, and yields a honey which rapidly granulates with a butter-like texture and colour. Packed originally in cans embossed with floral decorations (now in 300-kg drums), it soon won favour in Europe. No matter how large the crop, it has been sold out at premium prices every year. In the past decade Miel Carlota has developed another large operation in Vera Cruz on the east coast, where Dr. Oscar J. Barraza was the partner-in-charge (now the owner), and still another in Acapulco on the Pacific coast, with a huge modern plant for preparing bulk honey for the export market.

Meanwhile in Yucatán, the fascinating peninsula which was the home of the ancient Maya who kept stingless bees in log hives (17.31), a few beekeeping pioneers were doing well with *Apis mellifera* in the modern hives. In Mérida, the capital of Yucatán, three men (who themselves were keeping bees on a large scale) formed a partnership to build a receiving and refining plant for honey; they were Plinio Escalante Guerra, Adolpho Peniche Lopez, and Lic Manuel Rios Covian. The annual honey production of Yucatán was about 140 tons in 1945, but within the brief space of twenty years it became 15 000 tons or more. The 1971/72 crop reached 18 038 tons; and in 1972/73 it was 12 390.

Rich men who never saw the inside of a bee hive were investing in bees. Widows were mortgaging their houses to get some of this liquid gold. Wherever there was a road or a trail, there were soon bees, in hives that had been made in Mérida and equipped with frames and foundation also made there. It was a booming business.

The author had the privilege of carrying the first samples of Yucatán honey to Europe in 1949, when the import of honey first became possible again after the war. Germany was not allowed to buy honey as yet, nor were Italy, France, the Netherlands, or Belgium. Britain could not yet exchange sterling for dollars. But the Swiss could trade; in that year they started to buy Yucatán honey, and have since been buying about 1 000 tons annually.

Back in Yucatán the original partnership was meeting with competition. By the middle 1950s there were twelve exporters, most with honey-processing plants more or less of the same pattern: galvanized iron cylindrical receiving tanks sunk in the floor, holding 30 tons of honey, equipped with electrically driven agitators to secure uniformity of the product. Provision is made for straining, first coarsely, then (where required, and for the U.S. market in particular) finely through wire cloth with 80 meshes per inch (32/cm), after gently heating; the honey is cooled immediately after straining, to prevent heat damage (9.31). For special requirements, the wire cloth is replaced by an even finer nylon cloth.

Yucatán honey (now averaging 15 000–16 000 tons a year) goes to the world markets in beeswax-lined closed-head steel drums holding 300 kg. Some are reconditioned used drums from Mexico City; a few years back some were new ones made in Mérida by Sr Hector Medina V., proprietor of Miel Yucatán, S.A. Around 5–10% of the Yucatán crop goes to market in square tins holding 26 kg, each tin being in a carton case; this fetches $10 per ton more than honey in drums.

Most of the Yucatán crop is from two sources. (Species of *Ipomoea*— Convolvulaceae— vines that bloom in December, provide a third, minor, source of honey that is light in colour, of good flavour, but thin in body and thus often a trouble maker, 6.11.1.) One of the two main sources is a tall many-branched plant (Compositae) 2m high with yellow flowers, which goes by the Maya name *tah*; it blooms in January and is almost always dependable. The honey is extra light amber to light amber in colour, of good body and of pronounced flavour (as are most of the Compositae honeys), and quick to granulate. Some European buyers prefer it. The biggest part of the crop, however, comes from trees; some have spectacular yellow, white or lavender flowers, but the most important is a shrub-like tree with tiny greenish open-faced flowers, belonging to the Polygonaceae. It is *Gymnopodium antigonoides*, whose Maya name is *dzidzilché*. The honey is generally light amber in colour and of good body. It is one of the world's most aromatic honeys; when it is used in baking, the heat of the oven volatilizes the essential oils, esters, alcohols, and aldehydes which give honey its flavour and aroma, and the air has a fragrance like that of a summer garden full of flowers at eventide.

This *dzidzilché* honey of Yucatán has met with great favour around the world. If you eat honey in Germany, Switzerland, Britain, Belgium or the Netherlands, *pain d'épices* of France or *torrone* of Italy, it is likely that some of your nourishment is coming from Yucatán.

Table 14.22/1 gives official export figures for Yucatán from 1954/55 to 1972/73, with details of the importing countries. No official figures are available for the rest of Mexico, but from 1955 on, the average amount of honey exported can be estimated at 10 000 tons a year. Wulfrath believes that Mexico will produce 100 000 tons of honey a year by the end of the century. Prices of Yucatán honey provide an example of the way world prices changed in the early 1970s. The price per ton went to $450 in the fall of 1971, to $470 early in 1972, then to $750 in the fall of 1972 and $900 per ton in the fall of 1973; it remained there until May 1974.

14.23 Australia
Australia is now the world's fourth largest honey-exporting country, the People's Republic of China having moved into third place. Well arranged government statistics have been published since 1944; these show that on the average about 50% of Australia's honey production is exported, and they provide other interesting information (Table 14.23/1).

The 1948–1949 crop year* provided a record export of 15 000 tons, and no year since has come within 85% of this. There have been three crop years (including 1944) when honey exports were less than 25% of the 1948–1949 record, and there have been eight years with less than 50% of the record year. These facts exemplify the unique difficulties of the Australian beekeepers, whose principal sources of nectar are various species of eucalypts (Dyce, 1960). These trees, according to species, may flower only once in two, four or eight years and, depending on rainfall and other factors, may fail to blossom even in an expected year. Some species may be reliable for nectar two years out of three, some three years out of four, some one year out of two, but none every year.

As an important world source of honey, Australia must therefore be appraised as unreliable with regard to volume, because of the flowering biology of the plants from which much of the honey crop is derived. Another factor that weighs against Australian honey on the world market

* Australia's main honey-producing season lasts from November to February, so the 'crop year' spans parts of two calendar years.

Table 14.22/1

Honey exports from Yucatán (Mexico) to different countries, 1954-1971, in metric tons

The crop year is from November to October. The author estimates that Yucatán exports are approximately 50% of all exports from Mexico.

	Germany (B.R.D.)	U.S.A.	Britain	Netherlands	France	Belgium	Italy	Switzerland*	Denmark	African countries	Sweden	Lebanon	China	Jordan	Australia
1954/1955	757	2 902	—	227	206	232	—	—	—	—	—	—	—	—	—
1955/1956	1 653	1 464	—	661	275	526	100	—	—	—	—	—	—	—	—
1956/1957	4 260	1 615	—	1 067	509	437	135	—	—	—	—	—	—	—	—
1957/1958	4 164	1 945	370	653	376	284	19	—	—	—	—	—	—	—	—
1958/1959	5 754	1 377	128	1 030	515	442	230	99	—	—	—	—	—	—	—
1959/1960	4 749	2 177	312	1 973	353	550	167	23	10	39	—	—	—	—	—
1960/1961	5 390	2 410	711	1 681	104	453	321	—	75	10	17	15	27	2	2
1961/1962	7 751	1 799	834	1 293	738	1 122	306	—	124	43	5	2	—	—	—
1962/1963	4 558	841	667	1 815	454	322	695	—	77	—	39	4	—	—	—
1963/1964	8 241	1 610	1 367	1 882	582	645	77	—	146	38	10	1	—	—	2
1964/1965	6 086	2 510	1 005	1 475	353	307	200	—	33	—	—	—	—	—	—
1965/1966	10 411	2 931	1 802	1 765	260	511	705	—	463	—	—	—	—	—	—
1966/1967	5 154	2 755	1 470	1 360	155	187	221	—	91	—	10	—	—	—	2
1967/1968	12 853	4 417	1 179	2 034	222	314	849	—	50	—	10	—	—	—	2
1968/1969	6 012	3 870	3 244	2 127	472	523	885	—	105	—	—	—	—	—	2
1969/1970	6 145	2 314	2 386	1 252	—	618	38	—	100	—	—	—	—	—	—
1970/1971	2 946	917	1 556	369	—	184	87	—	—	—	—	—	—	—	—

* It is estimated that 90-95% of Yucatán honey entering the Netherlands is trans-shipped to Switzerland and consumed there.

is a 'musky' flavour that is generally associated with honey from eucalypts. (Their honeys, however, have nothing in common with eucalyptus oil, which is derived from only a few of the 600 or more species described by Penfold & Willis, 1961.) In 1967 Dr. Eva Crane undertook a study-tour of Australia at the request of the Honey Research Advisory Committee of the Australian Honey Board. After 1½ months spent visiting honey-producing regions and apiaries in six States, she told me that she could not identify the origin of this flavour, since she had not recognized it in any honey she had sampled from combs taken direct from the hive. In stressing the need for research on Australian honeys—much of which comes from plants growing only in Australia—Dr. Crane suggested that some of the honeys not being exported might well be more acceptable to the palate of consumers in importing countries than some that were sent overseas. One superb honey from Australia is leatherwood, *Eucryphia* (Eucryphiaceae, page 61), produced only in Tasmania.

Australians themselves prefer honeys with more flavour than the mild honeys that rate high on the world market. Most Australian honey for export is blended to conform with colour and flavour requirements of the various countries. In recent years the Australian government has reclassified the grading regulations; certain honey is now sold for export in retail packages; another classification is 'blending' honey, and a third is 'manufacturing'. The blending honey, now and before the new classification, goes in the main to Great Britain. West Germany became a customer in 1949, and in 1951 bought over 500 tons; in 1955, 1958, and 1962, it bought more than Britain (6 000, 5 000, 6 500 tons respectively). More recently German purchases have been dwindling, and Great Britain is again Australia's most reliable customer. Japan has recently come strongly into the picture as a buyer of Australian honey (Table 14.23/1).

The biggest shipping State for honey is now Victoria (Melbourne), with an edge over Western Australia (Perth) for second place, and followed closely by New South Wales (Sydney) and South Australia (Adelaide). As a result of close co-operation between government and producers, Australian honey goes to the world markets well prepared—carefully packed in small steel drums, and scrupulously well graded. Exporting is done both by co-operatives and private enterprise.

14.24 New Zealand

New Zealand, further south than Australia, and wholly in the temperate zone, is also a honey-exporting country. The principal product is white clover honey, but honeys from certain other floral sources that are exclusive to New Zealand have some interesting characteristics.

New Zealand honey is not very important from the world market point of view; the entire production currently averages around 6 000 tons a year, of which only about 1 000 tons are exported—the major export markets being Britain, Irish Republic, Japan, West Germany. What is important in relation to the international honey market is the way this relatively small export trade is managed. It is controlled by the New Zealand Honey Authority, which enforces rigid grading rules, prepares

Table 14.23/1
Honey production and exports from Australia,
1955-1973, in tons

Crop year	Production	Exports
1955–1956	15 387	12 126
1956–1957	18 062	5 727
1957–1958	14 413	8 788
1958–1959	14 499	4 997
1959–1960	20 340	6 121
1960–1961	15 983	7 512
1961–1962	19 487	11 335
1962–1963	14 589	11 946
1963–1964	20 378	8 419
1964–1965	18 786	6 121
1965–1966	17 851	7 247
1966–1967	15 696	5 826
1967–1968	19 624	5 083
1968–1969	12 982	5 467
1969–1970	21 907	6 560
1970–1971	19 126	10 013
1971–1972	20 240	8 834
1972–1973	18 083	7 965

detailed annual reports that anyone can understand, and—even with overheads weighing heavily against the small tonnage—gives New Zealand beekeepers about $US0·15 per lb ($0·33/kg) as an average for all grades, which is about 10% more than the beekeeper in the U.S.A. had been receiving up to 1970 under government price support. The Annual Reports of the New Zealand Honey Authority are available for consultation, and are still published each year.

14.25 China
The People's Republic of China is unquestionably a very large source of the world's honey supply. The quality has not been considered high,

but countries buying it say that this has improved steadily in recent years. We have no complete statistical data on the Republic's production or exports. United States Department of Agriculture Report FS 4-59 (1959) states that 3 336 tons were exported in 1956, and FHON 1-67 (1967) reports 13 900 tons for 1966—an increase of over 400% in ten years. Exports in 1968, 1969, and 1970 are reported as 19 909, 20 831, 16 760 metric tons respectively (U.S.D.A.). These huge quantities of honey go largely to West Germany and Japan.

14.26 Other exporting countries

There are other countries, primarily exporters of honey, that collectively provide a substantial part of the world supply. In the Western Hemisphere they are (in order of their export volume) Cuba, Guatemala, Chile, El Salvador, Dominican Republic, Costa Rica, and Jamaica. These countries together ship on average 13 000 tons a year or more.

In Europe, also in order of reported export volume, the following countries are collectively important to the world supply: Spain, Hungary, Yugoslavia, Rumania, France, and Greece. From some of these countries come world-famous varieties of honey—such as orange blossom from Valencia in Spain; acacia from Hungary, Rumania, and Yugoslavia; Gâtinais honey from France, and the legendary thyme honey from Mount Hymettus in Greece (Table 1.4/1, Labiatae). These countries together export about 3 500 tons each year towards the world supply.

14.3 EXPORT-IMPORT COUNTRIES

14.31 United States of America

Two countries both export and import significant amounts of honey: the U.S.A. and Canada. Except for trivial quantities imported in retail containers, the great bulk of honey imports into the U.S.A. is from Mexico and Argentina, very much smaller quantities coming from other parts of Central America and elsewhere. All imports of honey pay 1 U.S. cent per lb duty ($0·02/kg). Honey from Mexico (primarily Yucatán) and Central American countries is used almost entirely for industrial purposes in bakery and tobacco products. Argentina honey is bought exclusively for bottling, and is used interchangeably with U.S.A. 'white clover' in the packer's blend of table honeys. Except for war years, imports average less than 5% of the country's production.

Trade in foreign honey coming into the U.S.A. is greatly influenced by the 'Government Support Price' for honey. Since 1950, a honey producer in the United States has been able to obtain a government loan on his honey through his local bank, by following certain simple procedures; if he can do no better elsewhere, he can ultimately sell his honey to the Federal Government at a fixed price, based on quality, and his location in the country. The price is also related to a certain percentage of parity.* This safeguards U.S. producers against ruinous prices for honey, but when the world market price is low, the system attracts heavy imports, which compete with domestic production.

Prior to World War I the United States had a good honey export business to Europe. Britain was the largest buyer, followed by Germany, the Netherlands, and Italy. During the war and the period immediately afterwards, the world experienced its first great sugar shortage, and the price of honey then rose to 25 cents per lb in carload lots in the U.S.A. ($0·55 per kg).

Business increased after the end of the war, Britain taking large quantities of granulated Rocky Mountain clover honey—two 60-lb (27-kg) tins to a wooden case, Germany buying the same, but even more of the light amber alfalfa honey from California. The Netherlands took all the surplus buckwheat honey from New York State in wooden kegs (a container that has long since left the scene) holding 160 lb (73 kg); clover honey went to Italy too—it somewhat resembled the local acacia honey.

In the 1930s Germany placed a prohibitive tariff on honey, effectively shutting out imports. Depression prices, without government supports, drove honey prices to all-time lows. Honey was as cheap as sugar—sometimes cheaper—so Britain, the Netherlands, and Italy remained good buyers.

Then came World War II, and all was changed. Because of submarine warfare, no shipments of honey from the Western Hemisphere could reach Europe. Honey crops in 1939, 1940, and 1941 backed up, and markets were stagnated; Cuban honey reached the New York market at $3\frac{1}{2}$ cents a pound ($0·08/kg) duty paid. But by the end of 1941 the United States was in the war, and there was another sugar shortage, due mainly to poor crops in Cuba. So sugar and honey prices soared, but only briefly, because the United States Government put price controls on honey and sugar, and the latter was tightly rationed.

Beekeepers packed their honey in small containers, and thus received the full retail price; this only lasted until the end of 1946, because in 1947 sugar became abundant again, and prices returned to normal. But for the last six months of 1946, honey was taken out of price control while sugar was still scarce, and during this period honey in carload lots once again sold in the U.S.A. for 25 and even 30 cents a pound, as during World War I.

---

* This is a system of regulating prices, usually by government price support programmes, to provide farmers (in this case beekeepers) with the same purchasing power as in a selected base period. Honey is supported in the U.S.A. (along with such basic crops as wheat, cotton, and corn) because beekeeping is considered essential to American agriculture for pollination.

At the end of World War II Europe was a shambles, her cities devastated, her economy ruined. The export honey market for American countries was non-existent, except for Switzerland. But the Marshall Plan was soon in operation, and gradually the export market returned. The establishment of an export subsidy by the United States Government enabled honey producers to get their honey re-established in Europe. This export subsidy started in 1952; it was 5 cents a pound for the first year, 4 for the second, 3 for the third, and 2 for the fourth and final year. And it worked, for about 9 000 tons of U.S.A. honey was now going to Europe every year. The contact was well re-established.

Table 14.31/1 gives some official U.S. statistics for the import and export of honey.

Table 14.31/1
Honey production, exports and imports (U.S.A.), 1934-1973, in tons

Crop year	Production	Exports	Imports	Crop year	Production	Exports	Imports
1934	69 251	565	28	1954	96 913	12 147	4 281
1935	71 317	727	34	1955	113 297	8 584	3 970
1936	79 241	633	74	1956	95 770	8 576	2 036
1937	71 429	1 219	97	1957	108 346	8 916	1 722
1938	98 214	1 434	100	1958	117 989	9 138	1 783
1939	80 569	1 502	111	1959	109 748	5 606	2 397
1940	91 860	1 034	388	1960	115 284	3 295	6 156
1941	99 089	186	3 988	1961	122 363	3 013	3 559
1942	79 318	109	9 160	1962	121 781	6 960	2 603
1943	84 762	40	16 667	1963	133 640	11 265	875
1944	84 338	70	9 344	1964	127 554	3 354	3 031
1945	104 049	112	7 934	1965	124 215	6 160*	5 938*
1946	95 452	101	13 504	1966	112 564	14 433	9 538
1947	102 046	548	6 483	1967	101 500	7 582	11 446
1948	92 100	5 189	3 467	1968	90 946	3 684	7 680
1949	101 330	472	4 729	1969	128 182	4 485	6 693
1950	104 024	5 395	4 560	1970	105 755	3 624	4 657
1951	115 231	6 413	3 605	1971	93 785	3 422	5 193
1952	121 715	10 926	4 256	1972	97 341	1 861	17 578
1953	100 185	13 031	4 176	1973	107 985	7 985	4 854

* Estimates

14.32 Canada

Canada is also both an exporter and an importer of honey. Its boundaries encompass vast areas, and its honey production—limited by its long cold winters—comes mainly from the clovers (legumes). Over

H—13

the years the Canadians have done so well in teaching good beekeeping and the virtues of honey that they have large *per capita* production and consumption of honey. Thus although the average annual crop is over 14 000 tons, there is no acute marketing problem, and in many years more honey is imported than exported. In recent years, however, Canada has been exporting important quantities to Britain in retail containers. With this development, honey exports in Canada soared from an average of about 200 tons a year in 1955–1959 to almost 4 000 tons in 1965 and 11 000 tons in 1971, of which 40% went to Japan (Table 14.32/1).

Table 14.32/1
Honey production, exports and imports (Canada), 1955-1973, in tons

Crop year	Production	Exports	Imports
1955	11 175	38	2 704
1956	10 835	450	1 493
1957	14 309	208	3 008
1958	12 281	109	2 162
1959	14 074	386	2 022
1960	14 386	1 193	1 187
1961	15 651	434	264
1962	13 710	1 619	278
1963	18 813	1 942	1 619
1964	16 367	2 194	1 251
1965	21 945	3 527	1 930
1966	19 867	3 826	2 085
1967	20 391	1 931	1 188
1968	14 895	3 623	452
1969	24 232	2 180	408
1970	22 745	4 095	260
1971	22 894	10 980	311
1972	17 200	4 920	592
1973	24 792	9 847	291

Canada becomes a honey-importing country when her crops are short. Then she goes to the United States and Argentina to get honey similar to her own production. In the past decade, the U.S.A. has sold to Canada as much as 2 200 tons in one year (1958), and Argentina 500 tons (in 1965).

14.4 HONEY-IMPORTING COUNTRIES

Almost every country in the world, except those where it is prohibited by law, is an importer of honey, but there are only a few countries whose imports are large, running into thousands of tons a year. Except for Japan, all the important honey-importing countries are in Western Europe. They are dealt with here in their order of importance.

14.41 Germany (B.R.D.)

Argentina, Mexico, Australia, and China dominate the world honey export market: the Federal German Republic (West Germany) alone dominates the world import market in honey; some details are given in Table 14.41/1.

Table 14.41/1
Honey imports into Germany, 1925-1973, in metric tons

Year	Amount	Year	Amount	Year	Amount
1925*	4 482	1942	—	1958	34 516
1926	5 468	1943	—	1959	32 180
1927	7 834	1944	—	1960	37 670
1928	9 372	1945	—	1961	37 169
1929	9 852	1946	—	1962	48 686
1930	5 726	1947	—	1963	44 975
1931	4 668	1948*	4 709	1964	38 960
1932	4 469	1949	2 862	1965	48 667
1933	4 696	1950	5 482	1966	45 465
1934	4 680	1951	8 536	1967	45 064
1935	6 558	1952	15 511	1968	47 645
1936	5 763	1953	20 373	1969	45 156
1937	6 349	1954	27 358	1970	43 145
1938	7 036	1955	29 806	1971	46 779
1939	—	1956	26 753	1972	44 833
1940	—	1957	32 088	1973	46 198
1941	—				

* 1925–1938 figures are for the whole of Germany; 1948–1973 for West Germany (Bundesrepublik Deutschland) only.

In the five-year period 1961–1965, Germany imported much more than Europe + Japan + Canada + U.S.A., and more than *twice as much* as the rest of Europe combined. In 1965 imports of honey reached the staggering amount of 48 667 metric tons. Over 60% of this honey came from Mexico, China, and Argentina (in that order). In that year the *per capita* consumption of imported honey in West Germany was 0·84 kg; when this is added to the domestic production (11 000 tons), the annual *per capita* consumption becomes slightly over 1 kg. This is, from the incomplete statistics available, one of the world's highest rates of honey consumption. And the honey is used mostly as table honey, only a small percentage going into various kinds of cakes (15.22).

Germany today pays a duty on honey of 33%, but only 16% on honey from other Common Market countries.

14.42 Great Britain

Not close to West Germany, but leading the rest, Great Britain is now importing about 17 000 tons of honey a year, from Australia, Canada, Mexico, and Argentina. The *per capita* honey consumption has been notably increased by vigorous merchandizing campaigns in the press and on television, by Joseph Farrow and Sons of Peterborough, who market 'Gale's honey'. Import figures since 1950 are given in Table 14.42/1. The price per ton trebled between 1970 and 1973 (*Bee World*, 1974).

Table 14.42/1
Honey imports into the United Kingdom, 1950-1973, in tons

Year	Amount	Year	Amount	Year	Amount
1950	7 170	1958	7 041	1966	14 610
1951	7 002	1959	6 375	1967	12 933
1952	8 005	1960	8 458	1968	14 546
1953	6 987	1961	10 495	1969	15 040
1954	5 700	1962	9 799	1970	16 859
1955	8 228	1963	11 971	1971	19 824
1956	6 586	1964	5 318	1972	16 202
1957	6 004	1965	12 836	1973	17 202

In Britain honey is consumed almost entirely at the table. As far as the author knows there is no bread made with honey, and it is little used in cakes or biscuits. Also, Britain is unusual in prohibiting the use of additives in tobacco manufacture, so it cannot be used for this purpose (*see* 15.5).

14.43 France

France produces fairly large quantities of honey, some of which is among the world's finest. It exports a little, but its imports are limited: although free to import from certain countries where there are favourable trade balances (including U.S.A., which is France's best customer for champagne and other wines), French importers must get special licences for buying from most countries, including Argentina, Mexico, and Australia.

Nevertheless, as long as honey is used in the manufacture of *pain d'épices* and *nougat*, France will be an important honey importer (averaging perhaps 4 500 tons a year), except in years when the French honey crop is itself a bumper one, or when import licences are not issued.

14.44 Switzerland

There are many beekeepers in Switzerland in proportion to the population (4.21), and they know their beekeeping. But conditions are not conducive to the production of large honey crops, so—since the Swiss love honey—they import it in quite large amounts. Prior to World War II they bought much honey from Guatemala, but in 1949 they were introduced to honey from Yucatán, and most of their annual imports of 3 000 tons now comes from there.

The Swiss use substantial quantities of honey in baking, but by far the biggest portion of the honey of commerce is table honey. Duty on honey going to Switzerland is 600 Swiss francs ($156) a ton; until 1959 it had been 1 200 francs a ton, and by halving the duty the honey import business was doubled.

14.45 The Netherlands

Official statistics from the Netherlands (Table 14.45/1) reveal that the import honey business was larger in the 1930s than it has been since World War II. In 1939, 6 268 tons were imported, whereas since 1948 the highest annual import was 3 860 tons in 1965.

This country promotes honey to the consumer constantly, with attractive material designed to instruct six- and seven-year-olds with respect to honey. The Netherlands also has a large trade in honey which is imported and sold and forwarded to Central Europe, especially Austria and Switzerland, and to some extent the south of Germany. Prior to World War II the Netherlands bought 1 000 tons or more of buckwheat honey a year from New York, which was used to make honey cakes. Buckwheat is no longer grown in the U.S.A. except in inconsequential quantities, so buckwheat honey is no longer available; this may in part explain the lower Dutch imports of honey now than in the 1930s.

14.46 Belgium

In the five years 1961–1965, Belgian honey imports averaged about 2 000 tons a year. In this small country, the greatest promotional undertaking for honey is the work of Albert Florizoone, known familiarly as the Belgian Honey King. In the tiny village of Adinkerke—across the water from the historical beaches of Dunkirk—is Meli Park. Here, for a small admission fee, the public, including busloads of children from all over the country, come to see one of the finest collections of birds, bears, monkeys, seals, and many other animals, set out in a beautiful park. Provision is made in great comfort for people who bring their lunches, and they can get refreshments at low cost. And there is a free show on bees and honey, with the best colour films on bees and honey that are available. Literature on honey is free too, and Meli honey is on sale. So the now famous Meli honey dominates the Belgian market—the 'formula'

Table 14.45/1
Honey imports into the Netherlands, 1934-1973, in metric tons

Year	Amount	Year	Amount	Year	Amount
1934	3 264	1948	81	1961	2 272
1935	4 553	1949	263	1962	2 568
1936	4 492	1950	3 035	1963	2 816
1937	4 873	1951	1 969	1964	2 297
1938	5 113	1952	3 472	1965	3 860
1939	6 268	1953	1 874	1966	3 175
1940	2 907	1954	1 657	1967	2 912
1941	1	1955	1 410	1968	3 031
1942	—	1956	1 355	1969	2 973
1943	—	1957	1 222	1970	2 983
1944	—	1958	1 608	1971	2 934
1945	—	1959	2 293	1972	3 188
1946	245	1960	2 297	1973	2 737
1947	47				

is secret, but it is suspected that it incorporates honeys from legumes, acacia, and orange blossom, with Yucatán and Mexican varieties. What an intriguing combination of world honeys!

Belgium uses little honey in industry: some in honey cake and biscuits, and some in a very good sparkling honey wine. But almost all the imports go for table honey.

14.47 Italy

Italy produces substantial quantities of honey, but almost never enough for her own great needs. Her crop comes principally from the acacia tree (*Robinia pseudoacacia*): it is white, mild, and delicious, and very slow to granulate.

Italy might be self-sufficient for honey, were it not for the national confection called *torrone*, which is described in 15.23. So Italy imports about 1 200 tons of honey a year, at present mostly from Argentina and Yucatán in Mexico.

14.48 Austria

This country is enthusiastic about bees and honey, and chose as its motto for the XVI International Beekeeping Congress held in Vienna in 1956: *Österreich ein Bienenland*—Austria, a land of bees. The schools and experiment stations devoted to bees and honey stand witness to the country's interest. But, as with Switzerland, not enough can be produced within the country, and on average about 3 000 tons a year are now being imported.

Austrians too make honey cake but, like Germans, they consume most of their honey as table honey.

14.49 Scandinavian countries

Denmark, Sweden, Norway, and Finland are small honey producers, and relatively small consumers. Denmark, importing about 1 100 tons a year, buys more than the others put together.

Tariffs are very high in this area, and may thus be the restricting factor; if so, then reciprocal trade agreements might lead to higher imports. If honey were available at an attractive price, its consumption in these countries might well increase greatly.

14.4.10 Japan

The remaining honey-importing country is something of a sensation. Until as late as 1961, Japan imported no honey. It was still under U.S. Government influence, as a result of treaties signed after World War II. Gradually the breakfast of a bowl of rice was disappearing, and being replaced by the American-type breakfast—fruit juice and cereal, toast, honey, and coffee. By 1962 the Japanese industries were re-established, and business was starting to boom. Japan began to import honey, only 200 tons in that year, but—with rising wages—1 300 tons in 1963, 2 500 tons in 1964, 6 000 tons in 1965, and up to 25 000 tons by 1973 (Table 14.4.10/1). A revolutionary change had been made in eating habits.

The Japanese want mild white honey. They are keen merchants who know how and where to buy; at first they bought much of their requirements from Argentina, but China has been the major supplier for some years, with smaller amounts from Canada, New Zealand, and Australia. In maintaining her prosperity, Japan is destined to become one of the great honey importers and consumers of the world.

Table 14.4.10/1
Honey imports into Japan, 1963-1971, with details of country of origin at 4-year intervals, in metric tons
(Sumitomo Shoji Kaisha Ltd)

Year	Total imports
1963	1 339
1964	2 574
1965	6 240
1966	10 893
1967	14 358
1968	11 185
1969	11 661
1970	14 537
1971	16 358

Country of origin	1963	1967	1971
China (P.R.)	299	7 987	5 359
Canada	51	127	3 657
Argentina		3 405	3 295
New Zealand	172	2	1 254
Australia	386	87	1 099
Hungary	10	1 228	555
Rumania		667	354
U.S.A.	296	316	282
Spain	20	121	156
Cuba		20	101
North Korea	50	100	79
U.S.S.R.		217	57
Mexico		7	43
Israel			20
Honduras			17
West Germany			16
Guatemala			9
Taiwan (Formosa)	3	39	6

In 1963 and/or 1967 Bulgaria, Chile, Poland, South Africa, and the U.K. contributed small amounts (30 tons or less)

14.5 WORLD TRADING IN RETROSPECT

We have seen how honey, bought and sold as a valued food in ancient times, has risen to the status of a world commodity. There are obstacles to the further development of international trading in honey. Tariffs are one: the largest importing countries still impose high tariffs on a product which is grossly under-produced within the country for its own needs. Much would be gained by the complete removal of tariffs in these countries. Import regulations constitute another obstacle. Conditions laid down by Germany are the most stringent: unless honey has a certain minimum diastase content (5.71) and less than a certain hydroxy-methylfurfural content (5.84), it may not enter the country. Values outside the allowed limits are regarded as proof of overheating, although the basis for this distinction is not valid: some honeys still in the hive contain no diastase, and hydroxymethylfurfural contents may be high in honey that has not been subject to artificial heat, especially that produced in hot climates (9.31).

So world trading in honey has problems, but none is critical, and none without solution, provided that co-operation and open-mindedness between exporters and importers can be achieved.

CHAPTER 15

USES AND PRODUCTS OF HONEY

by R. B. Willson and Dr. Eva Crane

CHAIRMAN, R. B. WILLSON INC., NEW YORK, AND
DIRECTOR, BEE RESEARCH ASSOCIATION, RESPECTIVELY

*Cheese-cakes steeped most thoroughly
In the rich honey of the golden bee.*
EURIPIDES, CRETAN WOMEN (438 BC)

15.1 INTRODUCTION

In the long history of man, honey was used for many thousands of years before cane or beet sugar (Chapter 19). The origins of some of the present-day commercial products are therefore lost in antiquity. Others—and important ones—are new in the twentieth century. In his early history, man was almost entirely dependent upon honey for sweetness. Cane sugar became a direct competitor—and largely a successor—as the trade routes between Europe and the East were opened up (Baxa & Bruhns, 1967). There are, however, still a number of commercial and domestic products where honey is superior to sugar, because of flavour, texture, keeping qualities, and other factors.

The established uses of honey lie in many fields, including baking, confectionery, preserves, spreads and syrups, meat packing, tobacco manufacture, cosmetics, and a number of interesting minor applications.

In this chapter, commercial products made with honey are defined as those in which honey is an important ingredient, and which have become established in the market place. Many other products made with honey have been introduced into the market, but have not survived; on the other hand, some of the important uses of honey are not commercial in the above sense. Domestic uses are many and various.

Perhaps the most important commercial honey product that failed was ice cream. It had great sales appeal, and was delicious, but it ran into serious technical difficulties. The storage facilities of distributors are designed for ice cream made with cane sugar (sucrose). The invert sugars, glucose and fructose, of which honey is largely composed (Chapter 5), have much lower molecular weights than sucrose, so the freezing point of ice cream made with honey is lower than that of ice cream made with an equivalent amount of cane sugar. So in normal ice cream freezers, honey ice cream becomes soft and slushy, instead of remaining firm as sucrose ice cream does.

Another example is honey jelly. This has repeatedly been offered to the market, but has never caught on, because it is too bland in comparison with fruit jellies. Peanut butter made with honey sounds intriguing, but the product hardens, and the shelf-life is so short that it is commercially impractical. A combination of honey and cream was tried commercially, but it also did not survive.

15.2 MAJOR USES IN FOODS

15.21 Honey as a product in its own right
The major part of the world production of honey that is harvested (4.2), including the honey sold on world markets (Chapter 14), is eaten as table honey, not used as an ingredient of some other product. Primitive peoples have usually eaten honey on its own (19.2), but in general in the world today honey is either spread on bread, biscuits or crackers, or used directly to sweeten drinks (tea, coffee, fruit juice) or fruit or cereals. Of the estimated annual world consumption of honey (4.27), probably 90% is eaten directly as honey. The present chapter is, however, concerned with the remaining 10% which is used in other products, especially those manufactured commercially but also those made and used on a more domestic scale.

15.22 Honey in baking
The use of honey in baking is universal. Wherever baking is done, there are specialities made with honey, some of which have a very ancient origin. Beck (1938) discusses the honey cakes of the Ancient World, which were most highly valued. The Egyptians fed them to their sacred animals; in Rome *libum* was a sacrificial honey cake, possibly the root of the German *Lebkuchen*. Cheese-cakes baked with honey were praised by all Greek poets; cheese-cakes were glorified by Euripides and Aristophanes, and honey cake by Anacreon and Sophocles. In Rome, the honey cake *savillum* was eulogized by Cato as most savoury of all.

'Wafers made with honey' were familiar to the ancient Israelites, since this is how they described the manna they found in the desert after their flight from Egypt [*Exodus* 16: 3] (the manna itself may have been crystallized honeydew—cf. Section 2.13). *Baklava* is still made in Greece and the Near East: this consists of layers of flaky pastry with a rich filling of honey, almonds, butter, eggs, and spices (it is not a food for calorie watchers!). Among books that give interesting details are *The sacred bee* by Ransome (1937) and *Food and health* by Beck (1938), but the 1944 edition *Food and your health* by Beck & Smedley omits most of the historical information.

Possibly derived from the Roman *panis mellitus*, a great variety of spice cakes became famous in central Europe from mediaeval times onwards. In France and Belgium it is *pain d'épices*, in Germany *Lebkuchen*, *Pfefferkuchen*, etc.; they are flavoured with various spices. Nürnberg was (and is) a centre for their manufacture in Germany, and also Weissenberg. The *Lebkuchen* are baked in early summer, and removed from the oven dry and hard; by Christmas time—when they are eaten—because they are made with honey, they are ripe, soft, and chewy. In Budapest in Hungary there is still a shop devoted entirely to baking and selling these goods (*see* Figure 15.22/1—Plate 16); their manufacture is discussed in detail by Beliczay (1960), and the derivation of their names by Sturm & Hanssen (1961).

Czechoslovakia specializes in 'gingerbreads' of intricate design and with ornamented icing; they are likewise flavoured with spices, but not normally with ginger. Two recent books on them are by Bayerová *et al.* (1968) and Kopřivová (1961). In Poland Torún is still an important centre—the *décolletée* lady in Figure 15.22/2 (*see* Plate 16) will keep for twenty years if properly cared for. Descriptions of the industry, and detailed recipes, can be found in the many publications (e.g. *Européen*, 1962, Roski, 1968, Roussy, 1969). *La fabrication des produits alimentaires au miel* by Paillon (1960), which gives many large-scale recipes, quotes Roger Vaultier, archivist at the Bibliothèque Nationale: 'In France, *pain d'épices* was known from the time of the first Crusade. Agnes Sorel and Margaret of Navarre [*see* 15.7] valued it highly, but its use declined under Henri II, on account of a rumour that the Italians (who exported large quantities to France) mixed poison with it. It came back into favour under Louis XIV, and since then its popularity has remained high . . . We know that in the 1400s and 1500s *pains d'épices* were manufactured in many French towns . . . in Amiens in 1583 they were sold at a dozen for 7 *sols* [sous].'

Collections of wooden 'gingerbread moulds', in which the honey cakes were shaped, can be seen in various museums in the Netherlands; the mould in Figure 15.22/3 (*see* Plate 16) has a lady on the reverse side; German blocks and moulds are described by Hanssen & Hahn (1963). Bakers in Slovenia, the northern part of Yugoslavia, produce delicately executed 'gingerbread hearts', with elaborate designs moulded on top (Figure 15.22/4). Gingerbread houses are a speciality in Czechoslovakia (Figure 15.22/5). Today, any of these traditional honey cakes may be cheapened by being made with honey substitutes, but they cannot attain the delicacy of flavour of the genuine product, or the desired texture, unless made with honey.

In the Americas, all industrial uses of honey, including the extensive use of honey in baking, are confined largely to the United States and Canada. In the United States there is no area where bread made with honey cannot be bought, nor any place so remote that the nearest food market does not have graham crackers made with honey. Graham crackers are thin biscuits made of coarse (unbolted) flour, being named after Sylvester Graham, an American health-food promoter, 1794–1851. Honey in bread, ideally to the amount of 6% of the weight of the flour, enhances flavour, gives the bread a richer darker crust, and adds substantially to its keeping qualities. Honey in cakes and crackers (biscuits) improves flavour and gives additional sales appeal. Careful estimates indicate that in the U.S.A. alone some 25 000 tons of honey are used each year in commercial baking, bread and the graham crackers accounting for most of this huge quantity. There are, however, many other commercial baking specialities made with honey in North America, including honey buns, fruit cake, doughnuts, and cookies.

This industrial use of honey in baking has initiated, and is also derived from, research on the comparative behaviour of honeys and other sweetening agents during mixing, baking, and subsequent storage (Bailey & Lothrop, 1939, Miller *et al.*, 1960, Smith & Johnson, 1951, 1952, 1953*a*, 1953*b*, Nordin & Johnson, 1957). In general, the use of honey for up to one-third of the total amount of sugar is recommended. It improves moisture retention and thus increases storage life; it also tends to eliminate dryness and crumbliness in cakes, and to give them a richer flavour (Smith & Johnson, 1952). Mild honeys are better, except for fruit cake for which strong-flavoured honeys are satisfactory (Smith & Johnson, 1953*a*).

Claims that bread or cake contains honey can be tested by examining a sample for the presence of pollen grains, which are a normal constituent of honey (Chapter 7). Suitable techniques have been developed, adapted from those for the pollen analysis of peat (Louveaux, 1956*a*, Sturm & Hanssen, 1961).

There is a wealth of cookery books giving honey recipes from many countries. Many are listed in *Bee Research Association Bibliography No. 1* (1963), and *Bibliography No. 5* (1964) gives other useful information. The American Honey Institute has published a succession of recipe books since its foundation in 1928, *Old favorite honey recipes* was last issued in 1945 and *New favorite honey recipes* in 1947. The Australian Honey Institute has published *Secrets from the honey pot* (1954) and other material. Books from the U.S.A. include *Granny's honey and beeswax prescriptions* by Alice Cooke Brown (1957), Maxine Wilhelm's *Honey cook book* (1965), several from the Californian Honey Advisory Board by Monica Schafer (e.g. 1958, 1961, 1966), Mario Lo Pinto's *Eat honey and live longer* and Berto's *Cooking with honey* (1972). These books are largely, although not exclusively, interested in cakes, bread, confectionery, and other sweet foods.

The honey cookbook by Juliette Elkon (1955) has a wide range, and gives much attention to meat and vegetable cookery as well as to cakes, cookies, and pies. The author's European background has enabled her to present good and varied recipes from different parts of the world which will be widely acceptable in many countries, so that this book is outstanding.

Books from Czechoslovakia, Hungary, and France have already been mentioned. There is a Spanish book (de Vinuesa, 1965), one from Germany (Aisch, 1938) and Israel (Kornfeld, 1967), Poland (Bornus & Zalewski, 1962), and Yugoslavia (Perušić, 1959). None of these European books are in English, and Ambrose Heath's *Honey cookery* (1956), published in Britain, is based on Australian and New Zealand recipes. Numerous leaflets and brochures are available from beekeeping associations, departments, and institutions in many countries, and articles appear frequently in beekeepers' journals, especially around Christmas time.

15.23 Honey in confectionery

Evidence from primitive peoples suggests that man has always liked sweetmeats. So it is likely that confectionery made with honey pre-dates the written history of man, since the use of honey preceded that of cane sugar in most regions (*see* 19.42).

The relative cheapness and ease of use of cane sugar, and of the less sweet commercial glucose or corn syrup, have now pushed honey to an inconspicuous role in the confectionery field. There are, however, a few products—some of which are world famous—that cannot be made except with honey. They include *halvah* from Turkey, and *pasteli* of Greece, which is made from 50% honey, 40% sesame seed, and 10% sugar.

Italy has its traditional sweets made with honey (*see* e.g. Sherman, 1968, Zezzos, 1969). One Italian firm alone uses 300 tons of honey a year for manufacturing *torrone*, a sweetmeat made with almonds or other nuts and fruits, and white of egg; a New York firm uses 50 tons. An interesting result of emigration is that the traditional honey sweets are now made in other parts of the world than their country of origin, to satisfy the different cultural groups of immigrants.

The Spanish sweetmeat is *turrón*; ideally made from rosemary or orange honey, with almonds—its quality is given on the label as 1, 2 or 3 almonds (like stars for restaurants). Each Christmas some 8 000 tons of this are made, most of it in Spain, which has strict legislation governing standards for export (*Apicultura*, 1965, de Vinuesa, 1968). In Spain itself, *turrón* accounts for most of the *per capita* consumption of honey. The equivalent French delicacy is *nougat*; Montélimar became famous as the centre of its production after the almond tree had been introduced into the region in the 1500s (*Gazette apicole*, 1965). *Nougat* is still manufactured commercially in France (Gergeaux, 1961), and is likewise subject to official standards (Confiserie, 1964).

Confectioners in the United States have done more than any others in recent years to develop the use of honey in candies of various types. Its potentialities have been studied by Barth (1952), and instructions and descriptions published by Anderson (1958), Meineke (1967), Watson (1968), and others. The German journal *Kakao and Zucker* (1969) describes the manufacture of a butter-honey caramel.

One modern product now distributed about the world is a 'honey drop' made in Italy; this is the size of a small peach stone, with a hard outer shell of clear sugar enclosing a centre filled with liquid honey. Another is a honey cough drop made in the U.S.A. (*see* 15.6).

Altogether, in the world today, some 1 500–2 000 tons of honey are used annually in commercial confectionery, and an unknown further quantity in domestic production—for home consumption, for sale, or as gifts.

15.3 MINOR USES IN FOODS

15.31 Honey in breakfast cereals

Breakfast cereals of the cornflake type, sprayed with a sweetening agent, have been on the market for about twenty-five years, and are now well established. Honey is an ingredient in a number of these, but because it is hygroscopic, and a high degree of crispness is required in the product, 100% honey may well not be used (e.g. Fast, 1967). The annual world consumption of honey in these products is around 250 tons. A lot more honey is used domestically for eating with milk and cereal.

15.32 Honey spreads

Finely granulated honey, prepared by the Dyce process (Chapter 10), stays where it is laid down, and does not flow or drip like liquid honey. It is therefore suitable for use as a spread for bread, biscuits, crackers, sandwiches, and as a filling for cakes. Various ingredients such as butter, cinnamon, and fruits, can be blended with this granulated honey.

'Honey butter', the most successful of these spreads, has been established in the market in Canada and the United States for over forty years. Its formula is secret, but it probably contains about 55% 'white clover' honey and 45% dairy butter. One variant is made with cinnamon added, and others are constantly being devised; a new patent for honey butter was brought out by Finley & Hollowell in 1967. All spreads containing butter must be refrigerated for retail marketing.

Another promising development is the combination of reconstituted dried fruits with processed honey; alternatively juice or purée can be used (White, 1950). The best known of these honey-fruit spreads are made with apricots or strawberries. Methods for using sun-dried apricots, and freeze-dried strawberries and blueberries, are published by Berthold & Benton (1968a), who reported (1968b) that the honey-fruit spreads— with their sharper flavour—are enjoyed by many people who do not like honey. Graphs of monthly sales at one store show that during 1967 and 1968 honey-fruit spreads outsold liquid and creamed (finely granulated) honey combined; this was presumed to be because they were less sweet, offered a variety of flavours, and were less messy than liquid honey (Benton & Heckman, 1969). A maple-honey spread has been patented (Naghski et al., 1956), and in the U.S.A. a syrup made of honey, maple sugar, and cane sugar is on sale for dressing pancakes. Spanish markets offer syrups (arrope) made of fruit and honey.

Some 200 tons of honey are used annually for the various types of commercial honey spreads.

15.33   Honey in baby foods
For two generations baby foods have been a firmly established commercial product. Recommendations by pediatricians have helped them to develop into a multimillion dollar business in the United States alone. These prepared foods, packed in glass containers or tins, include a wide range of mashed and finely strained fruits, vegetables, and cereals. One large packer in the United States produces eight different varieties, and a substantial part of his business (and of his competitors') is exporting these baby foods all over the world. In the main, it is the fruits and cereals that are sweetened with honey, and about 100 tons of honey a year are used for the purpose.

15.34   Honey in meat packing
Many cooks have discovered how the flavour of a ham is enhanced if it is baked with honey. There are several meat packers in the United States —and possibly elsewhere—who use honey in some of their formulae, but these formulae are secret. The details seem to be so valuable from a sales point of view that the manufacturers will not divulge how the honey is applied, nor how much; see however 15.36. But at least 75–100 tons of honey a year are used for prepared meats in the United States.

15.35   Honey in preserves

There are many preserves (confitures) that can be made with honey, although the water content of honey can sometimes present difficulties, and honey jellies (Haynie, 1953) are in general too bland (15.1). In the commercial field, cane and beet sugar have taken over almost entirely, but many of the books mentioned in 15.22 give recipes which use honey for marmalades, jellies, jams, chutneys, and sweet pickles, and these are widely followed in beekeeping households.

One preserve made with honey is still world-famous as a commercial product: Bar-le-Duc currants. The town of Bar-le-Duc is the capital of the department of Meuse in north-eastern France, and has long been known for its jams and jellies. According to tradition, its early fame was based on a preserve made with red currants and honey, for which the seeds were removed from the fruit by hand. The currants are cooked in honey so that each remains whole; almost any small soft fruit may be used instead of currants.

Problems associated with the use of honey in fruit juices and preserves have been studied in Greece (Kodyne, 1962, 1967); in Germany a patent has been taken out for the use of honey in preserving finely divided fruit and vegetables (Schweigart, 1961); an Italian method for preserving plums and other fruit in honey is available (Muzzati, 1953).

Probably not more than 5–10 tons of honey a year are used commercially in preserves.

15.36   Dried honey

Dried honey, or honey concentrate, is made by spraying a film of honey on to a slowly revolving hot drum about 4 m long and 2½ m in diameter. At the completion of each revolution the film of honey, from which the water has been evaporated, is scraped off. Dried honey can also be produced by passing it through an evaporator under vacuum (Turkot *et al.*, 1960, Claffey *et al.*, 1961). When the pure honey concentrate is exposed to the air, it soon takes up moisture and sets into a hard and unmanageable mass. A practicable product can, however, be produced by blending the concentrate with about 55% of a starch or a non-hygroscopic sugar, or a combination of both (Glabe *et al.*, 1963, 1965, Shookhoff, 1957; also a Japanese patent, Tokyo Yakuhin Kaihatsu, 1969).

The stable powder thus produced is used in dry mixes for cakes and bread; it improves their flavour and texture, without the necessity of handling or packaging the honey as such (*Food Processing*, 1969). A coating of dehydrated granular honey can reduce shrinkage in meat products by as much as 19% (*Food Processing*, 1971). These types of honey product have been developed to suit modern manufacturing and marketing methods. At present the world consumption of honey in them is not more than 50 tons a year, but this amount will undoubtedly increase, and they may become one of the important commercial honey products.

15.37 Honey and milk

From very early times, milk and honey have been linked together as symbols of plenty. The land promised to the children of Israel was 'a good land and a large . . . flowing with milk and honey' (*Exodus* 3: 9). Hilda Ransome (1937) devotes a chapter to the ritual uses of honey and milk; she believes that their being offered to the dead is possibly one of the oldest forms of ritual, and quotes from Gilbert Murray's translation from Euripides:

> 'Milk of the mountain kine,
> The hallowed gleam of wine,
> The toil of murmuring bees.'

In the early Christian Church it was customary to present newly baptized persons with mixed milk and honey, and the blessing for the Whitsuntide baptisms in the Roman *Sacramentarium Leoninum* begins: 'Bless, O Lord, these thy creatures of the spring, of honey, of milk' and ends: 'Unite thy servants, O Lord, with the Holy Spirit, as here honey and milk are united, as a sign that heavenly and earthly substance is united in Christ Jesus our Lord.'

Coming to modern times, Spöttel's book *Honig und Trockenmilch* (1950), which is a valuable source of information about the earlier scientific literature on honey, was associated with a milk-honey product called Ho-Mi, manufactured in Germany (D.D.R.). It contained 65% dried milk, 25% honey, and 10% glucose. A book by Simonis (1965) in Germany (B.R.D.) also deals with milk and honey. In the U.S.A. Webb & Walton took out a patent for a dried honey-milk product in 1952, and in 1967 Torr took out a French patent for a somewhat similar product. Leomiel, on sale in France, is made from milk, honey, and predigested whole wheat.

The total world consumption of honey in these products is not at the moment of much consequence—perhaps 10 tons—but it may well increase.

15.38 Other honey food products

Honey has recently been introduced into yoghurt (Brown & Kosikowski, 1970). It is used in a coating sprayed on coffee beans during roasting (Nafe, 1966) to improve the flavour. It has been combined with oils from various fruit and nuts to provide a highly nutritious food (Bastet, 1969).

15.4 PRODUCTS OF HONEY FERMENTATION

These are dealt with in detail in Chapter 16, and only a few aspects of this age-old use of honey are mentioned here. In a paper on information collected by the Danish National Museum, Højrup (1957) reports finding residues from mead in a Bronze Age grave at Egtved, and in another funeral offering (two drinking horns from the first century AD) in a bog near Skudstrup. The early mediaeval Edda poems give various accounts of mead drinking at parties, and in 1590 24 tons of it were used at a princess's wedding; see Figure 19.41/4, Plate 26.

In Jewish communities in many parts of the world the mother of the household would, in the autumn, proceed to make a kind of mead with the aid of yeast and hops, for the next Passover. This drink is gradually phasing out in the West, as the old-fashioned Orthodox Jewish mother passes on, and a Kosher wine made with honey is taking its place.

Mead, made by fermenting honey (Chapter 16), is available in small quantities in many places in the world, but in few places does it warrant the status of a commercial honey product. The use of honey in wines is, however, wider than this. In 1970, a superficial survey of stores in up-state New York in the U.S.A. brought to light nine wines for sale that were made with honey—two of them imported from Japan.

Outstanding in the list of alcoholic honey products, and lending its own prestige to honey as one of its components, is Benedictine, the celebrated liqueur of Fécamp in Normandy (France). In South Africa a well known liqueur with a brandy base is made with honey, and others are discussed in 16.5. But the secrecy of the formulae used in these honey products is hard to break down. A new liqueur sweetened with honey, and with Scotch whisky as its base, has recently been produced in Aberdeen, Scotland; its name is Lochan Ora.

Further (aerobic) fermentation of honey has been used for the production of vinegar, from earliest times and in many parts of the world (see 16.33). But honey vinegar hardly qualifies as a commercial honey product under our definition.

The world consumption of honey for commercial alcoholic beverages is estimated at 100 tons for liqueurs and another 100 tons for wines. To this must be added perhaps 40 000 tons for the African honey beers (16.43). Their manufacture is the main use for honey among African peoples, although little is sold as a commercial product.

15.5 HONEY IN TOBACCO PRODUCTS

Honey is extensively used in tobacco products. In most countries, a sweetening agent of some kind is used in processing tobacco—usually cane sugar, molasses, corn syrup or honey. Honey has the particular advantage that its hygroscopic nature helps to keep the tobacco moist. Tobacco products depreciate in value in proportion to the rapidity with which they dry out. Moreover dry tobacco burns too hotly to be acceptable.

The Italian Government tobacco monopoly buys honey. It is known that manufacturers of cigarettes in Switzerland, the Philippines, Hong Kong, and the United States use honey but, here again, the formulae are secret. At least 2 000 tons of honey a year are used in these and other countries for processing tobacco, most of this in the U.S.A. In Britain the use of sweetening matter, or indeed of any other additive except water, is prohibited (Customs and Excise Act 1952, Part V, Paragraph 176).

A prominent American manufacturer of pipes for smoking imports French briar-root—from which the world's finest pipes are made—and, in preparing it for use, soaks the wood in a solution which includes honey. Here also the formula and process are secret.

15.6 HONEY IN PHARMACY

Honey has been regarded as a cure for many ills, from very early times. Books by Beck (1938), Ransome (1937), Jarvis (1958), Lerner (1963) give numerous details. A Sumerian tablet (Figure 19.31/1) dating from about 2000 BC has been found at Nippur in Iraq. Referred to as 'the oldest fragment of medical literature ever unearthed' (*Journal of the American Medical Association*, 1954), this tablet gives details of several prescriptions, in one of which honey and water were used for kneading up a powder.

The first book on honey in English, by John Hill (1759), was entitled *The Virtues of Honey in Preventing many of the worst Disorders; and in the Certain Cure of Several others; Particularly The Gravel, Asthmas, Coughs, Hoarseness, and a tough Morning Phlegm* . . . (19.513). It is worth noting that when cane sugar was rare and expensive it was also regarded as a panacea. Physicians agreed that sugar was an excellent remedy for constipation, flatulence, colic, and purges of the internal organs; its special virtue was as a laxative. But they recommended that sugar should not be used much, except in case of illness (Baxa & Bruhns, 1967).

Today honey is a component of many commercially manufactured pharmaceutical products. At least 200 tons are used in the world annually in various types of cough mixture; a cough drop incorporating honey and glycerine, made in the U.S.A., accounts for many tons of honey.

Another use is as a palatable sweetening agent in general pharmaceuticals. An extensive American survey of the stability of drugs in the presence of honey reported that 'most members of the taste panel considered the honey medicinals superior to the same medicinal made without honey' (Gennaro *et al.*, 1959). In England, Short discussed the subject in 1960. In the Soviet Union honey is commonly used as a base for ointments, and honey is used by itself for treating burns (e.g. Osaulko, 1953). In England Bulman (1955) and others have reported its successful use as a surgical dressing for open wounds, burns, and septic infections; it proved to be a more comfortable dressing than most, being —perhaps surprisingly—a non-adhesive. There is also interest in South Africa (Steyn, 1970).

Another use, reported from Switzerland (Buman, 1953), is against sickness resulting from radiation treatment; this effect seems to be due to the fructose in the honey (Fochem, 1954). The rapid utilization of fructose (Albanese *et al.*, 1952), and the increased rate of metabolism of alcohol in the presence of fructose, have led to the use of honey for sobering drunken patients (e.g. Martensen-Larsen, 1954).

Honey and fruit are the main natural sources of fructose, which has in the past been very expensive to produce, and therefore commercially unimportant. An International Symposium on the Clinical and Metabolic Aspects of Laevulose (1971, *see also Nature*, 1970) was held recently at the Royal Society of Medicine in London. At this Symposium developments were reported which should make it possible to produce cheaply large quantities of fructose from wheat starch. If this comes to pass, commercial uses of fructose will expand greatly, but honey will not have any obvious part in this expansion.

15.7 HONEY IN COSMETICS

Cosmetic uses of honey date from very early times. Beck (1938) refers to several famous women whose use of it is on record. Nero's wife Poppea, who employed a hundred slaves to attend her beauty, used honey and asses' milk as a face lotion, and the patrician women of Rome followed her example for many centuries. In France many famous women are known to have used honey extensively in their toilet preparations: in the 1400s Agnes Sorel, mistress of Charles VII; in the 1500s Margaret, wife of Henri II of Navarre; in the 1600s Mme de Sévigné; in the 1700s Mme du Barry, mistress of Louis XV. In England Queen Anne, to whom Warder dedicated his book on bees *The true Amazons* in 1712, was also reputed to use honey. In Italy, Marinello published a book in Venice in 1574, *Gli ornamenti delle donne* (A lady's adornments); this gives many details of the uses of honey in cosmetics at that time—for hair, skin, lips, hands, and eyes (*see* Chiavegatti, 1965).

Honey is still valued in cosmetics today, particularly for its emollient effect on the skin. Many beekeeping families use home-made lotions:

honey, egg yolk, and sweet almond oil; honey, lemon juice, and Eau de Cologne; honey, glycerine, alcohol, and lemon juice (Beck, 1938); honey, beeswax, and lard; honey, egg white, glycerine, and flour (Tonsley, 1969).

Commercially, the presence of honey in lotions, handcreams, etc. is commonplace; it is also used in facial masks, in preparations for tonic or relaxing baths, in hair conditioners and shampoos, and in toilet soap; honey soap is on sale in France, Germany, and Japan. In 1971 *Gazette apicole* (page 41) reproduced an early advertisement for a toothpaste based on honey—*Panacée dentifrice: Miel éthiopien*—which was awarded a royal warrant in 1844.

The use of honey and milk, referred to at the beginning of this section, has just been revived. The following is a quotation from Charles Revson appearing in a 1972 advertisement: 'I believe I have tapped a great new natural resource of beauty in 100% fat-free milk, rich with proteins . . . and moisturizing honey. Nothing I have seen gives skin such a look of vitality as these pure, natural organic ingredients.'

Cosmetics do not account for more than 25 tons or so of honey a year. On the other hand beeswax, another hive product, is an important component of many cosmetics—lipsticks, face creams, hair dressings, etc. It is difficult to get information as to quantities; cosmetics fetch higher prices if they can be described in terms of 'miracle' components, or if their contents are kept as trade secrets, than if it is made clear that they are based on well known materials, however wholesome.

15.8 OTHER USES OF HONEY

In the course of man's history, uses of honey have been legion, some having a logical basis—whether this was understood or not—and some having none. In the Ancient World, honey was used for feeding sacred animals (Ransome, 1937), and as an ingredient of many remedies for animal as well as for human ailments. Representative examples from the Middle Ages can be found in a manuscript written in the 1200s by a Provençal ecclesiastic, Daude de Pradas, entitled *Dels auzels cassadors* (Birds of the chase); the manuscript has been summarized in English (Schutz, 1945). Honey is frequently recommended for feeding and medicating falcons: to make meat attractive to them; as a binding agent for pills; to revive tired birds; in ointments, and especially to encourage the growth of new feathers in place of broken plumes.

Nowadays veterinary medicine uses more sophisticated materials, although honey is still a commonly applied 'home cure' for many minor ailments of animals. There is an outlet for larger amounts of low-grade honeys in stockfeeding—pigeons, fish, horses, steers, etc. (e.g. Bray, 1967, Riggs & Weaver, 1955).

In horticulture, honey has been used to stimulate root formation. It is reported to be effective in a $7\frac{1}{2}\%$ solution for cacao cuttings (Cabrera Villa & Soto Rosiles, 1962), and at 10% to enhance the effect of alpha-naphthalene-acetic acid in vine cuttings (Poma Treccani, 1950). It is also used for spraying on to fruit blossom (with or without hormones); the flowers then attract more pollinating bees (Brouchot, 1951).

CHAPTER 16

WINES FROM
THE FERMENTATION OF HONEY

by Dr. Roger A. Morse and Dr. Keith H. Steinkraus

PROFESSORS OF APICULTURE AND MICROBIOLOGY RESPECTIVELY,
CORNELL UNIVERSITY, U.S.A.

Section 16.43 by P. D. Paterson

C.I.D.A. BEEKEEPING PROJECT, KENYA

From the bonny bells of heather,
They brewed a drink long-syne,
Was sweeter far than honey,
Was stronger far than wine.
They brewed it and they drank it,
And lay in blessed swound
For days and days together
In their dwellings underground.
R. L. STEVENSON (1850–1894),
HEATHER ALE

16.1 THE HISTORY OF MEAD

16.11 Introduction: mead in Europe

Honey was the only concentrated sweet widely available in prehistoric times, and fermented honey may well have provided man's first common alcoholic drink, long before the cultivation of fruit or grain crops.

A drinking horn discovered under $2\frac{1}{2}$ metres of peat bog in north Germany, and dating from before AD 100, was found to contain pollen grains and yeasts, indicating that the horn had contained a fermented honey drink (Grüss, 1931, Betts, 1932b). This is one of the earliest items of material evidence of man's association with mead. Written and oral records existed very much earlier, however. In 1948, Gayre published a book which he described as 'an account of mead, metheglin, sack and other ancient liquors, and of the mazer cups out of which they were drunk, with some comment upon the drinking customs of our forebears'. He devotes a chapter to the difficult area of mythology, and quotes passages from many of the better known Greek and Roman writers including Plato, Plutarch, Theocritus, and Pliny. Gayre's notes document the fact that both alcoholic and non-alcoholic drinks made of honey were known, enjoyed, and apparently even worshipped, by certain of our ancient ancestors.

Mead was popular in central and northern Europe at least as early as 334 BC; Ransome (1937) states that when Pytheas, a contemporary of Alexander the Great, sailed to lands round the North Sea, he noted that people there ate honey and made a drink from fermented honey and grain. The early use of mead in Denmark, where it was a national drink, was reviewed by Højrup (1957). Bell (1962) states that the word 'mead' may be traced to Beowulf and the year AD 604. The linguistic history of the word is discussed fully in Chapter 18.

According to Ransome, Wulfstan found mead to be common in Esthonia in the ninth century, and in 1015 a fire in the eastern German city of Meissen was extinguished with mead because there was a lack of water. One of the best-known mead-making towns was Eger in Czechoslovakia, where in 1460 there were thirteen establishments producing mead.

In 1669 Digby's book appeared: *The Closet of the Eminently Learned Sir Kenelme Digbie Kt. Opened: Whereby is Discovered Several Ways for Making of Metheglin, Sider, Cherry-Wine. . . .* This book contains about a hundred recipes for mead, meath, meathe, metheglin, sack, and hydromel, as well as many recipes for fruit wines and cooking in general. This is not the first reference to early English mead making, but it is the most extensive.

Digby travelled widely in Europe and was well known in the English court—though, for a while, he was in prison for his religious beliefs. His recipes are likely to be fairly normal at the time in England. The honey-water ratio in most of Digby's recipes indicates that the resulting wine was sweet and had an alcohol content of about 12%. One recipe is an exception: 'Hydromel as I made it weak for the Queen Mother', where he used 18 quarts of spring water and one quart of honey, together with ginger, clover, and a sprig of rosemary. The mixture was boiled, skimmed, and a spoonful of 'ale-yeast' added. Digby indicates that this mead was satisfactory for drinking after about two months—a much shorter time than that recommended for most of his meads. In another recipe Digby recommends adding hops in fairly large quantity, but this is one of the few instances in which hops are used. Many of the concoctions contained much spice and herbs, strongly suggesting that Digby, as well as other mead makers of the time, was covering up faulty fermentations, or using the spices or herbs as sources of growth factors for the yeast. Certainly the lack of knowledge about the fermentation process is likely to have yielded some products containing more acetic acid than alcohol.

Typical of the recommendations in the 1700s and 1800s on mead making in England are those of Keys (1796). He recommended that between 3 and 4 lb of honey be added to a gallon of water. The use of 4 lb of honey per gallon of water (400 g/litre) would probably result in a residual sugar content of 10–20%; most of the meads appear to have been sweet. After the water and honey mixture had been prepared, it was boiled and skimmed. Boiling the diluted honey before fermentation aids in clarification of the mead. Less precipitate is formed, and undesirable micro-organisms are destroyed. Boiling was recommended by practically everyone during the seventeenth to nineteenth centuries; it leads to a cooked or baked flavour which may have been popular at the time.

After boiling, the honey-water mixture was then placed in a barrel. It was recommended that a slice of bread, which had been toasted hard on both sides and covered with fresh yeast and a little lemon peel, should be placed in the barrel on top of the liquid to be fermented. Keys suggested that, on occasion, mead might be flavoured with raspberries or currants or other suitable fruits.

The addition of 'ale-yeast' by Digby or 'yeast' by Keys was part of the ritual of mead making. The mead makers knew it was helpful, but it is certain that they did not understand why this was so. It was not until the mid-1880s that Pasteur discovered the significance of the yeast cell, and the fact that it was responsible for the production of alcohol and carbon dioxide. Yeast cells had been observed under the very early microscopes by several people, and references to them were published, but they were considered little more than curiosities.

16.12 Production of fermented honey drinks in the Americas
The honeybee (*Apis mellifera*) is not native to North or South America but was introduced by colonists, probably first in 1638. However, several species of bees of the genera *Trigona* and *Melipona*, known as 'stingless bees' or meliponins, are native to South and Central America (and Africa), and some of these store quite large quantities of honey. Much information on these bees was published by Schwarz (1948). For instance Columbus, on his first voyage to the Western Hemisphere, found honey in Cuba which must have been produced by *Melipona beecheii fulvipes*, the only species of *Melipona* in Cuba at the time. Schwarz states also that the explorer Gomara found apiculture in an advanced state on the Yucatán Peninsula of Mexico in 1578. The amount of honey stored varies from species to species, and 2 kg per colony would be considered a good yield. One unusual nest of *Melipona* in Brazil yielded 45 kg of honey in 1930. Sections 17.3 and 19.39 give further information.

Several authors have stated that an alcoholic drink was common amongst the natives of Central America at the time the Spaniards landed.

The records indicate that the honey was diluted with water and fermented, either by itself or together with fruit and corn. At certain religious events it was consumed in large quantity. Indians in America north of Mexico were less fortunate: with the exception of those in the north-eastern part of the continent who made maple syrup, they had no ready source of sugar. There is no evidence that maple syrup was used to make an alcoholic drink, but this may reflect a shortage of the raw material more than anything else.

The introduction of the European honeybee to the Americas has led to a reduction in the use of the stingless bees for honey production, but alcoholic drinks are still made today by Indians in Central and South America from honey, using any source available.

16.13 The decline of mead making

Gayre (1948) stated, as others have, that for all practical purposes mead or honey wine was the national drink in England and certain other northern European countries for many centuries. One or two thousand years ago there were fewer people in continental Europe, and in Britain, than today; there were probably more colonies of bees, so more honey was probably available *per capita*. Gayre believes that the decline of mead making in the 1700s and 1800s resulted from: (1) a shortage, and as a result an increase in the price, of honey; (2) a desire for sweet wines —as witness the large imports of ports, sherries, and madeiras into England in the 1800s. It is difficult to believe that the second reason is a very important one, for it is easy to make a mead which is either sweet or dry.

It seems far more probable that the poor quality of mead was responsible for its decline. Yeast is mentioned in many of the early recipes, but it was added without understanding of its function. The yeasts present in ripe honey all belong to the genus *Zygosaccharomyces*, osmophilic yeasts which grow only in saturated sugar solutions. When honey, or any other medium containing yeasts of this genus, is diluted to less than 50% sugar, *Zygosaccharomyces* will not grow.

It is therefore important to introduce live wine yeasts into the diluted honey-water mixture; this is an ideal medium for the growth of many undesirable micro-organisms, which will multiply if not suppressed by yeast growth. (The opposite is true of most ripe fruits, especially grapes, whose surface is covered with yeast cells.)

Excellent mead can be made without the addition of spices and herbs, just as good meat is not in need of improvement. The strength, as well as the quantity, of many of the materials added in these old recipes— either before or after the fermentation—strongly suggests an attempt to mask a poor or faulty product.

16.14   From Pasteur to the present time

It was not until 1866, when Pasteur published his book *Études sur le vin*, that there was any serious scientific study of wine. This was about the time that men were arguing about spontaneous generation; Darwin's *Origin of species* had been published only a few years earlier. Pasteur's book on wine was followed by one on beer in 1877, and this provided further information on the fermentation process.

At that time there was little interest in honey wine, and not much was published on the subject until the 1900s. Vinson (1907), and later Fabian (1926), both in America, wrote on honey vinegar; others in Europe were writing on the same subject. Reading between the lines, it seems likely that Vinson and Fabian were interested in the production of honey wine rather than honey vinegar, but they were writing during the period of interest in Prohibition. In England, Bancks (1905) also wrote on honey vinegar. The important fact in these papers is the recognition that honey, when diluted with water, is not a satisfactory medium for fermentation; hence the recommendation to add various salts: ammonium phosphate, sodium phosphate, ammonium chloride, etc. The authors very carefully differentiated between the alcoholic fermentation and the acetic fermentation (the first of which must precede the second), so that the discerning reader could also separate the two processes.

In the 1930s papers by Filipello & Marsh (1934) and Fabian (1935*a*) in America, Vouloir (1935) in France, and a number of authors in England (Castle-Turner, 1933, Betts, 1932*a*, Harland, 1932, etc.) indicated a wide and renewed interest in the general subject. More important, it was recognized that a healthy fermentation could be conducted only if the proper medium were provided.

The experiments of Filipello & Marsh in 1934 were much more sophisticated than previous trials. As well as adding nutrients, they recommended the addition of citric acid, and they pointed to the clarification problem which exists with most honey wines. They also undertook preliminary experiments with a honey brandy. Part of the interest in honey wine in the 1930s developed because of surpluses of honey in various parts of the world at that time; these surpluses disappeared in the late 1930s and during World War II.

A flurry of articles, and the outstanding book by Gayre (1948), followed the war. These included the conservative approach of Brother Adam (1953) as well as interest in a more rapid fermentation, for instance by Bergeret & de Castro (1943*b*) in South America, Dennis (1954) in England, Palmer-Jones (1953) in New Zealand, Hocking (1958) in Canada, and Morse (1953*a*) in the United States. Morse (1961) prepared a bibliography concerned with the general subject of mead making.

During the period from 1900 to date, Vouloir (1935) was not the only

French author on honey wine, but his book is perhaps one of the best. Alphandéry (1931, pages 361–364), and several authors in *l'Apiculteru* and other French journals, wrote about the virtues of honey wine, and about the necessity of adding certain nutrient factors prior to fermentation.

The use of hops in manufacturing honey wine is specifically mentioned in U.S. law. The addition of hops is a very old technique and can be traced to early English references. The flavour and aroma of hops are quite popular, as is shown by the consumption of beer and ale. In addition, hops contain tannins which help to precipitate proteins that cause cloudiness. Hops also contain compounds which help to make beer and similar drinks more stable biologically (Prescott & Dunn, 1959). In honey wine, hops might also contribute certain nutrient factors for the yeast.

One argument amongst honey wine producers around the world, that will not be easily resolved, is whether or not to add nutrients to the honey-water mixture. Most people agree that the addition of some amount of acid (even if only the juice or rind of lemon) enhances the flavour of honey wine. We know that yeasts can synthesize their own vitamins; also, because yeast cells which die in the fermentation process break down organically, certain elements are released which may be re-used in the same medium by living yeast cells. It is possible to produce a honey wine with 10–12% alcohol without adding salts and acid, given a sufficient amount of time. The major danger with such a prolonged fermentation, especially where a certain amount of head-space or air-space is left in the barrel, is that the wine is susceptible to contamination by a multitude of organisms for a long period of time. Also autolysis, the slow disintegration and breakdown of yeast cells in the fermenting medium, produces undesirable flavours. Some people do not find them disagreeable, but most authorities on wine believe that it is well to separate the new wine from the lees (dead yeast cells) as soon as possible; a rapid fermentation is thus recommended for obtaining the best product.

16.2 RECENT STUDIES

We have seen in 16.1 that the quality of many meads in the past was not generally very good, and that this may account for the inclusion of flavouring agents in many of the recipes. Adams & Niesen (1963*a*, 1963*b*) summarized the situation in the early 1960s quite well: 'Mead, however, has not gained acceptance as a beverage. Although sweet tasting due to the presence in it of unfermented sugars, mead has had a generally undesirable flavour. The poor flavour of mead beverages has been attributed in important part to the excessive period of time necessary for fermentation of honey to become complete.'

Much of the recent research on honey wines (Crowther, 1960, Adams & Niesen, 1963*a*, 1963*b*, Maugenet, 1964, Steinkraus & Morse, 1966, Morse & Steinkraus, 1971) has been directed towards speeding up the fermentation with the aim of improving the flavour.

16.21   Various types of mead
Mead can be made from very nearly any sort of honey, and the mead produced retains many of the characteristics of the honey from which it was derived.

Experience has shown that light honeys yield meads similar to light grape wines such as sauterne. Dark honeys of stronger flavour should perhaps be used primarily for making 'honey ales', or bock-type beer drinks. The low natural acidity in light honey meads makes them ideal for the production of sherry flavour, either through subsequent flor yeast fermentations or baking processes, or a combination of the two. It has been our experience that the average American taster prefers mead from a light honey to that from a darker honey.

Light honey meads fermented to dryness, with an alcohol concentration in the range 10–11%, may be refermented in the bottle to yield good quality sparkling meads or champagne-like beverages.

Mead offers considerable potential for the production of distilled spirits. The brandy obtained retains just a hint of the honey from which it was originally fermented.

16.22   Factors influencing the fermentation
In Japan, Nakayama *et al.* (1961) studied the fermentation of several types of honey. They concluded that regular wine yeasts of the *Saccharomyces cerevisiae* species were best for fermentation of honey where the concentration of fermentable sugars was less than 15%. For fermenting honey where the sugar concentration was above 15%, they reported that the osmophilic yeasts present in honey were better. Steinkraus & Morse (1966) had no difficulty fermenting diluted honey (25% sugar) to alcohol concentrations of 12–15% by using *S. cerevisiae* type wine yeasts. Generally the osmophilic yeasts are not as good producers of alcohol as the wine yeasts.

Nakayama & Koike (1965) reported that they could ferment diluted buckwheat honey with a wine yeast to a delicious mead with 11·5% alcohol, in a period of one month. They added 2% yeast extract when they wanted the fermentation to continue to about 10% residual sugars. We would consider a 30-day fermentation too slow for development of the best flavour.

Other papers by the Japanese group include one by Nakayama *et al.* (1966) which deals with the prevention of browning in mead by use of sulphite, and others (Kushida *et al.*, 1963, Nakayama *et al.*, 1962*a*, 1962*b*, Kushida *et al.*, 1961) which report results of research in which honey was added to grape juice and fermented. Generally, the honey contributed a characteristic flavour to the grape wine.

Crowther (1960) found that it was essential to add nitrogen to the diluted honey base to stimulate fermentation. He used diammonium phosphate. Maugenet (1964) reviewed the problems involved in the successful fermentation of honey, which contains mainly fermentable carbohydrate and is deficient in nitrogen, minerals, and other factors that stimulate yeast growth and fermentation. He recommended the addition of 250 mg/litre of diammonium phosphate to the diluted honey base. Maugenet preferred honey fermentations leaving a residual sugar content of 5–10% in the mead. Using a temperature of 20–25°C (68–77°F), his fermentations required 20–30 days.

Adams & Niesen (1963*a*, 1963*b*) emphasized the importance of keeping the yeast cells in suspension (by slow agitation) in order to increase the rate of fermentation. Steinkraus & Morse (1966) were able to get very rapid fermentations by selection of proper yeast strain, and addition of appropriate yeast growth factors, without agitation. Unless growth factors are added, it is not unusual for light honey fermentations to require 6 months or longer.

Steinkraus & Morse (1966) studied the factors influencing the rate of fermentation of honey. They found that honeys vary considerably in their fermentability. Light clover honey (*Trifolium*) was more difficult to ferment, and required more additives—vitamins, minerals, and nitrogen—than did a dark honey from buckwheat (*Fagopyrum*) or goldenrod (*Solidago*). Not all batches of clover honey reacted in the same way: some lots permitted a very slow fermentation, but others allowed practically no fermentation without the addition of supplementary growth factors.

With the addition of growth factors, it was found that all honeys tested could be fermented rapidly, yielding meads characteristic of the types of honey from which they were derived, and lacking certain unpleasant flavours believed to be due to long fermentations resulting from the absence of sufficient yeast growth factors.

Clover honey yielded a mild-flavoured, 'sauterne-like' mead, which was acceptable to the majority of (American) tasters participating in organoleptic tests—there was only a hint of clover honey flavour. Buckwheat honey yielded a dark-coloured, rather heavy-flavoured mead, which was less acceptable than clover mead. Other honeys yielded meads reflecting their particular flavour characteristics.

Additives used are referred to below as Formulae I and II.

Formula I		Formula II	
Component	*Weight in g*	*Component*	*Weight in mg*
ammonium sulphate	1·0	biotin	0·05
potassium phosphate	0·5	pyridoxine	1·0
(K$_3$PO$_4$)		*meso*inositol	7·5
magnesium chloride	0·2	calcium pantothenate	10·0
sodium hydrogen sulphate	0·05	thiamin	20·0
citric acid	2·53	peptone (Roche)	100·0
sodium citrate	2·47	ammonium sulphate	861·45
Total	6·75	Total	1000·00

Formula I added at rate of 6·75 g/litre   Formula II added at rate of 0·25 g/litre

The most rapid fermentation of a clover honey base was produced by the use of Formulae I and II combined. Using 25% solids and a 0·4% yeast inoculum, the alcohol content reached 12% in less than 2 weeks. Formula I alone, containing mineral salts and sources of ammonia, nitrogen, and phosphate, permitted a slightly less rapid fermentation. The addition to clover honey of only ammonium sulphate and (tri-basic) potassium phosphate (the two most important yeast foods in Formula I) resulted in about the same rate of fermentation as the complete Formula. Formula II, containing vitamins, was much less effective in promoting the fermentation of clover honey in the absence of Formula I. Honey entirely from clover (unifloral) permitted practically no fermentation, but Formula II on its own stimulated the fermentation slightly. In other lots of clover honey, with more admixture from other sources, Formula II on its own failed to stimulate the fermentation rate. The control (no additives) showed no fermentation after 18 days and reached an alcohol content of only 6% in 54 days.

Buckwheat honey, without additives, fermented much more rapidly than clover honey. This could suggest that dark honeys contain more natural yeast nutrients than light honeys. In buckwheat honey, Formulae I + II again produced the most rapid fermentation, but Formula II by itself resulted in some stimulation. Formula I by itself gave more rapid fermentation of buckwheat honey than did Formula II alone. Ammonium sulphate added to buckwheat honey slightly stimulated the rate of fermentation. Separate addition of the other factors in Formula I showed no positive effect upon the rate of fermentation.

The pH of the honey influenced the rate of fermentation considerably, and the pH of different lots of honey varied (*see* 5.42). Initially, as part of Formula I, 5 g citric acid per litre was added, and the pH was often below 3·0. By adding citrate in the form of 2·53 g citric acid and 2·47 g sodium citrate, the diluted honey was buffered more closely to a favourable pH, so less adjustment was necessary. Fermentation of clover honey progressed much more slowly at pH 2·9 than at pH 3·7. It was found that a pH of 3·7–4·6 was desirable for a honey fermentation; 3·7 was high enough to permit rapid fermentation, but still low enough to inhibit growth of undesirable bacteria.

A 10% (by volume) inoculum resulted in a more rapid initial fermentation than 0·4%. However, fermentation from the smaller inoculum generally caught up with that from the larger inoculum after 10–12 days, so there was no advantage from using the larger inoculum. At a low temperature, however, such as 13°C (55°F), a 10% inoculum produced a more rapid fermentation than a 1% inoculum.

It was found that yeasts varied considerably in their ability to ferment honey. One yeast strain, widely used for grape juice and champagne fermentations, did not yield as much alcohol as another yeast which has been widely used for honey fermentations. It was also observed that some yeasts caused the formation of more haze than others. Since haze is a serious problem with honey, it is well to avoid the use of such yeasts.

It was concluded that with added growth factors, a selected yeast, a proper pH, and a temperature of 24–27°C (75–80°F), even light-coloured honey such as clover can be fermented to an alcohol content of 12–13% by volume in about 2 weeks, without agitation.

Although more rapid fermentations can be achieved at higher temperatures, there is a belief among some wine makers that the use of lower temperatures and a longer fermentation results in a better flavour. The truth of this has not yet been demonstrated with clover meads. However, when speed of fermentation was less essential, we used a fermentation temperature of 18°C (65°F).

16.23   A pilot plant procedure

Based upon our laboratory studies (Steinkraus & Morse, 1966), a pilot-plant procedure was developed for the production of clover honey mead in approximately 40-gallon lots in 55-gallon oak barrels [1 gallon = 3·785 litres]. The process was designed to yield a dry, light, almost colourless mead, devoid of harsh or bitter flavour, with good stability in the bottle. The process consists of the following steps:

1. Clover honey is diluted to 21% solids with water. Crystallized honey is heated to 60–65°C (140–150°F) to facilitate solution.
2. The following yeast nutrients are added in grams per U.S. gallon (3·785 litres) of diluted honey:

citric acid*	18·9
ammonium sulphate	4·65
potassium phosphate (K_3PO_4)	1·9
magnesium chloride	0·7
peptone	0·1
sodium hydrogen sulphate	0·2
thiamin	0·02
calcium pantothenate	0·01
*meso*inositol	0·0075
pyridoxine	0·001
biotin	0·00005

The above additives are approximately equivalent to the addition of Formulae I and II (16.22).

3. The pH is adjusted to 3·7–4·0 with sodium hydroxide or hydrochloric acid.
4. When cooled to about 27°C (80°F), the 40-gallon batch is placed in a 55-gallon barrel, inoculated with 0·5% by volume of actively growing yeast culture, and sealed with a bubbler.
5. The mead is fermented at 18°C (65°F).
6. The mead is allowed to age in the barrel for about 6 months.
7. It is then decanted and filtered through Celite 503 or a similar filter-aid.
8. Total acidity is adjusted to 0·6% with citric or tartaric acid.
9. The mead is pasteurized at 63°C (145°F) for 5 minutes and bottled while hot.

The above process yields a dry mead with an alcohol content of about 12% by volume. Starting with an initial solids content of 25%, the resulting mead has an alcohol content in the range 14–15%.

16.24 Sparkling mead

For a sparkling mead, it is preferable to use a clover honey base, diluted to about 18–19% solids at the start of fermentation. Formulae I and II are added to the diluted honey. This yields a mead (cuvée) with an alcohol content of about 10% by volume, which is better for champagne manufacture than a cuvée with higher alcohol content. The steps in making the sparkling mead are as follows.

Sucrose (2% by weight) is added to the cuvée, with the following yeast nutrients (per U.S. gallon, 3·785 litres):

| peptone | 100 mg |
| thiamin | 20 mg |

* or preferably citric acid 9·57 g plus sodium citrate 9·34 g, which requires less subsequent adjustment of pH.

calcium pantothenate	10 mg
inositol	7·5 mg
ammonium sulphate	861 mg
pyridoxine	1·0 mg
biotin	0·05 mg

This is essentially Formula II, added to ensure sufficient yeast vitamins for the fermentation in the bottle.

The cuvée is inoculated with yeast (7% by volume), and the inoculated cuvée bottled in champagne bottles, sealed with metal caps, and incubated at 18°C (65°F). Within 21 days, the alcohol content should increase to 12% by volume, and carbon dioxide pressure to 4·8 kg/cm²; (68 psi) at room temperature. Standard disgorging procedures are used to remove the yeast cells. The sparkling mead produced from clover honey has a pleasant flavour and retains its carbonation very well. There is a greater tendency to form a head than is usual in grape champagnes.

16.25   Sherry mead
Clover honey mead can be made into a light sherry by refermenting it with a flor sherry yeast (Steinkraus & Morse, 1966). Acetaldehyde, a measure of sherry flavour development, rose from 48 to 190 mg/litre in 48 hours by passing the mead through a glass column filled with ceramic tile pieces inoculated with the yeast. The initial alcohol content of the mead was 10·4%, which was increased (fortified) to 13% by volume before commencing the sherry fermentation. The sherry produced from mead had a definite flor sherry flavour that many tasters preferred to the original mead flavour. Because of its low acidity, mead offers an excellent base for the production of sherry.

Higher levels of acetaldehyde have been obtained recently by using an agitated, aerobic process, and a honey wine base of 10·2% alcohol. Acetaldehyde reached 600–700 mg per litre in 6–7 days at 25°C (77°F) (Steinkraus & Morse, 1973).

16.3   SPECIAL PROBLEMS

16.31   Ageing
Ageing clover honey mead in charred oak barrels for 6 months produces a very smooth product, which shows a greater stability than freshly fermented mead.

16.32   Clarification
Mead poses a rather difficult problem as far as stability is concerned. There are distinct hazes and precipitates that form—sometimes early, sometimes later, sometimes when the mead is chilled, and sometimes

when the mead comes into contact with air. The hazes have not been adequately studied, and their chemistry needs attention.

To produce a completely stable product, it is desirable that the mead be decanted off the lees as soon as fermentation is complete. The mead should then be held at 1·5°C (35°F) for 5–7 days to precipitate cold precipitable fractions. The mead can be decanted, leaving the precipitate-containing fraction behind, or it can be passed through a continuous centrifuge. If a centrifuge is used, the temperature must be maintained at about 1·5°C to prevent resolution of the haze, with resultant instability in the product.

Before bottling, the mead should be filtered using a filter-aid such as Celite 503. There should be sufficient Celite and sufficient recirculation through the filter to make the product crystal clear.

The product should be heated to 77°C (170°F) to ensure that the mead does not contain any heat-precipitatable materials. If it does not, the mead can be pasteurized by heating to 77°C and bottled while hot.

Steinkraus & Morse (1973) found that boiling the diluted honey before fermentation precipitates unstable matter, and thus leads to a much more stable honey wine.

16.33 Honey vinegar
Several authors (Gray, 1933, Fabian, 1928, 1935a, Kelty, 1920, Morse, 1953b) have discussed the production of vinegar from honey. They have obviously been aware of the delicious flavour and aroma of honey vinegar which distinguish it from others.

In our laboratory, clover honey mead was circulated through a glass column filled with oak wood shavings inoculated with *Acetobacter aceti*. Total acid rose from 0·6% to 5·1% in 30 days at room temperature. Under optimal conditions, the process would require at most 1 or 2 days. The honey vinegar produced had an excellent aroma that could make it very attractive as a specialty vinegar on the commercial market.

16.4 WORLD MEAD PRODUCTION

16.41 Mead production in the Americas
Orlando Muñoz of Costa Rica produced honey wine from 1948 to 1964. However, the increased price of honey, especially influenced by the European market, together with the production of sugar wines in other Central American countries, forced him to eliminate the production of mead, as well as the fermentation of honey in combination with various fruit juices. At the present time we are not aware that any mead is being produced commercially in Central or South America.

Three companies are listed by *Wines and Vines* (1971) as producing honey wine in North America in 1970; one is in New York City, one in London, Ontario, and the third in Ohio. It is estimated that 20 tons of honey a year are used for making honey wine in New York City; most of the honey wine is produced for the Kosher trade. During the past few years a small quantity of mead has been imported into the United States from England and elsewhere.

16.42   Europe

Despite the fact that Europe was the original mead-making area, little mead is made in western Europe today. At least one company is now making mead in England (see 16·41). In France, Maugenet (1964) reports that the commercial production of mead is not over 3 000 hectolitres, and that it is essentially a product made for home use only.

The eastern European countries have retained their greater interest in the subject. In Poland, the honey co-operative at Kraków makes two types of mead, using the traditional methods of past centuries. *Dwojniak* is made with equal weights of water and honey and is fermented with an osmophilic yeast. The alcohol content is around 16%, but there is residual sugar which makes the mead quite sweet. *Dwojniak* is aged 5–7 years in 4 000-litre wooden vats. For *Trojniak*, $\frac{1}{2}$ kg honey per 1 kg water is used; its alcohol content reaches $12\frac{1}{2}$%, and it is still slightly sweet. *Trojniak* is aged for 3 years or more. It is very highly regarded, especially if made from cornflower (*Centaurea*) honey. Special strains of yeast (Málaga type) are used, which are maintained at a microbiological institute.

16.43   The traditional making of honey beer throughout
tropical Africa   by P. D. Paterson

The brewing of beers which include honey as a partial or total ingredient has for centuries been a part of the way of life for many African tribes. Beer brewing is universal among the tribes of central and eastern Africa and Ethiopia and is practised to varying degrees on the west coast. Neither honey beer nor other beers are normally found in north-eastern Kenya or in Somalia; this is attributed to the Muslim influence, which is absolute in Somalia but varies in different parts of West Africa. It is interesting to note that Islam in general is to be found in desert and semi-desert areas. These, by their nature, have a low potential for honey production; but the devout Muslim is a staunch abstainer even in the more fertile areas.

African honey beer is unlike mead in that it is a long drink; it is never left to mature, but is consumed within a day or two of commencement of fermentation. The method of brewing has a considerable number of variants, as do the social circumstances in which the beer is drunk.

Honey—either brewed, or for the purpose of brewing—is frequently used for bartering, for village compensations or fines, for celebrating feasts, and as an award for special accomplishments. Its most important use, perhaps, is for bridal prices and marriage ceremonies. In some tribes the drinking of beer is confined to the old men, who may also be the most enthusiastic beekeepers. Young women are often discouraged from drinking lest they behave improperly. (Some interesting observations have been made by Seyffert (1930).)

Brewing may take place in the home, for domestic purposes, or in the village beer halls. The principles are the same, although home brews for special occasions, such as weddings, are likely to be of a rather stronger quality. Any readily available container would be used as the brewing receptacle; at home this might be a calabash, and in the beer hall an old wooden cask. Tins and drums are also used.

In most parts of Africa there is little or no frame-hive beekeeping, and facilities for efficient separation of honey from the comb are not generally available. The honey used for brewing has therefore not undergone any processing after it has been collected from wild nesting places or from the traditional barrel-type hives. It includes a mass of broken-up combs, with pollen and bees included. Contrary to many reports, it is not usual for the beekeeper to mix brood combs into the honey for brewing, although he will frequently remove the brood comb from the hive for direct consumption; it is a valuable source of protein in many areas where malnutrition is the problem rather than starvation. Brewers are normally very reluctant to use any honey that has been separated from the wax, as they believe that crude honey makes the best beer. Perhaps the pollen content of the combs acts as a yeast nutrient, and the waxy layer that forms on the surface of the brew may perhaps also assist fermentation.

Beers of many varieties may include honey as an ingredient, but usually 'honey beer' is either brewed as a pure honey beer, or sugar is the basic ingredient (in the form of juice crushed from the cane, or jaggery, or even refined sugar), and the honey is added with the sugar. The higher the proportion of honey the better; a typical 'commercial' village brew of the latter type might have a recipe as follows: 60 lb honey, 240 lb sugar, with water added to make a total of some 50 gallons (27 kg, 108 kg, 250 litres). To the brew is often added the loofah-like fruit of the muratina (Kikuyu) or sausage tree, *Kigelia aethiopica* (Bignonaceae). Twenty to thirty slices of the fruit may be put into a 50-gallon barrel, and after fermentation is complete they will be removed for use in the next brew. These slices of muratina are supposed to give strength and flavour to the beer, but they may well be the means whereby yeasts are transferred from one brew to the next; cultured yeasts are never used in a traditional beverage. When fermentation is completed, the brew is strained through long bags of woven reeds, or some more modern device; the waxy residue is either discarded or rendered for wax, and the muratina is put out in the sun to dry, and is used again for the next brew either days or weeks later.

Various beers are distilled, for the production of such drinks as Nubian gin. Distillation is generally illegal, and takes place under very crude conditions, in dirty hideouts on river banks. The product is often a poisonous drink containing methyl alcohol, which can lead to total blindness in those who drink it.

Honey beers are considered to be the best of all beers, and this use of honey accounts for the consumption of almost the entire crop in tropical and subtropical Africa. The fermentation process has been little studied, but Rossi (1959) has published an account of the yeasts of Ethiopian *tej*; *see also* 19.23.

16.44   Other countries

Interest in mead-making appears sporadically in various parts of the world. Articles, giving results of independent investigations, have appeared for example in Canada (Crowther, 1960) and New Zealand (Palmer-Jones, 1953), but despite this there is little progress towards commercial production.

16.5   HONEY IN HIGHER ALCOHOL DRINKS

The product of distillation may be high proof, which for practical purposes would be pure alcohol, or low proof, which would contain many fusel oils, aldehydes, acids, and esters. These materials, called 'congeners', give many distilled spirits their individual flavour. Some of the world's distinctive honeys might very well be candidates for a product in this category; we have, however, not found any literature on the subject. Apparently honey is not used as a base for any distilled spirits—such as whisky or gin.

Honey is added to several cordials or liqueurs, some of which have an international reputation. Best known, probably, is Drambuie, which has no imitator, and which is made from Scotch whisky, heather honey, and several herbs. In Ireland, a liqueur called Irish Mist is made of Irish whiskey, heather honey, and herbs. In Poland a similar drink is called Krupnik; its base is Polish whisky and honey. Svoboda (1935) has described Czech liqueurs. In the United States, an American Krupnik is produced, and also Forbidden Fruit, a cordial containing citrus fruits and American honey. Several lesser known cordials also use honey as all or part of their sweetening agent.

SECTION 5

OTHER ASPECTS
OF HONEY

CHAPTER 17

HONEY FROM OTHER BEES

by Dr. Eva Crane

DIRECTOR, BEE RESEARCH ASSOCIATION

But, for the point of wisdom, I would choose
To know the mind that stirs between the wings
Of bees and building wasps. . . .

GEORGE ELIOT,
THE SPANISH GYPSY (1868)

17.1 INTRODUCTION

The 'Recommended European regional standard for honey' of the Codex Alimentarius Commission (1969; *see also Bee World*, 1970) defines honey as 'the sweet substance produced by honeybees from the nectar of blossoms or from secretions of or on living parts of plants, which they collect, transform and combine with specific substances, and store in honey combs'. European honey, and indeed most of the world production, is obtained from the honeybee *Apis mellifera*. Preceding chapters of this book are concerned almost entirely with this honey.

There are, however, three other species of honeybee (17.2)—which some authors subdivide into many more (Maa, 1953)—and many other social bees that also store honey. The Meliponinae or stingless bees (17.3) have been the basis of rich beekeeping cultures in tropical America, and have also been exploited for honey in Africa, Asia, and Australia. Many bumble bees (17.4) form summer colonies in temperate zones, and honey from them may be worth harvesting in some circumstances. Certain tropical wasps and other insects produce honey or similar stores (17.5). It is interesting that as long ago as 1500–1300 BC, in Ayurvedic times in India, eight varieties of honey were recognized—produced by various bees, by wasps, and by insects working in ant-hills (Joshi & Godbole, 1970).

There have been comparatively few scientific studies on these honeys so far, but recent work by Burgett (1973) on their glucose oxidase activity and inhibine content is reported in 8.2.

There are also a few direct uses of nectar by man (17.6), without any insect intermediary.

17.2 HONEY FROM OTHER HONEYBEES (*APIS*)

17.21 Distribution and characteristics of the honeybees

Two *Apis* species, *mellifera* and *cerana*, build their combs in the dark, and can therefore be kept in hives. The other two, *dorsata* and *florea*, both tropical species indigenous to south-east Asia, build a single comb in the open; they cannot be 'kept' in an enclosed hive, and their honey is normally harvested from nests in the wild. The physiological basis for these behaviour characteristics—on which the world's beekeeping industry depends—is not completely understood; it has, however, been established that a clustered swarm of *Apis mellifera* will not commence comb building unless the light intensity falls below a certain threshhold (Morse, 1965); comb building will then be continued and extended, irrespective of the illumination. Presumably the same is true of *Apis cerana*. Some other circumstance must initiate comb building in *Apis dorsata* and *Apis florea*.

The various races of the honeybee *Apis mellifera* evolved in Europe, Asia, and Africa. There are no honeybees indigenous to the New World. European bees were first taken to different parts of the Americas in the 1600s, or possibly the 1500s, and to Australia in 1822 and New Zealand in 1842 (Crane, 1963). Tropical Africa has its own subspecies, one of which (*Apis mellifera adansonii*) was introduced to Brazil in 1956.

Figure 17.21/1 gives some idea of the distribution of the different *Apis* species; *see also* Table 17.21/1. In south-east Asia, including the Indian sub-continent, and stretching northwards through China and Japan and the extreme east of the U.S.S.R. (Ussuri), the native honeybee is *Apis cerana*. This bee is very similar to *Apis mellifera*, but is a distinct species (Ruttner, 1969), and the two cannot interbreed. There are, or were (*Bee World*, 1971a), many races and strains of this bee, e.g. the Indian, Chinese, Japanese, bee (*A.c.indica*, *cerana*, *japonica*). In the more northerly and the higher areas, *Apis mellifera* has been successfully introduced in recent years, and can give higher honey yields than *Apis cerana*; in the Indian Punjab introduction is well under way on an experimental basis (Atwal & Sharma, 1968). Further information on the *Apis* species can be obtained through Bee Research Association Bibliography No. 8 (1967). Goyal (1974) discusses interspecific competition.

All early descriptions of honey from hives in Asiatic countries relate to honey from *Apis cerana*, not *Apis mellifera* (Table 17.21/1).

17.22 Characteristics of *Apis* honey

Even within the species *Apis mellifera*, there are variations in honey according to the plant visited. Caucasian bees (*A.m.caucasica*), with long tongues, get nectar from plants with a deep corolla that shorter-

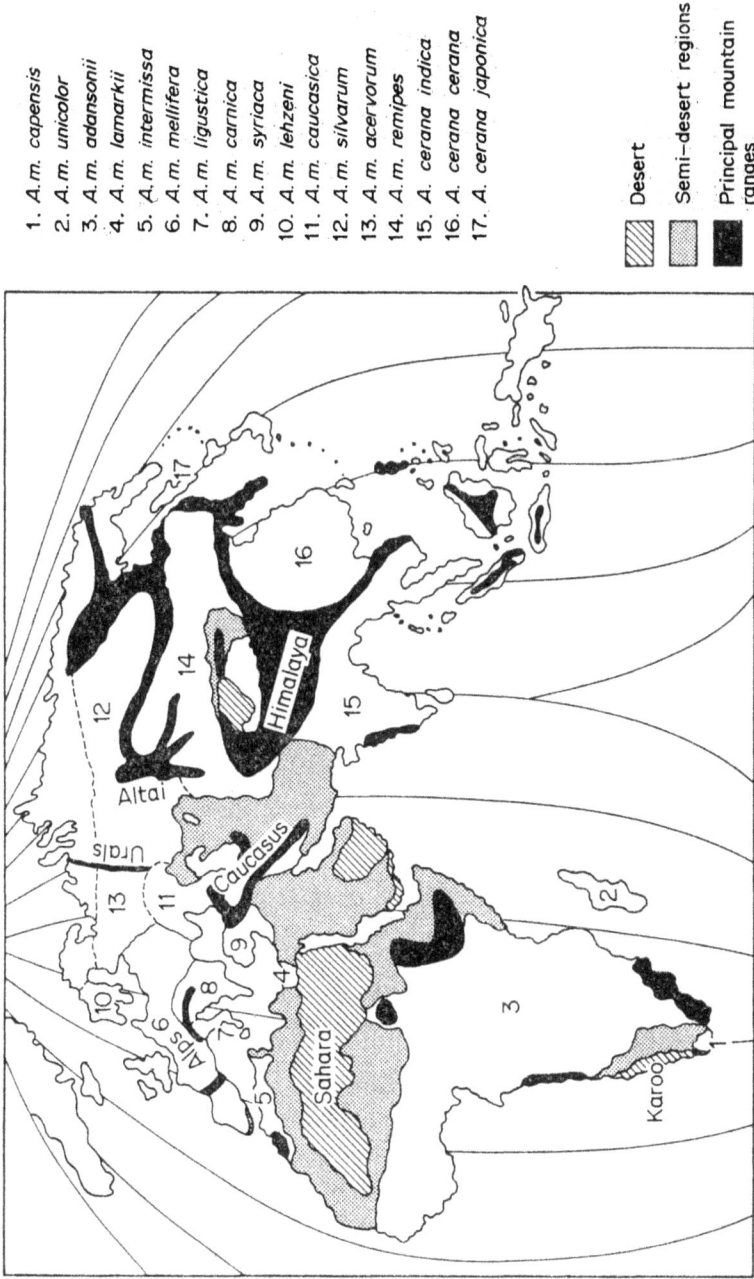

1. A.m. capensis
2. A.m. unicolor
3. A.m. adansonii
4. A.m. lamarkii
5. A.m. intermissa
6. A.m. mellifera
7. A.m. ligustica
8. A.m. carnica
9. A.m. syriaca
10. A.m. lehzeni
11. A.m. caucasica
12. A.m. silvarum
13. A.m. acervorum
14. A.m. remipes
15. A. cerana indica
16. A. cerana cerana
17. A. cerana japonica

Desert

Semi-desert regions

Principal mountain ranges

Figure 17.21/1 Distribution of the four *Apis* species (after Kerr, 1960). Except for the recent introduction of *Apis mellifera adansonii* into South America, the New World countries have only *Apis mellifera* (*mellifera, ligustica, carnica, caucasica*).

tongued bees cannot reach. Italian bees (*A.m.ligustica*) build very large summer colonies, which can get more honey from a crop then in flower, whereas the more frugal Carniolan bees (*A.m.carnica*) are better able to exploit and store honey from an earlier crop.

Northern races tend to leave a space between the honey in the cell and its wax capping. The resulting combs look white and dry and are therefore favourites for prizes at shows. Southern races are inclined to fill the cell to the brim, and their cappings look darker, being wet with honey underneath. Gubin (1953) compared the thermal expansions of honey and

Table 17.21/1
Transition from *Apis cerana* to *Apis mellifera* for honey production in different parts of Asia

Country	A. mellifera first introduced	Species now used in hives	Proportion of marketed honey derived from A. cerana
Japan	1876[e]	almost entirely *A. mellifera*; *A. cerana* still in Kyûshû in 1957[a]	none since 1923[b] (or earlier)
China	1925[d]	*A. cerana* in some areas; elsewhere replaced by *A. mellifera*	appreciable amounts from both species
India	1919[f]	all *A. cerana* except a few experimental colonies	all[†]
Pakistan	*see* India	all *A. cerana*	all[†]
S.E. Asia	1958[c]*	*A. cerana*; *A. mellifera* well established in some parts	most[†]

* Philippines, probably earlier elsewhere
[†] Some also from *Apis dorsata* (honey hunting, not beekeeping)
[a] Akahira & Sakagami (1958)
[b] *Bee World* (1925)
[c] Morse & Laigo (1969)
[d] Oschmann (1961)
[e] Sakagami (1959)
[f] Sharma (1960)

beeswax, and on this basis suggested that the space left by the northern bees was *necessary*, in order to prevent the cells becoming deformed or bursting in cold winter weather, when the temperature of the colony's food store might drop from the summer level of 30° to say 0°C (86°–32°F).

Honeys from Italian and 'Africanized' bees in the same apiaries in Brazil were recently compared, because of complaints that the introduction of *A.m.adansonii* (17.21) had led to the sale of more watery honey.

There was, however, no evidence that honey of 'Africanized' bees (mean relative density 1·436) was significantly different in water content from that of Italian bees (1·420) or of others of European origin (1·455).

Variations between the nectar foraging behaviour of different species of honeybee have been investigated to some extent. Studies by Lindauer (1956, 1957) and F. G. Smith (1958b) respectively show certain differences in foraging range and dance behaviour. We may assume that *Apis mellifera* gets most of its forage within 3 000 m of the hive, and that the same is true of *Apis mellifera adansonii* in Africa. *Apis cerana indica* flies less far, say 700 m (Lindauer) or rather more (Smith). *Apis cerana japonica* has been found to collect nectar from less prolific sources than *A.mellifera* (Akahira & Sakagami, 1958). *Apis dorsata* seems fairly similar to *A.mellifera*, but the tiny *A.florea* keeps within 350–400 m of its nest. *Trigona iridipennis* (17.31), also studied by Lindauer in Ceylon, had a smaller foraging range still, only about 100 m.

These variations in foraging behaviour will have their effects on the plant sources of the honeys produced. A comparative study of foraging habits of the three tropical *Apis* species in Pakistan showed less difference between the nectar sources they use than between the pollen sources (Latif *et al.*, 1958; *see also* Goyal, 1974).

Section 1.1 discusses many other factors that affect honey production, some of which also affect the composition and properties of the honeys produced. Table 17.22/1 sets out most of the information available on the gross composition of honeys from *Apis cerana*, *A.dorsata*, and *A.florea*, with two sets of figures on *A.mellifera* honey for comparison; the first is the average of the 490 samples in Table 5.2/1, and the other represents a single composite sample from the same place as the first *Apis dorsata* honey quoted.

Little work has yet been done on honeys from the other three species of *Apis*, compared with the detailed studies on *A.mellifera* honey described in Chapter 5, and even less work has been done on their physical properties (cf. Chapter 6). Accepting this, Table 17.22/1 shows that, in broad outline, variations between the gross composition of honeys produced by different *Apis* species are due less to interspecific differences between the bees than to other factors, especially climate and plant origin—both depending on the geographical region in which the honey is produced. Perti & Pandey (1967b) report that the water content of *A.dorsata* honey collected in Nainital, India, was about 17% in the summer months and 26% in the rainy season. Mallik (1958) found a variation from 16·6 to 26·4% in the water content of 40 samples of 'genuine honey' (presumably from *Apis cerana*) in different parts of India.

The collection of honeydew as well as nectar will materially affect the

Table 17.22/1

Composition of honeys produced by the four *Apis* species and some other species of bees

Some figures have been rounded off. In the final column, f=as formic acid; g=as gluconic acid; m=meq/kg. In the previous column, p=protein (not nitrogen). All results are in percentages, except L/D, meq/kg in the final column, and sugars marked t which represent percentages of total sugars. *See* 5.3 for validity of entries under sugar and dextrin.

Author and Country	No. Samples	Water	Total reducing sugars	Glucose (dextrose)	Fructose (laevulose)	Sucrose	L/D	Dextrin	Ash	Nitrogen	Total Acid
Apis cerana											
Latif *et al.* (1956) Pakistan	5	16·4	73·5	32·5	42·8	2·12	1·32		—		0·11f
	8	14·3–21·2	69·8–76·9	27·1–34·2	39·0–53·9	1·09–2·75	—		0·11–0·32		0·07–0·16f
Mitra & Mathew (1968) Calcutta, India	30	20·5	72·3	33·9	38·9	1·3	1·15	0·28			—
		17·5–24·0	66·9–77·4	28·8–37·7	33·3–42·9	0·0–7·9	0·95–1·42		0·03–0·52		—
Nair *et al.* (1950) Travancore, India	1	25·2		33·3	28·9	5·98	0·87	1·13	0·73	0·59p	0·28f
	1	28·4		36·4	31·1	1·93	0·86	0·71	0·37	0·15p	0·18f
Phadke (1962) Mahabaleshwar, India	9	16·1–19·6	66·5–78·1	28·2–38·3	35·7–42·5		1·06–1·36	0·50–2·14	0·014–0·048	0·39–0·14p	0·09–0·17f
Giri (1938) Madras, India	12	19·2	75·0	35·7	39·3	0·60	1·10	0·10			
		16·2–22·1		34·2–39·2	36·8–40·5	0·3–1·0		0·03–0·46			
Phadke (1967a) all India	80	20·9±2·0	70·2±2·9	33·4±2·8	36·5±2·5	—	1·10±0·12	1·97±0·93	0·19±0·05	0·56±0·15p	3·19±0·5f
Phadke (1967b) Mahabaleshwar, India	64*	17·2–19·1	—	31·2–38·3	35·2–43·3	—	1·03–1·34	0·67–1·92	0·11–0·25	0·04–1·11p	0·07–0·17f
Apis mellifera											
White *et al.* (Table 5.2/1) U.S.A.	490	17·2	69·5	31·3	38·2	1·3	1·21	1·50	0·17	0·041	0·57g 7·1m
Minh *et al.* (1971) Philippines	1	20·7	64·7	35·3	29·4	1·27	0·83	0·22	0·18	0·042	26·1m

Apis dorsata

Source	n										
Minh et al. (1971) Philippines	7	27·8 (23·4–35·9)	59·6	29·0 (23·2–36·7)	30·7 (27·9–33·1)	1·50 (0·08–2·84)	1·06	0·81 (0·36–1·20)	0·17 (0·06–0·32)	0·08 (0·04–0·13)	40·2m (25·5–81·8m)
Latif et al. (1956) Pakistan	5	16·2	69·2	27·0	42·2	1·43	1·56	—		0·16f	0·19f
Mitra & Mathew (1968) Calcutta, India	14	23·5 (19·0–27·1)	69·9 (65·7–75·9)	35·0 (30·5–37·5)	35·0 (31·7–39·2)	0·27 (0·0–1·3)	1·00 (0·87–1·15)		0·26 (0·06–0·80)		0·19f (0·09–0·40f)
Nair et al. (1950) Travancore, India	1	27·8	62·0	35·3	26·7	2·41		0·37	0·49	0·91p	0·39f
	1	31·0	63·6	35·4	28·2	0·56		0·31	0·38	0·99p	0·29f
Phadke (1968) India	20	20·9 (18·9–24·2)	69·5 (—)	32·1 (29·8–33·8)	37·4 (34·6–39·9)		1·17 (1·02–1·27)	1·57 (1·02–2·17)	0·39 (0·19–0·52)	1·07 (0·86–1·32)	0·25f (0·11–0·34f)

Apis florea

Source	n										
Latif et al. (1956) Pakistan	5	17·4	62·9	28·3	40·4	1·8	1·42				0·18f
Nair et al. (1950) Travancore, India	1	23·8	67·9	34·2	33·6	0·60	0·99	0·54	0·41	0·98p	0·22f
Phadke (1968) India	5	16·5 (16·0–17·5)	71·2 (—)	32·3 (28·2–35·6)	38·9 (36·9–41·3)		1·21 (1·09–1·31)	8·67 (6·00–13·6)	0·73 (0·68–1·04)	1·14p (0·78–1·45p)	0·27f (0·21–0·33f)

Trigona irridipennis

Source	n										
Nair et al. (1950) India	1	29·1		33·3	28·5	2·37	0·86	0·83	0·55	0·16p	0·27f
Phadke (1968) India	5	24·1 (22·8–25·3)		20·1 (13·7–22·1)	32·3 (27·5–34·9)		1·58 (1·38–2·00)	5·9 (5·1–7·0)	0·52 (0·48–0·54)	0·78p (0·59–1·14p)	0·23f (0·18–0·32f)

Various meliponins

Source	n										
Maurizio (1964) Brazil	5	—		27·9–45·9t	49·0–54·8t		1·07–1·96				

Bombus spp.

Source	n										
Maurizio (1964) Europe	14	—		5·3–53·5t	37·1–79·4t	0–3·3t	0·74–11·42				

* Averages for 8 samples of each of 8 unifloral honeys are considered

honey composition, but few workers give valid information on plant origin. Phadke (1962) is an exception, and he was able to show, for instance, that *A.cerana indica* honey from *Carvia callosa* (Acanthaceae) is thixotropic. Other studies have been made by Kalimi & Sohonie (1964*a*, 1964*b*, 1965) on unifloral honeys in Mahabaleshwar, and by Latif *et al.* (1958) on plants foraged by three *Apis* species in West Pakistan.

Undoubtedly differences will come to light as investigations are made on honeys produced by different species of bees within the same *foraging areas*. Vorwohl (1968*a*) has already been able to compare enzyme activity of honeys from *A.mellifera* and *A.cerana* colonies in Germany, kept in the same place. Both spring and summer *A.cerana* honeys showed diastase activities around half as high as the corresponding *A.mellifera* honeys (5.71); and their water contents were higher. Diastase is used to indicate quality or authenticity of honey, so low values due to natural factors are of economic significance. Invertase, unlike diastase, has an important function in bees (5.72), and honeys from the two species in Germany had fairly similar invertase contents.

The amino acid content of honey shows certainly interspecific differences. The major amino acid in *Apis mellifera* honey is normally proline, and Maslowski & Mostowska (1963) and Bergner & Körömi (1968) have shown that most of the proline originates from the bee. Proline was not detected in two of four samples of *Apis cerana* honey by Kalimi & Sohonie (1964*a*). Recent work has shown that in honey from *Apis cerana* the proline content is low, and there is a larger quantity of an unidentified amino acid (Davies, 1974).

17.3 HONEY FROM STINGLESS BEES (MELIPONINAE)

17.31 Information about stingless bees

Stingless bees do not, as might be thought, provide the basis of painless beekeeping. They are a group of exclusively tropical social bees which have only vestigial stings, but which defend their colonies in other—equally effective—ways. Although a few species are timid, others bite, tickle, or burn the skin with a caustic fluid, or generally demoralize the person who disturbs their nest by crawling into his eyes, ears, nose, and hair.

Figure 17.31/1 shows the present distribution of stingless bees in America, Africa, Asia, and Australia. The two main genera are *Trigona* and *Melipona*, the latter being larger bees. A number of species have been quite extensively 'domesticated' in the American tropics and subtropics. The important report by Schwarz (1948) gives much information on

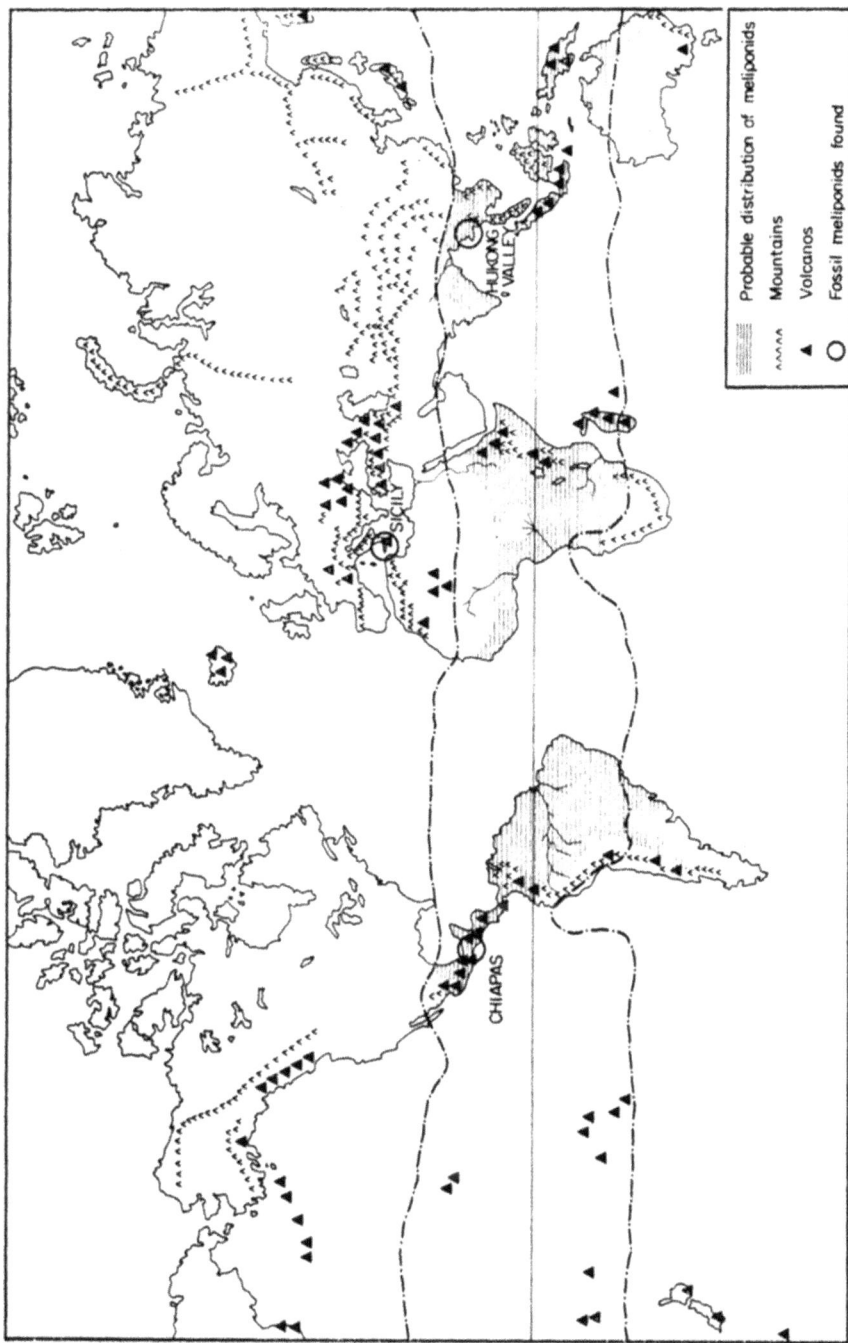

Figure 17.31/1 Distribution of stingless bees (meliponins) in relation to the boundaries of the tropical region (-·-·-·-) and other features (after Kerr and Maule, 1964).

stingless bees. Nogueira-Neto has written a short introductory study (1951), and a detailed book (1953, 1970), explaining how they can be kept in hives. Figure 17.31/2 (*see* Plate 17) shows a dried gourd used as a hive in Brazil; Bennett (1965) noted how colonies of *Melipona favosa phenax* are housed in similar gourds in Panama. Portugal Araújo (1954, 1955, 1957) has promoted the use of stingless bees in Angola in Africa. In Asia and Australia little or no attempt has been made to keep them in hives.

17.32   Honey from stingless bees
Yields vary from species to species, being often a kilogram or less each year, more rarely 10 to 20 kg. The nests are not composed of parallel combs like those of honeybees, so frame hives cannot be adapted for use by stingless bees, nor can the normal centrifugal extractor be used for the honey; it is usually pressed out (Nogueira-Neto, 1970), but a suction-pump extractor has also been devised (*Abelhas*, 1970).

Unlike honeybees, many of the meliponins do not confine their foraging to nectar, pollen, and honeydew. Sugars, fruits, and resins may be used and—what is less acceptable for honey destined for human consumption —some species also collect unhygienic material, for instance fluids from dead animals, and excrement, sweat, and urine; Schwarz (1948) and Nogueira-Neto (1953, 1970) give details. In spite of this unpleasant characteristic of some of the stingless bees, their honey is greatly prized in regions where it is produced, and often commands a higher price than honeybee honey. This may be because the honey is believed to have special medicinal properties.

Nogueira-Neto mentions the honeys of a number of species, describing their flavour as pleasant (*saboroso*) or not, and reporting many to be acid.

17.4   HONEY FROM BUMBLE BEES (*BOMBUS*)

17.41   Bumble bee colonies
The bees discussed in 17.2 and 17.3 form permanent colonies, and stores of honey are a continuing advantage to them. Most of the species are tropical or subtropical, but *Apis mellifera* and some races of *Apis cerana* are the only group of insects that has solved the problem of surviving cold winters as a colony, by remaining in a nearly inert cluster in the hive or other cavity which constitutes the nest. These bees can therefore live outside the tropics, and with man's aid have spread over most of the earth's surface which supports vegetation (19.511).

Bumble bees are widely distributed in the Northern Hemisphere, and some species live even beyond the Arctic Circle. There are also species as far south as Tierra del Fuego in South America. Bumble bees are not indigenous to Australia or New Zealand, but were successfully introduced there in the last century. In the tropics, bumble bees form permanent colonies; in the temperate regions, where they are more common, they solve the winter problem in a more primitive way than honeybees. A colony is built up in the spring by a queen which has overwintered alone, probably in an underground hibernaculum (Alford, 1971). Towards the end of the active season the colony produces new queens, which mate before overwintering; the colony, as such, dies out.

Since they build only temporary summer colonies, bumble bees do not store much food, but they and their nests may be sought out by country people. In *A Midsummer Night's Dream*, written in 1595, Bottom says to Cobweb: 'kill me a red hipt humble-bee on the top of a thistle; and, good monsieur, bring me the honey-bag'; this presumably refers to *Bombus lapidarius*. Gilbert White's *Natural history of Selborne* refers to an idiot-boy who was a specialist in this form of hunting around 1750. 'In the summer he was all alert, and in quest of his game in the fields, and on sunny banks. Honey-bees, humble-bees, and wasps, were his prey wherever he found them: he had no apprehensions from their stings, but would seize them, . . . disarm them of their weapons, and suck their bodies for the sake of their honey-bags.'

An ethnological study of bee hunting in the Carpathian mountains in eastern Europe (Gunda, 1968) includes a description of the hunting of bumble bee nests, and also of their transference to clay pots called 'bumble-houses', which were taken to the village for subsequent use (Figure 17.41/1).

17.42 Bumble bee honey

Although honeybee honeys have been described and studied for so long, as recently as 1959 an authoritative book on bumble bees could only report that 'little seems to be known about the composition of "honey" found in bumblebees' combs. . . . Bumblebee honey may also [i.e. like honeybee honey] be very concentrated but to what extent its sugars have been similarly converted appears to be unknown' (Free & Butler, 1959). Research has, however, since then provided interesting information about bumble bee honey (*see* Crane, 1972b).

Knee & Medler (1965) studied eleven nests of four North American species, and found that two types of honey were stored. One was 'thick honey', containing 70–87% sugar or more; the percentage in honeybee honey in the same area at the same time was 83–84%. This thick honey was stored in empty cocoons near the centre of the nest. Sladen (1912) reported that in a favourable season all the vacated cocoons, which may

amount to over 400, would be filled with thick honey and sealed over
with wax. The other type, 'thin honey', was stored in peripheral wax
honey pots which may be temporary structures (Plath, 1934); it contained
only 52–42% sugar or even less. At any one time a nest might contain
both, or one, or neither type of honey. Bumble bees have apparently
never been observed to collect water, and the thin honey may perhaps
be used to dilute the thick honey for use as larval food.

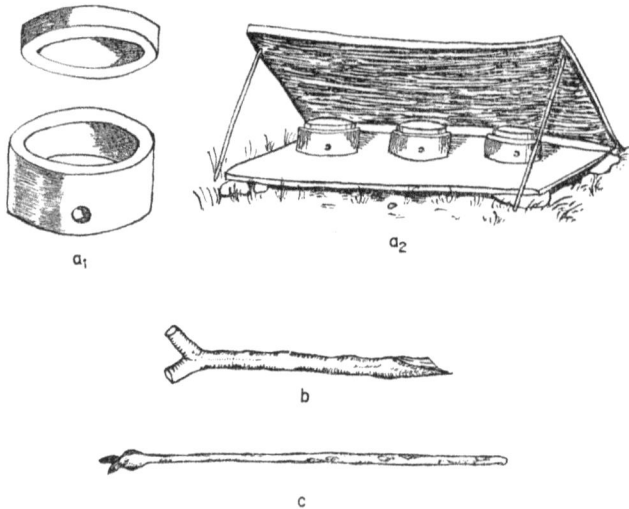

Figure 17.41/1

a_1 A clay 'bumble house' from Firtosváralja, a Hungarian
village in the Hargita Mountains of Transylvania,
Rumania.

a_2 Three bumble houses in position under a shelter.

b A stick for digging out bumble bee nests (Zalabaksa,
Göcsej region, Hungary).

c A stick used by Australian aborigines (Cape York) for
digging *Trigona* nests out of trees.

a and *b* from Gunda (1968), *c* from a photograph (E.
Crane) taken in the South Australian Museum at Adelaide,
all redrawn by Celia Hart.

In Canada 'thin' bumble bee honey has been sought out and examined
(Spencer *et al.*, 1970), as a likely source of a certain yeast isolated from
flowers, which was of potential commercial interest. In 189 samples of
bumble bee honey, 5 species of yeasts were found (compared with 19 in
flowers), the commonest being *Torulopsis bombicola* and *Candida* species.

The main sugars of honeybee honey are fructose and glucose (laevulose
and dextrose); these are present in fairly similar amounts, with laevulose

predominating (5.31). However, an L/D ratio of 2 would be considered very high; even for tupelo honey (*Nyssa aquatica*) the ratio is only 1·75 (Pryce-Jones, 1944). But in a study of European bumble bee honeys, Maurizio (1964) found much higher laevulose contents. Single samples from different *Bombus* species gave the following L/D ratios:

B.hortorum	1·34	B.lucorum	3·90
B.lapidarius	1·52	B.terrestris	7·09, 8·21
B.muscorum	3·22	B.agrorum	11·42

Results for other examples from *Bombus* species not identified varied from 1·12 to 5·21, except for one sample with a ratio 0·74. The high L/D ratio was associated with a high melezitose content of the samples, many of which were largely derived from honeydew (2.132).

Maurizio (1964) also compared the pollens in honeys from the bumble bees with those in honeys from honeybees in the same area. Bumble bee honeys frequently contained red clover pollen, but there was also a great variety of other pollens, showing that bumble bees visit many of the food resources available to them, and not only those designated 'bumble bee flowers' by scientists.

17.5 HONEY FROM OTHER INSECTS

17.51 Other bees

The three groups of social bees already discussed—Apini (17.2), Meliponini (17.3) and Bombini (17.4)—are closely related (Michener & Michener, 1951, 1974). Research in the past twenty years has brought to light much evidence on nesting and colony structure in the distantly related subsocial bees such as *Halictus* and *Allodape*. There is, however, too little information on the honeys of these bees to warrant a discussion of them at present; Burgett's work is referred to in 8.2.

17.52 Social wasps

Most social wasps are tropical; outside the tropics they make temporary summer colonies, like the bumble bees. But instead of wax, paper is the usual building material, made by collecting and chewing fibres of weathered wood and similar materials. The books by the Micheners (1951) and Richards (1953) give details of the mode of life of tropical social wasps. Most of the nests are hung in trees; some contain up to 15 000 individual insects. Certain species, notably of *Nectarina*, *Brachygastra*, and *Polybia*, are regular collectors of nectar, and stock their nests with honey, as honeybees do. The combs are parallel to each other, but usually horizontal instead of vertical. A few of the species have been more or less 'domesticated'. Evans & Eberhard (1970) think it likely that most, if not all, social species of wasps collect and consume nectar or honey. Nectar seems to be a major part of the diet of *Polistes* larvae, and in *Polistes*, at any rate, methods of evaporating water from nectar to make honey are similar to those used by honeybees (2.222). Nectar droplets are regurgitated on to cell walls, and in autumn a wasp may be seen sucking in a drop of nectar, and then slowly exuding it again. By evaporating some of the water, an individual can store an increasingly concentrated sugar solution in the gut. (*See also* Schremmer, 1972.)

In Central and South America, various Indian tribes collect honey from nests of honey wasps, some of which are very aggressive, whereas others are not and are easily robbed of their honey. Vellard's study (1939) describes the dependence of the Guayaki Indians in Paraguay on wasp honey, as well as on honey from meliponins (17.3). Both the Indians and Paraguayans collect a fair quantity of rather acid but pleasant flavoured honey from various *Nectarina*, and especially *Polybia*. One of the most highly valued and most common honeys is from the aggressive *Polybia scutellaris*. This honey granulates fairly easily; samples analysed by Santolaya & Gentile (1953) had a high solid content, with only 17·5% water. The question of sucrose inversion in wasp 'honey' needs more study.

Vellard's descriptions of honey hunting (17.3) includes the comment that the Paraguayans and Indians regard honey of meliponins, and more especially of wasps, as very intoxicating, but that it actually has 'stupefying properties'; also that some honeys are left untouched, being regarded as toxic.

17.53 Ants and syrphids
There are also ants that store honey, but they are unable to build storage vessels as bees and wasps do. Instead, the honey is stored within the bodies of some of the young workers of the colony, such a worker being known as a 'replete'. Returning foragers empty their crop contents by feeding the replete, whose abdomen becomes greatly distended; in the end the ant is so incommoded that it can only hang motionless from the roof of the underground nest. It may remain there for many months, but will regurgitate the honey store from its crop when stimulated to do so by other worker ants.

Honey ants are found in parts of Africa, America, and Australia. The American honey ant (*Myrmecocystus mexicanus hortideorum*), which occurs from Mexico City as far north as Colorado, may have 300 repletes in its colony (Michener & Michener, 1951). Honey-filled repletes are relished by Mexican Indians. In Central Australia, aborigines dig out the nests of the local honey ant, *Meliphorus inflatus* (see Figure 17.41/1c), and obtain the honey from the repletes by biting off the abdomen. (The distribution of the honey ants does not overlap that of *Trigona*, so only one or the other is available to aborigines.) The honey is slightly acid, and samples analysed by paper chromatography contained fructose and glucose. Reducing sugars were estimated as 59% of the total weight, and the L/D ratio was only 0·67 (Badger & Korytnyk, 1956). A report by Wetherill in 1853, that the Mexican [American] honey ant produces honey which is almost a pure fructose solution, is discounted.

'Honey' in the crop of the syrphid fly *Lasiopticus seleniticus* was found to have a sugar spectrum fairly similar to that of honeybee honey, except that 10–24% of the total sugar was maltose; the L/D ratio was 0·98–1·39 (Maurizio, 1964). Here, and also with the ants, one might question whether the material should be included under the definition for 'honey', since it is not matured and stored, as honeybee honey is.

17.6 USES OF NECTAR BY MAN

In general, nectar can only be used as a source of sugar by man if bees are available to act as intermediaries, because—unlike man—they have the anatomical structures which can serve as micro-manipulators for extracting the nectar. Flowers yielding much nectar have certainly been utilized in many parts of the world. From a mission school on Elcho Island, off the north coast of Australia, a letter recently reported: 'the *Grevillea* flowers are now [August] so full of nectar that they ooze; the children love them, and come to school smelling strongly of them'. The diet of Australian aborigines contains very little sugar, and they will eat flowers of *Cassia* and *Hakea*, and possibly some of the eucalypts, valuing their sweetness. In the Nullarbor Desert of Western Australia, north of Rawlina, aborigines pick flowers of *Grevillea excelsior*, crush them, and shake the nectar into water in bowls made out of wirras (galls on *Hakea multilineata*). The liquid is drunk after being left to ferment for 24 hours. It seems to be the only alcoholic drink known to be used by these Australian aborigines (*see* Crane, 1967*b*).

Edlin (1951) says that the true old heather ale in the Scottish highlands was made by steeping the nectar-bearing flowers in water, which was afterwards used to flavour the drink (*see* the quotation on page 392, also Section 16.5).

CHAPTER 18

THE LANGUAGE OF HONEY

by D. E. Le Sage

LECTURER IN GERMANIC PHILOLOGY,
UNIVERSITY OF SHEFFIELD

Every bee's honey is sweet.
GEORGE HERBERT,
OUTLANDISH PROVERBS (1640)

18.1 THE INDO-EUROPEAN LANGUAGES

This chapter presents a comparative study of the terms for 'honey', 'mead', 'wax', and 'honeycomb' in the Indo-European languages. A fuller discussion of the methods used here, and of the detailed structure of the Indo-European language family, has been published in an article on certain other apicultural terms (Le Sage, 1974).

The Indo-European languages form the world's largest single linguistic family, which covers almost the whole area of our western civilization (Europe and both Americas) and goes beyond it well into western and central Asia (India, Persia, Afghanistan, etc.). The main branches dealt with here, and their constituent individual languages, are listed below; for further information *see* Lockwood (1969).

(i) *Indo-Iranian* (also called Aryan): Consideration of this branch is here restricted mainly to Sanskrit, a well documented language rich in beekeeping terms, possessing in Indian culture roughly the same status as Latin in Europe, and now extinct as a spoken language. This branch includes also Hindi, Urdu, Bengali, Gujarati, etc.; Persian, Kurdish, Pashto (spoken in Afghanistan); and so on. The Iranian sub-branch (especially Persian) has been infiltrated by the vocabulary of Arabic, which is not an Indo-European language.

(ii) *Hellenic:* This branch is represented only by Greek—in all the stages of its history.

(iii) *Romance:* Latin—the parent language—with its modern descendants Italian, Spanish, Portuguese, French, and Rumanian.

(iv) *Celtic:* Irish, Scottish Gaelic, Manx; Welsh, Cornish, Breton.

(v) *Germanic:* Gothic (the oldest recorded representative, but not identical with the parent Germanic language); English, High and Low German, Dutch; Icelandic, Swedish, Danish, Norwegian. Modern (High) German is descended from Old High German, English from Old English (or Anglo-Saxon), and the Scandinavian group from Old Norse. These three plus Gothic represent the earliest recorded stages in the disintegration of the parent Germanic language.

(vi) *Baltic:* Lithuanian and Latvian are the only modern survivors of this group. Estonian is not an Indo-European language.

(vii) *Slavonic:* Old Church Slavonic (analogous to Gothic in that it is the oldest recorded Slavonic language but is not the same as the parent or Common Slavonic language); Russian, Polish, Czech, etc.

Occasional reference is made to other languages in the group: Hittite, Tocharian (once spoken in Chinese Turkestan), and Albanian. Only the last is still a living language.

When we speak of 'Indo-European languages', compare their similarities and contrast their differences, we are tacitly (if not explicitly) assuming the one-time existence of a single parent language from which they are all descended and which was spoken at a certain point in history as a discrete linguistic unit, albeit perhaps with strongly differentiated dialects. Indo-European philologists still differ on important issues, but that a linguistic unity existed in some form at some time as the parent language of a large number of modern tongues is not doubted. Eastern Europe is one of current favourites for its location; and the third millenium BC can be roughly postulated as the period in which it had a unified existence. One must be careful not to speak in terms of an Indo-European people or race, since no research to date has enabled us to associate the undeniable existence of this language with a single ethnic or national group.

Similarly, we should not attempt to draw cultural conclusions from the evidence of language. Thus, for example, merely because the Germanic languages all have a word for honey which is based on a root quite unconnected with the rest of Indo-European honey-terminology, we are not entitled to say that the Germani themselves did not know the substance honey when they first became heirs to an Indo-European dialect. Words appear in and disappear from languages for a wide variety of reasons; Le Sage (1974) gives a more detailed description of the potential pitfalls implied here. We shall therefore be concerned entirely with *linguistic* description, with statements about the words which a given language possessed. The nearest we shall come to extra-linguistic comment is the occasional self-evident conclusion drawn from the etymology of a word about the way in which its users conceived and looked upon the thing designated; for instance both German *Wachs* (wax) and *Honigwabe* (honeycomb) contain a hidden reference to the process of weaving, and thus may be said to point to close observation of bee behaviour (18.3).

Finally, the very fact that we are comparing several languages descended from a common ancestor means that our approach will be essentially historical as well as comparative. Frequent reference will be made to extinct languages including Indo-European itself. It will usually be necessary to quote only the roots in the latter, rather than whole words. In accordance with normal practice, all words and roots which are reconstructed by the usual and well tried methods of historical and comparative linguistics are preceded by an asterisk (*).

18.2 THE WORDS FOR HONEY AND MEAD

As might be expected, there is greater unanimity among the Indo-European languages over the word for 'honey' than for 'bee'. Honey was presumably the reason why the bee first became interesting to man, and we should expect to find more common cognates for this valuable substance than for its producer, which might be variously designated as the 'buzzing fly', the 'honey maker', or even something initially unanalysable deriving from a taboo (Le Sage, 1974). It will soon become clear that 'honey' and 'mead' must be dealt with together.

The very small number of separate roots behind the various Indo-European words for honey is remarkable indeed. With the special exception of Germanic, only two primaeval honey terms provide all the Indo-European languages with their attested terminology: *melit-* and *medhu-*. *Medhu-* occurs in virtually all Indo-European languages, and in one (eastern) section its descendants denote both 'honey' and 'mead' (or other alcoholic drink) without distinction, whereas in the other (western) section *medhu-* means mead only, and honey is denoted by a derivation of *melit-*. *Melit-* is correspondingly absent from the first section. Although it seems obvious that in common Indo-European *melit-* signified honey and nothing else, it is difficult to know exactly what meaning to ascribe to *medhu-*: this is because of its great variety of derivatives. In the western group it came at an early stage to denote mead, an alcoholic drink fermented from honey, and was at all times distinguished from the word for honey itself. In the east, however, *melit-* appears to have been absent almost from the beginning, and *medhu-* has expanded its scope to take in a large number of connected concepts: honey, mead, sweetness, intoxicating liquor, etc. This process

* *See* explanation at end of 18.1.

went furthest in the Indo–Iranian branch: Platts (1960) lists words from both ancient Sanskrit and the later Hindi. His definition of Sanskrit *mádhu* (page 1016) is: 'Honey; the juice or nectar of flowers; anything sweet; mead; sugar; liquorice;—sweetness;—a spirituous liquor obtained from the blossoms of the *Bassia latifolia*; any sweet intoxicating drink; wine; spirituous liquor;—water [!] . . .'. The Hindi word *madh* shows just as much semantic diversity: 'Honey; the juice or nectar of flowers . . . the grape;—wine; ardent spirit; intoxicating liquor or drug; intoxication . . .'. Selected words from other Indo–Iranian languages show that the same diversity of meaning continued within that branch: Nepalese has *maha* = honey, while the Gypsy language of Armenia has *mahl* = wine or brandy; Sinhalese has *mihi* = honey, and in the Kharosthi inscriptions of Chinese Turkestan we find *masu* = wine.

Clearly at an early stage in the Aryan branch **melit-* (if indeed it survived the Primitive Indo-European period in that area) was felt to be redundant due to the extension in semantic scope of **medhu-*, which —in view of the range of senses it subsequently developed—possibly had an original meaning something like 'sweet liquid' or 'sweetness' in Indo-European.

We can see the same phenomenon in the Slavonic and Baltic languages, though in these **medhu-* has not extended its scope to such a great degree. Old Church Slavonic had *medŭ* probably meaning both honey and mead, and many of the modern Slavonic languages similarly make no basic lexical distinction between the two substances: both Russian and Polish can use the simple *mjod* and *miód* respectively—for either honey or mead, but it is possible in both languages to avoid ambiguity by the addition of *napítok* in Russian and *pitny* in Polish (both meaning 'drink'), so that a distinction is made (in for example Russian) between *mjod* and 'the drink *mjod*'. On the other hand, modern Czech can distinguish explicitly between *med* = honey and *medovina*, literally 'honey-wine'. This word must be regarded as an independent Czech development, since the 'western' language (according to our east-west distinction drawn above) most closely in contact with Czech throughout its history, viz. German, has never designated mead as a type of wine.

There is an interesting case of borrowing in the Baltic languages which can be established by careful use of historical phonology. Latvian has— in line with the rest of the 'eastern' languages—one word for both honey and mead—*medùs*, while Lithuanian calls honey *medùs* but mead *mìdus*. On the face of it this is puzzling: *medùs* is the regular development of Indo-European **medhu-*, but according to Lithuanian sound-laws the *i* in *mìdus* ought never to have developed. We must look for a solution in the realm of borrowing, and in particular from a language which (a) belongs to a branch that is 'western' as defined, i.e. made a lexical distinction between honey and mead; (b) regularly shifted Indo-European short e to i; and (c) was at some time in contact with Lithuanian. Gothic fits all these requirements, for apart from the linguistic aspects we know that the Goths settled for a while on the south Baltic coast when they first migrated from their Scandinavian homeland. And while, because of the fragmentary nature of its preservation, Gothic has no attested word *midus, there is enough evidence here to justify our reconstructing it in this form, and further identifying it as the origin of the Lithuanian term.

Before leaving the 'eastern' branch, brief mention of a few other languages will help to complete the picture. The only 'eastern' language to distinguish lexically between mead and honey (though not on the *medhu-/melit- basis) is Avestic, the Old Iranian language of the Avesta or sacred book of the Zoroastrians. In this we find madu- = 'wine' beside paēnaēna = 'made from honey' and suggesting an unattested form *paēna = 'honey'. This appears to come from the same root as the Lithuanian and Iranian words for milk (pienas and payah respectively; see Le Sage, 1974), and seems to indicate that the Old Iranians looked upon honey as 'bees' milk'. Modern Persian calls honey asal which is borrowed—like so much of its vocabulary—from a totally unrelated Arabic word ('asal). But Persian keeps the root *medhu- in mai = wine.

In view of its geographical position it is not surprising that Tocharian has mit (from *medhu-) for honey, and no derivative of *milit-. It is conceivable that similar words in some non-Indo-European oriental languages are derived from this: cf. Japanese mitsu, Chinese mi and Sino-Korean mil, all meaning honey.

Other non-Indo-European languages which can be seen with more or less certainty to have borrowed some form of *medhu- are Hungarian with méz = honey, and Finnish which has mesi = honey beside a native word sima also meaning mead. As in Slavonic, however, the beverage can be conveyed more explicitly by a composite word: simajuoma meaning honey-drink.

Turning now to the 'western' branch we find a clear distinction drawn at all times between honey and the drink fermented from it. The original Indo-European lexical distinction was apparently between *medhu- (probably first meaning any sweet liquid or drink and then specifically mead, which would often have been sweeter than beer or wine, the only other alcoholic drinks of such apparent antiquity), and *melit- restricted to honey only. As we have seen, the latter root disappeared very early from eastern Indo-European languages.

If the *medhu-/melit- hypothesis is correct, then Primitive Celtic would seem to have preserved the pair longer than most other branches: Welsh has mêl and medd, Breton mel and mez, Cornish mel and mēth; in the other (so-called Goidelic) branch of Celtic we find Irish with *mil* and *miodh*, but Scottish Gaelic has done away with the **medhu-* root and speaks only of *mil* = honey (genitive *meala*) and *leann meala* = mead, literally honey-beer. The Welsh word *meddyglyn* which gives us *metheglin* (defined in the *Oxford English Dictionary* as a 'spiced or medicated variety of mead peculiar to Wales') seems, contrary to appearance, to derive not from *medd* = mead, but from Latin *medicus* + Welsh *llyn* = liquor, i.e. doctored drink.

Mendelsohn (1965) says of metheglin: 'The term is especially used of mead flavoured with mace, cinnamon, cloves, and pimento. Hydromel seems to have been identical.' But he offers no proof of the latter assertion, and this brings us to Latin and Greek in which we find evidence that the two cultures concerned had lost a taste for mead, though it is from these that we get the word *hydromel*.

There is extensive reference in both literatures to the bee (*apis*, *mélissa*) and its main product (Latin *mel*, Greek *meli*), but mention of honey-based drink is so sporadic as to give the impression that it was a rarity. During the classical period of each language there was no word of separate etymology used to denote only mead. Latin lost the root **medhu-* altogether and Greek preserved it only in a poetic word for wine—*méthu*. Indeed it seems that at an early stage in Greek the stem *meth-* came to denote alcohol generally, for we find quite common use of words such as *méthusis* = drunkenness, *méthusos* = drunk and *methúō* = I am drunk. The same stem is the basis of our 'methylated spirit'.

As to the drink mead itself, the Greeks seem to have regarded this as alien, barbaric, having its proper place in distant antiquity, and they called it *melíteion*. Plutarch in the early part of his *Life of Coriolanus* says: 'men at the first beginning did use acorns for their bread and honey (*melíteion*) for their drink' (North's translation, 1906). Juvenal (Satire VI, 10) also designates the acorn as the staple diet of pre-civilized man, but he does not mention honey products in the same context.

There were, however, other drinks containing honey known to the Greeks—apparently made by mixing honey into some other liquid. *Melíkraton* was one such substance. The second element is derived from the verb *keránnumi*, to mix, dilute. In the *Odyssey* (X, 510) this liquid is mentioned as one of a number of suitable libations to the dead.

A similar dilution is the Latin *mulsum*, probably from **mel-sos* = honeyed, sweetened and formed after the analogy of *salsus* = salted. Cicero in *De Oratore* (II, 70) mentions a certain incompetent lawyer who was advised to cure a hoarse throat with *mulsum frigidum*, rendered in the Loeb translation as 'chilled wine and honey'.

But a fermentation of honey diluted with water was just as foreign to the Romans as to the Greeks, as witness the recipe given by Pliny the Elder (*Natural History* XIV, 113) for *hydromeli* (borrowed from a late Greek word) which he says is *nusquam laudatius quam in Phrygia*.

This *hydromeli*, however, is the basis of all the subsequent Romance words for mead: Italian *idromele*, French *hydromel*, Spanish *hidromel* and also *aguamel* (with *agua* = water substituted for the Greek first element). Rumanian has *hidromel* but also *mied* clearly borrowed from a neighbouring Slavonic language, like much of its vocabulary. All the Romance words for honey are directly descended from the Latin term (Table 18.4/1).

If mead yielded in popularity to other drinks in the classical world of the Mediterranean, in the Germanic north it was as much favoured as beer. All the oldest native Germanic literature contains reference to the substance which is designated in all cases by a derivative of **medhu-*: Old Norse has *mjoðr* (whence Danish *mjød*, Swedish *mjöd*); Old English called the drink *meodu*, Old High German *metu* from which comes modern German *Met*, but German also uses the tautological word *Honigmet* (= honey-mead; cf. apple cider). Dutch has *mee*. As we have seen above, Gothic most probably used the word *midus*.

Gothic is also the only Germanic language to contain unequivocal evidence that Germanic at first preserved the **medhu-/melit-* pair: in Mark 1, 6, in which John the Baptist is said to be in the wilderness eating 'locusts and wild honey', the Gothic translation has *milip haipiwisk*, heath honey (*þ* is pronounced like English *th* as in 'thin'). Now although this is the only place in which the word occurs, we can be sure that it is not a loan formation from the Greek (*méli*) because of the form of the word: the changes from Indo-European *e* to *i* and from *t* to *þ* are characteristic of Gothic; had the word been borrowed it would probably have taken the form **maili(t)* or *mailei(t)*—cf. Gothic *Aileisabaiþ* for Greek *Elisabeth*.

But if earliest Germanic preserved the **medhu-/melit-* distinction, all the separate Germanic languages apart from Gothic lost it at an early stage (except for a few residual obscure compounds of **melit-* dealt with below). It would seem that after the Goths left the Germanic ancestral lands the Germani coined a new word for honey which in Primitive Germanic was something like **hunaga-*. This appears to be based on an Indo-European root **keneko-* meaning golden-yellow (cf. Greek *knēkós* = pale yellow, tawny; Latin *canicae* = bran; Sanskrit *kāñcana* = gold). It is extremely difficult to say what semantic changes had taken place in that stage of Germanic to create the need for a new honey-word in the structure of the language, because of the total lack of written records from the period. But the new word maintained its semantic position with great tenacity, and has remained the basis of all Germanic terms for honey to this day (Table 18.4/1). Alongside the other two words quoted

Figure 17.31/2 (see page 420) Old colonial house in São Paulo Province, Brazil, with gourd hive for stingless bees sheltered by the roof of the porch. (photo: Eva Crane)

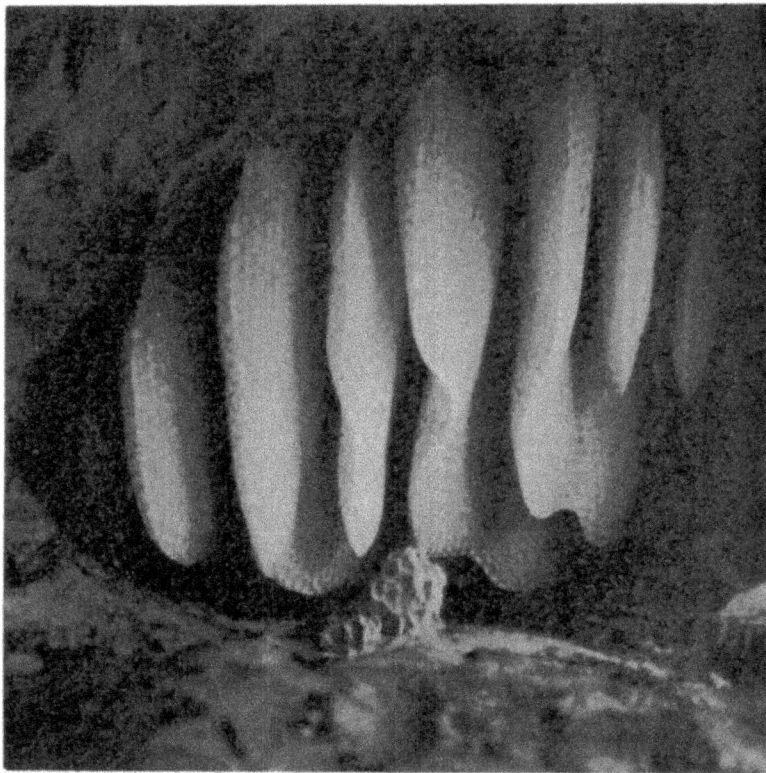

Figure 19.11/1 (see page 440) An empty honeybee nest in a hollow tree, viewed from below. A honeybee nest would have looked like this as far back as the Ice Ages. (B.R.A. Collection)

PLATE 17

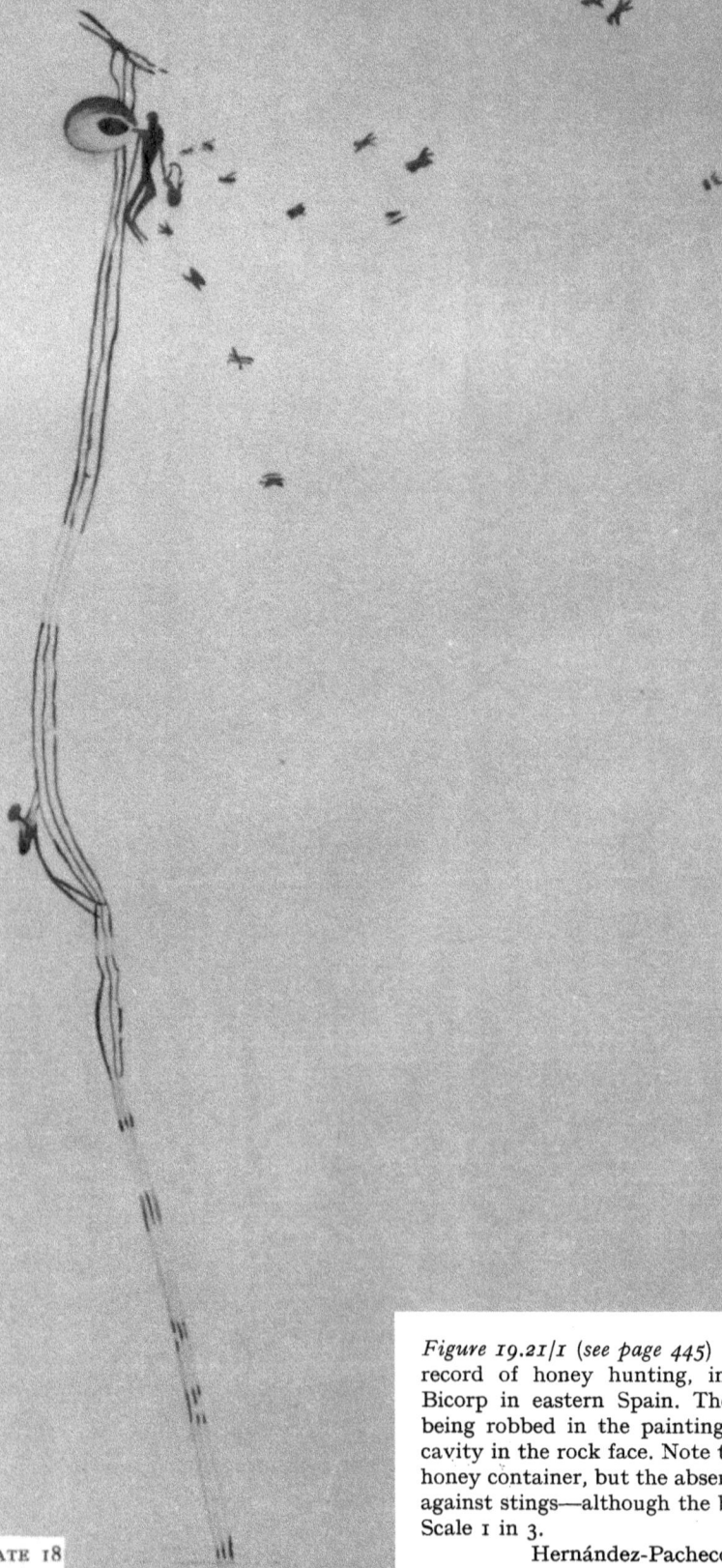

Figure 19.21/1 (see page 445) The earliest known record of honey hunting, in a rock shelter at Bicorp in eastern Spain. The bees' nest that is being robbed in the painting is actually a small cavity in the rock face. Note the use of ladder and honey container, but the absence of any protection against stings—although the bees are well alerted. Scale 1 in 3.   (from a painting by Hernández-Pacheco, B.R.A. Collection)

PLATE 18

Figure 19.21/4 (see page 446) Honey hunting equipment from two African tribes today. The sisal rope (B68/19) and honey barrel of wood and leather (B68/17) are from the Tharaka tribe, Kenya. The shallow basket to the left (B71/80) is used for honey collection in Mali; the gourd honey vessel (B71/81) from the same source is decorated with a representation of a Mali hive.   (B.R.A. Collection)

Figure 19.21/5 (see page 448) Rock painting of a nest of stingless bees by Australian aborigines, who prize the honey highly. Secure Bay, near Darwin.
(photo: Dr. I. M. Crawford, Western Australian Museum)

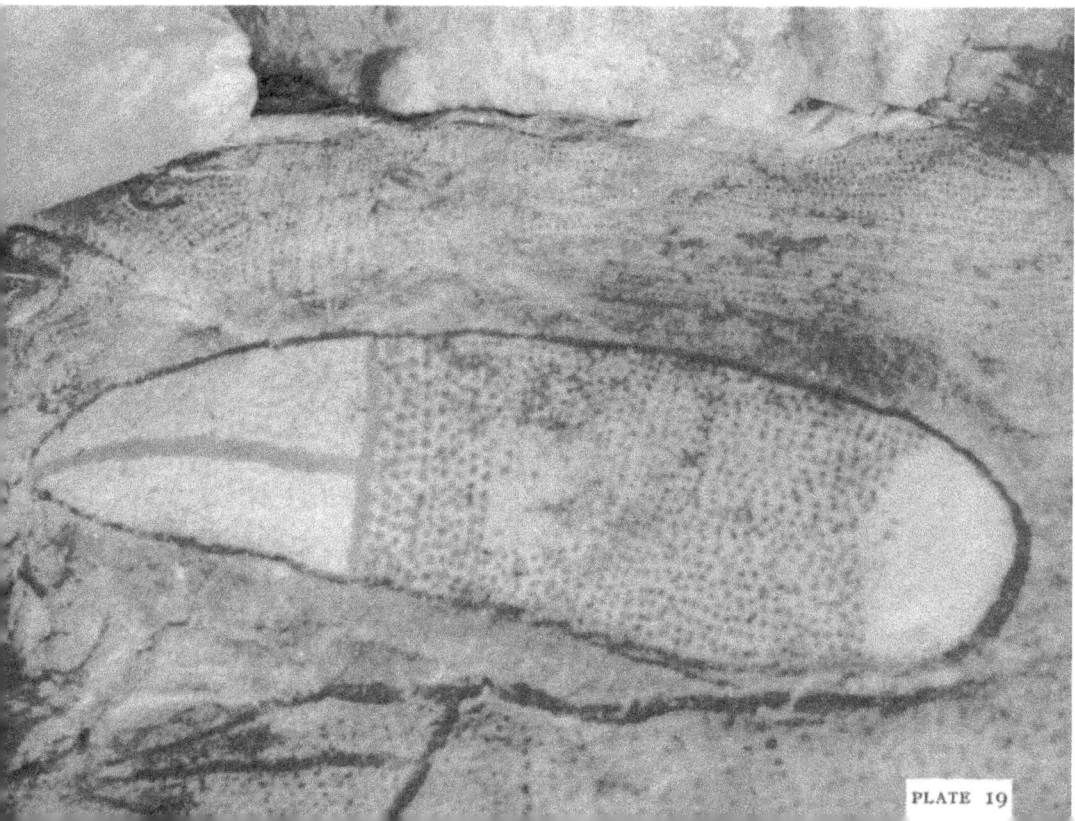

PLATE 19

Figure 19.22/2 (see page 449) Honey harvesting in a German forest, from honeybee colonies in the trees, owned and worked by beekeepers. The operator on the right (F.153) stands on a ladder; he is removing the hive door with his left hand and holds a knife in his right. He appears to be wearing a veil. The man on the left (Fig. 152) is slung from a sort of bosun's chair; he has already removed the hive door and is operating a cutting implement with his left hand. Below are two vessels on the ground between which an assistant stands ready to receive the comb to put into them. Fig. 151 represents the traditional *Zeidler* (forest beekeeper) with bow and arrow. Fig. 156 is a honey bag, and Fig. 157 a swarm catcher; just below, the owner's mark, a cross, similar to that in Figure 19.22/1 opposite.

(from *Das wesentlichste Bienen-Geschichte und Bienen-Zucht* by J. G. Kränitz, 1774)

PLATE 20

Figure 19.23/1 (see page 450) Bark hive (Tanzania), suspended by a forked stick; the hive would be lowered on a rope for honey collection. (photo: F. G. Smith)

ure 19.21/2 (see page 445) An early
tograph, probably taken in northern
e Province, South Africa, showing
oney hunter's ladder in position.
oto: Marloth; by courtesy of Alex-
er McGregor Memorial Museum,
aberley)

ure 19.23/2 (see page 450) Primitive
e made of mud, cow-dung and
ken straw, used today in the High
aien mountains of Ethiopia. The
s fly from a hole at the right-hand
. Compare the shape of this hive
h those portrayed in Egypt around
BC (PLATE 24).
ught from the High Simien in 1972
John Rea, now No. B72/15 in the
A. Collection.

ure 19.22/1 (see page 449) Owner's
rk on a tree in a Polish forest that
tains a honeybee nest. Twentieth
tury. (photo: S. Kirkor)

PLATE 21

Figure 19.23/3 (see page 451) Apiary of Greek movable-comb hives. A frame of one hive has been removed by its top bar; the other top-bars are covered with mud/dung and cannot be distinguished. The protective straw hackles are similar to those represented in two honey pots in PLATE 30.

(photo: G. Georgantas, 1956)

Figure 19.31/1 (see page 454) Sumerian clay tablet that contains the earliest known written reference to honey (in a prescription), *c.* 2000 BC. (Photograph of reproduction B58/17 in B.R.A. Collection)

Figure 19.32/2 opposite (see page 455) Wall painting in Tomb 101 at Thebes (*c.* 1450 BC) showing honeycombs, with other foods, being offered to Pharaoh. Bees are hovering over the combs.
This drawing is reproduced as the frontispiece of *Beekeeping in antiquity* by H. M. Fraser (1931).

PLATE 22

PLATE 23

Figure 19.32/1 (see page 455) Relief in the tomb of Pa-bu-Sa at Thebes (*c*. 625–610 BC). The person in the lower scene is probably a beekeeper getting honey from clay or mud hives at the back; compare Figure 19.23/2. In the upper scene the honey is being poured, and possibly strained, into a second vessel.

PLATE 24

above, Finnish also has *hunaja* for honey which is clearly a loan-word from Germanic, and because of its extremely conservative nature Finnish has preserved the Primitive Germanic form of the word almost intact.

Other Indo-European languages containing a **melit-* word for honey are Armenian with *melr*, Hittite with *milit* and Albanian with *mjaltë*. It is possible that the last-mentioned is a loan-word from Latin (Hamp, 1966), but the connection is doubted by Walde (1954, Vol. II, page 62).

18.3 THE WORDS FOR COMB, WAX, AND HONEYCOMB

Because of overlapping terminology, it is also necessary to take the other two main items of investigation—comb and wax—together. As might be expected, the further away one gets from the first aspects of man's contact with honey, and the more specialized the notions become, the less etymological correspondence there is between the appropriate lexical items in each language. The great variety of words for honeycomb, which gives an interesting insight into the way in which each linguistic community saw the object concerned, makes it impossible to postulate any original Indo-European designation for it. And the position is not much simpler with 'wax'.

There are two basic Indo-European roots connected with wax: one is the basis of Latin *cera*, etc., and the other of English *wax* and its cognates. Of these possibly only the first originally meant wax. It produces Greek *kērós* (=wax) and *kērion*, a simple derivative by suffix meaning honeycomb. There is no attempt here to describe the shape or nature of the comb, and *kērion* is thus somewhat exceptional among the terminology concerned, which tends to be explanatory or descriptive.

The Latin word *cera* may be a loan-word from the Greek, or it may be an independent derivation from the same root. All the modern Romance words are descended from it: French *cire*, Italian *cera*, Spanish *cera*, etc., and in the specific sense 'beeswax' they are usually further qualified (*d'api*, *de abejas*, and so on).

In Celtic all the words for wax are definitely loan-words from Latin, and in the Goidelic branch 'honeycomb' is expressed by a mere circumlocution with the word for wax as its basis. The borrowing from Latin apparently took place during the Roman occupation of Britain and found its way to Ireland (cf. Lewis & Pedersen, 1937, pages 56, 58). Irish now has *céir* and *cíor mheala* (*mheala* is the genitive of *mil*); Scottish Gaelic has *céir* and *cìr-mheala* but also the expression *cìrchuachag*, literally woven wax, which is thus conceptually very similar to German *Honigwabe;* see below.

Welsh calls wax *cwyr*, and honeycomb is either *dil mêl*, honey structure, or *crwybr gwenyn*, bees' comb. *Crwybr* also means froth or foam, and probably the honeycomb was so named because of its similarly cellular appearance. Cornish for wax is *cōr*, and for honeycomb *crÿben mel*. *Crÿb* in Cornish is a (toothed) comb or cock's comb, and *crÿben* the crest of any bird. Cornish is thus the only language other than English so to designate the bee's cellular construction; the *Concise Oxford Dictionary of English Etymology* (Oxford, 1966) includes under 'comb': 'flat cake of cells of wax made by bees (*an exclusively English use*, the origin of which is doubtful), late OE in *hunigcamb*' [my italics]. Clearly a loan-formation cannot be ruled out in respect of the Cornish term. English 'comb' comes from an Indo-European word **gombhos* meaning tooth. Again some attempt to describe the appearance of the object would seem to be the principal reason for the usage. The Welsh for (toothed) comb is *crib* cognate with Cornish *crÿben*, but not apparently related to Welsh *crwybr*.

The majority of words for wax in the Indo-European languages are cognate with the English word (German *Wachs*, Russian *vosk*, Lithuanian *vãškas*, etc.), but this word is further—and most interestingly—analysable. The Indo-European root is **weg-* which apparently meant to weave or plait: cf. Sanskrit *vāgurā* = net or snare, and English *wick* so named from its having been spun or plaited. Thus it would seem that the substance wax was in the first place named from the fact that it was 'woven' or built up into a fabric resembling woven material. Notice, moreover, that in German, when the original meaning of 'wax' had become forgotten so that the word merely called to mind the substance and not its manner of production, the 'woven' nature of the waxen structure was again that aspect which gave its name to the structure or comb itself: Old High German calls the honeycomb *wabo*, cognate with our words *web* and *weave*, and this usage is preserved in the modern German *Honigwabe*. On the way Middle High German coined another term on the same theme: *wift*, which is clearly identical in formation to English *weft*.

In the other Germanic languages, however, we get evidence of a different conception being attached to the honeycomb. All the members of the group have cognates of 'wax' (Old Norse *vax*, Danish *(bi-)voks*, etc.), but 'comb' is more variously designated. The Gothic word is not attested. Old Norse has *hunangsseimr*. This is the same as German *Honigseim* which means virgin honey. Kluge (1960) lists under *Seim* (page 716) a number of cognates including Greek *haîma*, 'blood' and a Westphalian dialect word *siemern* 'to trickle'. This is perhaps the meaning of the original root: other derivations are Old Norse *simi*, 'sea' and Old English *siolop*, 'lake'. Thus in Old Norse *hunangsseimr* we have a semantic transference from the fresh dripping honey to the whole comb from which it issues.

The later Scandinavian languages have not kept this word, but speak either of a 'waxen cake' (Swedish and Icelandic *vaxkaka*) or of a 'wax tablet' (Norwegian *vokstavle*, Danish *vokstavle* also *-kage*). This word *tavle* is borrowed from mediaeval Latin *tabula*, '(writing-) tablet' or 'board'. Thus these languages all demonstrate the tendency mentioned earlier for each linguistic community to develop individual and often explicitly descriptive terms for this object.

The Slavonic languages all designate wax by a word which is cognate with the Germanic terms: Old Church Slavonic has *voskŭ*, Russian *vosk*, Polish *wosk* and so on. While the organic relationship between these words and the Germanic ones is obvious, the suggestion—at one time maintained—that they were borrowed from Germanic is now no longer credited; *see* Vasmer (1953). The metathesis of the original *-ks-* to *-sk-* is a Balto-Slavonic development. Lithuanian has *vãškas* and Latvian *vasks*.

For honeycomb the Slavonic languages, like the Goidelic, mainly use a compound based on 'wax': cf. Russian *voshchina*, Polish *woszczyna*. The tendency to import an element of graphic description into these words is seen in the alternative Polish term *plaster woszczyny* (cf. also the Czech *plást*). Baltic diverges from Slavonic here and appears to use the ancient wax-words *korỹs* (Lithuanian) and *kãres* (Latvian) which have come to mean only 'honeycomb' and not the substance from which it is made. The connection with Greek *kērós*, etc., would seem obvious, and it is maintained by Pokorny (1959). It would also appear to be confirmed by Armenian *xorisx* = honeycomb which Hübschmann (1962, Vol. I, page 455) links with the Greek word. But Fraenkel (1962) prefers a postulated relationship with Lithuanian *kárti*, 'to hang', implying—presumably—that the word for honeycomb is derived from an original emphasis at the lexical level on the suspended construction of freely built combs. Since there are no apparent phonological obstacles to the former (*korỹs* = *kērós*) theory, this one seems logically preferable.

Returning now to Latin, whose term for comb is in no way connected with 'wax' or any attempt at circumlocutive description, we find another sadly obscure word—*favus*. This occurs *passim* in Virgil's famous fourth *Georgic*, but is given its most explicit—almost prosaic—description by Varro (*De Re Rustica* 3, 16, 24): 'favus est quem fingunt (sc. apes) multicavatum e cera, cum singula cava sena latera habeant'. From attested usage it was clearly the most usual word for the object. Walde (1954, Vol. I, page 469) admits the uncertainty of its origin, but gives a pointer to the most likely solution: since the problem of establishing cognates is the *a* of the stem-syllable, he suggests that this is an analogical development from an original *o* (thus **fovus*). This is a not uncommon change: cf. *cavus*, *-um* (as, for example, in the quotation from

Varro above) which in the ordinary way would have become *covus; and
the verb *faveo* ('I favour') which should strictly be *foveo—see* Pokorny,
page 453. If this supposition is correct, it provides a plausible derivation
from the same root (Indo-European *bheu/bhou-) which gives (*inter alia*)
German *bauen*, to build, and English *booth*. The semantic connection
between these and 'honeycomb' is not difficult to establish.

Of the individual Romance languages only Italian, Portuguese and—
surprisingly in view of its long detachment from the rest of Romance—
Rumanian have retained a direct descendant of this word: they have,
respectively, *favo*, *favo de mel*, and *fag* or *fagure*. Rumanian can also say
turta de miere = honey-cake.

Spanish for honeycomb is *panal*, which looks deceptively like English
panel and thus to be derivable from the same notion which gave rise to
Danish *tavle*. But in fact the Spanish word comes from Latin *panis*,
bread. Corominas (1961) says that the use of *pan(al)* to denote an
aggregate of various substances goes back to the thirteenth century 'en
particular pan de cera'. Spanish also has a word *bresca* of somewhat
obscure origin of which Corominas (1954) says: 'voz prerromana, prob-
ablemente céltica'.

The French *rayon* (*de soleil*) and *rayon* (*de miel*) are etymologically
quite unrelated. The first goes back to Latin *radius*, while the second is
a loan-word from Germanic. The word from which it is taken (in the
language of the Franks) has been reconstructed as *hrata. This gives Old
French *rei*, and indeed for a while in later French the word remained as
raye. Dutch (technically known as New Low Franconian and descended
from one of the old Frankish dialects) calls honeycomb *honingraat* with
the last element identical to the French word. Other (albeit unlikely
seeming) cognates are Russian *krósno*, loom, and English *roost* (part of a
henhouse). Again a descriptive element seems present.

Turning finally to India, we find that the word *mádhu* forms the basis
of Sanskrit honey-vocabulary in regard to 'comb' as well as many other
concepts (Le Sage, 1974). This tendency to view all things connected
with honey from the point of view of the honey itself is reflected
especially in the word for wax: contrast the Germanic words, which
designate the substance as something woven by the bee and thus testify
to a close observation of the insects' behaviour, with the Sanskrit *madhu-
ukkhista*, which literally means 'remainder of honey'. *Madhusishta*, also
meaning 'wax', has the same semantic analysis.

Honeycomb in Sanskrit is *madhu-patala* of which the second element
can mean a cover, veil, cataract (on the eye), and also a basket or hive.
In Urdu both honeycomb and hive are also called *madhukosh* = honey-
container.

Altogether the Sanskrit words appear to testify to a long period when

Table 18.4/1

Words for honey, mead, wax, and honeycomb in the main Indo-European languages

	Honey	Mead	Wax	Honeycomb
Greek	mḗli	melíteion, hydromeli	kērós	kēríon
Latin	mel	hydromel(i), mulsum	cera	favus
Italian	miele	idromele	cera (d'api)	favo
Spanish	miel	hidromel, aguamel	cera (de abejas)	panal
Portuguese	mel	hidromel, aguamel	cera (amarella)	favo de mel
French	miel	hydromel	cire (d'abeilles)	rayon (de miel)
Rumanian	miere	hidromel, mied	ceara	turta de miere
Gothic	miliþ	*midus		—
Old Norse	hunang	mjǫðr	vax	hunangsseimr
Danish	honning	mjød	(bi-) voks	tavle, vokskage
Norwegian	honning	mjød	(bi-) voks	(voks-) tavle
Swedish	honung	mjöd	(bi-) vax	vaxkaka
Icelandic	hunang	mjöður	bývax	vaxkaka
Old English	hunig	medu, meodu	weax	hunigcamb
English	honey	mead	wax	honeycomb
Old High German	hona(n)g	metu	wahs	wabo
German	Honig	Met, Honigmet	(Bienen-)wachs	Honigwabe
Dutch	honing	mee, mede	(bijen-) was	honingraat
Old Church Slavonic	medŭ	*medŭ	vosku	—
Russian	mjod	mjod (napítok)	vosk	voschina
Polish	miód	miód (pitny)	wosk	woszczyna, plaster woszczyny
Czech	med	med, medovina	vosk	plást
Lithuanian	medùs	mìdus	vaskas	korŷs
Latvian	medus	medus	vasks	kāres
Irish	mil	miodh	céir	clor mheala
Scottish	mil	leann meala	céir	cir-mheala
Welsh	mêl	medd	cwyr	dil mêl, crwybr gwenyn
Cornish	mel	mêth	côr	cryben mel
Sanskrit	mádhu	mádhu	madhusishta etc.	madhupaṭala

apiculture was not practised as an advanced branch of husbandry, but in which the bee and its honey were none the less known and exploited. This would explain the 'honey-centred' vocabulary connected with so many aspects of beekeeping.

18.4 SUMMARY

Before listing the various words in Indo-European languages for the substances and objects in question, it will be as well to recapitulate briefly the findings of our investigations in these areas.

The attested Indo-European languages have for the most part only two roots giving rise to the various honey and mead terms in each, respectively *melit-* and *medhu-*. At an early stage the Indo-European languages fall into two groups according to how they treated each of these roots: in one group—represented now by languages mainly in the west —the derivatives of *medhu-* came to designate either 'honey-wine' or alcohol in general, while *melit-* was restricted entirely to 'honey'. In some branches—such as Italic—*medhu-* died out altogether: the apparent lack of interest in mead among the Romans would appear to be a contributory factor in this development. In Greek *méthu* remains a designation for alcoholic liquor of some sort, while its derivatives denote various aspects of the alcoholization of the individual. In the other (eastern) branch of Indo-European *melit-* disappeared at a pre-historic stage and *medhu-* expanded its scope to cover both honey and the drink fermented from it. One sub-branch of the western group—Germanic—retained the root *melit-* in its earliest stages and in some fossilized compounds, but soon replaced it by a root apparently meaning 'golden yellow'.

Greek *kērós* and Latin *cera* are perhaps based on the original Indo-European word for wax; they are not—as yet—further analysable. All Romance words are taken from the Latin term, and so are those in the Celtic languages—by borrowing. In Germanic, *wax*, etc. originally meant 'the woven substance'. The same root is found—apparently not due to borrowing—in Slavonic and Baltic.

The semantic analyses of the words for honeycomb are many and varied: a full list is given in Table 18.4/1. Many words contain an element referring to wax, others continue to indicate the weaving or building process; yet others describe the shape of the structure. In these words— as in those for hive—independent formations in each linguistic community are evident.

Table 18.4/1 sets out the lexical findings of this chapter. Only the main branches of Indo-European dealt with are listed; Indo-Iranian, as already indicated, is represented only by Sanskrit. Equivalents of 1 000 bee terms in twelve of the languages have been published (Crane, 1951–1971).

CHAPTER 19

HISTORY OF HONEY

by Dr. Eva Crane

DIRECTOR, BEE RESEARCH ASSOCIATION

Where bees are there is honey.
JOHN RAY (1670)

19.1 HONEY WITHOUT MAN

19.11 Honey production

Books have been written on historical aspects of beekeeping and man's relations with bees in many parts of the world, including Britain (Fraser, 1958), North America (Pellett, 1938), Germany (Bessler, 1886), Switzerland (Sooder, 1952), the Ancient World (Billiard, 1900, Fraser, 1951), and Russia (Galton, 1971). Bodenheimer's *Geschichte der Entomologie* (1928/29), Ransome's *The sacred bee* (1937), and Glock's *Die Symbolik der Bienen* (1891) provide access to further valuable source material. Far fewer books have concentrated on historical aspects of honey; of the more recent of these, Lerner's *Aber die Biene nur findet die Süssigkeit* (1963) is worth special mention.

These and other publications are concerned more with man's production and use of honey than with the history of honey from the natural history or ecological aspect, as a substance produced by certain insects and exploited by various other animals. Yet honey (and its source-material, nectar) are of considerable significance in environmental biology, and man has been present to enjoy honey only for a small fraction of the long period of its production.

Bees, which were thus producing honey long before man appeared on the earth, can still do this as well without man's intervention as with it, in many areas. Leftn o their own, diseased or weak colonies are more likely to die out than are healthy strong ones, but colonies with stored honey will not be systematically robbed of it, as they are by the honey hunter or beekeeper. The population of hive bees in any given area may or may not be increased by the care of beekeepers, depending on a number of factors. On the other hand beekeepers have enormously increased the total world population of honeybee colonies (19.51). Honeybees did not reach the New World until emigrants from Europe took them there, after 1500 or 1600. The interesting thing is that the bees throve very well in the New World, often better than in their native country, and even today there is better bee forage in the New World than in the Old, as is clearly shown by Table 4.27/1.

Flowering plants producing nectar and pollen, and insects that forage on them, had already evolved by the Jurassic/Cretaceous periods 150–100 million years ago; the first mammals appeared about the same time. Bees came later; the first (solitary) bees and the first monkeys developed in the Eocene, 50–25 million years ago. The honey-storing social bees, and the higher arthropods, developed more recently still. They have existed for perhaps 20–10 million years, since the Miocene, and man himself for one (or a few) million years, since the Pleistocene.

Honeybees (*Apis*) were present, and probably widespread, in all the regions of the Old World where there were flowering plants, and none of these regions was completely isolated from the rest by a barrier of desert or sea that bees could not cross. The *Apis* genus is presumed to have originated in Asia, perhaps in the region of Afghanistan. It became differentiated into the four species recognized today, probably between 20 and 10 million years ago. Two of the species, *Apis dorsata* and *Apis florea*, build a single comb in the open, and will not nest in an enclosure; they have never succeeded in spreading beyond the tropics of south-east Asia. The other two species, *A. mellifera* and *A. cerana*, nest in a cavity, and in some strains of these species the colony acquired the ability to survive a winter of dearth and cold by remaining almost inactive in a cluster sheltered within its cavity. These species were able to spread northwards, *A. mellifera* west of the Himalayas into the Middle East, Europe, and Africa; *A. cerana* east of this great mountain range into China, Japan, and adjacent regions.

Colonies of these honeybees nested in hollow trees, in crevices, in rocks, and in holes in the ground; Figure 19.11/1 (*see* Plate 17) shows such a nest, viewed from below. As they still do today, the bees collected and stored pollen, collected nectar, and made and stored honey. Honeybees could survive anywhere in the Old World where there were plants that provided sufficient stores to last the colonies through the next dearth period. Some of these plants are still virtually unchanged: Fossel (1970) discusses some of the berry-bearing *Vaccinium* in this context. Regions habitable for honeybees in general exclude deserts, and areas at high latitudes where the (cold) dearth period is too prolonged, but not much else. Hansson (1955) has suggested that the northern boundary of hazel (*Corylus avellana*) and other woodland trees gives a good approximation to the northernmost limit of survival; Straka (1970) gives relevant maps for different periods. In tropical regions the dearth period is generally determined not by temperature but by rainfall (either wet or dry season), and the honeybees there may survive it by migrating rather than by clustering.

During the successive glaciations known as the Ice Age—which started something less than 1 million years ago and recurred at intervals until 20 000 years ago—bees, with other animals in the Northern Hemisphere, were driven southwards. Certain areas remained unfrozen, and served as isolated 'retreat' areas for long periods. The important 'retreats' for honeybees were the Iberian, Appenine, and Balkan peninsulas, and the south Caucasian region. Here *Apis mellifera* became differentiated into the races *mellifera, ligustica, carnica*, and *caucasica* respectively. As the ice retreated, the first of these races extended its range to northern Europe (the barrier of the Pyrenees does not quite reach the sea); the second was largely prevented from spreading by the Alps; the third moved up into lower parts of the eastern Alps; and the fourth remained fairly stationary. Ruttner (1952) has set out this history in some detail.

Another limiting factor to honeybee survival must have been suitable nesting places. Populations spread to adjacent areas by swarming, a 'daughter' colony establishing itself in a cavity of a size that is acceptable; adequate shelter from the weather, and an entrance small enough to be protected against robbers, are likewise important for survival.*

Although there were no indigenous honeybees in any part of the New World—the Americas and Australasia—meliponin bees and social wasps stored honey in some of the tropical parts of these and other continents (17.3, 17.52).

All these insects produced honeys that might well be hard to distinguish from the honeys produced today, except in so far as the distribution of plants has changed. And however cleverly man exploits the bees, these are no more 'domesticated' today than before man existed. There has been some genetic selection, both intentional (bee breeding) and unintentional (e.g. by bush fires, by man's destruction of the less aggressive colonies to take their honey, or by utilization of swarms), but the nature of honey is unchanged because the life habits of the bees, and their enzyme chemistry, are unchanged.

19.12 Exploitation of honey by animals
In a large part of the Old World, honeybees (*Apis*) have been producing honey from flowering plants (1.1), and probably also from honeydew (1.2), for 10–20 million years. These bees were absent from the New

* The cavity-size factor can be significant. European honeybees *Apis mellifera* have been in South America for several centuries (19.51), but did not penetrate the Amazon valley. After the African subspecies *Apis mellifera adansonii* had been introduced to Brazil in 1956, hybrids spread quite rapidly, by swarming, into areas not occupied by European races of *Apis mellifera*. This spread seems to have been largely due to the fact that swarms would occupy smaller nesting cavities than those acceptable to the European bees present in the area before the 1960s (National Academy of Sciences, 1972, *Bee World*, 1972/73).

World (19.11), but stingless bees were producing substantial amounts of honey in tropical and subtropical areas there. Then, as now, honey was stored for the bees' future use; then, as now, the honey was sometimes taken instead by various mammals—although not yet by man—and also by a few marsupials, by birds, and by other insects. (Many birds with names such as honeyeaters or honeycreepers suck *nectar* from flowers (Keast, 1968), and a fifth of the world's birds are involved in flower pollination (Fisher & Peterson, 1964). The brush-tongued lorikeets crush the flowers rather crudely to get the nectar; the even less specialized eating of flower tissue with nectar and pollen still survives in some species.)

We can get an idea of behaviour patterns of animals that actually braved bees' nests to get honey or other food by observations on similar animal marauders today. As far as is known, no one has yet made a special study of this subject, but the observations quoted here are likely to be representative of events in the 10 million years or more before man appeared and developed honey hunting (19.21) into a more deliberate pursuit.

Many animals enjoy sweet foods, and some are willing to suffer considerable inconvenience, and many stings, in getting honey from bees' nests. Honey produced by the bees in the London Zoo is used for the animals that especially like it.

The mammals best known as honey eaters are various species of bears, and the honey badger. Perry (1970) describes the behaviour of the sloth bear (*Melursus ursinus*) in India, which in the hot season is 'on the lookout for the combs of the small forest-bee [*Apis cerana indica*] in hollow trees, and also for the huge combs of the large rock-bee [*Apis dorsata*], which hang in clusters from the branches of trees or from the undersides of rocks. The latter he knocks down to the ground—possibly deliberately, for in the darkness the bees tend to buzz around the site of the comb rather than fly down in pursuit of the raiding bear—but the combs of forest-bees cannot of course be knocked down. They must be taken the hard way, and a Himalayan black bear can be heard bawling with the pain of the stings implanted in his muzzle, but nonetheless persisting in his "honey-combing".' Perry (1970) quotes V. K. Arseniev's description of a black bear's behaviour in the Manchurian taiga. There was a colony of bees in a tall lime tree that was growing almost flush against a rock. 'He was standing on his hind legs. The stones prevented him thrusting his paw into the hole in the tree. He was not a patient bear, and was shaking the tree with all his might and growling with irritation. A cloud of bees was humming round him, stinging him on the head. He kept rubbing his head with his paws, uttering a high-pitched little scream, rolling on the ground, and going back once more to his work.' Finally the bear ran to the tree, climbed to the top of it and, 'putting his back against the rock, began to push the tree with all his might. Presently the tree gave a rending sound, split and crashed to the ground. That was just what the bear wanted. All he had to do now was to open the nest and collect the honey.'

Bears have adapted their honey hunting to cope with man-made hives, upon occasion even in apiaries surrounded by an electric fence (Crane, 1966).

Honey hunting by the ratel or honey badger (*Mellivora capensis*) in southern and tropical Africa has often been reported to be done in 'partnership' with the honey-guide, a bird belonging to the parasitic family Indicatoridae. The apparent guiding behaviour, which leads a mammal (including man) to a bees' nest, is restricted to the greater honey-guide *Indicator indicator* and a few other species. It was investigated in detail by Friedmann (1954, 1955), and the commonly held belief that the bird actually guides the mammal to the bees' nest so that the nest is opened, and the bird can get at the honey, is no longer tenable. What seems to be true is that the so-called guiding behaviour is released in the bird by the sight or sound of a ratel, baboon or (away from a village) a human. As long as the mammal 'follows' the bird—i.e. remains within its sight or hearing—the guiding behaviour continues, except that it is suppressed by the sight or sound of bees. If, as is probable, this happens near a bees' nest, the mammal is likely to locate and open the nest. The bird will take some of the comb, to which it now has access, eating honey along with the rest of the comb contents, pollen, and brood. Wax and insects are the main food of the honey-guide, which is most unusual in being able to digest beeswax, an ability shared by only a few other animal species, such as the wax moths *Galleria*, *Achroia*, and *Phormia*. Friedmann & Kern (1956a, 1956b) have shown that in the honey-guides this is due to the presence in the gut of *Micrococcus cerolyticus* and the yeast *Candida albicans*.

Very few other birds brave bees' nests to take the contents, and these may well be feeding on brood or pollen rather than on honey. This is certainly true of woodpeckers, of which the green (*Picus viridis pluvius*) especially can do widespread damage to hives by boring holes in them. Some birds have developed special adaptations which protect them from stings. A honey buzzard (*Pernis apivorus*), for example, is able to ignore bees that crawl over its face when it is digging their nest out of the ground (Willis, 1972) 'because the feathers that cover this part of the head are small and close-fitting, forming an armour that is apparently impregnable to wasp or bee stings'. (It is not known if any other members of the Accipitridae have such sting-proof feathers immediately in front of the eyes; the common buzzard (*Buteo buteo*) does not.)

There is less information about marsupials, which did not evolve in the same regions of the world as honeybees. The tiny honey possum (*Tarsipes spenserae*), about the size of a pygmy shrew, feeds mainly on nectar, especially from *Banksia*.

From the viewpoint of man's liking for and use of honey, the greatest interest attaches to other primates. Observations have recently been published on baboons, gorillas, and chimpanzees.

In discussing the antiquity and evolution of the co-operation between the honey-guide bird and mammals referred to above, Friedmann (1955) says 'in Northern Rhodesia, Major E. L. Haydock was told by his native collectors that the bird also calls to baboons and monkeys, and that occasionally the baboons do follow it, but the monkeys never do. A corroborating and wholly independent bit of evidence came to me from the Cape Province, where Mr. Trevor McKenzie Crooks told me that one morning around Christmas time near Uitenhage he saw a baboon (one of a troop) opening a wild bees' nest with a greater honey-guide in attendance, chattering from a perch a little way up in a tree close to the hive. Crooks watched the procedure from a distance. The baboon first made a clearing by repeated short dashes towards the hive, which was low down in a soft-barked tree, until it had made more or less of a path. Then the baboon backed up to the tree and reached in with its hind foot, grabbed a piece of comb, and dashed off about 30 yards where it dropped it. Then, when the bees had calmed down, it came back to the piece of comb and cleaned it of bees by wiping it in the sandy ground, and frequently wiped its hands in the sand to get rid of the bees.'

Schaller (1965) was told by natives that gorillas frequently raid the nests of wild bees, but around Kabara (near the Congo-Uganda border) he did not find them doing so; 'several gorillas once climbed up on a hollow log which contained bees, but the apes seemingly ignored the combs which were clearly visible'.

Merfield & Miller (1956) report on the use of stick tools by chimpanzees to obtain honey: 'each ape held a long twig, poked it down the hole and withdrew it coated with honey'. On the other hand Reynolds & Reynolds (1965), watching chimpanzees in the Budongo Forest where there were numerous bees' nests, saw no signs of interference with them. Deschodt (1969) gives an entertaining description of a honey raid by a group of some twenty chimpanzees near Lake Kivu (west of Lake Victoria). A large chimpanzee got the combs from a crevasse in a granite wall, and threw what he did not eat to the others. All got stung, but all seemed to enjoy the party—except the babies, who snuggled their vulnerable faces against their mother's chest.

The scene is now set for honey hunting by primitive man, other primates being seen to rob bees' nests, devising a way of getting the comb partially free from bees, and even using a tool to extract the honey (compare the digging stick in Figure 17.41/1).

19.2 EXPLOITATION OF HONEY BY PRIMITIVE PEOPLES

The behaviour pattern of honey hunting by primitive tribes can also be deduced from observations on similar groups today. It seems likely that this pattern developed independently, and in a fairly similar way, among peoples in many parts of the Old World. This development concerned the two species of honeybees that nest in a cavity—*Apis mellifera* in Europe and Africa and *Apis cerana* in Asia (17.21), and two other species that build a single comb in the open and cannot be hived—though Thakar (1973) has just succeeded in hiving *Apis dorsata* by using unorthodox techniques. Most of the honey from this bee, and from *Apis florea*, is still harvested by honey hunting.

In some regions a gradual transition took place from honey hunting to ownership of wild nests (19.22), then to transport of the wild nests to an 'apiary' near the dwelling (19.23), and finally to the construction of containers specifically for bees:hives. In the tropics of the New World a similar and parallel development took place, but the bees there were the social Meliponinae (17.31).

The date and duration of these phases of honey-getting varied widely in different places, and the author believes that the whole sequence occurred independently in many regions, although details must have differed according to the habitat of both bees and man. We shall never know the whole story, but recent studies in archaeology and prehistory have greatly extended the range of evidence available.

19.21 Honey hunting

Man's earliest honey hunting probably followed a similar pattern to that of other mammals. Notable advances by man were the use of a ladder to reach the nest, and of a container to hold the honey combs; the earliest evidence so far available for these is a palaeolithic rock painting in eastern Spain, dated by Hernández-Pacheco (1924) around 7000 BC (Figure 19.21/1—*see* Plate 18). In *The honey hunters of Southern Africa* Guy (1972b) describes the history of the use of ladders and climbing pegs, which he has found in Bushman rock paintings in the Drakensberg Mountains in Natal, and still in place in a baobab tree in Mozambique; he also shows an early photograph (Figure 19.21/2—*see* Plate 21).

Another, more sophisticated, development among primitive people was the use of smoke for 'subduing' the bees. A rock painting in Southern Africa which portrays this (Figure 19.21/3) is one of about eighty

Bushman paintings showing bees, bee colonies, and honey hunting which were discovered and recorded recently by Pager (1971, 1973). The fact that this is the only one showing a honey hunter using smoke, whereas several show him climbing a ladder that leads to a nest, suggests that smoke is a relatively late development. Honey hunting equipment in use in the 1960s is shown in Figure 19.21/4 (*see* Plate 19).

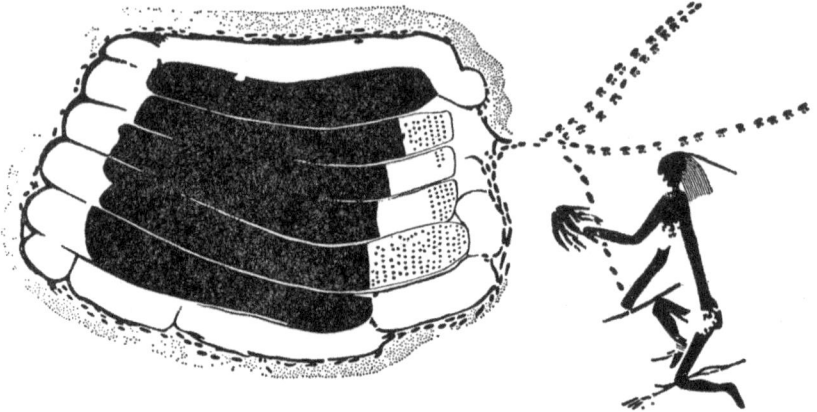

Figure 19.21/3 Rock painting of a honey hunter near the Toghwana Dam, Matopo Hills, Rhodesia. It is the only such painting known where the use of smoke is shown. Scale 1 in 6. Compare the view of the combs (from below?) with 19.11/1.   (From a copy by H. Pager)

Honey-getting from *A. dorsata* is vividly depicted in an Indian rock painting (Gordon, 1960).

The primitive Onges people are a negrito (pygmy) tribe who live on Little Andaman Island in the Bay of Bengal. They still lead a palaeolithic existence—hunting and gathering, but not growing crops, and Cipriani (1966) has described their very primitive honey getting: 'The Onges make no attempt to protect their bare bodies while they are extracting honey. . . . I watched an Onge from the ground with my field glasses as he climbed up a tree to a *dorsata* nest and saw him blowing on the bees round about the trunk as he went up. As he approached the nest they huddled round it in a protective cloud, hiding it completely from view. Suddenly, as the Onge's face came within twenty centimetres of the boiling, humming mass, the bees seemed to shrink back as he blew on them. Not one moved to attack as he gouged out the combs with his bare hands, throwing handfuls of bees out into the air. And then the whole cloud of bees gathered into a swarm and left the tree and the nest to the interloper. The Onge stripped lengths of green bark from the lianas growing up the tree, and then broke the nest up. Tying the combs to his back with the strips of bark he lowered himself down to the ground, swinging like an ape from liana to liana.'

In parts of Asia *Apis dorsata* nests under rock overhangs; Harrer (1953) describes getting honey from such a nest in Tibet. 'This honey taking is a very risky adventure as the bees hide the honeycomb under projecting rocks of deep ravines. Long bamboo ladders are dropped, down which men climb sometimes two or three hundred feet, swinging free in the air. Below them flows the Kosi and if the rope which holds the ladder breaks it means certain death for them. They use smoke balls to keep the angry bees away as the men collect the honeycomb, which is hoisted up in containers by a second rope. For success of this operation perfect and well rehearsed combination is essential, as the sound of shouts or whistles is lost in the roar of the river below. . . . I much regretted that I had no cine camera with which to take a picture of this dramatic scene.'

Several anthropologists have more recently been able to take such films. James Woodburn included a long sequence on a honey-hunting expedition in *The Hadza: the food quest of an East African hunting and gathering tribe* (1965); (Bee Research Association, 1973). Alan Root directed *Baobab*, shown by BBC-2 in 1971, which has a sequence on honey hunting using the honey-guide bird (19.12). *The pygmies*, made by Herbert Ricz, in the tropical forest of the Congo (Brazzaville), also has a short sequence.

Other recent anthropological studies include detailed first-hand accounts of the seasonal cycle of primitive honey hunting, and the accompanying rituals and dances. Turnbull (1966, 1968) gives a vivid description of the African Mbuti pygmies who live by hunting and gathering in the Ituri forest between the Congo River and Lake Albert. He includes (1966, pages 303–307) a song relating a legend which, translated, begins:

'I went to the forest to collect honey.
I went, I went, I went.
I went very far, beyond the big river.
I heard the sound of bees. I saw up high the place of honey.
I said, "Nobody has enclosed the tree with vine, it is mine; it is my honey." [*see* 19.22.]
I sharpened my axe, very sharp indeed.
I cut a vine. I fanned my fire, and put the fire in my basket.
I began to climb; I climbed, I climbed.
The honey was very far. It was honey, real honey, not *apuma*.
I reached it. I fastened the vine, I sat. I put fire into the hole.
I blew. I got much smoke, I drove out many bees.
I chopped, I chopped, I chopped.'

Mwaniki (1970) deals with the Embu in Kenya; Woodburn (1965, 1968) describes the Hadza, east of Lake Eyasi in Tanzania. The bees robbed by these tribes were the often aggressive *Apis mellifera adansonii*, and the same is true for the Batwa, primitive forest dwellers near the Rwanda–Uganda border studied by Schaller (1965):

> 'A Batwa . . . discovered bees hovering by a small hole in a tree about twelve feet above ground. Immediately the Batwa wedged a sapling against the trunk and climbed up. They lit a dry, rotten piece of wood and blew the heavy smoke into the opening. They made little progress pecking away at the bark with their spears in an attempt to enlarge the hole and decided to return the following day. During the next attempt their spears were fitted with a chisel-like point. The chips flew as they chopped, and they were coughing and spitting in the dense smoke. One Batwa reached into the tree and hauled out the combs dripping with golden honey. They ate and ate, laughing and swatting at the bees that buzzed angrily around them. Some combs held white grubs and these were eaten too. Only the wax was spat out. The Batwa had honey all over their arms and chest, and the bees landed on their skin and stung them. Then they licked their fingers and arms and stuffed the extra honey into the small calabashes they carried.'

Allchin (1966) refers to various peoples in both Africa and Asia, and *Australian Bee Journal* (1970) to Tasman's 1642 description of the now extinct natives of Tasmania, who hunted honey of stingless bees (17.3). Coon (1972), in a brief survey of primitive honey hunting in a number of regions of the world, says that in one part of Tasmania the sap of *Eucalyptus gunnii* was also harvested by these Tasmanians, who cut holes in the trunk. The sap sometimes dried into a white paste which was eaten. The 'sweet white manna' from certain eucalypts mentioned by Daisy Bates (1938) as food of Australian aborigines might have been the exudate from man-made holes or injuries (Basden, 1965), or the lerp produced in a rather similar way to honeydew (2.13) by lerp insects (Basden, 1966).

Nests of stingless bees are occasionally shown in rock paintings by Australian aborigines, one of which is reproduced in Figure 19.21/5 (*see* Plate 19).

19.22 Ownership of scattered honey-producing colonies

The position with regard to ownership is a remarkable one. Over widely separated areas in the world, nests of species of bees from which honey could be harvested each season seem to have been the only thing individually owned among hunting-gathering peoples. A nest of bees is not mobile as animals are. Honey seems to have had a very special significance, and a strong magico-religious character, for many peoples who were hunter-gatherers—and this was man's way of life for over 90% of the total period of his existence (Lee & Devore, 1968).

Morris (1967) regards a 'sweet-tooth' as something typically primate-like, and offers the rationalization that fruits were a main food of primates, and that these usually become sweeter as they ripen and are more suitable for consumption, so man's 'ape ancestry' expresses itself in seeking out specially sweet substances. Current evidence suggests that the liking for sweet foods—or at any rate its indulgence—has varied greatly among different peoples, and also that honey was valued for other reasons than its sweetness. The subject is discussed further at the end of 19.23.

As examples of peoples who have had a specially strong predilection for honey, Turnbull (1966) says that no amount of alternative foods, even meat, can reduce the passion for honey shown by the Mbuti, African pygmies. Cipriani (1966), whose description of honey hunting in the Andaman Islands is quoted in 19.21, says: 'When one of the Onges finds a nest of bees he puts a mark on the rock or tree as a sign of ownership. None of the others would then dare to take possession of it, for fear of punishment by the group, but when the time comes to take the honey everyone helps in the hard work of getting at the nest and opening it up.'

As and when hives were brought into use (19.23), these were always 'owned', as were crops and domesticated animals. Crosse-Upcott (1956) has studied the intricacies of hive ownership and transfer among the Ngindo in what is now Zambia. The hives are scattered among woodland and thickets, and there is a fairly close link here with the ownership of wild colonies.

Honey was not hunted only in the tropics. The forests of Northern Europe were rich in honey, and enough contemporary documents still exist to provide a picture of the state of affairs in the Middle Ages. Ownership was easily established by cutting a sign in the trunk of a tree (Figure 19.22/1—see Plate 21), and was defended by law. Dorothy Galton's study of Russian sources (1971) provides a coherent picture of honey hunting in the woods round Moscow, which would be similar to that in other forests, such as that shown in Figure 19.22/2—see Plate 20. 'For the 1400s and 1500s there is much evidence provided both by foreign observers and internally of the great extent of tree beekeeping in the Moscow lands. Bee walks were often mentioned in wills and deeds of the time, the lands often being left to monasteries. Whether trees had bees in them or not, they were not allowed to be felled, for bee-less holes could be used later by swarms. In a charter of 1565 the people of a village were forbidden to cut down bee trees with or without bees. There were frequent disputes about ownership of and rights in bee forests. . . . The Code of Ivan Grozny (1550) laid down fines for destroying and robbing bee trees.' Rights of ownership run like a refrain through accounts of honey hunting and the care of occupied (or empty) nesting sites for bees.

19.23   Honey from apiaries

Without man's intervention, cavities in trees and rocks are the usual sites for honey production. A swarm of bees (*Apis mellifera* or *A. cerana*) will occupy any cavity of a suitable size (19.11); if sites are plentiful, it may be able to select a cavity that is at a favoured height, suitably sheltered, dry and warm, does not get overheated, and whose entrance is small enough to be easily defended. If there is a population pressure on such cavities, swarms may occupy holes in the ground, or build their combs under the large leaves of dense vegetation such as that of banana or sugar cane. Comb building is not *initiated* unless the light intensity is low but, once initiated, it is continued even if the illumination is increased (Morse, 1965).

Hives probably came into being in two ways. Firstly, when man acquired tools that could separate off the part of a tree trunk containing a natural bees' nest, he could take this near his dwelling for protection, or attach it to the trunk of another tree that suited his purpose. Secondly, where competition by the bees for nesting sites was severe, swarms would take over man-made receptacles for their nests. Suitable receptacles included clay water pots—often left empty lying on their sides in the Mediterranean and Middle East—and baskets of wicker or cane or, when cereals were cultivated, of coiled straw. Usually the wooden tree-trunk hives were left vertical (Figure 19.22/2), and the word *Bienenstock* (*Stock* = tree-trunk) is still used in German for a hive. The successors of these hives in Europe north of the Alps, and throughout the New World, are still 'vertical' hives. Usually the basket hives were inverted for use by bees, like the skeps still remembered in rural parts of Europe. The clay pot or pipe hives were generally horizontal, and it may well be that the horizontal position of the traditional log and bark hives throughout tropical Africa (Figure 19.23/1—*see* Plate 21) was taken over from the horizontal clay hives of Ancient Egypt (*see* 19.32 and Figure 19.32/1), as a result of communication along the Nile Valley. The hive in Figure 19.23/2 (Plate 21) is egg-shaped, made of the humblest materials—mud, cow-dung, and broken straw. It is used today in the High Simien mountains of Ethiopia, and it could well be a primeval link with some of the earliest hive types.

The above remarks greatly over-simplify the complex development of world hive types, a subject which has yet to be studied (however, *see also* Crane, 1974). The most significant outcome for the history of honey is that in all the types of hive mentioned, the bees attached their combs to the inner surface of the hive, and the beekeeper could obtain the honey only by breaking or cutting the combs away. Among primitive peoples the combs were eaten, more or less as described in 19.21, or put in water to produce an alcoholic drink (16.1). Separation of honey from the combs by straining and pressing is a later development, dealt with in 19.41. Extraction of honey centrifugally from the (framed) combs of modern hives (3.4, 19.52) is hardly more than a century old (19.52).

A unique advance was made in Greece, where some of the wicker hives were used open end up instead of down like other skeps (Crane, 1963*b*, Georgantas, 1957). Over the open top a series of bars was laid, each bevelled on the underside. Each bar was made with its width equal to the natural spacing between the midribs of adjacent honeybee combs, about 4 cm (1½ inches), and the bees built their combs down from the bevel under the bars. This upright wicker hive, the *anastomo cofini*, was fairly large, and wider at the top than the bottom, and *the bees did not attach their combs to the side walls* (Figure 19.23/3—*see* Plate 22). Combs could thus be lifted out and inspected, and those full of honey removed expeditiously. Evidence has not yet come to light to show when these hives were first used; some believe that Aristotle (384–322 BC) could not have made the observations on bees recorded in his *Natural History* without them. News of the hives first reached the western world in 1682, when Sir George Wheler described them in his book *A journey into Greece*.

This movable-comb hive has recently been adapted for use in Rhodesia, and a trough hive—with sloping sides and bars across the top all of the same length—has been used in development programmes in some other African countries (Crane, 1971).

It is sometimes taken for granted that honey, 'man's first sweetener', was obtained from hives and eaten by primitive peoples in large quantities, being the precursor of cane or beet sugar. This is unlikely to be true. Honey was, and is, a useful food source in times of famine in any community (Crosse-Upcott, 1958). Among many peoples it has been highly valued and regarded as a magico-religious substance. Morris (1967) regards a 'strong positive response to sweet-tasting objects' as a particular characteristic of the higher primates (19.12). But there is much evidence that in ancient civilizations (19.3) honey was regarded as a medicine rather than as a food. And in many areas, especially in Africa, honey was used not to eat, but to produce a drink that contained alcohol (16.43)—from the fermentation of the honey sugars (16.2), probably aided by other materials in the combs such as pollen and yeasts. Seyffert (1930) has made a study of the subject, and the map in Figure 19.23/4 is from his book. During a visit to Ethiopia in 1973 I visited the country's largest honey market, in Addis Ababa, which was taken up with stacks of bags, made of two goat skins sewn together and holding 50 kg or so of broken honey combs and their contents, harvested from primitive hives. The honey would be sold not for eating, but for making *tej*, a light drink with an alcohol content varying from say 8 to 13%, which is drunk throughout Ethiopia (Rossi, 1959, Rea, 1974).

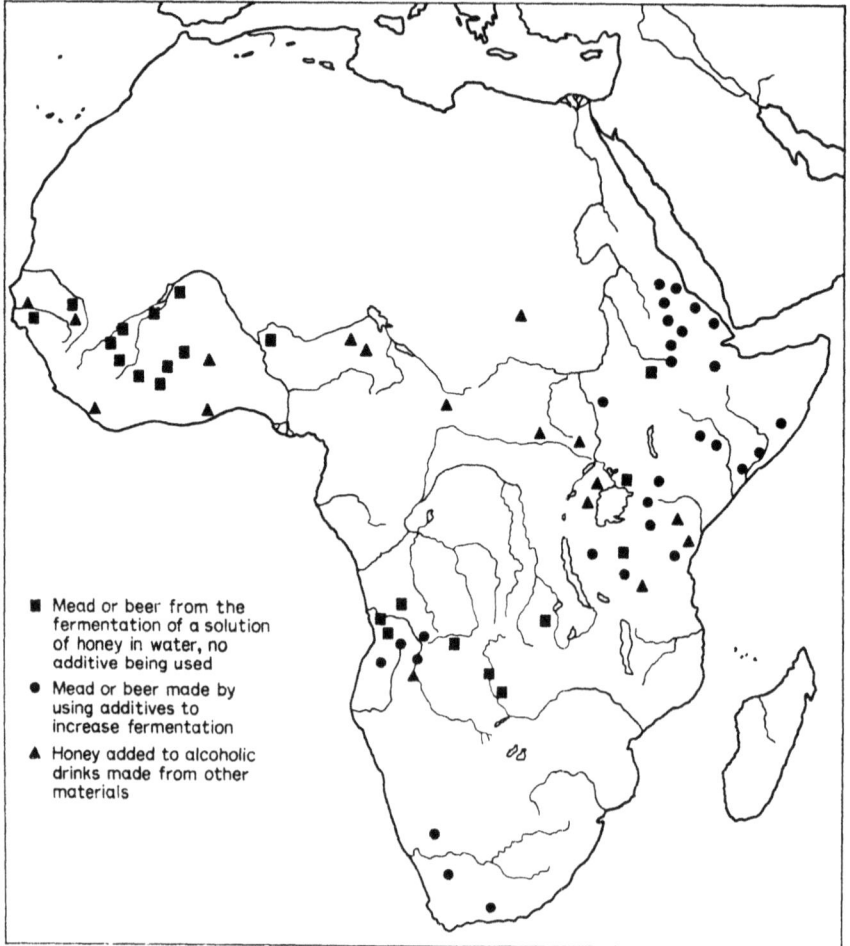

Figure 19.23/4 Map of Africa showing the distribution of different types of alcoholic drinks made with honey; *see also* 16.43.   (Seyffert, 1930)

The human craving for sweetness (Nordsiek, 1972) is thus by no means universal. It has been largely developed during the present century, among the so-called advanced peoples of the world, and is an outcome of the supersedure of honey by cane sugar (19.42). The world production of sugar increased nearly sevenfold, from about 8 million to 54 million tons, between 1900 and 1964 (Pyke, 1972).

19.3 HONEY IN ANCIENT CIVILIZATIONS

So far the evidence used has been mainly pictorial, or derived from recent studies on animals and primitive peoples. For the ancient civilizations there is a wide range of scattered graphic records, and many more will doubtless be discovered and interpreted as time goes on. We begin to get information about various uses to which honey was put in diet, pharmacy, and ritual, the varieties that were preferred in different regions, laws relating to honey, and standards imposed for its treatment and sale.

Much evidence points to the widespread importance and magico-religious significance of honey in the many ancient civilizations that rose and fell and were superseded by others. The sections 19.31–19.39 are used to present a little of what is known, with emphasis on recent findings rather than what is already available in earlier books. For those who want to read more widely, the following are recommended:

The sacred bee by Hilda Ransome (1937)
The concept of the sacredness of bees, honey and wax in Christian popular tradition by A. E. Fife (1939)
Beekeeping in antiquity by H. M. Fraser (2nd ed. 1951)
Geschichte der Bienenzucht by J. G. Bessler (1886)
Materialien zur Geschichte der Entomologie bis Linné by F. S. Bodenheimer (1928/29)
Traité de biologie de l'abeille ed. R. Chauvin, Vol. V (1968c)
Food in antiquity by D. & P. Brothwell (1969)
Honey and health by G. D. Beck (1938)—not the 2nd edition (1944)

There are pitfalls for those searching contemporary written records and their translations, in that the word rendered as 'honey' may have various meanings: honey from bees; honeydew or its crystallized form known as manna (but 'manna' also has other meanings); a concoction made from dates, or some other fruit syrup; or in fact anything sweet—and nectar similarly. Unless the context makes clear a connection with hives, bees, or honeycomb, caution is warranted.

There is one set of pictorial records relating to honey that may perhaps be a link between the earliest known rock painting of honey hunting (Plate 18) and the civilizations in the Middle East. A complex of buildings which follows a systematic plan has recently been excavated at Çatal Hüyük in Anatolia (Mellaart, 1963, see also Brentjes, 1964); and among the many decorations are several painted designs that Mellaart believes depict honey-comb and brood cells, and bees foraging on flowers. These have been assigned about the same date as the rock painting, and were thus made several thousand years before the earliest written records concerning honey (19.31).

19.31 The early civilizations of Mesopotamia and Anatolia

History begins at Sumer is the title of a book by Kramer (1958). Around 5000 BC, Sumerians from Central Asia settled in the plain by the Tigris and Euphrates, in a region known then as Sumer but after 2000 BC as Babylonia. The Sumerian language was already in use in 5000 BC, and by 4000 BC it was being written on clay tablets; this is the earliest writing known.

Two Sumerian passages that still survive are the first known written references to honey. One is on a tablet giving recipes for drugs and ointments (Figure 19.31/1—*see* Plate 22), which has now been translated by Kramer & Levey (*Journal of the American Medical Association*, 1954). It was found at Nippur in Iraq, and dates from the Third Dynasty of Ur, around 2100–2000 BC. Among the directions for prescriptions is the following: 'Grind to a powder river dust . . . and [word destroyed], then knead it in water and honey, and let [plain] oil and hot cedar oil be spread over it'. The other passage, quoted by Kramer (1958), is perhaps the earliest recorded love song. It celebrates the Sumerian king Shu-Sin and his bride, and includes the antiphonal lines:

'Bridegroom, dear to my heart,
  Goodly is your beauty, honeysweet,
Lion, dear to my heart,
  Goodly is your beauty, honeysweet.'

The Babylonians continued to use honey in medicine; they also used it in rituals (Leibovici, 1968*b*), and it is referred to in the Code of Hammurabi, around 1800 BC (Deerr, 1949). The Hittites had then already attacked and plundered Babylon; by 1500 BC the Sumerian language, although still written, was no longer spoken, and by 1400 BC the Hittite Empire was the strongest kingdom in western Asia. What is known as the Hittite Code (about 1300 BC) was discovered at Boğasköy in Anatolia in 1906–7. Bodenheimer (1942) discusses in detail the translation of passages on bees and honey. Among the laws regulating prices, paragraph 181 deals with honey: 'one tub (*zimmittani*, value unknown) of honey costs 1 shekel of silver'; a tub of butter or lard had the same price—and so did a sheep. Two paragraphs relate to the theft of bees and hives:

'91. The theft of a bee-swarm formerly carried the fine of 1 pound of silver. Now the thief is fined 5 shekels of silver only. His house is the guarantee for the payment of the fine.

92. The fine for the theft of 2 or 3 bee-hives was formerly exposure to bee-stings. Now the thief pays only 6 shekels of silver. The theft of an empty bee-hive is punishable by a fine of 3 shekels of silver.'

Here we have clear evidence of beekeeping, a hive—literally a 'bee house'—being either occupied or empty. Moreover the reference to earlier, higher fines suggests that bees had been kept in hives for a long time. Since the fine for stealing a domestic animal was usually the restitution of six animals, the 5-shekel fine means that a swarm was worth roughly a shekel—as much as a sheep. A hive was worth over half as much, so it was probably a specially constructed container (Bodenheimer, 1942).

The Euphrates river rises in eastern Anatolia, but it was another seven centuries before hives were used for honey production lower down the Euphrates valley, in Sukhi. An inscription of about 630 BC reads: 'Bees which collect honey, which no man had seen since the time of my fathers and forefathers, nor had brought to the land of Sukhi, I brought from the mountains of the Khabkha tribe, and I put them in the garden of Gabbari-ibri. . . . They gather wax and honey. The preparation of wax and honey I understand and the gardeners understand it. Whoso may arise later may ask the oldest in the land: "Is it true that Shamash-resh-usur, Governor of Sukhi, brought the honey bee to the land of Sukhi?" '

19.32 Egypt

The best-known early civilization developed in Egypt, and a number of pictorial records are available to show how honey was obtained from hives, and how it was used. Around 3400 BC the king of Upper Egypt joined his kingdom to Lower Egypt. From the First Dynasty (3200 BC) the bee was the sign of the king of Lower Egypt, and drawings and engravings of a single stylized bee are frequently found in tombs and on statues, wall paintings, etc. The coffin of Men-kau-ra (Fourth Dynasty, c. 2500 BC) in the British Museum is one of many early examples. The treatment of honey is shown in a relief from the Temple of the Sun built by Ne-user-re of the Fifth Dynasty at Abusir, c. 2400 BC, which Armbruster (1921) has discussed in detail. A much later relief in the tomb of Pa-bu-Sa at Thebes dating from the reign of Psamtek I, around 625–610 BC, shows the hives more clearly (Figure 19.32/1—see Plate 24). Their shape is of interest in comparison with that of the hive of mud, cow dung, and straw from the High Simien in Ethiopia shown in Figure 19.23/2—see Plate 21.

Much honey was used in Ancient Egypt, by the priests for religious rites, for feeding the sacred animals, and in many ceremonials (Figure 19.32/2—see Plate 23). As in other early civilizations, corpses were sometimes preserved in honey (Leibovici, 1968a, Ransome, 1937). The tomb of Rekhmara, vizier of Queen Hatshephut and her stepson King Thothmes III (c. 1450 BC) has a relief showing men baking honey cakes. There is one record that Rameses III made gifts of 20 800 and 1 040 jars of honey for the Nile god, and another 7 050 jars for making cakes. This would represent some 20 tons of honey—a year's harvest from perhaps 5 000 hives. A marriage contract runs: 'I take thee to wife . . . and promise to deliver to thee yearly twelve jars of honey' (Glock, 1891, page 126).

The Ashmolean Museum in Oxford has a papyrus manuscript dating from the Nineteenth Dynasty (c. 1300 BC) in which a scribe complains of the inadequate returns of honey made by two beekeepers. One of them had already been dismissed but was continuing his job; and the scribe urges his superior to take firm action.

Evidence of migratory beekeeping to exploit a series of honey flows (3.34) is provided in a 'petition from beekeepers' to Zenon, an official who lived in the Fayum around 250 BC (Ransome, 1937, page 27). The beekeepers were desperate for donkeys—that had been promised but not sent—to move their hives away from an area due for imminent irrigation flooding. It is often stated that in ancient Egypt hives were moved up the Nile each year at the end of October, when flowering started, and put on rafts which were floated downstream to Cairo, keeping pace with the blossom continuously until February, when the flowering (and the rafts) reached Cairo, where the honey was sold. This practice was described in detail by a French traveller De Maillet in 1740 (Ransome), and may indeed have been customary for many centuries.

Many other details and references to bees and honey in Ancient Egypt can be found in Kuény (1950) and Leclant (1968).

19.33　The Hebrew kingdoms

Jewish writings have received much notice by those interested in bees, especially Bodenheimer (1960). In particular, the description of the promised land 'flowing with milk and honey' has been quoted and elaborated upon.* R. J. Israel (1972) is one of several who have tried to put the biblical references in a reasonable perspective. There are none to beekeeping, but several to 'honey from the rocks' (Psalm 81, 16) or 'wild honey' (Matthew 3, 4), and also to honey-comb—so certainly bees' honey was known. Samson's collection of honey out of the young lion's carcass is comparable with other descriptions of primitive honey hunting (19.21) 'He turned aside to look at the carcass of the lion, and he saw a swarm of bees in it, and honey. He scraped the honey into his hands and went on, eating as he went. When he came to his father and mother, he gave them some, and they ate it' (Judges 14, 8–9). It may have needed more bravery to get the honey than to kill the lion, for the honeybees

* This and other biblical passages are quoted from the *New English Bible* (1970).

native to the region were an aggressive race, referred to in describing tiresome enemies, for instance 'The Amorites . . . like bees they chased you' (Deuteronomy 1, 44).

R. J. Israel suggests that the honey mentioned in the *Bible* and the *Talmud* as a *plentiful* commodity was a preparation made from dates (*see* page 453), whereas 'honey' that was rare and)valuable was honey from bees. Thus Psalm 19, 9–10: 'The Lord's decrees are . . . more to be desired than gold, pure gold in plenty, sweeter than syrup or honey from the comb'.

Blum (1955), who discounts the date-product interpretation, regards the honey sent by Jacob to the ruler of Egypt (Genesis 43, 11) as probably the earliest reference to comb honey: *dvash nachoth*. Another is in 1 Samuel 14, 25–27: 'There was honeycomb in the country-side; but when his men came upon it, dripping with honey though it was, not one of them put his hand to his mouth for fear of the oath [Saul had forbidden eating until nightfall]. But Jonathan had not heard his father lay this solemn prohibition on the people, and he stretched out the stick that was in his hand, dipped the end of it in the honeycomb, put it to his mouth and was refreshed.' This use of a stick is a widespread primitive way of getting honey for immediate eating (cf. 19.12).

It has been stated that the Dead Sea scrolls provide evidence that the Essenes were beekeepers around 100 BC (La Fay, 1956). Allegro (1956) disputes this, and supposes that the honey they ate was from wild colonies; it was certainly bees' honey, since the Zadokite fragments include a passage: 'Let no man defile his soul with any living being or creeping thing by eating of them, from the larvae of bees (in honey) to all the living things that creep in water'. This instruction carries an interesting suggestion of an early use of straining in preparing honey for use.

The *Talmud*, a great Rabbinical thesaurus which grew up between say 0 and AD 600, continued to allow honey, but not bees, to be eaten. Moreover, in the *Talmud*, honey was not considered unclean if left uncovered, as were water, wine, and milk. There are various references to medicinal uses of honey, as well as to beekeeping practices (Springer, 1954).

19.34 China
Although the Chinese is the oldest living civilization in the world, and there are records of silk production in 2640 BC, information about honey is hard to come by. Kellogg (1968) makes the interesting points that a distinct term for honey did not appear in the written Chinese language before the fourth century BC, and that at some earlier period the only form of sweet mentioned was maltose, produced by soaking rice or wheat

grains until they fermented and then cooking them to extract the maltose. In the first century AD honey was listed among the main imports into China from the West; it was bartered for silk in Samarkand. The idea that honey was an imported product is supported by the fact that there are four distinct words for honey in the Chinese language, all of them merely 'sound words' from other languages. One of Kellogg's students discovered a manuscript from AD 232 which referred to honey production:

> 'In the process of rearing wild honeybees for domestication at home, the people make a wooden box in which several holes are bored. Inside of this box they place some honey and catch the bees. These bees, after a day or so, will fly out and bring materials for making honey. The amount made depends upon the prosperity of the year.'

Such boxes were still in use in Foochow, and Kellogg includes photographs of them.

Honey was greatly prized, and used as a medicine; an official who retired in AD 500 was granted from the Emperor a quart of white honey a month as his 'drug of immortality'. Honey was described not by plant source or geographical region, but according to the site of the bees' nest it was obtained from. 'Ground honey' came from nests in the dry soil of the north; in the south the soil is damp, and there 'wood honey' came from nests in wood (but *see* the 1727 reference below). In Szechuan Province nests in bamboos gave 'bamboo honey'. Honey from rock crevices and caves was 'stone honey'. Chang Hwa, in the Tsin Dynasty (AD 200–317) wrote: 'In far off mountains and out-of-the-way countries of the South honey and wax are found. These stick to dangerous and perpendicular cliffs which are impossible to climb. When one wishes to get the honey he must sit in a chair or in a basket which is tied to ropes and lowered by others over the face of the cliff.' Chen Chuan Chi in the Yuan Dynasty (AD 1229–1368) wrote somewhat similarly, and explained how honey was obtained from trees: 'Where the nests are in high trees that cannot be climbed, the people use long poles to pierce the nests and collect the dripping honey in a vessel.' During the same Dynasty it was reported of one new Magistrate in Chekiang Province that he 'opened to all people the privilege of collecting honey from the cliffs of the district, which all former governors had taken as their exclusive right'. On the other hand, another Magistrate 'stopped the people from collecting wild honey as many had met their death by venturing to climb the dangerous cliffs for it'. This information was published by Kellogg & Chung-Chang Tang (1963); it may be compared with that on page 447.

Osborn (1952) quotes Sicien H. Chen of the Chinese Academy as reporting that the earliest authentic record of honey as food can be traced to the later part of the Chow dynasty about 500 BC, and of beekeeping to the early part of the East-Han dynasty (AD 25–150). Ku P'o wrote a short poem in praise of honey in AD 276–324, and honey is known to have been mixed with opium around AD 600–700 (Ransome, 1937). In 1958 Dr. J. Needham drew my attention to the existence of *Fêng Chi*, an essay on beekeeping by Wang Yuan-Chih who lived about AD 1200. Fan Tsung Deh in Taiwan was able to find a copy of *Fêng Chi*, and kindly translated it into English. The instructions for taking honey (by cutting out the combs, as elsewhere) are: 'when cutting away the bees' honey, do not cut out too much, or they will become hungry and weak, but do not cut out too little, or they will become lazy'. Dr. Needham had not come across further relevant information by 1973.

The *Chinese Encyclopaedia* (1727) has about 30 pages on honey, including both bees' honey and wood honey, a term which is here used for sweet juices from dates, sugar cane, and so on (*see* page 453).

Nemoto published a paper in 1961 on honey as recorded in ancient literature of the Orient and the Far East, and discussed various important historical data from the point of view of the relation between culture and pharmacy. His study confirms what we have already seen, that honey is one of the oldest foods, sweets, and drugs used by man.

19.35 Early civilizations in India

The first city communities in the Indus valley were established by perhaps 3000 BC. The *Rig-Veda*, the oldest of the sacred books of India, probably compiled between 3000 and 2000 BC, contains many references to both honey and bees. These and other sources show how old are some of the beliefs about honey that have been held by various Indo-European peoples through succeeding centuries. For instance Ransome (1937) quotes from the *Atharva-Veda* and the *Grihya-Sutra* respectively:

'O Asvins, lords of Brightness, anoint me with the honey of the bee, that I may speak forceful speech among men.'
'I give thee this honey food so that the gods may protect thee, and that thou mayst live a hundred autumns in this world' [birth-rite for a male child].

Honey from bees was used in many rituals and ceremonies, and in pharmacy. We know that there was trade in honey by 1000 BC, because this was expressly forbidden to the Brahmans. But there seems no evidence of the use of hives for producing honey in the writings of ancient India (Canus, 1886, Ransome, 1937).

Joshi & Godbole (1970) have searched the Sanskrit literature of Ayurvedic times for references to honey. Susruta, a famous surgeon *c.* 1400 BC recognized eight varieties of honey, produced by different insects (honeybees, stingless bees, wasps) from various plant sources. Specific medicinal properties were attributed to each variety—some were cooling, others cured asthma and coughs, others cured skin diseases. Charak, a physician who lived a century or two later, also described the different medicinal properties of honeys according to their source, but he was more concerned with their internal use. He regarded freshly collected honey as a tonic, which did not remove phlegm as easily as ripe honey, and which was a mild laxative; ripe honey was slightly constipative. Their approach was not so unlike that of Sir John Hill in his book on honey written 3000 years later (19.513).

19.36 Greece

The famous city states, Sparta, Corinth, Athens, and Thebes, flourished around 900–600 BC. Then came the 'golden age', with the growth of Athenian democracy (600–300 BC) and finally (to 140 BC) the decline of Athens and the spread of Greek civilization to the Near and Middle East in the wake of the armies of Alexander the Great. During this period the foundations of natural history were laid. Bees were studied—along with other animals and plants—for their own interest rather than for their exploitation, and thought was likewise given to the nature of honey.

Aristotle's *Natural history* (344–342 BC) contains a variety of direct observations on honey: the bee collects the juices of flowers with its tongue, and carries them to the hive; bees gather honey from every flower that has a calyx; honey is collected in the stomachs of the bees and regurgitated by them into the cells; when first collected, honey is like water, but in about twenty days it attains consistency; its source (if it is thyme honey) can be recognized from its taste. Aristotle's conclusion was, disappointingly, that honey is deposited from the atmosphere. Nevertheless, some of his observations are interestingly near the truth and, together with those on other aspects of bee life, suggest that he must have watched bees at work on their combs. This could have been so if the movable-comb hive *anastomo cofini*, found in Greece in the 1600s (19.23), was already in use in Aristotle's time. We shall probably never know. Fraser said in 1931 that 'if Aristotle knew these hives he would have seen more of the interior of a hive than the average eighteenth-century English skeppist'. But in 1950 he went much further. 'This hive was evidently in existence in the time of Democritus (*obiit* 361 BC) . . . The advice to sprinkle the insides of the covers of the hives with water and to open the hives at break of day can refer to no other ancient hive, and if, as Armbruster thinks likely, this hive was invented between the days of Homer and Aristotle, this latter's accurate knowledge of the interior of a hive is comprehensible.'

Apart from the possibility that some of the hives used in Ancient Greece had movable combs, we know curiously little about hives in use there, and no contemporary illustrations of them survive, such as those of Ancient Egyptian hives. However, archaeological findings are now coming to light. During excavations at Isthmia, Kardara (1961) found a near-cylindrical clay vessel dating from around 300 BC which may well have been a hive used horizontally; it is 63 cm long and 17–30 cm in diameter, the wider end being the open one. More recently still, in excavations at Vari and Trachones, clay vessels with cylindrical extensions, which are likely to be hives of a more elaborate type, have been found by Jones, Graham & Sackett, with Geroulanos (1973). In general terms these date from 400–200 BC.

These hives may provide a direct link with the earlier Egyptian hives, and with the primitive hives still used in other parts of the eastern Mediterranean—Crete and Malta, for example. Fuller evidence on this intriguing subject, and its implications, will be published elsewhere (Graham & Crane, 1975).

Honey is frequently referred to in the literature of Ancient Greece, and much has been written about these references, by Fraser (1951), Ransome (1937), Moulé (1910), Turnow (1893), and others. The many references in the *Iliad* and the *Odyssey* have been well explored. Ransome quotes from Athenaeus (*c.* AD 230): 'bread and honey were the chief food of the Pythagoreans according to the statement of Aristoxenes, who says that those who eat this for breakfast are free from disease all their lives'. Herodotus (*c.* 482–428 BC) comments on honey made by bees, and other honey made by the skill of man—for instance 'out of the fruit of the tamarisk'. The term honey denoted a wide range of sweet foods (*see* page 453).

Strabo lived from 63 BC to AD 25, and his *Geographica* (republished 1960) has sundry references to honey. That from Hybla is excellent, but that from Hymettus the best. Other honeys mentioned were from Hyrcania and other parts of Asia including Colchis, now in Georgia, U.S.S.R. In Europe, honeys produced in what are now parts of Italy, Spain, and Switzerland were worthy of his comment, and also Thule— which may have been part of Shetland or Norway. Strabo reported that there were reeds in India 'that produced honey' (*see* 19.42), and that the Babylonians 'bury their dead in honey'. This last statement may be based on a confusion with the use of wax for embalming; honey was upon occasion also used for preserving bodies, in Egypt, in some parts of Asia, and probably elsewhere (Forbes, 1966). Forbes, who has collected together much information on honey in the *Hellenic and Hellenistic world*, quoted two descriptions of the quality a good honey should possess. The first is Greek, by Dioscorides (*c.* AD 50):

'That best liked is sweet and sharp, of a fragrant smell, of a pale yellow colour, not liquid but glutinous and firm, and which in the drawing does leap back to the finger. . . . The spring honey is the best, then that of the summer, but that of the winter being thicker is the worst . . .'

The other is Roman, by Pliny (AD 23–79), quoted by Forbes:

'It is always of the best quality where it is stored in the calyces of the best flowers. This takes place at Hymettus and Hybla in the region of Attica and of Sicily, which are sunny localities and on the island of Calydna. But at the start it is honey diluted as it were with water, and on the first day it ferments like must and purifies itself, while in the twentieth day it thickens and then is covered with a thin skin which forms the foam of the actual boiling.' [*Nat. hist.* XI. 11ff]

Plutarch (*c.* AD 100) records a report that, among the treasure Alexander took from the Persians, was 5 000 talents' weight of purple from Hermione which had kept its colours fresh and lively after 190 years' storage, because honey was used in the purple dyes.

19.37   Rome

Rome conquered Italy and some other Mediterranean areas around 400–200 BC, and Greece by 100 BC. By AD 200 the Greco-Roman culture had been spread throughout the Roman Empire, which then included the whole of the Mediterranean region and much of Europe to the north. This Empire declined as barbarian invasions increased during the next 200 years; Rome itself was sacked by Goths and Vandals in the fifth century.

During the Roman expansion, for the first time much was written about beekeeping and honey production, and much of this has survived. The Roman literature has been discussed in some detail by Billiard (1900), Clark (1942), and Fraser (1951). Columella's writings have been dealt with by Becker (1891), his and Pliny's by Klek & Armbruster (1921), and Virgil's by Hérouville (1926/27) and Whitfield (1956); Tilly (1973) has published selections from Varro's works. The general picture one can build up includes owners of bees who were interested in them, observant of them, and able to write about them. If the owner did not himself do the beekeeping work, this was the duty of the *mellarius*, who knew how to handle the bees and to take the honey from them. Virgil, who lived from 70 to 19 BC, is the best known writer on bees, beekeeping and honey, and his fourth *Georgic* is devoted largely to these subjects. The English translation of the *Georgics* by C. Day Lewis (1940) is used below; there are many others, and many Latin editions available.

Although Virgil's fourth *Georgic* opens with the line: 'Next I come to the manna, the heavenly gift of honey', he was far more interested in the bees and their doings than in the honey. He himself was attracted by the scents of different plants, and believed correctly that the bees were too: 'The work goes on like wild fire, the honey smells of thyme'. There is a lovely description of the spring-time development of a colony of bees as the flowers come into bloom (lines 51–63):

... 'When the golden sun has driven winter to ground
And opened up all the leagues of the sky in summer light,
Over the glades and woodlands at once they love to wander
And suck the shining flowers and delicate sip the streams.
Sweet then is their strange delight
As they cherish their children, their nestlings: then with craftsmanship they
Hammer out the fresh wax and mould the tacky honey.
Then, as you watch the swarm bursting from hive and heavenward
Soaring, and floating there on the limpid air of summer—
A vague and wind-warped column of cloud to your wondering eyes:—
Notice them, how they always make for fresh water and leafy
Shelter. Here you shall sprinkle fragrances to their taste—
Crushed balm, honeywort humble.'

Columella (*c*. AD 50) spells out the method for getting honey, which was probably similar to that in other early civilizations:

'But whatever be the number of honeycombs that are harvested, you should make the honey on the same day, while they are still warm. A wickerwork basket or a bag rather loosely woven of fine withies in the shape of an inverted cone, like that through which wine is strained, is hung up in a dark place, and then the honeycombs are heaped in it one by one. But care must be taken that those parts of the waxen cells, which contain either young bees or dirty matter are separated from them, for they have an ill flavour and corrupt the honey with their juice. Then, when the honey has been strained and has flowed down into the basin put underneath to catch it, it is transferred to earthenware vessels which are left open for a few days until the fresh product ceases to ferment; and it must be frequently skimmed with a ladle. Next the fragments of the honeycombs, which have remained in the bag, are handled again and the juice squeezed out of them. What flows from them is honey of the second quality and is stored apart by itself by the more careful people, lest any of the honey of the best flavour should deteriorate by having this brought into contact with it.' [*De re rustica IX* 15.12, *see* Forbes (1966).]

A large Roman honey storage jar could be a *cadus* or an *urceus*. A smaller two-handled jar used for honey is shown in Figure 19.37/1 (*see* Plate 25). In several countries that were once within the Roman Empire (e.g. Italy, Britain, France), a children's game survived until recently—and may indeed still survive—known as the honey-pot game. In one version of this, a child stands with his or her hands under the armpits, making two handles, and two other children swing the 'honey-pot' child in order to weigh it. It has been suggested (Duruz, 1953) that the game was a survival from Roman times, and this could well be so. Opie & Opie (1969), who describe many variants of the honey-pot game, stress the unchanging nature of many children's games.

Recent excavations at the Greek city of Paestum, south of Naples in Italy, uncovered a small building without doors or windows (Kolb, 1973), which contained eight ornate bronze vessels filled with a soft yellow substance said to be honey 'at least 2 500 years old'. (A sample sent to me contained clover pollen grains, and was probably residue from straining a honey-wax mixture, after the application of heat (Inglesent, 1974).)

In Roman cookery honey was added to sweetmeats, pease pottage, and various dressings for root and salad vegetables. The honey was used to correct excessive sharpness in salad dressings and sauces containing vinegar. One such dressing contained honey, oil, vinegar, and powdered mustard. Virgil mentions the use of honey 'to soften your wine's harsh flavour', and this would have been desirable or necessary when the wine was poor. Fresh fruit, vegetables, and meat were sometimes immersed in honey to preserve them (Wilson, 1973).

19.38 Arab-Muslim civilizations

The Arabs were a Semitic race who built up a vast empire between AD 600 and 800. By 732 this empire extended over the Middle East, the whole of North Africa, Spain, and eastwards beyond what is now Iran. The Arabs carried with them the Muslim religion, founded by Muhammad the Prophet who lived from AD 571 to 632. A later writer, Ibn Magih, quotes Muhammad as saying: 'Honey is a remedy for every illness, and the *Koran* is a remedy for all illnesses of the mind, therefore I recommend to you both remedies, the *Koran* and honey.' Sura 16 of the *Koran* is entitled 'The bee'; it records that, in addressing the bee, the Lord spoke of 'houses in the mountains, and in the trees, and of those materials wherewith men build hives for thee' (1892 English translation). So honey was presumably collected from both wild colonies and colonies in hives devised for them. An Arabian hive hewn out from wood (Figure 19.38/1— *see* Plate 25), and a gourd containing honey (Figure 19.52/4a, Plate 30) exemplify what could well have been in use in Muhammad's lifetime. A valuable detailed discussion of bees, honey, and wax in the Muslim world

Figure 19.37/1 (*see page 464*) Romano-British honey jar. Larger two-handled jars have been found in various parts of the Roman Empire, some stamped with the weight of honey they contained (*see* Duruz, 1953). 12 × 9 cm. Compare the Spanish jar in Figure 19.52/4a.   (B.R.A. Collection B61/1)

Figure 19.38/1 (*see page 464*) Hive hewn from wood, in the Radfan mountains of Arabia. In use in the 1960s, but probably similar to hives used for the last 1 000–2 000 years. 116 × 45 × 19 cm.

(B.R.A. Collection B68/16)

Figure 19.41/1 (*see page 468*) Miniature drawing from a thirteenth-century English manuscript, Bodleian 764 (172C/2).
Against a background of gold-leaf and an ornamented blue sky, bees collect nectar from a green and a blue flower. The stylized hive is striped in red and gold.

iud in psalterio secin hebreos. miltii abies
domus eius.

PLATE 25

A	warm winter	stormy summer	corn dear	crops good garden fruit fairly good	honey plentiful	youths will di
B	friendly winter	wet summer	corn dear	great injuries & thunderstorms	grape harvest good	there will be wailing
C	black winter	turbulent summer	wet autumn	crops good much garden fruit	honey plentiful	youths and women will die
D	warm winter	good summer	good autumn	crops good garden fruit	honey plentiful	good grape harvest

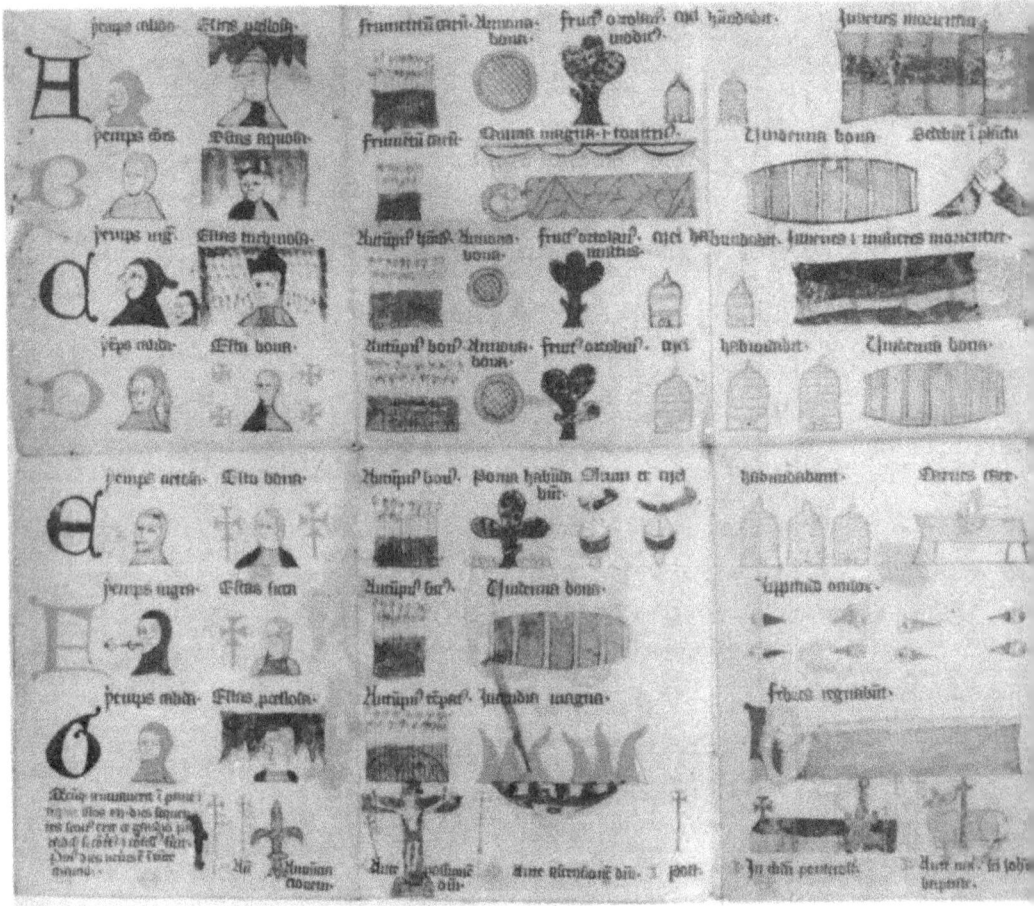

E	windy winter	good summer	good autumn	fruit will be plentiful	oil and honey plentiful	meat is scarc
F	black winter	dry summer	dry autumn	grape harvest good	soreness of eyes	
G	warm winter	stormy summer	mild autumn	great fires	fevers will prevail	

PLATE 26

disease of livestock	there will be fights	great robberies		some news about kings	
kings will die		bees will die	sore eyes	soldiers will fight	many demons
livestock killed by lightning	great fires		danger on the sea		they will work at the grape harvest
women will die in childbed		hay in danger		some news about kings	

livestock will die and perish		crops will vary dear flax	there will be peace	rivers will flood
much honey, i.e. abundant	children will die	there will be starvation in various places		earthquakes
honey plentiful	high price of flax [i.e. flax scarce]	old men will die		much hay

Figure 19.41/3 (see page 470)   English illuminated calendar drawn around 1370 (Bodleian Rawl. 937), predicting honey harvests with other crops and events.

Figure 19.41/2 (see page 469)   Detail from an Exultet Roll showing Italian bee-keepers *c.* AD 1000 harvesting honeycombs from a long wooden hive and collecting them in a bowl, which possibly served as a strainer since another vessel is held below it.   (photo: Vatican Library)

Figure 19.41/4 (see page 471)   Wall painting in Ørslev church, Denmark, made around 1350, showing two princesses welcoming guests with a drink such as mead.

PLATE 27

Figure 19.52/1 (see page 481) Vertical section cut through a Langstroth movable-frame hive. The 'bee-space' can be seen above and on the right of the frames, which are separated by a piece of metal queen-excluder. (photo: British Museum (Natural History))

Figure 19.52/2 (see page 482) Early centrifugal honey extractor, to hold a single frame. The container with the frame in it is spun rapidly round the rod. (photo: Rev. G. H. Hewison, B.R.A. Collection)

Figure 19.512/1 (see page 478) A Stewarton hive 'in full work'. (From the Rev. E. Bartram's *The Stewarton, the hive of the busy man* (1861))

Figure 19.51/2 (see page 476) Bee hunter's box used for 'hunting' 250 bee trees in Massachusetts, U.S.A., between 1915 and 1953. A bee alighting on the honeycomb (*left*) was trapped by closing the lid over it; it was finally released in the place selected, by manipulating the gates and lid on the right. (B.R.A. Collection B53/135)

PLATE 28

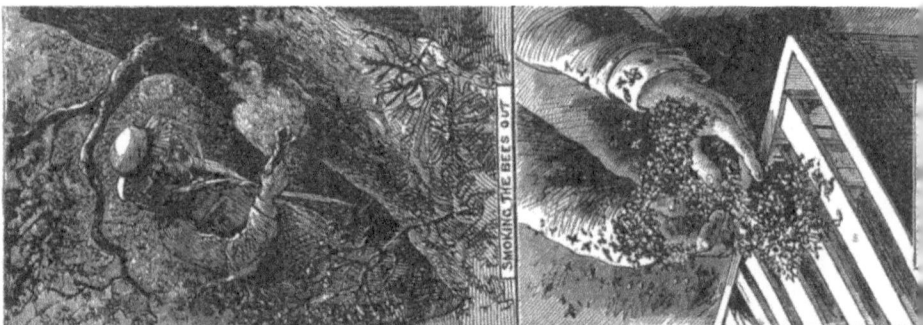

SMOKING THE BEES OUT

FOLLOWING A BEE TO THE HIVE

Figure 19.51/1 (see page 476) Part of a sequence showing a bee hunt in North America. The scenes show how a bee is followed to the hive, the limb containing the nest sawn off the tree, and the honey gathered into a wooden tub. A meddlesome boy is attacked by the bees; the honey is strained from the combs, and brood combs are transferred to wooden hives.

(from *The Graphic*, 16 April 1881, B.R.A. Collection)

PLATE 29

Figure 19.52/4 (a–d) (see page 486) A selection of honey pots in the B.R.A. Collection. Descriptions run from left to right.

a Gourd of honey purchased in a country market in Arabia in 1968. The triangular opening has been resealed with wax. Diameter 10 cm. (B68/10)
Unglazed pot used for marketing honey in Madrid, Spain, 1961. Height 13 cm. Compare the two-handled Roman honey pot in Figure 19.37/1. (B69/52)

b Traditional Russian carved wooden bear with honey pot 'Natural'nyi med' c. 1930. Height 33 cm. (B71/78)
Heavy black clay pot from Rwanda (1968), used for holding honey combs in the home (a mud hut). The lid is of banana leaves; it is a replacement made in Kenya after the original had been eaten by animals. Diameter 20 cm. (B69/63)

c Small French pharmaceutical pot, faience decorated in pink, 'Miel Roséat', date unknown. Height 7 cm. (B68/56)
French honey pot 'Miel du Gatinais', date unknown. Height 18 cm. Gâtinais is an old region lying between Ile-de-France and Orleans, famous for its sainfoin honey. (B52/57)
English honey pot of yellow slag glass, c. 1830. Height 13 cm. (B69/27)
Similar type to pot on extreme left, decorated in blue. Height 8 cm. (B68/4

d Reproduction (1972) of a silver honey pot designed by Paul Storr in 1800. Height 12 (B73/5). Blue jasper ware honey pot designed by Wedgwood and Sons for R. B. Will in 1959, ornamented with bees and a wreath of orange and clover blossoms, in wh relief. Height 12 cm. (B68/35). Honey pot of Queen's ware, made for R. B. Willson 1952 by Wedgwood and Sons, using a mould made about 1765. Height 11 cm. (B52/

PLATE 30

is given by Fahd (1968), who quotes many references to honey in Arabic prose and poetry.

The influence of the Muslim religion now extends over the Middle East, Iran, Afghanistan, and into Mongolia, and over North Africa to about 10° north of the equator. This has a special significance as far as honey is concerned, in that the religion forbids the use of alcohol. So in Muslim areas honey cannot be put to its traditional use in Africa south of the Sahara—the production of mead or honey beer. The honey produced is therefore eaten, bartered or sold. Interestingly, the Christian areas of Ethiopia remain one of the main strongholds of the use of honey for fermentation (*see* 16.43, also page 452).

In Spain, Muslim influence was maintained by the Moors (Arabs and Berbers from north-west Africa) until their final expulsion in 1492. Honey was a familiar and highly valued commodity there. Ibn-al-Awam, a Moorish writer who lived at Seville in the twelfth century, wrote a book on agriculture, including bees (*Bee World*, 1932). Much of the information in it was derived from Aristotle; but there is also evidence of other Arabic writers on bees whose works have been lost. It was the Arabs, and later the Moors, who 'built the shining bridge which spans the Dark Ages': they passed on to the rest of Europe the learning of Ancient Greece, including what was known about bees and honey, still mainly derived from Aristotle's writings.

Before moving on to honey in mediaeval Europe, a group of early civilizations in the New World must be considered—where honey came from a different group of bees.

19.39 Amero-Indian civilizations

Honeybees are not indigenous to any part of the New World; they were introduced by settlers from Europe. So they did not reach the Americas before the sixteenth or seventeenth centuries. The ancient American civilizations had no *Apis* honey as the ancient Old World civilizations did. However, in the tropics of Central and South America another group of social bees evolved, and these also build nests and store honey. This group, the Meliponinae or stingless bees, has many genera and species, and honey from a number of these has been used by man for a very long time. In some regions honey was a comparatively important food, and in many more it was much used in medicine; Schwarz (1948) gives many details; *see also* 17.3 and Favre (1968).

When Columbus landed on Cuba on his first voyage to the New World in 1492, he noted 'a variety of honey' among the natural assets of the island; Schwarz (1948) believes that this must have been from *Melipona beecheii* var. *fulvipes*, the only stingless bee in Cuba. It may well have been wild honey, but some of the 'sundry kinds' of honey encountered

H.—16

by Hernando Cortés and his conquistadores in Mexico in the 1520s were almost certainly from 'domesticated' bees of the same species. They found a well established honey production, which was surprisingly large in view of the tiny surplus obtained from each colony. The Mayas were the greatest honey producers of all the Central Americans, and around 1570 Bishop Diego de Landa reported that they were paying tributes of beeswax both from their hives and from wild colonies they sought out in the jungle.

In 1562 this same bishop collected and burned hundreds of Maya books in an attempt to stamp out their learning. Among the few Maya manuscripts which he did not destroy, and which can now be seen in the Biblioteca Nacional in Madrid, are the *Codices Troano, Tro-Cortesianus*, and *Mendoza*, which provide interesting evidence of Maya beekeeping before Columbus. Ransome (1937) reproduces some of the hieroglyphs and figures, which show bees, honeycomb, honey jars, and vigorously fermenting mead. The bees had their own gods in the shape of large bees, who looked after them in all their activities. Representations of these gods can be seen in various places. The British Museum has a Totoniac vase decorated with a very benign bee god; in the Museum at Mérida (Willson, 1953) is what appears to be a Maya bee god complete with log hives and honeycomb. The ancient Maya beekeepers held two special feasts. One was in the month *Tzec* (October/November), with the object of increasing the honey crop; this ceremony ended with a great drinking of honey wine. The other was in the month *Mol* (December/January) to persuade the gods to provide flowers for the bees.

Beekeeping with stingless bees was practised in many other parts of the American tropics. Honey and beeswax were valued commodities to the Aztecs, who collected them as tribute from subservient tribes in neighbouring regions. According to Sahagún, the most important medicinal and tonic beverage among the Aztecs contained powdered cacao beans, vanilla, honey, and pepper (Jürgen, 1962). Cortés found honey and beeswax on sale in the Aztec markets in 1519. Several species besides *Melipona beecheii* are known to have been kept in hives in different parts of Mexico, and when honeybees (*Apis mellifera*) were introduced, they too flourished there.

Thus, although America was without honeybees until well after Columbus's voyage in 1492, honey was part of the economy of its ancient civilizations in the tropical and subtropical areas. Moreover, these have a tradition of both honey hunting and beekeeping which goes back without interruption to the time before recorded history—as in Europe. Vellard's book *Une civilisation du miel* (1939) provides a detailed account of the life of the Guayaki Indians in Paraguay as he knew it, which may be indicative of life soon after man first moved into the region. The Guayakis are hunters, but on a lowly level using only arrows (even for fish), and not having traps or nets. They prefer meat when they can get it, but honey is the basis of their diet. Although there might be no other provisions in a camp, Vellard always found honey; one group of fifteen people had seven large vessels, holding at least 40 litres altogether. The honey is pressed out by hand, and carried from camp to camp in special baskets of plaited sedge (*Carex*), made water-tight by coating them with wax from the nests of stingless bees. These vessels are egg-shaped, and are called *gaytü-hü* (water nest). The Guayakis also make pots by mixing clay with wax, but have no baked pottery, and so cannot cook food by boiling. The heaviest of the many types of rope they make is used when climbing trees to reach the bees' nests (cf. Figure 19.21/4, Plate 19, from Africa).

Vellard notes that the Guayakis could not adapt themselves to 'civilization', and their circumstances now are not known to the present author. The bee and honey vocabulary in Guarani, of which the Guayaki language may well be a pure form, was described in 1954 by W. Braun.

There are many folk-stories relating to honey and honey hunting in the stingless-bee region, some of which are remarkably similar to those in the honeybee-regions of the Old World. Lévi-Strauss (1966, 1973) has recently analysed a number of them.

19.4 HONEY IN MEDIAEVAL EUROPE

The uses of honey for food, for drink, for medical and various preservative purposes, and in magico-religious rites, started with primitive man and continued after the ancient civilizations collapsed. Honey was, after all, widely available, palatable, and easily preserved—if not so easily collected from the bees that made it. This section continues the history in Europe, where many contemporary records, both written and pictorial, have survived. In other parts of the Old World (but not yet in the New) honey was also harvested from bees, but change was much slower, and records are scanty.

19.41 Evidence from contemporary records

For the thousand years before printing, we rely largely on manuscript material for depictions of daily life. From the sixth century onwards, increasing numbers of illuminated European manuscripts survive, with monochrome line drawings or with designs and scenes in brilliant colours, made even richer by the application of gold-leaf. The Bodleian Library in Oxford has recently brought together representations of bees in its own collection of manuscripts (*Bee World*, 1971b); a common motif is a group of flowers with bees flying from them to a hive (skep) nearby. One interesting miniature in a thirteenth-century English manuscript (Figure 19.41/1—*see* Plate 25) shows two bees inside the flowers, one deep in a corolla; this may well be the earliest representation of a bee in relation to nectar in a flower, from which honey will be made. (Bee-keeping scenes are few, and those showing honey collection, such as Figure 19.41/2, Plate 27, even fewer.)

A ninth-century Irish manuscript includes a poem about the hermit Marbán, half-brother of King Gúaire of Connacht. As well as giving an indication of the hermit's meatless diet, the poem is rich in appreciation of nature. Bees and honey are there along with the rest—herons, sea-gulls, speckled badgers, and beech mast.

'I have a bothy in the wood—
none knows it save the Lord, my God;
. . . .
but alone I live quite happy.
. . . .
Eggs in clutches, and God gives mast,
   honey, heath-pease;
. . . .
A cup of mead from noble hazel
. . . .
Bees and chafers, gentle humming
. . . .
I thank the Prince who so endows me
   in my bothy.'

The English translation, by James Carney, is quoted by Bieler (1966).

The ancient Irish and Welsh laws provide many references to bees and honey; these laws were unaffected by Roman influence, so the references give glimpses of life in Europe from much earlier times. Bees were in fact chosen as an example to illustrate many early laws, and Roman Law regarded wild bees as *res nullius*, not necessarily owned by the owner of the land they were over or on (Ellis, 1926), as they were in early English and Norman law.

The Ancient Laws of Ireland (Harris *et al.*, 1865–1901) were codified by St. Patrick in AD 438–441. One passage concerns the food of fostered children; fostering was a common practice, not a result of deprivation. All children were fed on 'stirabout'. Children of inferior grades were fed to sufficiency on stirabout of oatmeal on buttermilk or water, eaten with stale butter. Sons of chieftains got stirabout of barley meal on new milk, with fresh butter. But the sons of kings were fed stirabout of wheaten bread upon new milk, taken with honey. 'A man's full meal of honey' was one of the penalties prescribed for stinging by bees. Ireland must have been good bee country at this time: of three types of honey storage vessels used, 'a man could raise the smallest over his head when it was full of honey, the middle one up to his breast, and the largest to his waist only' (Fraser, 1958). There is mention of an Irish ship partly laden with honey, and the trade seems to have been two-way, because at marts set up on the coast of South Wales, 'Norse merchants maintained a brisk trade in Welsh slaves, horses, *honey*, malt, and wheat in exchange for Irish (or Irish-imported) wines, furs, . . . butter and coarse woollen cloth' (Williams, 1941, quoted by Bowen, 1972, and referring to the period after AD 989).

In Wales, honey-rents were quite common, although they were later commuted to cash payments. The Gwentian Code valued bees by the swarm, at 4*d* to 24*d*, and ascribes their worth less to their utility than to the fact 'that bees originated in Paradise, but when man sinned in Eden the bees fled, whence arose the blessing of God upon them, and this blessing made it incumbent for candles made of their wax to be used whenever mass was sung'.

The holiness of bees is exemplified in the hymn used in the Roman church when blessing the paschal candle on Holy Saturday. Various copies still exist of illuminated manuscripts used for the paschal service, made around AD 1000 in Italy, especially at Monte Cassino. These manuscripts are called Exultet rolls, from the first word of the hymn in praise of the candle, and they provide us with almost the only illustrations of beekeeping and honey-getting at that time (*see* Figure 19.41/2—Plate 27). This paschal hymn probably dates from the fifth century (Dockery, 1958), and praises the bees 'who produce posterity, rejoice in offspring, yet retain their virginity'. In regions where the Christian church was established, the widespread keeping of bees and production of honey was, in fact, related to this choice of beeswax for church candles, which in turn was based on a misconception about the method of reproduction in honeybees.*

In England, Domesday records compiled between AD 1000 and 1200 abound in records of bees and honey (Burtt, 1930, Fraser, 1958). Dues were payable in honey, but never in wax, so they would be of pre-Christian origin. They were measured in sextars, from which Fraser

---

* In 1771 Anton Janscha gave a clear account of the mating between a queen and drone, which is necessary before the queen can lay eggs that give rise to workers or other queens. But even in the 1950s there was resistance in some quarters to the concept of anything more promiscuous than a single monogamous union, although by then it had been established beyond doubt that a queen honeybee was likely to mate with ten or more drones within a short space of time.

argues that the honey must have been liquid, but some of it could well have been the general 'scoopings' from a hive, as in 19.21.

A calendar of prognostications compiled in England around 1370 throws interesting light on the importance attached to honey production. the relevant part has recently been published for the first time (Crane, 1972a) and is reproduced here (Figure 19.41/3—see Plate 26): honey is the only crop for which a forecast is made for each of the seven years in the Dominical cycle. The picture one gets is of a hard life for man and beast in the fourteenth century, but of good conditions for the bees in six years out of seven. This could well be a fair assessment.

A housewife's diary written near Chepstow in 1796 (Hughes, 1964) includes a vivid account of killing the bees and harvesting their honey, which could well apply to many previous centuries:

'September 28th
John comes in to say we must take the honey from the bees, so he to the making of sulfur papers, which he do put too near the fire, it flaring up did burne his fingers, thereby he did drop all on my clean hearth stone, and did dance about like a bee in a bottle. I did put some butter on his finger to stop the smarting thereon, but he did make a mighty fuss. Sarah (the maid) did scrape up the sulfur from the hearth stone and cleaned it, but it be stained and do smell very nasty. Later carter's wife did come and make a goodly pile of the papers, and so we now all ready to take the honey on to morrow.

September 30th
We did have a busy time taking the honey from the bees yester night. Me and Sarah and carter's wife did have to do it all, John saying his fingers being very sore from the burns.

Sarah did dig a big hole in the ground for each skep, wherein we did put a sulfur paper which we did set alight, and put the skep of bees on the top, the smell of the sulfur do kill the bees, and so we do get the honey there from. It do grieve me to kill the poor things, being such a waste of good bees, to lie in a great heap at the bottom of the hole when the skep be took off it, but we do want the honey, using a great lot in the house for divers times. Carter's wife did fall backwards and sat in a skep of bees, which did make a great buzzing and did send her youping out of the garden, at which Sarah did laff so heartily to see carter's wife holding up her gown while jumping over the cabbidges, that she did nearly do the same thing, at which I quite helpless to reprove her, laffing myself at the carter's wife spindly legs abobbing up and down among the vegetables. She back anon, with a mighty big nose where a bee had sat on it, and we to the out house with the honey skeps, there to leave them till sure all the bees be gone.

We shall break the honey combs up and hang it up in a clean cotton bag to run it through then we shall strain it divers times and when clear put the pots reddie to use. The wax we do boil many times till it be a nice yaller colour and no bits of black in it, when it can be stored for use for the polishing and harness cleaning.'

This quotation does not include one important subsequent operation: steeping honey-combs in water so that they ferment to make mead. A simple domestic description of this is given by Hartley (1964), and much information will be found in Chapter 16. Mead making was an important industry in mediaeval Europe, and mead was a common drink north of the regions where wine could be grown; Gayre (1948) has written a well documented historical account of mead and related drinks such as metheglin, mulsum, pyment, clarre, hippocras, and bracket. He also describes the horns, mazers, and mether cups out of which it was drunk; Birch (1959) gives further examples. A wall painting in Ørslev church in Denmark made around 1350 (Figure 19.41/4—see Plate 27), shows two princesses welcoming guests with a drink. Højrup (1957) reproduces this mural when referring to a passage from the mythological Edda poem in which an envoy is greeted with a goblet of mead. There are comparable passages in *Beowulf* (AD 700), for instance:

'She to Beowulf
braceleted Queen,
Noble-minded
the mead-bowl bore.'

We have now reached a period where records give some idea of yields of honey, and of its price. Walter of Henley's *Husbandry*, referring to England in the 1400s, says that if honey is taken from a hive every other year, the yield should be 2 gallons from each hive (Lamond, 1890); this would be about 14 lb or 6 kg a year. On the other hand, Harrison's *Description of England* (1587) quotes Olaus Magnus (1555) as saying that 'in Podolia, which is now subject to the King of Poland, their hives are so great and combs so abundant that huge boars [bears] overturning and falling into them are drowned in the honey before they can recover and find the means to come out'.

Records kept in the mid-1500s at Ingatestone Hall in Essex (England) show that the local charcoal-burner who worked in Writtle Great Wood kept bees, and sold the honey to his landlord Sir William Petre at the rate of 2 gallons [13 kg] for 5s. 5d; pigeons were then 18d. the dozen and geese 5d. the piece. The detailed accounts of food purchases, and the menu lists (Emmison, 1964) exemplify the small part that sweetening played in the diet. Until cane sugar, and then beet sugar, were available as a cheap food, the *per capita* consumption was much smaller than the current figures given in 4.27; this subject is discussed further in 19.42.

King Edward I and Queen Eleanor bought honey in 1281 at Conway Fair in Wales; for 1s. 4d. they got so much that a cart was hired to carry the honey away to Rhuddlan Castle. Honey was sold at Conway Fair every year until 1954, when the harvest was so bad there was none to sell (Fraser, 1958).

Records quoted from the British Isles could be matched by others of similar interest from other parts of Europe. For instance account books of hospitals in Frankfurt am Main, Munich, and Augsburg still exist, covering between them the period 1493-1732. They show (Bindley, 1965) that honey was eaten almost only in periods of fasting, but that it was also used liberally for the Christmas baking (*see Lebkuchen* in 15.22).

Contemporary records exist in plenty, although collecting and inter-preting them is a slow process. They could form the basis of an interesting study of honey in mediaeval Europe such as that published in 1971 on honey and wax in Russia during the past thousand years (Galton).

The general picture that emerges is of honey as a specific food and medicine (not yet as a 'sweetening agent') and as a source of alcoholic drinks, widely produced in small apiaries. Colonies from which honey was harvested were killed in the process. To some extent, the level of production of honey was determined by demands for beeswax candles for church and domestic use. The custom of eating many sweetened foods had hardly arisen, and the honey consumption *per capita* would probably have been less than 5–10% of the consumption of sugar today (Table 4.27/1).

Contemporary records of honey yields and prices, over a number of years, are available: e.g. for various German towns, 1493-1732 (Bindley, 1965); for the Rhone valley in Switzerland, 1578-1586 (Bindley, 1967); for France, 1201-1800 (Avenal, quoted by Deerr, 1949). Swiss yields were only about 2 kg per hive. The German prices exemplify the effects of the Thirty Years' War (1618-1648); in 1622 especially there seems to have been a honey famine everywhere, and prices were 2½–8 times those in 1621. Table 19.41/1 shows (A) some of the fluctuations in the price of honey over the years, and (B) the gradual reduction in the cost of white cane sugar imported from the east, from a luxury costing ten times as much as honey to a price nearer parity (19.42).

19.42 The introduction of cane sugar
Sugar—'the honey from reeds'—had been known throughout antiquity as a supplement to the limited production of honey (Forbes, 1966). Sugar-cane seems to have been introduced to southern China around

200 BC. In India it was readily available; the cane was sucked or chewed, and later—after perhaps AD 300—extracted with water. The Ancient Greeks and Romans knew of the existence of cane sugar. Dioscorides called it 'a kind of concreted honey . . . like in consistence to salt, and

Table 19.41/1

A. Average prices of honey and white sugar in England, 1400-1600, in pence per pound (Deerr, 1949)

Decade ending	1410	1420	1430	1440	1450	1460	1470	1480	1490	1500
Honey	1·17	1·08	1·08	1·13	1·23	1·13	1·11	1·23	1·15	1·34
White sugar	24·00	——	——	24·00	24·00	14·25	14·83	8·67	6·50	4·29

Decade ending	1510	1520	1530	1540	1550	1560	1570	1580	1590	1600
Honey	1·32	1·27	1·64	——	——	——	——	3·40	——	——
White sugar	3·18	6·21	6·75	7·82	11·02	13·93	9·40	17·23	17·10	19·10

B. Average price of honey in France, 1200-1800, as a percentage of the price of sugar (calculated from Deerr, 1949)

Percentages in brackets relate to English prices during decades within the same period as the French, calculated from Table A.

	%		%
1376–1400	9	1551–1575	34
			(20)
1401–1425	10 (5)	1576–1600	7
1426–1450	9 (5)	1626–1650	13
1451–1475	5 (8)	1651–1675	59
1476–1500	11 (14, 18, 31)	1701–1725	26
1501–1525	9 (42, 20)	1726–1750	30
1526–1550	24 (24)	1776–1800	65

brittle to be broken between the teeth, as salt is. It is good for the belly and the stomach being dissolved in water and so drank. . . .' To Pliny, sugar was a kind of honey, and only used as a medicine.

Deerr (1949) and Baxa & Bruhns (1967) have written extensive treatises on the history of sugar. They indicate the trade routes by which cane-sugar reached Europe from the east, and later from the west, and much more of interest. The first mention of sugar recorded in Germany was in the love lyrics of the troubadours around 1205; in the epic *Parzival* it is referred to as something eaten by the ladies. In England, the Bishop of Hereford's Household Roll for 1289 lists sugar among purchases from dealers in Hereford and Ross-on-Wye; it was classified with spices. In Cologne, where as much as 9 tons were imported annually in the 1490s, small 'cruyde' were fine spices such as nutmegs and cloves, and coarse 'cruyde' included figs, dates, rice, and sugar; 'cruyde' denoted condiments. Sugar was regarded as a medicine as well as a condiment; but there were recommendations that it should not be used much except in case of illness, when small amounts might be beneficial. In succeeding centuries sugar is mentioned more and more frequently, in poetry, in lists of commodities, and in various other documents. But for many centuries it remained an expensive foreign delicacy, and until well after 1500 it cost very much more than honey, in France at any rate (Table 19.41/1).

Morus (1954) makes the interesting point that the shift from honey to cane sugar was the first recorded supersedure of an animal product by a vegetable product. It was by no means the last, and many plant products have in turn been superseded by mineral ones—for instance, fats, fabrics, containers.

The extent of the dislodgement of honey by sugar can be gauged by present total world production. The very rapid increase in the consumption of cane sugar is a recent phenomenon; it is deprecated by many, such as Yudkin (1972), who reports a 50-fold increase in the last 150 years and quotes the following figures for annual world production, in million tons:

1850	1890	1900	1950	1972
$1\frac{1}{2}$	5+	11+	35	70+

Table 4.27/1 gives a comparable recent figure for honey production—0·6 million tons, about 1% of the sugar production. For the U.K., which has better long-term statistics than any other country, annual sugar consumption just over 200 years ago was around 2 kg *per capita*; now it is over 50 kg. A level of 2 kg could well be similar to honey consumption in the Middle Ages. It is commonly said that honey was then 'the only sweetener' for most people, but sweetness was not at all a common characteristic of foods. The general consumption of sweet foods and drinks in many countries is a concomitant of the growth of the sugar industry, and has little to do with honey. Papers by Nicol (1971) and Nordsiek (1972), as well as Yudkin, provide further reading.

19.5 HONEY FROM 1500 TO THE PRESENT DAY

In England, the dissolution of the monasteries in 1537 reduced the demand for beeswax candles, and hence indirectly the production of honey (19.41). Sugar was still much more expensive than honey, but gradually the price differential decreased, until sugar was no longer a costly spice or medicine, and could be used as a foodstuff. For several more centuries honey was produced in home-made hives, at almost no cost, among the rural population. And most of the population were rural until the industrial revolution drove them to the towns. Sugar was then more easily transported than honey to the growing urban populations. So the roles of honey and sugar were exchanged: sugar—instead of honey —became the cheap, easily obtained sweetener, and honey—instead of sugar—became the exotic, more expensive and desirable luxury. The speed of this change varied greatly from place to place.

Throughout Europe there were corresponding changes, with variations, their speed likewise varying within each country, as well as between countries. In some areas local production of sugar from beet accelerated the change to sugar. In some others, ease of production of honey and other factors kept down the price of honey to about the same as that of sugar even to the present century; it was true in Spain in 1960 (Crane, 1961). By and large, however, consumption of honey as a food is now a feature of affluent countries and, within them, of the higher income groups—a complete reversal of the situation in the Middle Ages (19.41).

19.51 Honey in the New World as well as the Old: 1500-1850

The honeybee belonged to the Old World—to Europe, Africa, and Asia. Prior to 1500 there were neither wild colonies nor hives of honeybees in the Americas, Australia, and New Zealand. But, like the dog, the honeybee had accompanied man on most of his major migrations, and some of the early settlers in each part of the New World took hives of bees with them.

19.511 *Honey production moves to the New World*

Unfortunately early records of the establishment of honeybees in the Americas are scanty. The east coast of North America was their first landfall: in 1622 they were in Virginia, and by 1648 honey was abundant there. In 1640 a municipal apiary was established in Newbury, Massachusetts; since the expert in charge of the apiary became the town's first pauper, the honey yields were probably poorer than further south. In 1644 honeybees were recorded in Boston, Massachusetts, imported from England; *see* Adams, 1921.

Once a colony of bees had survived the Atlantic sea passage, the bees could themselves colonize any area with adequate nesting sites and bee forage, where the winter was not too long or harsh. The bees often moved inland ahead of the settlers, especially towards the south and west; a hurricane is reputed to have carried them over the Allegheny Mountains. By the American Indians, these 'white man's flies' were justifiably dreaded as heralding the arrival of pale-faced intruders; white clover, which spread similarly, was known as the 'white man's foot':

> 'Whereso'er they move, before them
> Swarms the stinging fly, the Ahmo,
> Swarms the Bee, the honey-maker;
> Wheresoe'er they tread, beneath them
> Springs a flower unknown among us,
> Springs the White Man's Foot in blossom.'
>
> *Hiawatha* by H. W. Longfellow (1855)

Washington Irving's *Tour on the Prairie*, published in 1832, includes the story of a bee hunt which throws light on the situation at that time; it was reprinted in Cotton's book (1842). 'It is surprising in what countless swarms the Bees have overspread the far West within but a moderate number of years. The Indians consider them the harbinger of the white man, as the buffalo is of the red man; and say that, in proportion as the Bee advances the Indian and the buffalo retire.' Honeybees were taken to the west coast of North America in the 1850s, both in ships and overland at the rear of covered wagons; honey crops in the west surpassed even those further east.

Bees had probably become fairly common throughout the eastern part of North America by 1800 (Nelson, 1967). In all but the harshest north, the settlers thus had a supply of honey. They kept their bees in hives, usually hollowed tree trunks or constructed from four planks of wood nailed together. In the many areas of rich bee country, wild colonies in trees (or in caves or buildings) could accumulate large honey reserves. Honey hunting was a profitable undertaking, and there were many exciting stories of the chase and huge finds of honey (Figure 19.51/1— see Plate 29). The nests were located by finding a foraging bee, trapping it (Figure 19.51/2—see Plate 28), releasing it, and watching its line of flight. Releasing bees at other points served to give several 'lines', which together located the nest. Edgell (1949) and Slater (1969) have published books on this type of honey hunting.

As and when honeybees were introduced in Central America, they largely supplanted the stingless bees as major honey producers. Honey bees were already in Florida in 1763, when Spain ceded its land there to England, and refugees took them from Florida to Cuba. There the colonies 'multiplied with a celerity which appeared incredible. The hives yield four crops every year and the swarms succeed each other without interruption'. This passage is quoted by Beck (1938) from a report by M. Montelle.

In Cuba, Barbados, and other sugar-growing areas, the planters waged war on the bees, which foraged on the cut cane and in the sugar refineries. They collected large honey crops from indigenous plants, some of it being of very good quality, and extended their territory rapidly.

We now know that honeybees were not introduced to Brazil until 1839 (Nogueira-Neto, 1962); they are said to have reached Chile and Peru in 1857 (Beck). Dates quoted for the introduction and spread of honeybees into the various countries of Central and South America are often at variance. The whole story, which is of considerable environmental and economic interest, still needs unravelling. Meanwhile the first honeybees had been landed in Australia in 1822 (Morgan, 1956–58), and W. C. Cotton took the first consignment to New Zealand in 1842. Cotton gives detailed directions for transporting bees on the long sea voyage from England.

These countries, also, gave wonderfully rich honey yields, and the foundations were laid for the immense honey production that followed. It is little more than a hundred years since the nectar of all five continents has been utilized for honey production.

19.512 *Other developments*

In Europe, the period 1500–1850 was characterized by a considerable growth of knowledge about the fundamental facts of the life of bees, and by modest developments in beekeeping techniques, some of which gave better honey and also avoided killing the bees to get it. In particular, various types of extension to a hive were devised to form a separate honey-chamber, where the queen was unlikely to lay eggs, so that the honey was uncontaminated by brood rearing. A straw skep could be extended *at the top* by a smaller skep (a cap) or a glass jar (a bell), above a hole. Alternatively an 'eke' was placed *under* the skep; this was a straw cylinder of the same diameter as the skep and say 10 cm high. The bees built new comb down into the eke, which was sliced off with a long knife. Hives made of wooden boards came into use, of many designs; some were collateral, with boxes *at the sides* for honey storage. As with caps, the bees' access to these boxes was through holes small enough to deter the queen from entering them.

Honey-getting was a laborious and messy affair for the beekeeper, little changed from Roman times. Having driven the bees from the honey-chamber or the hive, he next cut the combs out of it. John Hill's book published in 1759 (*see* 19.513) sets this out neatly, in words not very different from Columella's (19.37). 'The virgin Honey being such as runs out of it self, when it is set in a position for that purpose; and the common Honey such as is pressed out of the combs with violence. This also is of two kinds, some having been pressed without heat and some with: of these the former is by far the better; but both are much inferior to the pure virgin kind' (cf. Section 9.3).

In certain areas, the more affluent customers would buy the complete honey-chamber—straw cap or glass bell. The *Gardeners' Chronicle* for 16 August 1856 reported, for instance: 'There is now to be seen in the shop window of Mr. Cory, of Trumpington Street, Cambridge, a glass of honey containing the extraordinary weight of 96 lbs. net. The glass is hexagonal, about two feet six inches high, and the diameter of a common cottage hive. The empty glass was placed upon the top of a strong stock, and it was filled perfectly with the finest virgin honey in seven weeks. Not a speck or blemish is visible in the combs from top to bottom.' Systems of this sort saved the beekeeper much work, and there were various attempts to design special honey-chambers that could be sold as comb already packaged. One of the most successful and elegant systems was that of Robert Kerr, a Scottish cabinet maker of Stewarton in Ayrshire, who lived from 1755 to 1840. He devised an ingenious hive, which became known as the Stewarton (Struthers, 1951). It consisted of tiered octagonal boxes (Figure 19.512/1—*see* Plate 28) each being fitted with fixed top-bars and intermediate wooden slides. A honey-super was sold with its contents of say 20 pounds [9 kg] of honey. Stewarton supers won many prizes at honey shows; in one class at the First International Honey Show at the Crystal Palace in London in 1874, six of them shared the first prize, their owners receiving 6s. 8d. each—not a large reward for sending the honey 400 miles from Scotland.

19.513   *The first book on honey*
The first book on honey—and the only one in English until the present century—was published in London in 1759 by Sir John Hill, a Covent Garden apothecary. Its title was *The Virtues of Honey in Preventing many of the worst Disorders, and in the Certain Cure of Several others. . . . Direction of the Manner of taking it for the Cure of Consumptions: To which is prefix'd An account of the Origin and Nature of Honey, its various kinds, English and Foreign. . . .*

The author shows much salesmanship, and makes his statements with confidence, whether they are true or false. In Chapter 1, 'Honey, tho' we owe it to the bee, is originally a vegetable substance. . . . The quantity each flower affords is so very small, that human industry never cou'd collect it; but these little insects are indefatigable: their numbers make the general store considerable, and we are supplied by them for all our occasions. The Bee makes no change in the nature of the Honey: such as she receives it from the flower, such she delivers it to the Hive; and the thing is the same in all respects. . . .' Chapter II: 'Besides what we have thus of our own produce [English honey], there is very fine Honey of different kinds brought from abroad, in general the Honey which is commonly sold is so bad; partly from a dishonest mixture of flour and other ingredients, and partly from the ill manner wherein it is procured from the combs, that it wou'd be advisable for every one who depends upon its virtues either to settle a correspondence with some honest person who keep Bees; or to purchase that which is brought from abroad.' 'The spring Honey is vastly best, and shou'd be the only kind used as a medicine. The Bees are in their full vigour when they collect this, and the first flowers of the year from which they gather it, are in their glory.' Again in Chapter III: 'None exceeds the pure spring Honey of England. . . . This Honey will be of the thickness of a syrup, no more; and of a very pale amber colour, quite clear, of a fragrant smell, and in taste of a delicate sweetness with a sharp or biting quality upon the tongue.' Narbonne honey also receives high praise; it is collected in 'the beginning of August, when the lavender has done flowering.' It is equated to Hymettian honey, the favourite of Ancient Greece. Then 'The second Honey in estimation among the ancients, was that of Hybla . . . we find it had a singular fragrance, not accute and piercing as the Hymettian, but soft and delicate.' Wild thyme, the source of Hybla honey 'we have also wild in England. . . . On the left side of the road leading from Denham to Rickmansworth, a little more than a mile from the house of the late Sir George Hill, there is a dell or hollow just within the gate of a field, where gravel many years since was dug: the whole surface is now covered with wild Thyme, which we call Mother of Thyme, whose fragrance scents the evening air to a great distance. Half a quarter of a mile off, there is a farm house, where Bees are kept; and they have an easy and quiet passage over the fields to this dell. The Honey of their hives is perfectly Hyblaean: paler than any I have seen of English produce, and of a delicate sweetness, heightened by this mild aromatick flavour.'

Hill gives full directions for using honey to cure the gravel and stone, asthmas, coughs, hoarseness, and a tough morning phlegm. Taking honey is only part of the regime which, for consumption, includes 2–3 hours' riding every day, gentle purging, and issues (blood-letting), as necessary.

Another London honey promoter was Richard Hoy, whose Honey Warehouse was at 175 Piccadilly, a house near Fortnum and Mason's shop which still sells honey in great variety. In 1788 Hoy published a four-page leaflet, of which only one copy is known, giving *Proper directions how to manage Bees in Hoy's Octagon Box Bee-hives*; Plate 15 shows the introductory engraving. Like John Evelyn in the seventeenth century and Robert Kerr in the nineteenth, Hoy regarded an octagonal shape as especially suitable for bees. Hoy was knowledgeable; his bees came home 'with yellow Bee-bread round their Thighs, which they live on during the Winter; the Honey they carry Inside of them'. He was both a realist and a romantic: 'At the time of Year when the Lime-trees begin to blow, which is about the last ten Days in June, until the 18th or 20th of July . . . that is the Season when the Bees lay in their Store of Honey; at this Time you will see two or three Bees standing before the Mouth of their Hive, drumming their Wings, and sounding their Trumps for Joy of their Harvest coming Home'. The trumpeter bees would be fanners, creating a current of air through the hive and thus helping to ripen the honey.

Chapter XI of Hill's book strikes a note in accord with today's concern with land utilization and environmental resources. It is headed 'Of the quantity of Honey that might be made in England', and discusses the 'quantity annually lost, by neglect among us. Every flower yields Honey, and every spot of ground throughout the kingdom produces these; tho' 'tis in so few places that we raise the industrious insects, that would amass it for our service. Farmers know the profit of bees is very considerable. . . . Wherever bees are not kept nature has given Honey in the flower in vain; it perishes unused, and this is plainly thro' the whole kingdom, except a few small spots. . . . We import a great deal, and it is sold at a considerable price, while there is prodigious waste at home by our neglect. Indeed the price, the excellence, and the possible quantity, all plead for its becoming a more general concern.'

A similar plea for the greater use of bee forage for honey production was made nearly a century earlier by John Evelyn, the English diarist. His great work on gardening, *Elysium Brittanicum*, was written between 1650 and 1700 and ran to 900 manuscript pages. The section on bees is the only part that has been published (in 1966), and here Evelyn says 'I never kept 20 stalls [hives], and usually take but half; yet do I value my Wax and Honey worth 20 nobles at the least. Now if he that is valued but as the tenth part of a parish, at most, can make so much, what may the rest? What may the County? What may the Nation?' Evelyn might well have added 'What may the World?', if he could have foreseen events and developments two centuries later, as set out in the next section.

19.52   The revolution in honey production techniques: 1850-1900
By 1850, honey was being produced by bees, and harvested by man, over almost the whole world. The only large area of land that still lacked the bees to produce honey was Siberia, which was not settled until the present century; it now includes some of the best honey country in the U.S.S.R.

Since the 1600s, beekeepers in various countries—Evelyn among them—had been thinking and experimenting, trying to devise hives and fitments that would allow more control over the bees, and easier removal and handling of the honey combs. This is not the place to relate in detail the story which culminated in 1851; it has been done elsewhere, and Johansson & Johansson (1967) have set the record straight in several important ways. What is significant for the history of honey is that L. L. Langstroth in the U.S.A. devised a hive—described in his book *The hive and the honeybee* (1853)—which has since formed the basis of modern beekeeping throughout the world. A tribute by T. W. Cowan after Langstroth's death in 1895 is quoted by the Johanssons: 'The opening of the hive at the top, the perfect interchangeability of the movable combs, and the lateral movement of the frames, have given the bee-keeper the most perfect control over his bees, and have more than justified Langstroth's expectations when he wrote the note in his diary in 1851, that "The use of these frames will, I am persuaded, give a new impetus to the easy and profitable management of bees".'

The Stewarton (19.512) and other hives were opened from the top, but gave little control over the bees. The Greek hive (*anastomo cofini*) gave some control, having movable combs, but the combs hung from bars of different lengths and so were not interchangeable, and they had no frames. The Swiss naturalist François Huber devised a hive with movable frames, but these were hinged and thus had no lateral freedom of move-ment; the hive was useful for observation, but not for honey production.

In Langstroth's movable-frame hive and its successors (Figure 19.52/1 —*see* Plate 28), the top-bar of each frame has a short lug at each end to support the frame as it hangs in the hive box, with a 'bee-space' (about 6 mm) between the frame and the hive sides, and between the frame and other frames in the boxes above and below. The spacing between adjacent frames in the same box is such that, when a frame is fully built with comb, there is also a bee-space between comb surfaces. The bees build parallel combs, attached to the frames (which support them) but not to the hive surfaces; the importance of the support will appear shortly. To ensure that combs were in practice built parallel, and within each frame, a device used in bar-hives was adopted: the underside of the top-bar of the frame was bevelled, and then smeared with beeswax; this encouraged the bees to build their comb down from the bevelled edge.

In 1857 Johannes Mehring in Germany carved wooden plates which could be used to press out beeswax sheets imprinted with hexagons. Such a sheet, which came to be known as 'foundation' (Johansson & Johansson, 1969), could be fitted into a frame. Provided the hexagons were of the right size, the bees accepted them as cell bases for their comb, and they then literally 'drew out' the wax ridges between the hexagons to form part of the cell walls. Many stronger substances, especially plastics, have been tried as substitutes for beeswax in foundation (Johansson & Johansson, 1971), but with only moderate success. Methods were, however, devised for strengthening the foundation by embedding wires across it which were secured to the frames.

The strengthening was necessary because the use of a frame made it possible to get the honey out of the combs by centrifugal force after the wax cappings over the cells had been sliced off with a sharp knife. The first centrifugal honey extractor is credited to Major F. Hruschka in Austria in 1865. Early extractors held one frame and were spun by hand (Figure 19.52/2—*see* Plate 28). Larger models held four frames in a square, or more disposed similarly; the speed of rotation (and of honey extraction) was increased by the use of gears. In the present century electric motors gave higher speeds, and setting the frames radially instead of tangentially enabled 50 or more to be accommodated at one time in a large extractor (*see* 9.24).

The production and sale of 'extracted' honey brought its own problems. Buyers felt they had no assurance that they were buying pure honey, as when they bought honey in the comb. Sometimes their fears were indeed justified, and the quality of other sweet foodstuffs was likewise suspect because of adulteration. Each country in turn was faced with this problem, and legislation was usually introduced to prohibit the sale of a more expensive product adulterated with a cheaper one. Figure 19.52/3 reproduces a petition drawn up in the United States in 1878, by a committee of which Charles Dadant was Chairman; Pellett (1938) says that it went to Congress with more than 30 000 signatures, and that it was referred to the Committee on Ways and Means, but was never reported for action. Some countries, including South Africa, still lack adequate legislation, and adulterated honey may be sold there. Information on current honey legislation in many countries is presented in Chapter 13.

A development that gave the buyer comb honey as the bees made it, but in a much smaller package than the honey-super, was originated in 1857 by J. S. Harbison of California: the 'section honey-box'. The modern production of section honey, and of its successors cut-comb honey and chunk honey, is described in Chapter 11. Thin unwired beeswax foundation is commonly used, but this is not in principle necessary, nor is the use of a movable-frame hive, although section honey post-dated its development. It is, however, necessary to keep the queen out of the honey-chamber, which was much simplified by the use of a 'queen excluder', perfected in 1865 by Abbé Collin in France; the excluder is a flat sheet or grid with slots wide enough for worker bees to pass through easily, but not wide enough for the larger queen.

Between 1850 and 1900 there were also advances in ways of 'subduing' bees to manipulate the frames, and of getting bees out of a honey-chamber which was ready for harvesting. Two significant events, both

To the Honorable Senate and House of Representatives of the United States :

Your petitioners respectfully represent to your honorable body:—

1. That the sweets now in use in the United States, including cane-sugar, maple-sugar, syrups, candies, jellies, honey, etc., are often adulterated with glucose, and sometimes are manufactured entirely of it.

2. That this glucose is manufactured from corn starch, by boiling the starch with sulphuric acid, (oil of vitriol), then mixing with lime. The glucose always retains more or less of sulphuric acid and lime, and sometimes it has copperas, sucrate of lime, etc.

3. That seventeen specimens of common table syrups, were recently examined by R. C. Kedzie, A. M., Professor of Chemistry in the Michigan State Agricultural College, at Lansing. Fifteen of these proved to be made of glucose; one of the fifteen contained 141 grains of sulphuric acid, (oil of vitriol), and 724 grains of lime to the gallon; and another, *which had caused serious sickness in a whole family*, contained 72 grains of sulphuric acid, 28 grains of sulphate of iron, (copperas), and 363 grains of lime to the gallon.

4. That the American people are pre-eminently a sugar-eating people. The consumption of sugar, by each individual in our country, is shown by statistics to be about 40 pounds a year. It is seen at once that the adulterators of sugars and other sweets, not only cheat our people in the quality of what they consume, since glucose contains only from 30 to 40 per cent. of sugar, but injure also the public health, by selling under false names, an article injurious to health.

5. It is as much the right and duty of Congress to enact laws against such frauds in food as it is to enact laws against frauds in money, for if the counterfeiters of money injure the public wealth, the counterfeiters of food injure the public health.

In view of the above facts your petitioners earnestly request your honorable body to decree that the adulteration of sweets, and the sale of such adulterated products, are crimes against the people, and to enact laws for the suppression of this illegal business.

And your petitioners will ever pray.

Figure 19.52/3 Petition to the United States Senate and Congress, 1878   (by courtesy of Professor R. A. Morse)

in the U.S.A., were T. F. Bingham's production of a bellows smoker that could give a controlled stream of cool smoke (1877), and E. C. Porter's sale of a 'bee-escape' (1891). The bee-escape is a metal contrivance inserted in a board that is put below a honey-chamber due for removal from a hive. It has one or more channels through which a bee can pass down out of the honey-chamber, for instance by pushing between two delicate springs; and the device prevents a bee returning up into the honey-chamber. Various types which have no moving parts have been developed in the present century; chemical repellants were, and are, also used (3.35).

The second half of the nineteenth century brought about more developments in methods of exploiting bees to get their honey, and for handling the honey, than the whole period before 1850. North America was the main centre for these developments, partly because the honey potential of the newly pioneered regions there was so great. By contrast, the honey potential of the developed regions of the Old World had been exploited to some extent for many centuries, and the undeveloped regions were not ready for a honey industry. In Australia and New Zealand honeybees had only recently arrived, and the main growth of their honey industry was yet to come.

News of the American developments soon reached European countries. The speed with which they were adopted (or, more usually, adapted) often depended on individual contacts and actions on—or reactions against—new ideas. Two examples must suffice. Developments in continental Europe owed much to Charles Dadant, who emigrated from France to the U.S.A., settling in Illinois in 1863. Dadant's beekeeping experience led him to believe that Langstroth's brood-chamber was too small. He wrote extensively in European bee journals promoting movable-frame hives; as a result, variants of the Dadant hive with its large brood-chamber have been used ever since in many parts of Europe, and in the U.S.S.R. In England, the movable-frame beekeeping was introduced by T. W. Woodbury of Exeter, who devised a movable-frame hive on Langstroth's principles, but with an insulating layer of straw incorporated in the sides of the hive boxes and the cover. This was soon after 1858, when Woodbury received a copy of Langstroth's second (1857) edition; the Bee Research Association has this book, inscribed 'T. W. Woodbury Esq. from L. L. Langstroth'. The hives developed in Britain deviated in various details from Langstroth's, but the Langstroth hive was adopted in most other English-speaking countries and in some others, and it is today the most important hive on a world scale.

Between 1850 and 1900, three interlinked features assumed an increasing importance in honey-producing countries: beekeeping journals, beekeeping associations, and honey shows. A few beekeeping journals and associations had been formed before 1850, but after the introduction of movable-frame beekeeping such institutions had a new and important function: to spread knowledge of the new 'rational beekeeping'. There were many exciting new ideas to pass on, to discuss, and to argue about. Once centrifugally extracted honey was produced, much activity was directed towards setting standards of cleanliness and quality in the finished product, and here the honey shows played an important part.

In Britain, for example, a meeting in 1874, attended by various beekeepers, resolved:

> 'That these gentlemen now present constitute themselves a society to be called "The British Bee-Keepers' Association", whose object should be the encouragement, improvement, and advancement of bee-culture in the United Kingdom, particularly as a means of bettering the condition of cottagers and the agricultural labouring classes, as well as the advocacy of humanity to the industrious labourer—"the Honey Bee".'

The humanity referred to—which had already been advocated by some for many years—consisted in harvesting honey only from a hive (or part of one) after the bees had been removed from it, instead of killing the bees to take the honey, as described in 19.41.*

The first general meeting of the Association was held at the Crystal Palace in September 1874, during the first Great Exhibition of Bees and their Produce, Hives, and Bee-Furniture, arranged at the instigation of the *British Bee Journal* which was started by private enterprise in May 1873. The six Stewarton supers referred to in 19.512 were described as 'the pride of the show' in the *Journal*, which also commented on the 'beauty of the magnificent supers exhibited by T. W. Cowan, Esq., of Horsham'. Cowan's book (1928) describes the excitement of this exhibition, and of many others held in the first fifty years of the British Bee-Keepers' Association. In 1874 the number of cottager competitors was disappointing, but '. . . the cottager had not been instructed, and it was hoped that the clergy would be interested in the movement'. Use of a straw skep supered with a box of sections, which is illustrated, would presumably be within the capabilities of a cottager.

Interest at honey shows soon moved to 'extracted honey'; in 1886, at the Royal Show in Norwich, the Princess of Wales 'was graciously pleased to . . . accept . . . the first prize extracted honey . . . and expressed her desire for some jars covered with wickerwork': the uninteresting

---

* An early method was to 'drive' the bees out of the hive or honey chamber. A skep, say, was inverted, and driving irons were used to fix an empty skep above it at an angle (mouth to mouth); the lower skep was thumped rhythmically and the bees crawled out of it into the upper one.

character of a squat glass honey jar was recognized early. The use of clear glass was nevertheless of great merit in making the contents visible. Competition between jars of honey at shows had a notable effect in raising the standards of cleanliness, until a jar with the smallest speck or bubble of froth could not win a prize. Once this level was reached, it would seem to have been more useful to take advantage of scientific methods of assessing the composition, properties, and quality of honey. But even today, when commercial institutions put much effort into achieving objective methods, the judging of honey at shows is still largely subjective (*Bee World*, 1963).

The sale of liquid honey gave rise to a multiplicity of designs for ornamental honey pots for table use. Most early honey containers had been utilitarian, made from gourds, skins, wood, or earthenware, but ornamental pots certainly predated extracted honey. In England in the 1750s honey tureens were listed in Chelsea sales among porcelain table wares; these would have been for honey in the comb. Hughes (1973) gives an interesting account of English honey pots made between 1750 and 1850; most were in the shape of a coiled-straw skep, of silver or other metal with a glass lining, or of porcelain, earthenware, or pressed glass. Paul Storr was the most famous designer in silver, and in the late 1960s one of his pots was sold for over £900 at a London auction. Hughes' statement that the skep honey pot was on the way out by 1851 is hardly true. Such pots are still on sale today, although their design is in general much cruder than that of the elegant Georgian ones. Some examples are shown in Figure 19.52/4—*see* Plate 30.

By 1900 the change-over to the production of 'extracted' honey from movable-frame hives was largely completed among the more advanced beekeepers in the more advanced countries. But much honey was still produced in fixed-comb hives, even in parts of advanced countries. The comb was sold as such, or cut out and drained or pressed, as in earlier centuries. Two personal reminiscences from England serve as examples. Charles Beer (1973) recalled his start as a beekeeper when he was a boy in the 1890s in Cambridgeshire. 'At the end of the summer, as was the practice in those days, I suffocated the bees with sulphur fumes. The bees had not quite filled the box. I carefully cut out all the combs, placed them on a dish, covered them with a cloth and took them to a head gamekeeper's wife. I asked her three shillings for the lot.' Flora Thompson (1939) writes about Queenie, the old cottage beekeeper in Lark Rise. Queenie 'had one great day every year, when, every autumn, the dealer came to purchase the produce of her beehives. Then, in her pantry doorway, a large muslin bag was suspended to drain the honey from the broken pieces of comb into a large, red pan which stood beneath.'

19.53 Towards modern honey production: 1900 onwards
The history of honey during the present century has largely been told in the preceding chapters. The stage is world-wide, and records become more prolific each decade.

Techniques for honey production (Chapter 3) have become more mechanized and, for this and other reasons, more specialized. Good road systems, and vehicles suitable for carrying hives, led to the development of migratory beekeeping on a scale undreamt of earlier. Thousands of colonies could be moved on from one crop to another during the flowering season. As and when labour became too expensive, mechanical loading devices were adapted which enabled one man to 'manage' a thousand hives or more. The apiary (a truck load of hives) became the unit, not the hive, and the days when the beekeeper 'knew' each of his colonies were past.

Removal of honey from the hives has been facilitated by the use of repellants—and latterly of a forced stream of air—to make bees leave the honey-supers when the honey is to be harvested. Uncapping the combs by hand with a knife became too slow, and power-operated mechanical uncappers were devised. Just as the apiary superseded the hive as the unit, so the full complement of frames in a honey chamber has begun to supersede the single frame, and honey extractors have been designed into which the complete boxes of frames are loaded—uncapping may even be done box by box, the frames being fixed in the boxes.

The above developments have been the concern of large-scale commercial operators who produce the bulk of the honey that appears on the world market. Meanwhile the domestic and side-line beekeepers have taken advantage of such innovations in equipment and techniques as suited small-scale operations. Recently the fact that their honey receives minimal handling and heating has made it especially desirable among a growing section of the public, as explained below.

During the present century honey processing (Chapters 9–12) has been continually elaborated and mechanized; as a result, the packer can now provide liquid or granulated honey with a shelf-life acceptable to supermarkets. (Honey itself will keep almost indefinitely, but liquid honey would not usually stay liquid for long without pre-treatment.) In recent years there has been a widespread swing away from processed foods towards 'natural' ones, and the demand for honey as the bees made it is steadily increasing. Honey in the comb is natural honey *par excellence*; the 'virgin honey' of the Middle Ages, dripped through a cloth bag, has not returned to the market, but unheated and even unstrained honey (entirely clean, but with particles of wax cappings and perhaps some pollen) is in increasing demand.

In earlier decades much effort was put into the large-scale production of honey that could be maintained as a standard type by judicious blending. Such honey is now getting increasing competition from honeys offered according to source of origin. The source may be geographical; Canadian prairies, Alpine meadows, or Mexican jungle, would in principle be possible as source-descriptions of this type. Alternatively, the source may be defined by floral origin—acacia, lavender, heather, orange blossom, clover, etc. This is a much more difficult task if there is to be no misrepresentation. What proportion of a honey must have come from, say, lavender for it to be called lavender honey? There are problems of both analysis (Chapter 7) and legislation (Chapter 13). Also, many plant names have different meanings in different countries. *Acacia* is properly the name of a genus of African trees, but 'acacia' honey exported from Eastern Europe comes from a different tree altogether—*Robinia pseudoacacia*, known in North America not as acacia but as black locust. In Britain clover honey comes from *Trifolium* (usually white clover, *Trifolium repens*), but in North America clover honey may mean honey from legumes in general.

World honey production (Chapter 4) has increased during the century, and the relative importance of different regions has shifted notably. World trade in honey has been through various periods of expansion and depression, as explained in Chapter 14. The situation today is, however, unprecedented. A general shortage of honey began to develop in 1970; this increased the world price by 8%, and there were further sharp rises in 1971. By 1972 the price of honey from all origins had more than doubled in the major importing countries; prices were still rising in 1973 (F.A.O., 1973). The history of honey thus closes on a most satisfactory note for those who work with bees in order to harvest their honey. As for the bees, the message given on the endpapers of this book is as true now as when the first honey hunter robbed a bees' nest: 'so we the bees make honey, but not for ourselves'.

BIBLIOGRAPHY

References are listed in alphabetical order of author(s) and, for the same author(s), in year order. Publications are written in the language of the title quoted, except where otherwise indicated, e.g. *In Russian*. An English translation is added of some titles that are in less common languages.

The letter B at the end of a reference indicates that the publication is in the Library of the Bee Research Association; the letter T indicates that this Library has an unpublished English translation. Numbers such as 252/55 refer to an English summary in the B.R.A. journal *Apicultural Abstracts*; this example is abstract number 252 in the 1955 volume. The letter L (as in 501 L/72) indicates a listing in the journal by title only, or with a short note. *Apicultural Abstracts* has been published since 1950.

Abelhas (1970) Extractor de mel para meliponas. *Abelhas* 13(151): 82 only
[B, 501 L/72

Abramson, E. (1953) Ein Vergleich verschiedener Methoden für die Bestimmung des Wassergehaltes in Honig. *Mitt. Geb. Lebensmittelunters. u. Hyg.* 44(6): 468–471
[B, 252/55

Achert, O. (1912) Über die Inversion von Saccharose durch Bienenhonig. *Z. Unters. Nahr.-u. Genussmittel* 23: 136–139

Adam, Brother (1953) Mead. *Bee Wld* 34: 149–156; *Repr. BRA No. M12* [B, 64/55

Adams, G. W. (1921) Incidents in Massachusetts Colony prior to 1654. *Am. Bee J.* 61(7): 276–278
[B

Adams, S. L. & Niesen, G. V. (1963*a*) Honey beverage and process for making it. *U.S. Pat.* No. 3,100,705

Adams, S. L. & Niesen, G. V. (1963*b*) Beverage from honey and process for making it. *U.S. Pat.* No. 3,100,706

Adcock, D. (1962) The effect of catalase on the inhibine and peroxide values of various honeys. *J. apic. Res* 1: 38–40
[B, 916/64

Aganin, A. V. (1965) [Determination of maximum water content of honey by the method of suspended drops.] *Trudy saratov. zootekh. vet. Inst.* 13: 316–321. *In Russian*
[B, 197/71

Agthe, C. (1951) Über die physiologische Herkunft des Blütennektars. *Ber. schweiz. bot. Ges.* 61: 240–274
[B, 36/52

Aisch, F. (1938) *Ich koche mit Honig* (Innsbruck: Vereinsbuchhandlung und Buchdruckerei AG) 2nd ed.
[B,T

Akahira, Y. & Sakagami, S. F. (1958) Zum gegenwärtigen Zuchtzustand der japanischen Honigbiene in Kyûshû, Süd-Japan. Studien zur japanischen Honigbiene, *Apis indica cerana*, Fabr. II. *Z. Bienenforsch.* 4(5): 87–96 [B, 348/58

Albanese, A. A. & 4 others (1952) Utilization and protein-sparing action of fructose in man. *Metabolism* 1(1): 20–25
[B, 84/54

Alford, D. V. (1971) Egg laying by bumble bee queens at the beginning of colony development. *Bee Wld* 52(1): 11–18; *Repr. BRA No.* M59
[B, 311 L/72

Allchin, B. (1966) *The stone-tipped arrow. Late Stone-age hunters of the tropical Old World* (London: Phoenix House)

Allegro, J. M. (1956) Private communication

Allen, M. Yate (1937) *European bee plants and their pollen* (Alexandria: Bee Kingdom League)　　　　　　　　　　　　　　　　　　　　　　　　　　[B

Alphandéry, E. (1931) *Traité complet d'apiculture* (Paris: Berger-Levrault)　　[B

Amaral, E. (1957) Honey bee activities and honey plants in Brazil. *Am. Bee J.* 97(10): 394–395　　　　　　　　　　　　　　　　　　　　　　　　[B, 291/58

American Honey Institute (1945) *Old favorite recipes* (Madison: American Honey Institute)　　　　　　　　　　　　　　　　　　　　　　　　　　　　[B

American Honey Institute (1947) *New favorite honey recipes* (Madison: American Honey Institute)　　　　　　　　　　　　　　　　　　　　　　　　　[B

Ammon, R. (1949) Der Ursprung der Diastase des Bienenhonigs. *Biochem. Z.* 319: 295–299　　　　　　　　　　　　　　　　　　　　　　　　　　　　[31/51

Anderson, E. J. (1947) Controlled crystallization of honey. *J. econ. Ent.* 40(1): 55–57　　　　　　　　　　　　　　　　　　　　　　　　　　　　　　[B

Anderson, E. J. (1958) Honey candies. *Progr. Rep. Pa agric. Exp. Stn. No.* 186: 4 pp.　　　　　　　　　　　　　　　　　　　　　　　　　　　[B, 5a140/60

Anderson, L. D. & Atkins, E. L., Jr. (1968) Pesticide usage in relation to beekeeping. *A. Rev. Ent.* 13: 213–238　　　　　　　　　　　　　　　[B, 556/69

Anderson, R. H. & Perold, I. S. (1964) Chemical and physical properties of South African honeys. *S. Afr. J. agric. Sci.* 7: 365–374　　　　　　　　[B, 561/66

Andrásfalvy, M. (1969) Quantitative pollen analysis of Hungarian honeys. *Acta agron. hung.* 18(3/4): 297–306　　　　　　　　　　　　　　　　[B, 1038/71

A.O.A.C. (1970) *Official methods of analysis.* 11th ed. See also Horwitz, W. (1960, 1965)

Aoyagi, S. & Oryu, C. (1968) Honeybees and honey. III. Yeasts in honey. *Bull. Fac. Agric. Tamagawa Univ. No.* 7/8: 203–213　　　　　　　　　　[B, 195/72

Apicultura, Madrid (1965) Legislación. Normas reguladoras de las exportaciones de turron. *Apicultura, Madrid* (155): 16–17　　　　　　　[B, see 805 L/69

Apiculture in Western Australia (1964) Honey colour grading. *Apic. W. Aust.* 1: 20–21　　　　　　　　　　　　　　　　　　　　　　　　　　　　[B

Arai, J., Akiyama, K., Sakai, S., Doguchi, M., Suzuki, N. & Ogawa, S. (1960) Some notes on Japanese honey. *Jap. Bee J.* 13(4): 101–107　　　　　[B, 105/61

Argentina. Laws & Statutes (1971) *Código alimentario argentino* (Buenos Aires: Ministerio de Bienestar Social). honey (Art. 782–784), artificial honey (art. 1035), labelling (art. 220–246)

Aristotle. *Natural history.* Oxford translation (1910) ed. D'A. W. Thompson *et al.* (Oxford: University Press)

Armbruster, L. (1921a) Vergleichende Eichversuche an Bienen und Wespen. *Arch. Bienenk.* 3(7): 219–230　　　　　　　　　　　　　　　　　　[B

Armbruster, L. (1921b) Bienenzucht vor 5000 Jahren. *Arch. Bienenk.* 3(1/2): 68–80　　　　　　　　　　　　　　　　　　　　　　　　　　　　[B

Armbruster, L. (1931) Die Biene im Orient: I. Der über 5000 Jahre alte Bienenstand Aegyptens. *Arch. Bienenk.* 12(5/6): 221–273　　　　　　　　[B

Armbruster, L. & Jacobs, J. (1934/35) Pollenformen und Honigherkunft-Bestimmung. *Arch. Bienenk.* 15(6/7): 277–308 (1934); 16(1, 2/3): 17–106 (1935); also *Bücher Arch. Bienenk.* 2: 1–122 (1934/35)　　　　　　　　　　　[B

Armbruster, L. & Oenike, G. (1929) Die Pollenformen als Mittel zur Honigherkunftsbestimmung. *Bücherei für Bienenkunde* 10 (Neümunster: Wachholtz)　[B

Aso, K., Watanabe, T. & Yamao, K. (1960) Studies on honey. I. On the sugar composition of honey. *Tôhoku J. agr. Research* 11: 101–108　　[B, 91/59

Atwal, A. S. & Sharma, O. P. (1968) The introduction of *Apis mellifera* queens into *Apis indica* colonies and the associated behaviour of the two species. *Indian Bee J.* 30(2): 41–56 + 1 plate [B, 585/71]

Auclair, J. L. (1958) Honeydew excretion in the pea aphid, *Acyrthosiphon pisum* (Harr.). *J. Insect Physiol.* 2: 330–337 [B, 825 L/63]

Auclair, J. L. (1959) Feeding and excretion by the pea aphid, *Acyrthosiphon pisum* (Harr.) reared on different varieties of peas. *Ent. exp. appl.* 2: 289 only [B, 139 L/64]

Auclair, J. L. (1963) Aphid feeding and nutrition. *A. Rev. Ent.* 8: 439–490 [B, 473 L/64]

Auerbach, F. & Borries, G. (1924) Die Bestimmung der Trockenmasse echter Honige. *Z. Unters. Nahr.- u. Genussmittel* 48: 272–277

Austin, G. H. (1953) Maintaining high quality in liquid and recrystallized honey. *Can. Bee J.* 61(1): 10–12, 20–23 [B, 301/53]

Austin, G. H. (1958) Maltose content of Canadian honeys and its probable effects on crystallization. *X Int. Congr. Ent. 1956* 4: 1001–1006 [B, 191/64]

Australia. Department of Primary Industry (1964) Standards for export honey. Memorandum CV.64/6, Melbourne 26.2.64

Australian Bee Journal (1970) Dinna . . Giraga . . Warrul. *Aust. Bee J.* 51(3): 12 [B

Australian Bee Journal (1972) Budding statistics. *Aust. Bee J.* 53(1): 5–6 [B

Australian Honey Board (1965) *About Australian honey; 100 cake, biscuit, meat, dessert and party recipes* (Sydney: Australian Honey Board) [B

Australian Honey Institute (1954) *Secrets from the honey pot* (Melbourne: Australian Honey Institute) [B, 312/55]

Austria. *Codex alimentarius austriacus* (1957) Section B3: Honig und Kunsthonig (Vienna: Bundesministerium für Soziale Verwaltung)

Auzinger, A. (1910a) Über Fermente im Honig und den Wert ihres Nachweises für die Honigbeurteilung. *Z. Unters. Nahr.- u. Genussmittel* 17: 19(2): 65–83 [B, T

Auzinger, A. (1910b) Weitere Beiträge zur Kenntnis der Fermentreaktionen des Honigs. *Z. Unters. Nahr.- u. Genussmittel* 19(7): 353–362 [B, T

Averza, J. (1968) Food legislation in Central America. Chapter 7 from *The safety of foods* ed. J. C. Ayres *et al. AVI*

Bacon, J. S. D. & Dickinson, B. (1955) The origin of the trisaccharide melezitose. *Biochem. J.* 6: xv–xvi [B

Bacon, J. S. D. & Dickinson, B. (1957) The origin of melezitose: a biochemical relationship between the lime tree (*Tilia* spp.) and an aphid (*Eucallipterus tiliae* L.). *Biochem. J.* 66: 289–299

Bacon, J. S. D. & Edelman, J. (1950) The action of invertase preparations. *Arch. Biochem.* 28: 467–468

Badger, G. M. & Korytnyk, W. (1956) Examination of honey in Australian honeyants. *Nature, Lond.* 178: 320–321 [B, 390/57]

Bahr, L. (1939) Honig und Gesundheit. *Schweiz. Bienenztg* 62(12): 684–689 [B

Bailey, L. (1952) The action of the proventriculus of the worker honeybee, *Apis mellifera* L. *J. exp. Biol.* 29(2): 310–327 [B, 264/53]

Bailey, L. (1954) The filtration of particles by the proventriculi of various aculeate Hymenoptera. *Proc. R. ent. Soc. Lond.* (*A*) 29(7/9): 119–123 [B, 67/55]

Bailey, L. H. (1949) *Manual of cultivated plants* (New York: Macmillan Co.) [B

Bailey, L. H. & Lothrop, R. E. (1939) Honey in commercial bread baking. *Bakers' Tech. Dig.*: 23–25 [B

Bailey, M. E., Fieger, E. A. & Oertel, E. (1954) Paper chromatographic analyses of some southern nectars. *Glean. Bee Cult.* 82: 401–403, 472, 474　　[B, 156/55

Balogh, E., Szöcs, J. & Molnár, V. (1964) Cercetări experimentale privind influenţarea alcoolemiei prin unele substanţe alimentare [Experiments on the effect
of some foods on the blood-alcohol level]. *Revta med., Turgu Mures* 10(1): 49–51
+ 2 plates　　　　　　　　　　　　　　　　　　　　　　　　　　　[B, 331/66

Bancks, G. W. (1905) *The production of vinegar from honey* (Dartford: Perry)
15 pp., 4th ed.　　　　　　　　　　　　　　　　　　　　　　　　　　[B

Barbier, E. C. (1951) La sécrétion de nectar chez les eucalyptus. *Rev. franc. Apic.*
2: 529–532, 533–559　　　　　　　　　　　　　　　　　　　　　　[B, 10/52

Barbier, E. C. (1958) Examen pollinique de quelques miels unifloraux. *Annls
Abeille* 1(2): 73–76　　　　　　　　　　　　　　　　　　　　　　　[B, 281/59

Barbier, E. C. (1963) Les lavandes et l'apiculture dans le sud-est de la France.
Annls Abeille 6(2): 85–159　　　　　　　　　　　　　　　　　　　[B, 336/64

Barbier, E. C. & Pangaud, C.-Y. (1961) Origine botanique et caractéristiques
physico-chimiques des miels. *Annls Abeille* 4(1): 51–65　　　　　　[B, 509/62

Barbier, E. C. & Valin, J. (1957) Détermination de la couleur des miels. *Annls
Falsif. Fraudes* 50(587/588): 400–411　　　　　　　　　　　　　　[B, 220/60

Barbosa da Silva, R. M. (1967) Densidade do mel das abelhas 'Africanizadas'.
Bolm Ind. anim. (São Paulo) 24: 219–222　　　　　　　　　　　　[B, 193/72

Barschall, H. (1908) Über das Molekulargewicht des im Koniferenhonig vorkommenden Dextrins. *Arb. Kais. Gesund. Amt.* 28: 405–419

Bartels, W. & Fauth, A. (1933) Beobachtungen bei der Untersuchung californischer
Honige. *Z. Unters. Lebensmittel* 66(4): 396–407

Barth, O. M. (1969) Pollenspektren einiger brasilianischer Bienenhonige. *Z.
Bienenforsch.* 9(9): 410–419　　　　　　　　　　　　　　　　　　[B, 1012/70

Barth, O. M. (1970a) Análise microscópica de algumas amostras de mel. 1. Pólen
dominante. 2. Polen acessório. 3. Pólen isolado. *Anais Acad. bras. Cienc.* 42:
351–366, 571–590, 747–772　　　　　　　　　　　　　　　　　　[B, 795-7/72

Barth, O. M. (1970b) Análise microscópica de algumas amostras de mel. 4. Espectro
polínico de algumas amostras de mel do Estado do Rio de Janeiro. *Revta bras.
Biol.* 30(4): 575–582　　　　　　　　　　　　　　　　　　　　　　[B, 798/72

Barth, O. M. (1970c) Análise microscópica de algumas amostras de mel. 5. Melato
('Honeydew') em mel de abelhas. *Revta bras. Biol.* 30: 601–608　　[B, 805/72

Barth, O. M. (1971a) Análise microscópica de algumas amostras de mel. 6. Espectro
polínico de algumas amostras de mel dos estados da Bahia e do Ceará. *Revta bras.
Biol.* 31(4): 431–434　　　　　　　　　　　　　　　　　　　　　　[B, 684/73

Barth, O. M. (1971b) Mikroskopische Bestandteile brasilianischer Honigtauhonige.
Apidologie 2(2): 157–167　　　　　　　　　　　　　　　　　　　　[B, 430/73

Barth, W. (1952) Report on the use of honey as an ingredient in confectionery.
Cornell University: unpublished dissertation, 24 pp.　　　　　　　　[B, 63/55

Bartlett, B. R. (1962) The stabilization of relative humidity with honey in closed
systems. *J. econ. Ent.* 55(1): 149–150　　　　　　　　　　　　　　[B, 882/63

Basden, R. (1965) The occurrence and composition of manna in eucalyptus and
angophora. *Proc. Linn. Soc. N.S.W.* 90: 152–156　　　　　　　　　[B, 670/71

Basden, R. (1966) The composition, occurrence and origin of lerp, the sugary
secretion of *Eurymela distincta* (Signoret). *Proc. Linn. Soc. N.S.W.* 91: 44–46
　　　　　　　　　　　　　　　　　　　　　　　　　　　　　　　[B, 672/71

Basden, R. (1968) The occurrence and composition of the sugars in the honeydew
of *Eriococcus coriaceus* (Mask.). *Proc. Linn. Soc. N.S.W.* 92(3): 222–226
　　　　　　　　　　　　　　　　　　　　　　　　　　　　　　　[B, 669/71

Basden, R. (1970) A note on the composition of the lerp and honeydew of *Euca-
lyptolyma maidenii* Froggatt. *Proc. Linn. Soc. N.S.W.* 95(1): 9–10　[B, 724/72

Bastet, P. (1969) Produit alimentaire à base de miel. *Fr. Pat. No.* 1,572,373
[B, 529 L/72

Bates, D. (1938) *The passing of the aborigines* (London: John Murray) [B

Bates, F. J. *et al.* (1942) Polarimetry, saccharimetry and the sugars. *Circ. U.S. natn. Bur. Stand. No.* C440

Battaglini, M. [B.] & Battaglini, M. (1973) Sulle caratteristiche della frazione glucidica del nettare di alcune specie di cultivar di alberi da frutto. *Simposio Internazionale di Apicoltura, Torino 1972*: 71–76 [B, 421/74

Battaglini, M. [B.] & Bosi, G. (1973) Ricerche comparate sulla natura dei glucidi di alcuni mieli monoflora e dei rispettivi nettari. *Simposio Internazionale di Apicoltura, Torino 1972*: 123–129 [B, 476/74

Battaglini, M. & Ricciardelli d'Albore, G. C. (1971) Lo spettro pollinico di alcuni mieli della Sicilia. *Annali Fac. Agrar. Univ. Perugia* 26: 277–297 [B

Battaglini, M. & Ricciardelli d'Albore, G. C. (1972) Contributo alla conoscenza dei mieli monoflora italiani. *Annali Fac. Agrar. Univ. Perugia* 27: 219–224 [B

Battaglini, M. [B.] & Ricciardelli d'Albore, G. C. (1973) Differenziazione dei mieli italiani e stranieri in base allo spettro pollinico. *Simposio Internazionale di Apicoltura, Torino 1972*: 96–111 [B, 471/74

Baumgarten, F. & Möckesch, I. (1956) Über die papierchromatographische Auffindung freier Aminosäuren im Bienenhonig. *Z. Bienenforsch.* 3: 181–184
[B, 391/57

Baxa, J. & Bruhns, G. (1967) *Zucker im Leben der Völker. Eine Kultur- und Wirtschaftsgeschichte* (Berlin: Verlag Dr. Albert Bartens) [B, 479/70

Bayerová, V., Janíčková, E., Svobodová, H. & Pavlik, I. (1968) *Zužitkování medu v domácnosti* [Uses of honey in the home] (Prague: Státní zemědělské nakladatelství) [B

Bealing, F. J. (1953) Mould 'glucosaccharase': a fructosidase. *Biochem. J.* 55(1): 93–101 [B

Beck, B. F. (1938) *Honey and health* (New York: Robert M. McBride & Co.) [B

Beck, B. F. & Smedley, D. (1944) *Honey and your health* (New York: Robert M. McBride & Co.) [B

Beck, B. F. & Smedley, D. (1971) *Honey and your health* (London, New York & Toronto: Bantam Books) ix + 207 pp. Reprint of 1944 ed. [B

Becker, H. (1891) Bienenzucht und Bienenkenntnis der Griechen und Römer nach Columella. Nördlingen: Dissertation

Becker, J. & Kardos, R. F. (1939) Über den Vitamin C-Gehalt von Honig. *Z. Unters. Lebensmittel* 78: 305–308

Bee Research Association (1963) Selected list of publications on uses of honey. *Bibliogr. Bee Res. Ass. No.* 1: 7 pp. [B, 415 L/64

Bee Research Association (1964) Selected list of publications on the composition and properties of honey. *Bibliogr. Bee Res. Ass. No.* 5: 8 pp. [B, 647 L/64

Bee Research Association (1967) Honeybees other than *Apis mellifera*. *Bibl. Bee Res. Ass. No.* 8 [B, 66/68

Bee Research Association (1973) *World list of films on bees and beekeeping* (London: Bee Research Association) [B, 81/74

Bee World (1925) Japanese beekeeping. *Bee Wld* 7(3): 44–45 [B

Bee World (1932) A twelfth-century bee book. *Bee Wld* 13(11): 129 [B

Bee World (1963) Honey quality. *Bee Wld* 44(2): 45 [B

Bee World (1970) Recommended European regional standard for honey. *Bee Wld* 51(2): 79–91 [B, 508 L/72

Bee World (1971a) *Apis cerana* at risk (Editorial). *Bee Wld* 52(4): 141–142
[B, 332 L/72

Bee World (1971*b*) Illuminated manuscripts. *Bee Wld* 52(2): 80–81   [B, 311 L/73

Bee World (1972/73) Experts assess the Brazilian bee problem. *Bee Wld* 53(4): 178–179; 54(1): 9–10   [B

Bee World (1974) Costly honey: cause and effect. *Bee Wld* 55(2): 41–42   [B

Beer, C. H. (1973) My earliest recollection of bees. *Cambridgeshire Circular* No. 106

Belgium. Laws & Statutes (1967) Arrêté royal du 20 juillet 1967 relatif au miel et aux produits similaires. *Monit. belge* page 9972 (22.9.67) [*See* next reference for methods of analysis.]

Belgium. Laws & Statutes (1969) Arrêté royal du 13 octobre 1969 fixant les méthodes d'analyses de référence pour l'analyse du miel. *Monit. belge* page 10 930 (18.11.69)

Beliczay, L. (1960) *A méz ipari feldolgozása mézes sütemények* [Industrial use of honey, including honey confectionery] (Budapest: Muszaki Könyvkladó)
                                                                               [B, 216 L/63

Bell, H. T. (1962) A short introduction to the history of mead and mead making. *Lect. cent. Ass. Beekprs:* 12 pp.   [B, 520 L/62

Benelux. Laws & Statutes (1963) [Recommendation of the Committee of Ministers of the Benelux Economic Union relative to the harmonization of the legislation regarding honey and similar products] (Brussels: Benelux Economic Union) Circ. M (63)21 *In French*

Bennett, C. F., Jr. (1965) Beekeeping with stingless bees in western Panama. *Bee Wld* 46(1): 23–24   [B, 62/66

Benton, A. W. & Heckman, R. A. (1969) Promotion of honey sales at the local level. *Am. Bee J.* 109(10): 382–383   [B, 764/70

Beran, F. (1970) Der gegenwärtige Stand unserer Kenntnisse über die Bienengiftigkeit und Bienengefährlichkeit unserer Pflanzenschutzmittel. *Gesunde Pfl.* 22(2): 21–31   [B, 179/72

Bergeret, G. & Castro, J. A. de (1943*a*) Curva de la amilasa en la técnica de Koch para la investigacíon del 'poder diastasico' en la miel de abejas. *Revta Asoc. Ing. agron. Montevideo* 15: 63–66

Bergeret, G. & Castro, J. A. de (1943*b*) Hidromiel. *Revta Asoc. Ing. agron. Montevideo* 15: 67–70

Bergeret, G. & Castro, J. A. de (1954) Hidromiel. *Apicult. am.* 1(1): 4–7 [B, 334/57

Bergner, K.-G. & Hahn, H. (1972) Zum Vorkommen und zur Herkunft der freien Aminosäuren in Honig. *Apidologie* 3(1): 5–34   [B, 422/73

Bergner, K. G. & Körömi, J. (1968) Zum Aminosäurengehalt von Honigen. *Z. Bienenforsch.* 9(5): 182–184   [B, 812/69

Berthold, R., Jr. & Benton, A. W. (1968*a*) Creamed honey-fruit spreads. *Fd Technol., Champaign* 22(1): 83–85   [B, 409/68

Berthold, R., Jr. & Benton, A. W. (1968*b*) The potential of honey-fruit spreads increasing honey sales. *Am. Bee J.* 108(6): 236 only   [B, 777 L/68

Berto, H. (1972) *Cooking with honey* (New York: Crown Publishers, Inc.)   [B

Bertullo, W. A. & Lembo, F. E. (1943) La miel como alimento. *Revta Asoc. Ing. agron. Montevideo* 15: 43–49

Bessler, J. G. (1886) *Geschichte der Bienenzucht* (Stuttgart: W. Kohlhammer)   [B

Betts, A. D. (1929) Das Aufnahmevermögen der Bienen beim Zuckerwasserfüttern. *Arch. Bienenk.* 10(8): 301–309   [B

Betts, A. D. (1932*a*) Mead. *Bee Wld* 13: 69–70   [B

Betts, A. D. (1932*b*). Nectar yeasts. *Bee Wld* 13: 115   [B

Beug, H.-J. (1961) *Leitfaden der Pollenbestimmung für Mitteleuropa und angrenzende Gebiete* (Stuttgart: Gustav Fischer Verlag)   [B, 3 L/72

Beutler, R. (1930) Biologisch-chemische Untersuchungen am Nektar von Immen-
blumen. *Z. vergl. Physiol.* 12: 72–176 [B

Beutler, R. (1936) Über den Blutzucker der Bienen. *Z. vergl. Physiol.* 24: 71–115
[B, T

Beutler, R. (1953) Nectar. *Bee Wld* 34: 106–116, 128–136, 156–162 [B, 30/55

Beutler, R. & Schöntag, A. (1940) Über die Nektarabscheidung einiger Nutz-
pflanzen. *Z. vergl. Physiol.* 28: 255–285 [B

Beutler, R. & Wahl, O. (1936) Über das Honigen der Linden in Deutschland. *Z.
vergl. Physiol.* 23: 301–331

B.I.B.R.A. (1963) German plastics legislation. *Fd Cosmetics Toxicol.* 1(1): 53–70

Bieler, L. (1966) *Ireland: harbinger of the Middle Ages* (London: Oxford University
Press)

Biino, L. (1971) Ricera di alcuni aminoacidi in due varietà di miele. *Riv. ital.
Essenze Profumi* 53(2): 80–84 [B, 791/72

Billiard, R. (1900) *Notes sur l'abeille et l'apiculture dans l'antiquité* (Lille: Bigot
Frères) [B

Bilozorova, E. I. (1964) [Nectar productivity of buckwheat in relation to its
variety and its treatment during growth.] *Bdzhil'nitstvo* (1): 75–78. *In Ukrainian*
[B, 122/69

Bindley, M. D. (1965) Some contemporary records of early honey production and
use. *Bee Wld* 46(1): 32–33 [B

Bindley, M. D. (1967) Swiss beekeeping four hundred years ago. *Bee Wld* 48(2):
71–72 [B

Birch, C. A. (1959) Mazers. *Bee Craft* 41: 3, 12–13 [B

Blanchard, P. H. & Albon, N. (1950) The inversion of sucrose; a complication.
Arch. Biochem. 29: 220

Blanck, F. C. (1955) *Handbook of food and agriculture* (New York: Reinhold)
[General regulations, pp. 862, 887; labelling, pp. 864, 888]

Bland, S. E. (1962) *Money in comb honey* (Regina, Sask.: Saskatchewan Dept.
Agriculture) 18 pp. [B, 380 L/63

Blomfield, R. (1973) Honey for decubitus ulcers. *J. Am. med. Ass.* 224(6): 905 [B

Blum, R. (1955) Imkerei im alten Israel. *Bienenvater* 76(10): 334–336 [B, 339/57

Bodenheimer, F. S. (1928–1929) *Materialien zur Geschichte der Entomologie bis
Linné* (Berlin: W. Junk) [B

Bodenheimer, F. S. (1942) *Studies on the honey bee and beekeeping in Turkey*
(Istanbul: Nümune Matbaasi) [B

Bodenheimer, F. S. (1960) *Animal and man in Bible lands* (Leiden: E. J. Brill)
pp. 78–79 [B

Boer, H. W. de (1931) The behaviour of diastatic ferments in honey when heated.
Bee Wld 12: 13–16 [B

Boer, H. W. de (1933) Het verband tusschen de chemische samenstelling en de
botanische herkomst van in Nederland gewonnen honig. *Chem. Weekbl.* 30(23):
401–408 [B

Boetius, J. (1948) Über den Verlauf der Nektarabsonderung einiger Blütenpflanzen.
Beih. schweiz. Bienenztg 2: 257–317 [B, T

Bond, D. A. (1968) Variation between tetraploid red clover plants in corolla tube
length and height of nectar. *J. agr. Sci., Camb.* 71: 113–116 [B, 143/70

Bonnier, G. (1879) Les nectaires. Étude critique, anatomique et physiologique.
Ann. Sci. nat. 6 Sér. Bot. 8: 5–212 (also Paris: Masson) [B, T

Borneck, R. (1959) Facts about beekeeping in France. *Bee Wld* 40(2): 29–37
[B, 401/59

Borneck, R., Gauthron, R., Guirante, F., Horguelin, P., Louveaux, J., Pedelucq, A.

(1964) Les techniques de conditionnement et de commercialisation du miel au Canada et aux U.S.A. *Annls Abeille* 7(2): 103–159　　　　　　[B, T, 184/66

Bornus, L. (1957) Topographical conditions as a factor in honey production. *Bee Wld* 38(6): 141–149　　　　　　　　　　　　　　　　　　　[B, 39/61

Bornus, L. & Zalewski, W. (1962) *Miód pszczeli w produkcji, obrocie i spożyciu* [Honey: its production, composition and uses] (Warsaw: Zakład Wydawnicto CRS)　　　　　　　　　　　　　　　　　　　　　　　　[B, 203 L/63

Borowska, A. (1971) *Capnophialophora pinophila* (Nees) comb. nov. *Acta Mycol.* 7(1): 99–103

Borowska, A. (1973) *Tripospermum pinophilum* (Neger) comb. nov. *Acta Mycol.* 9(1): 100–104

Borowska, A. & Demianowicz, Z. (1972) Grzyby na spadzi jodlowej. [Fungi on fir honeydew]. *Acta Mycol.* 8(2): 175–189　　　　　　　　　　　[B, 891/73

Borries, G. (1934) Untersuchungen über die Vorgänge beim Treiben des Honigs. *Z. Unters. Lebensmittel* 67: 65–75

Bosch, W., Fulmer, E. I. & Park, O. W. (1932) Studies on honeys and nectars. *Rep. Ia St. Apiar.* 52–56　　　　　　　　　　　　　　　　　　　　[B

Bowen, E. G. (1972) *Britain and the western seaways* (London: Thames & Hudson)

Bozhilova, El. D. & Anchev, M. E. (1967/68) [Pollen analyses of honey collected by bees in the district of Kjustendil-Znepole floristic region.] *Ann. Univ. Sofia Fac. Biol.* 62(2): 11–29 *In Bulgarian*　　　　　　　　　　　[B, 653/73

Braun, E. (1954) Equipment for processing honey. *Can. Bee J.* 62(5): 11–19
　　　　　　　　　　　　　　　　　　　　　　　　　　　　　　　　[B, 254/55

Braun, W. (1954) Heute gebräuchliche Namen für Bienen und Verwandtes aus dem Guarani. *Dusenia* 5(2): 68–70　　　　　　　　　　　　　[B, 269/55

Braunsdorf, K. (1932) Zuckerfütterungshonig und Diastaseherkunft. *Z. Unters. Lebensmittel* 64: 555–558

Bray, J. T. (1967) Let's sell honey to horses. *Glean. Bee Cult.* 95(8): 472–474
　　　　　　　　　　　　　　　　　　　　　　　　　　　　　　　　[B, 775 L/69

Brazil. Laws & Statutes (1946) Decreto-Lei No. 15 642 de 9.2.46. Secretaria da Saúde Pública 1952 [honey regulations, p. 40]

Brazil. Laws & Statutes (1967) Decreto-Lei No. 209 de 27.2.67 [quoted in *Int. Dig. Hlth Legisl.* 19(4): 711 (1968); includes regulations on labelling and on substitute foods]

Bremer, W. & Sponnagel, F. (1909) Über die Fiehe'sche Reaktion zur Unterscheidung von Kunsthonig und Naturhonig, unter Berücksichtigung der Reaktionen nach Ley und Jägerschmid. *Z. Unters. Nahr.- u. Genussmittel* 17: 664–667

Brentjes, B. (1964) Einige Bemerkungen zur Rolle von Insekten in der altorientalischen Kultur. *Anz. Schädlingsk.* 37(12): 184–189　　　　　[B, 434/66

Brewer, J. W. & Dobson, R. C. (1969) Seed count and berry size in relation to pollinator level and harvest date for the highbush blueberry, *Vaccinium corymbosum. J. econ. Ent.* 62(6): 1353–1356　　　　'　　　　　　[B, 518/70

Brice, B. A. *et al.* (1951) Permanent glass color standards for extracted honey. *U.S. Bur. Agric. & Industr. Chem.* AIC 307. May, 1951.　　　　　[B, 161/52

Brice, B. A., Turner, A., Jr., & White, J. W., Jr. (1956) Glass color standards for extracted honey. *J. Ass. off. agric. Chem.* 39(4): 919–937　　　　[B, 25/58

British Food Journal (1959) *Br. Fd J.* 61: 84

British Food Journal (1965) *Br. Fd. J.* 67: 41

British Standards Institution (1950) Honey grading glasses. B.S. 1656: 1950
　　　　　　　　　　　　　　　　　　　　　　　　　　　　　　　　[B, 34/51

British Standards Institution (1952) Grading of honey. *British Standard* 1920: 1952 [scheme explained in *Bee Wld* 34(2): 29–31 (1953)] [B, 1955 ed. 95/56

Brokensha, D., Mwaniki, H. S. K. & Riley, B. W. (1972) Beekeeping in Embu District, Kenya. *Bee Wld* 53(3): 114–123 [B, 91/73

Brothwell, D. & Brothwell, P. (1969) *Food in antiquity. A survey of the diet of early peoples* (London: Thames & Hudson) [B, 492 L/70

Brouchot, A. (1951) Un nouveau bienfait du miel. *Apiculteur* 95(5): 100–101 [B, 176/53

Brown, Alice Cooke (1957) *Granny's honey and beeswax prescriptions* (Rutland, Vermont: Tuttle Publishing Co.) [B, 235/58

Brown, C. A. (1960) *Palynological techniques.* (Baton Rouge, La)

Brown, G. D. & Kosikowski, V. (1970) How to make honey yogurt. *Am. Dairy Rev.* 32(4): 60–62 [B, 1043 L/71

Brown, S. S., Forrest, J. A. H. & Roscoe, P. (1972) A controlled trial of fructose in the treatment of acute alcoholic intoxication. *Lancet* II (7783): 898–900 [B, 434 L/73

Browne, C. A. (1908) Chemical analysis and composition of American honeys. *Bull. U.S. Bur. Chem. No.* 110 [B

Browne, C. A. & Zerban, F. W. (1941) *Physical and chemical methods of sugar analysis* 3rd ed. (New York: John Wiley & Sons, Inc.)

Brünnich, K. (1940) Irrtümer in der Bienenzucht. Wie entsteht aus dem Nektar Honig. *Schweiz. Bienenztg* 63: 163–167 [B

Bryan, A. H. (1908) Estimation of dry substance by the refractometer in liquid saccharine food products. *J. Am. chem. Soc.* 30: 1443–1451

Büdel, A. (1956) Das Mikroklima in einer Blüte. *Z. Bienenforsch.* 3: 185–190 [B, 72/58

Büdel, A. (1957) Das Mikroklima der männlichen Weidenblüte. *Z. Bienenforsch.* 4: 21–22 [B, 82/62

Büdel, A. (1959) Das Mikroklima der Blüten in Bodennähe. *Z. Bienenforsch.* 4: 131–140 [B, 313/61

Büdel, A. & Grziwa, J. (1959) Die Rolle des Wärmeübergangs bei Erwärmung des Honigs *Z. Bienenforsch.* 4(7): 149–150 [B, 270/61

Büdel, A. & Herold, E. (ed.) (1960) *Biene und Bienenzucht* (Munich: Ehrenwirth Verlag) [B, 4/61

Bulgarian Committee for Standardization (1957a) [Honey.] *Bulgarian Standard* BDS 2673: 1957: 4 pp. *In Bulgarian* [B

Bulgarian Committee for Standardization (1957b) [Honey: rules for sampling and methods of test.] *Bulgarian Standard* BDS 3050: 1957: 6 pp. *In Bulgarian* [B

Bulman, M. W. (1955) Honey as a surgical dressing. *Middx Hosp. J.* 55(6): 188–189 [B, 304/56

Buman, M. de (1953) M2 Woelm en gynécologie et en obstétrique. *Praxis* 42(14): 282 only [B, 203/54

Burgett, D. M. (1973) The glucose oxidase activity in stored food products of the social Hymenoptera. Cornell University, Ithaca: Ph.D. Thesis [B

Burtt, E. G. (1930) References to bees in the Domesday Book. Unpublished manuscript in the B.R.A. Library [B

Büttner, G. (1948) Über den Gehalt des Honigs an Bor. *Z. Lebensmittelunters. u. -Forsch.* 88: 573–576

Cabrera Villa, L. & Soto Rosiles, J. (1962) La miel de abeja en la propagación vegetativa de cacao. *Agricultura téc. Méx.* 2(1): 18–20 [B, 358/69

Callow, R. K. (1963) Chemical and biochemical problems of beeswax. *Bee Wld* 44(3): 95–101 [B, 657/64

Cameron, A. T. (1947) The taste sense and the relative sweetness of sugars and other sweet substances. *Technol. Rep. Ser. Sug. Res. Fdn No.* 9

Campbell, D. J. (1953) Beekeeping in the Netherlands. *Bee Wld* 34(2): 25–29
[B, 231/55

Campo, M. van (1954) Considérations générales sur les caractères des pollens et des spores et sur leur diagnose. *Bull. Soc. bot. Fr.* 101(5/6): 250–281

Campo, M. van (1967) Pollen et classification. *Rev. Palaeobot. Palynol.* 3(1/4): 65–71

Canada Department of Agriculture (1952) *The fruit, vegetables and honey act and regulations* rev. 1951 (Ottawa: Canada Department of Agriculture)

Canada. Laws & Statutes (1954) Department of National Health and Welfare, Office Consolidation of the Food and Drugs Act and of the Food and Drugs Regulations [with amendments to 1974: honey, Section B 18.025; general regulations, Part I, Section 4; labelling, Section B 01.004; additives, Section B 16.100, Table X]

Canada. Laws & Statutes (1967) Canada Department of Agriculture, Honey regulations. *Canada Gaz. Pt II* 101(19): 1536–1554   [B, 404 L/68

Canadian Government Specifications Board (1968) Standard for honey: pasteurized. *Canadian Standard* CGSB 32-GP-206a: 1968: 2 pp.   [B, 518 L/72

Canus (1886) Die Honigbiene im alten Indien. *Berl. ent. Z.* 30(1): 65–71   [B

Caquot, A. (1968) L'abeille et le miel dans l'Israel antique. Pp. 43–49 from *Traité de biologie de l'abeille*. Vol. V, ed. R. Chauvin (Paris: Masson et Cie) [B, 325/68

Carey, F. M., Lewis, J. J., MacGregor, J. L. & Martin-Smith, M. (1959) Chemical and pharmacological observations on some toxic nectars. *J. Pharm. Lond. Suppl.* 11: 269T–274T   [B, 135/63

Castle-Turner, W. S. (1933) Mead. *Br. Bee J.* 61: 512–514   [B

Cavanagh, D., Beazley, J. & Ostapowicz, F. (1970) Radical operation for carcinoma of the vulva. *J. Obstet. Gynaec. Br. Commonw.* 77(11): 1037–1040   [B, 187/73

Centre National du Commerce Extérieur. Service Agricole (1956) *Note sur le marché européen du miel* (Paris: published by the author) 18 pp.   [B

Ceriotti, A. & Delpino, L. P. (1930) Notas analíticas de mieles argentinas. *Revta Fac. Ciênc. Quím. Farm. Univ. nac. La Plata* 7(1): 83–91

Čermagič, C., Jankovič, A., Soldatovič, D. & Maksimovič, M. (1964) An analysis of Jugoslav honey. *XIX Int. Beekeep. Congr. 1963, Prague* 1: 127–133
[B, see 58L/64

Chandler, B. V., Fenwick, D., Orlova, T. & Reynolds, T. (1974) Composition of Australian honeys. *Tech. Paper, CSIRO, Australia* No. 38: 39 pp.   [B

Chataway, H. D. (1932) Determination of moisture in honey. *Can. J. Res.* 6: 532–547   [B

Chataway, H. D. (1933) The determination of moisture in honey by the hydrometer method. *Can. J. Res.* 8: 435–439   [B

Chataway, H. D. (1935) Honey tables, showing the relationship between various hydrometer scales and refractive index to moisture content and weight per gallon of honey. *Can. Bee J.* 43(8): 215 only   [B

Chaubal, P. D. & Deodikar, G. B. (1965) Morphological characterization of pollen grains of some major honey yielding plants of the Western Ghats (India). *Indian Bee J.* 27(1): 1–28 + 2 plates   [B, 136/70

Chauvin, R. (ed.) (1968a) Digestion et nutrition des adultes. Pp. 347–377 from Vol. I. *Traité de biologie de l'abeille* (Paris: Masson et Cie)   [B, 323/68

Chauvin, R. (ed.) (1968b) Action physiologique et thérapeutique du miel. Pp. 116–126 from Vol. III. *Traité de biologie de l'abeille* (Paris: Masson et Cie) [B, 397/68

Chauvin, R. (ed.) (1968c) Histoire, ethnographie et folklore from Vol. V. *Traité de biologie de l'abeille* (Paris: Masson et Cie)   [B, 325/68

Chiavegatti, S. (from Marinello, G.) (1965) Farmacopea, cosmesi e miel nell'Italia del XVI secolo. *Apicolt. Ital.* 32(2): 32–34 [B

Chinese Encyclopaedia, Great (1727) [Section on bees and honey.] *In Chinese* [B

Chistov, V. & Silitskaya, N. (1952) [Chemical composition of floral and honeydew honeys.] *Pchelovodstvo* 29(10): 14–16 *In Russian* [B, 68/57

Chudakov, V. G. (1963) [The composition and properties of honey.] *Pchelovodstvo* 40(7): 18–19 *In Russian* [B, 196/65

Chudakov, V. G. (1964a) [Characteristics of the components of the acid-base complex of domestic honeys.] *Trudy nauchno-issled. Inst. Pchel.*: 303–332 *In Russian* [B, 833/65

Chudakov, V. G. (1964b) [The acid-base complex of domestic honeys.] *Trudy nauchno-issled. Inst. Pchel.* 333–345 *In Russian* [B, 834/65

[Chudakov, V. G.] Čudakov, V. G. (1964c) [Composition and properties of honeys in the U.S.S.R.] *XIX Int. Beekeep. Congr. 1963, Prague* 1: 146–151; *Pchelovodstvo* 40(7): 18–19 (1963) *In Russian* [B, T, 196/65

Chudakov, V. G. (1966) [Diastase activity of honeys from the U.S.S.R.] *Trudy nauchno-issled. Inst. Pchel.*: 467–511 *In Russian* [B, 560/69

Cipriani, L. (1966) *The Andaman Islanders* (London: George Weidenfeld & Nicolson) [B, 702/66

Cîrnu, I., Tomescu, A. & Sănduleac, E. (1965) Contribuţii privind stabilirea metodei pentru prognoza culesurilor in apicultură [Establishing methods for predicting the honey flow in apiculture]. *Lucr. ştiinţ. Staţ. cent. Seri. Apic.* 6: 95–114 *In Rumanian* [B, 676/70

Claffey, J. B., Turkot, V. A. & Eskew, R. K. (1961) Estimated cost for producing dried honey commercially. *Publ. U.S. Dep. Agric. ARS* 73–33: 16 pp. [B, 907 L/63

Clapham, A. R., Tutin, T. G. & Warburg, E. F. (1962) *Flora of the British Isles* (Cambridge: University Press) [B

Clark, E. W. & Lukefahr, M. J. (1956) A partial analysis of cotton extrafloral nectar and its approximation as a nutritional medium for adult pink bollworms. *J. econ. Ent.* 49: 875–876 [B, 219/61

Clark, J. G. D. (1942) Bees in antiquity. *Antiquity* 16: 208–215 [B

Clinch, P. G. (1966) An improved intracerebral injection method for detecting tutin and hyenanchin in toxic honey. *N.Z. Jl Sci.* 9(2): 433–439 [B, 237/68

Cocker, L. (1951) The enzymic production of acid in honey. *J. Sci. Fd Agric.* 2(9): 411–414 [B, 218/55

Codex Alimentarius Commission (1969) *Recommended European regional standard for honey* (Rome: Joint F.A.O./W.H.O. Food Standards Programme) (CAC/RS 12–1969); reprinted in *Bee Wld* 51(2): 79–91 (1970) [B, 507 L/72

Codounis, M. I. *see* Kodyne, M. I., Kodoyne, M. I.

Confiserie (1964) Le nougat reçoit une définition professionnelle. *Gaz. apic.* 64(681): 19–21 [B

Cook, V. A. (1967) Facts about beekeeping in New Zealand. *Bee Wld* 48(3): 88–100 [B, 320/68

Coon, C. S. (1972) *The hunting peoples* (London: Jonathan Cape)

Cornejo, L. G., Rossi, C. O. & Dávila, M. (1971) Curva de aporte nectarífero mediante el uso de la colmena báscula. *Gac. Colmen.* 33(377): 260–261 [B

Corominas, J. (1954) *Diccionario crítico etimológico de la lengua castellana* (Bern: Francke Verlag)

Corominas, J. (1961) *Breve diccionario etimológico de la lengua castellana* (Madrid: Editorial Gredos)

Cotton, W. C. (1842) *My bee book* (London: J. G. F. & J. Rivington) [B

Cowan, T. W. (1928) *British Bee-keepers' Association Jubilee: History of the Associa-tion representing fifty years of beekeeping progress* (London: British Bee-keepers' Association) [B

Crane, E. (ed.) (1951, 1958, 1964, 1971) *Dictionary of beekeeping terms with allied scientific terms:*

Vol. I English-French-German-Dutch, with Latin index   [B, 74/52
Vol. II English-Italian-Spanish   [B, 310/63
Vol. III English-French-German-Czech-Polish-Russian, with Latin index
     [B, 542/64
Vol. IV English-Danish-Norwegian-Swedish   [B, 337/72
London: Bee Research Association (*Vol. III* Warszawa: Państwowe Wydawnictwo Polnicze i Lesne in association with BRA)

Crane, E. (1954) An American bee journey. *Bee Wld* 35(7): 125–137   [B

Crane, E. (1957) Second American bee journey. *Bee Wld* 38(12): 301–313   [B, 44/60

Crane, E. (1958) Bienenstöcke je 100 ha Agrarwirtschaftsfläche. Honig: Gesamtproduktion und Ertrag je Stock. Pp. 65–66 from *Agrarwirtschaftsatlas der Erde* ed. B. Skibbe (Gotha: Hermann Haack)   [B, 142 L/64

Crane, E. (1960) Beekeeping in the People's Republic of China. *Bee Wld* 41(1): 4–8
     [B, 39 L/62

Crane, E. (1963*a*) A beekeeping visit to the Soviet Union. *Bee Wld* 44(2): 48–76
     [B, 752 L/63

Crane, E. (1963*b*) The world's beekeeping—past and present. Pp. 1–18 from *The hive and the honey bee*, ed. R. A. Grout (Hamilton, Ill.: Dadant & Sons) [B, 541/64

Crane, E. (1966) Canadian bee journey. *Bee Wld* 47: 55–65, 132–148   [B, 485/67

Crane, E. (1967*a*) Personal communication

Crane, E. (1967*b*) Unpublished information from A. R. Main and A. E. Guest   [B

Crane, E. (1971) Frameless movable-comb hives in beekeeping development pro-grammes. *Bee Wld* 52(1): 33–37   [B, 423 L/72

Crane, E. (1972*a*) Honey crop forecasting in the Middle Ages. *Bee Wld* 53(2): 79–83
     [B, 83/74

Crane, E. (1972*b*) Bumble bee honeys and others. *Bee Wld* 53(1): 38–39 [B, 1050 L/72

Crane, E. (1973) Honey sources in some tropical and subtropical countries. *Bee Wld* 54(4): 177–186   [B, 875/73

Crane, E. (1974) Directions in which bees build combs. *Bee Wld* 55(4): 153–155   [B

Crawford, I. M. (1968) *The art of the Wandjina* (Melbourne: Oxford University Press)

Cremer, E. & Riedmann, M. (1964) Identifizierung von gaschromatographisch getrennten Aromastoffen in Honigen. *Z. Naturf.* 19*b*(1): 76–77   [B, 920/64

Cremer, E. & Riedmann, M. (1965) Gaschromatographische Untersuchungen zur Frage des Honigaromas. *Mh. Chem.* 96(2): 364–368   [B, 183/66

Crosse-Upcott, A. R. W. (1956) Social aspects of Ngindo beekeeping. *J.R. anthrop. Inst.* 86(2): 81–108   [B, 283/61

Crosse-Upcott, A. R. W. (1958) Ngindo famine subsistence. *Tanganyika Notes* (50): 1–20   [B, 72/61

Crowther, R. F. (1960) Mead. *Rep. hort. Exp. Stn Prod. Lab. Vineland for 1950–1960* 98: 101

Cunningham, D. G. (1961) Facts about beekeeping in Australia. III. Tasmania. *Bee Wld* 42(7): 176–179   [B, 759 L/63

Curylo, J. (1961) Badania nad wplywem standaryzacji na β-amylaze i inwertaze w miodach oraz na zawartosc w nich 5-hydroksymetylofurfuralu [Effect of standardizing honey on its β-amylase, invertase, and 5-hydroxymethylfurfural contents]. *Pszczel. Zesz. Nauk.* 5(2): 65–74   [B, 211/63

Czarnowski, C. von (1952) Untersuchungen zur Frage der Nektarabsonderung. *Z. Bienenforsch.* 1: 171–173 [B, 100/53

Czarnowski, C. von (1953a) Über die Einwirkung unterschiedlicher Wasserversorgung auf die Nektarsekretion beim Boretsch. *Z. Bienenforsch.* 2: 85–91 [B, 139/54

Czarnowski, C. von (1953b) Die Nektarsekretion von *Sinapis alba* bei Einwirkung gestaffelter Stickstoffdüngung. *Arch. Geflügelz. Kleintierk.* 2: 135–140 [B, 14/54

Czarnowski, C. von (1954) Zur papierchromatographischen Blutzuckerbestimmung bei der Honigbiene. *Naturwissenschaften* 41: 577 only [B, 5/56

Dade, H. A. (1962) *Anatomy and dissection of the honeybee* (London: Bee Research Association) [B, 602/64

Dahlberg, A. C. & Penczek, E. S. (1941) The relative sweetness of sugars as affected by concentration. *Tech. Bull. N.Y. St. agric. Exp. Stn No.* 258

Dahlqvist, A. & Borgström, B. (1959) Characterization of intestinal invertase as a glucosido-invertase. II. Studies on transglycosylation by intestinal invertase. *Acta chem. scand.* 13: 1659–1667

Davies, A. M. C. (1974) The application of amino acid analysis to the determination of the authenticity of honey. 1. The amino acid analysis of honey from various sources. *J. apic. Res.* 13: *in press* [B

Dean, G. R. (1974) An unstable crystalline phase in the D-glucose-water system. *Carbohydrate Res.* 34: 315–322 [B

Deans, A. S. C. (1953) Honey sources. *Scott. Beekpr.* 29(1, 3): 4, 43 [B

Deans, A. S. C. (1957) *Survey of British honey sources* (London: Bee Research Association) [B, 195/65

Deerr, N. (1949) *The history of sugar* Vol. I, p. 6 (London: Chapman & Hall) [B, 95/50

Dehn, M. von (1961) Untersuchungen zur Ernährungsphysiologie der Aphiden. Die Aminosäuren und Zucker im Siebröhrensaft einiger Krautgewächse und im Honigtau ihrer Schmarotzer. *Z. vergl. Physiol.* 45: 88–108 [B, 660/68

Dehove, R. A. (1970) *La réglementation des produits alimentaires* (Paris: Commerce-Éditions). For honey and artificial honey *see* p. 469; for labelling *see* p. 492.

Delvert-Salleron, F. (1963) Étude, au moyen de radio-isotopes, des échanges de nourriture entre reines, mâles et ouvrières d'*Apis mellifica* L. *Annls Abeille* 6(3): 201–227 [B, 828/64

Demianowicz, A. (1957) Maximum exploitation of bee pasture. *Bee Wld* 38(6): 146–149 [B, 366/59

Demianowicz, A. & Demianowicz, Z. (1955) Nowe podstawy analizy pyłkowej miodów [A new method of pollen analysis of honey]. *Pr. Inst. Sadow. Skierniew.* 1: 185–195 [B, 305/58

Demianowicz, A. & Demianowicz, Z. (1956) Neue Grundlagen der Pollenanalyse des Honigs. *XVI Int. Beekeep. Congr. prelim. sci. Meet.* Vienna No. 395 [B, 243/56

Demianowicz, Z. (1957) O oblatywaniu przez pszczoly dwu odmian koniczyny czerwonej [Bee visitation to two varieties of red clover]. *Pszczel. Zesz. Nauk.* 1(2): 79–93 [B, 550/63

Demianowicz, Z. (1960) Zagadnienie miodów niezapominajkowych [Problems of forget-me-not honey]. *Pszczel. Zesz. Nauk.* 4(1): 43–48 [B, 413/64

Demianowicz, Z. (1961) Pollenkoeffizienten als Grundlage der quantitativen Pollenanalyse des Honigs. *Pszczel. Zesz. Nauk.* 5(2): 95–105 [B, 208/63

Demianowicz, Z. (1962a) Ocena jednoodmianowych miodów lipowych przy zastosowaniu metody współczynników pyłkowych [The evaluation of single-source lime honeys by the application of the pollen coefficient method]. *Pszczel. Zesz. Nauk.* 6(1): 25–45 [B, 927/64

Demianowicz, Z. (1962b) O miodach jednogatunkowych 6 przedstawicieli rodziny wargowych (Labiatae) [Some features of unifloral honey from six species of Labiatae]. *Pszczel. Zesz. Nauk.* 6(2): 75–80 [B, 412/64

Demianowicz, Z. (1963) Sur l'origine des macles d'oxalate de calcium contenues dans les miels de tilleul. *Annls Abeille* 6(4): 249–255 [B, 575/66

Demianowicz, Z. (1964) Charakteristik der Einartenhonige. *Annls Abeille* 7(4): 273–288 [B, 167/67

Demianowicz, Z. (1968a) Biologia kwitnienia i nektarowanie pięciu odmian rzepaku ozimego [Biology of flowering and nectar secretion of five varieties of winter rape]. *Annls. Univ.Mariae Curie-Skłodowska* 23E (19): 241-263 [B, 969/72

Demianowicz, Z. (1968b) Beitrag zur Pollenanalyse der Lindenhonige. *Z. Bienenforsch.* 9(5): 185–195 [B, 819/69)

Demianowicz, Z., Borowska, A., Dubik, G. & Pielka, J. (1972) Grzyby w spadziowych miodach jodłowych. [Fungi in conifer honeydew honey]. *Annls Univ. Mariae Curie-Skłodowska* 25E(15): 203–212 [B, 950/73

Demianowicz, Z. & Demianowicz, A. (1957) Nowe podstawy analizy pyłkowej miodów [A new method of pollen analysis of honeys]. *Pszczel. Zesz. Nauk.* 1(2): 69–78 [B, 654/63

Demianowicz, Z. & Hlyń, M. (1960) Porównawcze badania nad nektarowaniem 17 gatunków lip [Comparative investigation of nectar secretion in 17 species of lime]. *Pszczel. Zesz. Nauk.* 4(3/4): 133–152 [B, 91/62

Demianowicz, Z. & Jabłoński, B. (1959) Współczynniki pyłkowe 3 miodów jednogatunkowych z punktu widzenia analizy statystycznej [The pollen coefficients of three unifloral honeys, from the point of view of statistical analysis]. *Pszczel. Zesz. Nauk.* 3(1): 25–34 [B, 411/64

Demianowicz, Z. & Jabłoński, B. (1966) Nektarowanie i wydajność miodowa 4 gatunków roślin o drobnych kwiatach [Nectar secretion and honey yields of four species of plants with small flowers]. *Pszczel. Zesz. nauk.* 10(1/4): 87–94 [B, 712/67

Demianowicz, Z., Jabłoński, B., Ostrowska, W. & Szybowski, S. (1963) Wydajność miodowa ważniejszych roślin miododajnych w warunkach Polski [The honey yield of the main honey plants in Polish conditions. Part II]. *Pszczel. Zesz. Nauk.* 7(2): 95–111 [B, 854/64

Demianowicz, Z. & Ruszkowska, B. (1959) Gryka jako roślina pożytkowa [The nectar flow from buckwheat]. *Pszczel. Zesz. Nauk.* 3(1): 11–24 [B, 342/64

Demianowicz, Z. et al. (1960) Wydajność miodowa ważniejszych roślin miododajnych w warunkach Polski [The honey yield of the main honey plants in Polish conditions Part I]. *Pszczel. Zesz. Nauk.* 4(2): 87–104 [B, 853/64

Dennis, C. B. (1954) A background to mead making. *Lect. cent. Ass. Beekprs:* 14 pp. [B, 193/55

Deodikar, G. B. & Thakar, C. V. (1953) *A pollen study of major honey yielding plants of Mahalabeshwar Hills* (Poona: Village Industries Committee) 8 pp. [B, 185/55

Deodikar, G. B., Thakar, C. V., Phadke, R. P. & Shah, N. P. (1957) Thixotropy in honey of *Carvia callosa. Indian Bee J.* 19: 71–72 [B

Department of Health and Social Security (1969) Recommended intakes of nutrients for the United Kingdom. *Rep. publ. Health med. Subjects* No. 120 (London: H.M.S.O.)

Deschodt, C. (1969) Let me tell you about the chimps and the bees. *S. Afr. Bee J.* 41(1): 13 [B

Deshusses, J. & Gabbai, A. (1962) Recherche de l'anthranilate de méthyle dans les miels espagnols de fleur d'oranger par chromatographie sur couche mince. *Mitt. Geb. Lebensmittelunters. u. Hyg.* 53(5): 408–411 [B, 727/66

Detroy, B. F. (1966) Determining film coefficient for a viscous liquid. *Trans. Am. Soc. agric. Engrs* 9(1): 91–93, 97 [B, 190/70

Deutsche Lebensmittelbuch-Kommission (1972) Leitsätze der Deutschen Lebensmittelbuch-Kommission für Honig vom 16.3.1972 *GMinBl* 23(34): 610

Deutscher Imkerbund (1960) Richtlinien über die Gütebedingungen für deutschen Honig. Pp. 72–78 from *Das Honigbuch, die Gewinnung, Behandlung und Bewertung des Honigs* J. Evenius (Munich: Ehrenwirth Verlag) (1964) [B, 193/65

Dietz, A. (1966) Über den Einfluss der Umweltfaktoren auf den Tagesgang der Nektarsekretion. Universität Bonn: Dissertation.

Digby, Sir Kenelme (1669) *The closet of the eminently learned Sir Kenelm Digby, Kt. opened: whereby is discovered several ways for making of metheglin, sider, cherry wine* . . . (London: E.C. for H. Brome) Metheglin, meathe, hydromel, pp. 1–126

Dockery, J. (1958) A note on the Exultet rolls. Unpublished manuscript in the B.R.A. Library [B

Dold, H., Du, D. H. & Dziao, S. T. (1937) Nachweis antibakterieller, hitze- und lichtempfindlicher Hemmungsstoffe (Inhibine) im Naturhonig (Blütenhonig). *Z. Hyg. InfektKrank.* 120: 155–167 [B

Dold, H. & Witzenhausen, R. (1955) Ein Verfahren zur Beurteilung der örtlichen inhibitorischen (keimvermehrungshemmenden) Wirkung von Honigsorten verschiedener Herkunft. *Z. Hyg.* 141: 333–337 [B, 199/65

Dörrscheidt, W. & Friedrich, K. (1962) Trennung von Aromastoffen des Honigs mit Hilfe der Gas-Chromatographie. *J. Chromat.* 7(1): 13–18 [B, T, 895/63

Drawe (1908) Beitrag zur Dr. Fiehe'schen Reaktion auf Invertzucker im Honig. *Z. öff. Chem.* 14: 352

Drimtzias, N. (1968) Trachtpflanzen und Honige Griechenlands. *Z. Bienenforsch.* 9(5): 195–196 [B, 938 L/70

Duisberg, H. (1967) Honig und Kunsthonig. *Handb. Lebensmittelchemie* 5(1): 491–559 (Berlin, Heidelberg, New York: Springer Verlag) [B, 167/69

Duisberg, H. (1975) Wirkungen des Honigs auf den menschlichen Körper. In *Der Honig*, 2nd ed. by E. Zander & A. Maurizio (Handbuch der Bienenkunde Band 6) (Stuttgart: E. Ulmer Verlag) [in press]

Duisberg, H. & Gebelein, H. (1958) Über die Kontrolle von Erhitzungsschäden bei Honigen. *Z. Lebensmitt Untersuch.* 107(6): 489–501 [B, 349/59

Duisberg, H. & Hadorn, H. (1966) Welche Anforderungen sind an Handelshonige zu stellen? Vorschläge auf Grund der statistischen Auswertung von ca. 1600 Honig-Analysen. *Mitt. Geb. Lebensmittelunters. u. Hyg.* 57(5): 386–407
[B, 193/70

Duruz, R. M. (1953) The honey-pot game. *Bee Wld* 34(5): 90–93 [B

Duspiva, F. (1953) Der Kohlenhydratumsatz im Verdauungstrakt der Rhynchoten, ein Beitrag zum Problem der stofflichen Wechselbeziehungen zwischen saugenden Insekten und ihren Wirtspflanzen. *Mitt. biol. ZentAnst. Berl.* 75: 82–89

Duspiva, F. (1954a) Enzymatische Prozesse bei der Honigtaubildung der Aphiden. *Verh. dtsch zool. Ges.* 440–447 [B, 101/57

Duspiva, F. (1954b) Weitere Untersuchungen über stoffwechsel-physiologische Beziehungen zwischen Rhynchoten und ihren Wirtspflanzen. *Mitt. biol. ZentAnst. Berl.* 80: 155–162 [B, see 101/57

Dustmann, J. H. (1967a) Messungen von Wasserstoff-peroxid in Bienenhonig aus Edelkastanientracht (*Castanea sativa* M.). *Z. Lebensmittelunters. u. -Forsch.* 134(1): 20 only [B, 350/69

Dustmann, J. H. (1967b) Messung von Wasserstoff-Peroxid und Enzymaktivität in mitteleuropäischen Honigen. *Z. Bienenforsch.* 9(2): 66–73   [B, 784/67

Dustmann, J. H. (1971) Über die Katalaseaktivität in Bienenhonig aus der Tracht der Heidekrautgewächse. *Z. Lebensmittelunters. u. -Forsch.* 145(5): 294–295   [B, 1055/72

Dutcher, R. A. (1918) Vitamine studies. III. Observations on the curative properties of honey, nectar, and corn pollen in avian polyneuritis. *J. biol. Chem.* 36(1): 551–555

Dyce, E. J. (1931a) Fermentation and crystallisation of honey. *Bull. Cornell agric. Exp. Sta.* No. 528   [B

Dyce, E. J. (1931b) The crystallisation of honey. *J. econ. Ent.* 24: 597–602   [B

Dyce, E. J. (1933) Honey process and product. *Can. Pat.* No. 332,685   [B

Dyce, E. J. (1935) Honey process and product. *U.S. Pat. No.* 1,987,893   [B

Dyce, E. J. (1953) *Honey house equipment and methods* (Ithaca, New York: N.Y. St. Coll. Agric.) 4 pp.   [B, 274/54

Dyce, E. J. (1960) Beekeeping in Australia. *Glean. Bee Cult.* 88: 486–491, 505, 526–531   [B, 360/61

Dzialoszyński, L. & Kuik, K. (1963) Aktywność kwaśnej fosfatazy α-amylazy i katalazy w miodach z okolic torunia [The activity of acid phosphatase, α-amylase and catalase in honey from Toruń]. *Pszczel. Zesz. Nauk.* 7(1): 33–39   [B, 910/64

Echigo, T. (1970) Determination of sugars in nectar and honey, and the mutarotation coefficient of glucose, by gas liquid chromatography. *Bull. Fac. Agric. Tamagawa Univ.* 10: 3–12   [B, 889/73

Eckert, J. E. & Allinger, H. W. (1939) Physical and chemical properties of California honeys. *Bull. Calif. agric. Exp. Stn* No. 631   [B

Ecuador. Laws & Statutes (1963) Decree No. 462 of 16.9.63 [quoted in *Curr. Fd Addit. Legisl.* (77): abstract 1130; *Fd Drugs Cosmetics Law J.*: 609 (1964); *Fd Cosmetics Toxicol.*: 329 (1965)]

Edgell, G. H. (1949) *The bee hunter* (Cambridge, Mass.: Harvard University Press)   [B, 51/51

Edlin, H. L. (1951) *British plants and their uses* (London, New York, Toronto & Sydney: Batsford)

E.E.C. *see* European Economic Community

Egyptian Standards Institution (1963) [Honey.] *Egyptian Standard* EOS S355: 963: 11 pp. *In Arabic*   [B

Ehrhardt, P. (1961) Zur Nahrungsaufnahme von *Megoura viciae* Buckt., einer phloemsaugenden Aphide. *Experientia* 17: 461–463

Ehrhardt, P. (1962) Untersuchungen zur Stoffwechselphysiologie von *Megoura viciae* Buckt., einer phloemsaugenden Aphide. *Z. vergl. Physiol.* 46: 169–211   [B, 824 L/63

Ehrhardt, P. (1963) Untersuchungen über Bau und Funktion des Verdauungstraktes von *Megoura viciae* Buckt. (Aphidae, Homoptera) unter besonderer Berücksichtigung der Nahrungsaufnahme und der Honigtauabgabe. *Z. Morph. Ökol. Tiere* 52: 597–677

Ehrhardt, P. (1969) Untersuchungen zum Nahrungsbedarf einer Siebröhrensaugenden Aphide und der Bedeutung ihrer Endosymbionten für die Ernährung. *XXII Int. Beekeep. Congr., Munich*: 401–404   [B

Elkon, Juliette (1955) *The honey cookbook* (New York: Alfred A. Knopf)   [B, 143/56

Ellis, T. P. (1926) *Welsh tribal law and custom in the Middle Ages* 2 vol. (Oxford: Clarendon Press)

Elmenhorst, C. W. (1952) Beekeeping in Guatemala. *Bee Wld* 33(6): 93–96
[B, 261/53

Elser, E. (1924) Beiträge zur quantitativen Honiguntersuchung. *Arch. Bienenk.* 6: 118 only [B

Elser, E. (1928) Weitere Beiträge zur quantitativen Bestimmung der Aschenbestandteile des Honigs. *Z. Unters. Lebensmittel.* 55: 246–251

Emmison, F. G. (1964) *Tudor life and pastimes* (London: Benn)

Erdtman, G. (1966) *Pollen morphology and plant taxonomy. I. Angiosperms.* Off-set edition with addendum (New York: Hafner) [1952 ed. 43/54

Erdtman, G. (1969) *Handbook of palynology* (Copenhagen: Munksgaard) [B, 11/71

Erdtman, G. (1970) *World pollen flora* (Copenhagen: Munksgaard)
[B, 804 L–806 L/70

Erdtman, G. & Vishnu-Mittre (1958) On terminology in pollen and spore morphology. *Grana palynol.* 1(3): 6–9

Eschrich, W. (1961) Untersuchungen über den Ab- und Aufbau der Callose. *Z. Bot.* 49: 153–218

Eschrich, W. (1963a) Beziehungen zwischen dem Auftreten von Callose und der Feinstruktur des primären Phloems bei *Cucurbita ficifolia. Planta* 59: 243–261

Eschrich, W. (1963b) Der Phloemsaft von *Cucurbita ficifolia. Planta* 60: 216–224

European Economic Community (1974) European Economic Community Council Directive of 22 July 1974 on the harmonisation of the laws of the member states relating to Honey (74/409/EEC). *Off. Jl European Communities* No. L221: 10–14 (12.8.74)

Européen (1962) Le pain d'épices, friandise la plus ancienne du monde à son centre à Nuremberg. *Gaz. apic.* 63(664): 154–156 [B

Evans, H. E. & Eberhard, M. J. W. (1970) *The wasps* (Ann Arbor: University of Michigan Press) [B, 288/72

Evelyn, J. (ed. D. A. Smith) (1966) John Evelyn's manuscript on bees from *Elysium Britannicum* (London: Bee Research Association) [B, 632/67

Evenius, J. (1932) Eine Untersuchung pommerscher und mecklemburgischer Honige, insbesondere zur Prüfung der Trachtherkunft. *Pommersch. Ratgeber Bienenk.* 32(9/10): 241–246, 268–275

Evenius, J. (1933) Die Prüfung des Sedimentgehaltes norddeutscher Honige im Zusammenhang mit ihren chemisch-biologischen Eigenschaften. *Festschrift Zander:* 23–33 [B

Evenius, J. (1953) Die Strandaster (*Aster tripolium* L.) als Trachtpflanze. *Z. Bienenforsch.* 2(3/5): 112–116 [B

Evenius, J. (1958) Pollenanalyse und Begutachtung von sedimentreichen Honigen. *Annls Abeille* 1(2): 77–88 [B, 170/60

Evenius, J. (1960) Pollenanalytische Prüfung von Honigen aus der Bundesrepublik. *Dt. Bienenwirt.* 11(4): 76–80 [B, 139/61

Evenius, J. & Focke, E. (1967) Mikroskopische Untersuchung des Honigs. *Handb. Lebensmittelchemie* 5(1): 560–590 (Berlin, Heidelberg & New York: Springer Verlag) [B, 1005/70

Evenius, J. & Kaeser, W. (1970) *Das Honigbuch. Entstehung, Gewinnung, Behandlung, Lagerhaltung und Bewertung des Honigs.* (Munich: Ehrenwirth Verlag)
[B, 1009 L/71

Ewart, W. H. & Metcalf, R. L. (1956) Preliminary studies of sugars and amino acids in the honeydew of five species of coccids feeding on *Citrus* in California. *Ann. ent. Soc. Am.* 49: 441–447 [B, 56/58

Ewert, R. (1935) Die Förderung der Nektarausscheidung bei Raps, Buchweizen und Rotklee durch Kalidüngung. *Dtsch. Imkerführer* 9: 63–66   [B, T

Ewert, R. (1936) Honigen und Samenansatz des Rotklees. *Dt. Imkerführer* 10: 476–480   [B

Ewert, R. (1938) Das Honigen der Linden in Landsberg (Warthe) im Jahre 1937. *Dt. Imkerführer* 12: 85–87   [B

Ewert, R. (1940) *Das Honigen unserer Obstgewächse.* (Leipzig: Verlag Leipziger Bienenzeitung)

Faber, H. K. (1920) A study of the antiscorbutic value of honey. *J. biol. Chem.* 43: 113–116   [B

Fabian, F. W. (1926) Honey vinegar. *Circ. Bull. Mich. agric. Exp. Stn* 85: 13 pp. [B

Fabian, F. W. (1928) Honey vinegar—a profitable by-product. *Beekeeper* 36(8): 145–146   [B

Fabian, F. W. (1935a) Honey vinegar. *Ext. Bull. Mich. St. Coll. No.* 149

Fabian, F. W. (1935b) The use of honey in making fermented drinks. *Fruit Prod. J.* 14: 363–366

Fabian, F. W. & Quinet, R. I. (1928) A study of the cause of honey fermentation. *Bull. Mich. agric. Coll. Exp. Stn* No. 62   [B

Fabris, U. (1911) Über die Bestimmung des Wassers im Honig. *Z. Unters. Nahr.- u. Genussmittel* 22: 353–358

Faegri, K. & Iversen, J. (1964) *Textbook of pollen analysis.* (Copenhagen: Munksgaard) 237 pp. 2nd ed.   [B, 436/65

Faegri, K. & Ottested, P. (1949) Statistical problems in pollen analysis. *Univ. Bergen naturvit. Rek.* (3): 1–27

Fahd, T. (1968) L'abeille en Islam. Pp. 61–83 from *Traité de biologie de l'abeille.* Vol. V, ed. R. Chauvin (Paris: Masson et Cie)   [B, 325/68

Fahn, A. (1949) Studies in the ecology of nectar secretion. *Palest. J. Bot. Jerusalem Ser.* 4: 207–224   [B, 164/53

Fahn, A. (1953) The topography of the nectary in the flower and its phylogenic trend. *Phytomorphology* 3(4): 424–426   [B, 81/57

Fahn, A. & Rachmilevitz, T. (1971) Ultrastructure and nectar secretion in *Lonicera japonica*. Pp. 51–56 from *New research in plant anatomy* (London: Academic Press)   [B, 633/73

Fan, L. T. & Tseng, J. T. (1967) Apparent diffusivity in honey-water system. *J. Fd. Sci.* 32(6): 633–636   [B, 348/69

F.A.O. *see* Food and Agriculture Organization

Farnsteiner, K. (1908) Der Ameisensäuregehalt des Honigs. *Z. Unters. Nahr.- u. Genussmittel* 15: 598–604

Fast, R. B. (1967) Sugar-coated ready-to-eat cereals. *U.S. Pat.* No. 3,318,706   [B, 200 L/70

Favre, H. (1968) La symbolique de l'abeille et du miel en Amérique indienne. Pp. 121–143 from *Traité de biologie de l'abeille* Vol. V, ed. R. Chauvin (Paris: Masson et Cie)   [B, 325/68

Feder, E. (1911) Ein Vorschlag zur Prüfung des Honigs auf künstlichen Invertzucker. *Z. Unters. Nahr.- u. Genussmittel* 22: 412–413

Fedosov, N. F. (ed.) (1955) [*Beekeeper's encyclopaedia*] (Moscow: State Publishing House for Agriculture Literature) *In Russian*   [B, 79/58

Fedotova, T. K. (1957) [The catalase activity of honey]. *Nauch. Trudy Samarkand Inst. Sovet Torgovli* 8: 187–188. *In Russian*

Fehlmann, C. (1911) Beiträge zur mikropischen Untersuchung des Honigs. *Mitt. Geb. Lebensmittelunters. u. Hyg.* 2: 179–208, 220–261   [B

Fellenberg, T. von (1911) Viskositätsbestimmungen im Honig. *Mitt. Geb. Lebensmittelunters. u. Hyg.* 2: 161–178

Fellenberg, T. von & Ruffy, J. (1933) Untersuchungen über die Zusammensetzung echter Bienenhonige. *Mitt. Geb. Lebensmittelunters. u. Hyg.* 24: 367–392

Fellenberg, T. von & Rusiecki, W. (1938) Bestimmung der Trübung und der Farbe des Honigs. *Mitt. Geb. Lebensmittelunters. u. Hyg.* 29: 313–335

Ferreira, E. L. da S. (1970) *O mel e seus açúcares. Processos actuais de análise.* Universidade Técnica de Lisboa, Instituto Superior de Agronomia: Relatório final do Curso de Engenheiro Agrónomo [B, 186/73

Fiehe, J. (1908*a*) Eine Reaktion zur Erkennung und Unterscheidung echter Bienenhonige. *Z. Unters. Nahr.- u. Genussmittel* 15: 492–493

Fiehe, J. (1908*b*) Eine Reaktion zur Erkennung und Unterscheidung von Kunsthonigen und Naturhonigen. *Z. Unters. Nahr.- u. Genussmittel* 16: 75–77

Fiehe, J. (1931) Über Honigdiastase. *Z. Unters. Lebensmittel* 61: 420–427

Fiehe, J. (1932) Über die Herkunft der Honigdiastase. *Z. Unters. Lebensmittel* 63: 329–331

Fiehe, J. & Kordatzki, W. (1928) Beitrag zur Kenntniss der Honigdiastase. *Z. Unters. Lebensmittel* 55(2/3): 162–169 [T

Fiehe, J. & Stegmüller, P. (1912) Nachprüfung einiger wichtiger Verfahren zur Untersuchung des Honigs. *Arb. K. GesundhAmt.* 40: 305–356

Fife, A. E. (1939) *The concept of the sacredness of bees, honey and wax in Christian popular tradition* (Stanford University: Ph.D. dissertation) [B

Filipello, F. & Marsh, G. L. (1934) Experiments with mead. *Am. Bee J.* 74: 537–538 [B

Finley, R. D. & Hollowell, K. A. (1967) Honey-butter and process of manufacture. *U.S. Pat. No.* 3,351,472 [B, 778 L/68

Fischer, G. *et al.* (eds.) (1974) *Der Schweizerische Bienenvater: Lehrbuch der Bienenzucht* 15th ed. (Aarau & Frankfurt am Main: Sauerländer) [B

Fischer, H. (1890) *Beiträge zur vergleichenden Morphologie der Pollenkörner.* Universität Breslau: Dissertation [B

Fisher, J. & Peterson, R. T. (1964) *The world of birds* (London: Macdonald)

Fix, W. J. & Palmer-Jones, T. (1947) Control of fermentation in honey by indirect heating and drying. *N.Z. Jl Agric.* 74–75: 611–616 [B

Fix, W. J. & Palmer-Jones, T. (1949) Control of fermentation in honey by indirect heating and drying. *N.Z. Jl Sci. Technol.* 31A: 21–31 [B, 146/51

Fochem, K. (1954) Die Wirkung der Laevulose bei der Strahlenkrankheit. *Strahlentherapie* 93(3): 466–472 [B, 310/55

Focke, E. (1968) Das Pollenbild chinesischer Honige. *Z. Bienenforsch.* 9(5): 196–206 [B, 807/69

Food and Agriculture Organization of the United Nations (1969) Livestock numbers and products: Table 131, Production of honey. *F.A.O. Production Yearbook* 23: 414–415 [B

Food and Agriculture Organization of the United Nations (1973) World prospects for honey production. *World Anim. Rev.* (7): 7 pp. [B

Food Processing (1961) Mixing shaft of new design adapts Votator unit for rapid cooling, thorough mixing of highly viscous, granular product. *Fd Process.*: 2 pp. [B, 572/66

Food Processing (1969) Yield of baked foods increased with dry honey product. Flavor, texture of breads improved. *Fd. Process.* 30(5): 24–25 [B, 1044/71

Food Processing (1971) Dehydrated granular honey coating reduces shrinkage in meat products by as much as 19%. *Fd Process.*: 1 page only (Oct. 1971) [B, 809 L/72

Forbes, R. J. (1966) Sugar and its substitutes in antiquity. Chapter 2 (pp. 80–111) from *Studies in ancient technology*. Vol. V. (Leiden: E. J. Brill)   [B, 187/72

Fossel, A. (1955) Steirische Honigernte 1954. *Bienenvater* 76(7): 225–227   [B

Fossel, A. (1956) Steirische Honige. *Bienenvater* 77(5): 156–163   [B, 303/56

Fossel, A. (1958a) Die Bedeutung der Edelkastanie als Honigspenderin in der Steiermark. *Bienenvater* 79(2): 46–51   [B

Fossel, A. (1958b) *Iris sibirica*. Blütenbiologische und pollenkundliche Beobachtungen. *Z. Bienenforsch.* 4(6): 114–122   [B, 119/60

Fossel, A. (1960) Ein Sortenhonig der grossen Sterndolde (*Astrantia major*). *Bienenvater* 81(5): 146–148   [B, 331/61

Fossel, A. (1962) Kulturpflanzen als Nektarspender. *Dt. Bienenwirt.* 13(11): 330–333   [B

Fossel, A. (1966) Die Bedeutung der Heidetracht in Österreich. *Bienenvater* 87(8/9): 234–238   [B, 94 L/67

Fossel, A. (1968) Pollenersatzmittel im mikroskopischen Befund von Frühtrachthonigen. *Z. Bienenforsch.* 9(5): 206–211   [B, 822/69

Fossel, A. (1970) Seit der Eiszeit Heidelbeeren, Preiselbeeren und Rauschbeeren. *Bienenvater* 91(6): 179–184   [B

Fraenkel, E. (1962) *Litauisches etymologisches Wörterbuch*. (Göttingen: Van den Hoeck & Ruprecht)

Franzke, C. & Iwainsky, H. (1956) Über den Nachweis und die Bestimmung von Oxymethylfurfurol in Honigen und Kunsthonigen. *Fette Seif. AnstrMittel* 58(10): 859–862   [B, 199/59

Fraser, H. M. (1931) *Beekeeping in antiquity* (London: University of London Press)   [B

Fraser, H. M. (1950) The story of the progress of beekeeping before 1800. *Bee Wld* 31(5): 33–38   [B, 165/52

Fraser, H. M. (1951) *Beekeeping in antiquity* 2nd ed. (London: University of London Press)   [B, 22/52

Fraser, H. M. (1958) *History of beekeeping in Britain* (London: Bee Research Association)   [B, 715/65

Free, J. B. & Butler, C. G. (1959) *Bumblebees* (London: Collins)   [B, 287/59

Free, J. B. & Durrant, A. J. (1966) The dilution and evaporation of the honey-stomach contents of bees at different temperatures. *J. apic. Res.* 5(1): 3–8   [B, 662/66

Frei, E. (1955) Die Innervierung der floralen Nektarien dikotyler Pflanzenfamilien. *Ber. schweiz. bot. Ges.* 65: 60–114   [B

Frey-Wyssling, A. & Agthe, C. (1950) Nektar ist ausgeschiedener Phloemsaft. *Verh. schweiz. naturf. Ges.* 130 *Versammlung, Davos*: 175–176   [B, 35/52

Frey-Wyssling, A. & Häusermann, E. (1960) Deutung der gestaltlosen Nektarien. *Ber. schweiz. bot. Ges.* 70: 150–162   [B, 3/65

Frey-Wyssling, A., Zimmermann, M. & Maurizio, A. (1954) Über den enzymatischen Zuckerumbau in Nektarien. *Experientia* 10(12): 490–492   [B, 278/55

Friedmann, H. (1954) Honey-guide: the bird that eats wax. *Nat. geogr. Mag.* 105: 551–560   [B, 83/55

Friedmann, H. (1955) The honey-guides. *Bull. U.S. nat. Mus.* No. 208: 292 pp.   [B, 181/56

Friedmann, H. & Kern, J. (1956a) The problem of cerophagy or wax-eating in the honey-guides. *Quart. Rev. Biol.* 31(1): 19–30   [B, 217/57

Friedmann, H. & Kern, J. (1956b) *Micrococcus cerolyticus*, nov. sp., an aerobic lipolytic organism isolated from the African honey-guide. *Canad. J. Microbiol.* 2: 515–517   [B, 60/59

Frisch, K. von (1927) Versuche über den Geschmackssinn der Bienen. I. *Naturwissenschaften* 15(14): 321–327   [B

Frisch, K. von (1928) Versuche über den Geschmackssinn der Bienen. II. *Naturwissenschaften* 16(18): 307–315   [B

Frisch, K. von (1934) Über den Geschmackssinn der Biene. Ein Beitrag zur vergleichenden Physiologie des Geschmacks. *Z. vergl. Physiol.* 21(1): 1–156   [B

Fritzsche, J. (1832) *Beiträge zur Kenntnis des Pollens* (Berlin)

Führer, H. (1926) Über den giftigen Honig des pontischen Kleinasien. *Naturwissenschaften* 14: 1283 only

Fulmer, E. I., Bosch, W., Park, O. W. & Buchanan, J. H. (1934) The analysis of the water content of honey by use of the refractometer. *Am. Bee J.* 74: 208 only [B

Furgala, B., Gochnauer, T. A. & Holdaway, F. G. (1958) Constituent sugars of some northern legume nectars. *Bee Wld* 39: 203–208   [B, 46/59

Galli-Valerio & Bornand, M. (1910) Untersuchungen über die Präzipitine des Honigs. *Z. ImmunForsch. exp. Ther.* 7: 331–341

Galton, D. (1971) *Survey of a thousand years of beekeeping in Russia.* (London: Bee Research Association)   [B, 868/71

Gassparian, S. & Vorwohl, G. (1974) Vergleichende Qualitätsuntersuchungen an Iranischen Honigen. *Apidologie* 5(2): 177–190   [B

Gauhe, A. (1941) Über ein glukoseoxydierendes Enzym in der Pharynxdrüse der Honigbiene. *Z. vergl. Physiol.* 28(3): 211–253   [B

Gautier, J.-A., Renault, J. & Julia-Alvarez, M. (1961a) Recherche du sucre interverti dans le miel. Première partie: Étude critique des reactions de Fiehe et de Feder. *Annls. Falsif. Fraudes* 54: 177–193   [B, 663/63

Gautier, J.-A., Renault, J. & Julia-Alvarez, M. (1961b) Recherche du sucre interverti dans le miel. Deuxième partie: Étude critique de la chromatographie sur papier. *Annls. Falsif. Fraudes* 54: 253–260   [B, 663/63

Gautier, J.-A., Renault, J. & Julia-Alvarez, M. (1961c) Recherche du sucre interverti dans le miel. Troisième partie: Étude de la spectrophotometrie d'absorption dans l'ultra-violet et conclusions. *Annls Falsif. Fraudes* 54: 397–411   [B, 663/63

Gayre, G. R. (1948) *Wassail! in mazers of mead.* (London: Phillimore)   [B

Gazette Apicole (1965) Une friandise qui remonte au XVIe siècle: Le nougat de Montélimar. *Gaz. apic.* 66(703): 266–267, 260   [B

Gebel, W. (1968) Zur Problematik der Beurteilung von Bienenhonig. *Gordian* 68(2): 81–83

Geinitz, B. (1930) Die Entstehung des Tannenhonigs. *Arch. Bienenk.* 9: 308–318
                                                                                 [B

Geisler, W. (1962) *Vymezení včelařských oblastí po stránce bioklimatické a pastevní* [Demarcation of beekeeping regions from the bioclimatic and bee forage points of view]. (Prague: Ustav Vědeckotechnických Informací MZLVH) *In Czech.* Six detailed maps were published in association with this study   [B, 719/65

Geissler, G. & Steche, W. (1962) Natürliche Trachten als Ursache für Vergiftungserscheinungen bei Bienen und Hummeln. *Z. Bienenforsch.* 6(4): 77–92
                                                                           [B, 638/63

Génier, G. (1966) Le pollen des Ericaceae dans les miels français. *Annls Abeille* 9(4): 271–321   [B, 216/67

Gennaro, A. R., Sideri, C. N., Rubin, N. & Osol, A. (1959) Use of honey in medicinal preparations. *Am. Bee J.* 99(12): 492–493   [B, 107/61

Georgantas, P. D. (1957) The forerunner of the modern hive. *Bee Wld* 38(11): 286–289   [B, 83/59

Gergeaux, M. P. (1961) Fabrication du nougat. *Apiculteur* 105(3): 65–66 [B

Germany. Laws and Statutes (1927–1930) Verordnung über Honig vom 21. März 1930 (R.G.Bl. I, 101, 1930); quoted in Büdel & Herold (1960) Pp. 354–358; Evenius & Kaeser (1970) p. 94 ff.; Holthöfer *et al.* (1963) Vol. II p. 450; Duisberg in Zander & Maurizio (1975) p. 169 ff.

Germany. Laws & Statutes (1930*b*) Verordnung über Kunsthonig vom 21 März 1930 [quoted in Holthöfer *et al.* (1963) Vol. II, p. 470; Büdel & Herold (1960) p. 358]

Germany. Laws and Statutes (1960) Richtlinien des D.I.B. *See* Deutscher Imkerbund (1960)

Germany. Laws & Statutes (1971) Lebensmittel- und Bedarfsgegenstandsgesetz [This is so far a proposal only. Paragraph 18 deals with health claims]

Germany. Laws & Statutes (1972) Verordnung über die äussere Kennzeichnung von Lebensmitteln (Lebensmittel-Kennzeichnungs-Verordnung) vom 25.1.72 *Bundesgesetzblatt* Teil I: 86 (1972); *Deutsche Lebensmittel-Rundschau*: 120 (1972)

Gillette, C. C. (1931) Honey catalase. *J. econ. Ent.* 24: 605–606

Gilmour, D. (1961) *The biochemistry of insects* (New York & London: Academic Press)

Giordani, G. (1953) Le api, le frutta e l'uva. *Apicolt. Ital.* 20(2): 27–29 [B, 10/55

Giri, K. V. (1938) Chemical composition and enzyme content of Indian honey. *Madras agric. J.* 26: 68–72

Glabe, E. F., Goldman, P. F. & Anderson, P. W. (1963) Honey solids—a new functional sweetener for baking. *Bakers' Dig.* 37(5): 3 pp. [B, 333/66

Glabe, E. F. *et al.* (1965) Honey solids, functions of honey-starch complex in baking. *Bakers' Dig.* 39(2): 71–74 [1045 L/71

Glock, J. P. (1891) *Die Symbolik der Bienen und ihrer Produkte* (Heidelberg: Weiss) 411 pp. [B

Glukhov, M. M. (1955) [*Bee plants*] (Moscow: State Publishing House of Agricultural Literature) 6th ed. *In Russian* [B, 415/57

Glushkov, N. M. (1959) Facts about beekeeping in the U.S.S.R. *Bee Wld* 40: 169–172, 201–204 [B, 257/60

Goillot, C. & Louveaux, J. (1955) Études sur la sédimentation pollinique dans les miels fluides au repos. *Apiculteur* 99(5) *Sect. sci.*: 23–31 (1955); *Grana palynol.* (N.S.) 1(2): 90–98 (1956) [B, 183/56

Goldschmidt, S. & Burkert, H. (1955*a*) Über das Vorkommen einiger im Bienenhonig bisher unbekannter Zucker. *Hoppe-Seyler's Z. physiol. Chem.* 300: 188–200 [B, 197/56

Goldschmidt, S. & Burkert, H. (1955*b*) Die Hydrolyse des cholinergischen Honigwirkstoffes und anderer Cholinester mittels Cholinesterasen und deren Hemmung im Honig. *Hoppe-Seyler's Z. physiol. Chem.* 301: 78–89 [B, 198/56

Goldschmidt, S., Burkert, H., Helmreich, E. & Gramss, H. (1952) Über den cholinergischen Wirkstoff des Honigs. *Z. Naturf.* 7b: 365–367 [*see* 198/56

Gonnet, M. (1963) L'hydroxymethylfurfural dans les miels. Mise au point d'une méthode de dosage. *Annls Abeille* 6(1): 53–67 [B, T, 893/63

Gonnet, M. (1965) Les modifications de la composition chimique des miels au cours de la conservation. *Annls Abeille* 8(2): 129–146 [B, 573/66

Gonnet, M. (1969) Sur l'origine des substances phyto-inhibitrices présentes dans les miels de lavande. *C. r. hebd. Séanc. Acad. Sci., Paris* 268: 859–861 [B, 814/69

Gonnet, M. & Lavie, P. (1960) Influence du chauffage sur le facteur antibiotique présent dans les miels. *Annls Abeille* 3(4): 349–364 [B, 311/62

Gontarski, H. (1935) Leistungsphysiologische Untersuchungen an Sammelbienen. *Arch. Bienenk.* 16: 107–126 [B

Gontarski, H. (1948) Ein Vitamin C oxydierendes Ferment der Honigbiene. *Z. Naturf.* 3b: 245–249 [B, 41/51

Gontarski, H. (1951) Zur Analyse der Formbestandteile des Waldhonigs. *Z. Bienenforsch.* 1(3): 33–37 [B, 162/51

Gontarski, H. (1954) Fermentbiologische Studien an Bienen. I. Das physiko-chemische Verhalten der kohlenhydratspaltenden Fermente. (a) invertierende Enzyme. *Verh. dt. Ges. angew. Ent.*: 186–197 [B, 49/56

Gontarski, H. (1957) Eine Halbmikromethode zur quantitativen Bestimmung der Invertase im Bienenhonig. *Z. Bienenforsch.* 4(2): 41–45 [B, 345/59

Gontarski, H. (1960) Der Honig als Nahrung der Biene und seine Entstehung. Pp. 181–190 from *Biene und Bienenzucht*, ed. A. Büdel & E. Herold (Munich: Ehrenwirth Verlag) [B, 4/61

Gontarski, H. (1961) Über den Einfluss von Wassergehalt und pH des Honigs auf den Grad der Wärmeschädigung der Invertase. *Z. Bienenforsch.* 5(7): 191–198 [B, 653/63

Gontarski, H. & Hoffmann, I. (1963) Über Aufnahme und Verarbeitung der zuckerhaltigen Nahrung bei der Honigbiene. *Z. Bienenforsch.* 6(7): 184–198 [B, 589/64

Goodacre, W. A. (1926) Tannic acid in honey. *Agric. Gaz. N.S.W.* 37: 474–475

Gorbach, G. (1942) Zur Kenntnis der Stärkeverdauung durch die Biene. 2. Mitteilung in der Reihe: Ernährungsphysiologische Studien an der Biene. *Forschungsdienst* 13: 67–78

Gordon, D. H. (1960) *The pre-historic background of Indian culture.* 2nd ed. (Bombay: N. M. Tripathi (Private) Ltd)

Gösswald, K. & Kloft, W. (1961) Einblicke in das Staatenleben von Insekten auf Grund radiobiologischer Studien. *Imkerfreund* 16: 7–12 [B, 8c303(11)61

Gothe, F. (1914a) Die Fermente des Honigs. *Z. Unters. Nahr.- u. Genussmittel* 28(6): 273–286

Gothe, F. (1914b) Experimentelle Studien über Eigenschaften und Wirkungsweise der Honigdiastase sowie die Beurteilung des Honigs auf Grund seines Diastasegehaltes. *Z. Unters. Nahr.- u. Genussmittel* 28(6): 286–321 [T

Gottfried, A. (1911) Der Mangangehalt der Honige. *Pharm. Zentralhalle Dtl.* 52: 787–788

Gottfried, A. (1929) Die Formoltitration bei der Untersuchung von Honig. *Z. Unters. Lebensmittel* 57: 558–560

Goyal, N. P. (1974) *Apis cerana indica* and *Apis mellifera* as complementary to each other for the development of apiculture. *Bee Wld* 55(3): 98–101 [B

Graham, A. J. & Crane, E. (1975) Hives of the Ancient World. *Bee Wld* 56. *In press*

Gray, H. E. & Fraenkel, G. (1953) Fructomaltose, a recently discovered trisaccharide isolated from honeydew. *Science, N.Y.* 118(3063): 304–305 [B, 183/54

Gray, H. E. & Fraenkel, G. (1954) The carbohydrate components of honeydew. *Physiol. Zool.* 27(1): 56–65 [B, 50/55

Gray, J. (1933) Honey vinegar. *Am. Bee J.* 73: 353–354 [B

Gray, R. A. (1952) Composition of honeydew excreted by pineapple mealybugs. *Science, N.Y.* 115: 129–133 [B, 242/52

Griebel, C. (1930/31) Zur mikroskopischen Pollenanalyse des Honigs. *Z. Unters. Lebensmittel* 59: 63–79, 197–211, 441–471; 61: 241–306 [B, T

Griebel, C. (1938) Vitamin C enthaltende Honige. *Z. Unters. Lebensmittel* 75: 417–420

Griebel, C. & Hess, G. (1939) Vitamin C enthaltende Honige. *Z. Unters. Lebensmittel* 78: 308–314

Griebel, C. & Hess, G. (1940) Der C-Vitamingehalt des Blütennektars bestimmter Labiaten. *Z. Unters. Lebensmittel* 79: 168–171

Grout, R. A. (ed.) (1963) *The hive and the honey bee* (Hamilton, Ill.: Dadant & Sons) 556 pp. rev. ed. [B, 541/64

Grout, R. A. (1963) Extracting the honey crop. Pp. 303–322 from *The hive and the honey bee* ed. R. A. Grout (Hamilton, Ill.: Dadant & Sons) [B, 541/64

Gruch, W. (1957) Zur Frage der Giftwirkung gewisser Honige. *Z. Bienenforsch.* 4(2): 47–57 [B, T, 366/58

Grüss, J. (1931) Zwei altgermanische Trinkhörner mit Bier- und Metresten. *Prähist. Z.* 22: 180–191 [B

Gryuner, V. S. & Arinkina, A. I. (1970) [Carbohydrate composition, enzymatic and antibiotic activities of honey]. *Izv. vyssh. ucheb. Zaved. pishch. Tekhnol.* (6): 28–31 *In Russian* [B, T, 792–72

Gubin, A. F. (1945) [The Beekeeping Institute during the war: Honey in medicine.] *Pchelovodstvo* (1): 25–29. *In Russian* [B, 198/50

Gubin, V. A. (1953) [Why the cappings of northern and southern races of bees differ]. *Pchelovodstvo* 30(7): 14–15 *In Russian* [B, 13/54

Guglielmo, C. (1940) *Il miele e la cera nell'industria e nel commercio* (Pescara: Arti Grafiche) [B

Guillemin, A. (1825) Recherches microscopiques sur le pollen. *Mém. Soc. Hist. nat., Paris* 2

Gulyas, S. (1967) Zusammenhang zwischen Struktur und Produktion in den Nektarien einiger Lamium-Arten. *Acta biol., Szeged* 13(1/2): 3–10 [B, 405/72

Gunda, B. (1968) Bee-hunting in the Carpathian area. *Acta ethnogr. hung.* 17: 1–62 [B, 422/71

Gundel, M. & Blattner, V. (1934) Über die Wirkung des Honigs auf Bakterien und infizierte Wunden. *Arch. Hyg.* 112: 319–332 [B

Günther, F. & Burckhart, O. (1967) Bestimmung der sauren Gesamtphosphatase in Honig. *Dt. LebensmittRdsch.* 63(2): 41–44 [B, 574/67

Guy, R. D. (1971) A commercial beekeeper's approach to the use of primitive hives. *Bee Wld* 52(1): 18–24 [B, 992 L/72

Guy, R. D. (1972a) Commercial beekeeping with African bees. *Bee Wld* 53(1): 14–22 [B, 903 L/72

Guy, R. D. (1972b) The honey hunters of Southern Africa. *Bee Wld* 53(4): 159–166 [B, 192/74

Hackmann, R. H. & Trikojus, V. M. (1952) The composition of the honeydew excreted by Australian coccids of the genus *Ceroplastes*. *Biochem. J.* 51: 653–656 [B, 250/54

Hadorn, H. (1956) Beitrag zur Wasserbestimmung in Honig. *Mitt. Geb. Lebensmittelunters. u. Hyg.* 47(3): 200–204 [B, 301/57

Hadorn, H. (1961) Zur Problematik der quantitativen Diastasebestimmung in Honig. *Mitt. Geb. Lebensmittelunters. u. Hyg.* 52(2): 67–103 [B, 577/67

Hadorn, H. (1964) Enthalten Orangenblüten und Lavendelblütenhonige enzymhemmende Stoffe? *Annls Abeille* 7(4): 311–320 [B, 162/67

Hadorn, H. & Kovacs, A. S. (1960) Zur Untersuchung und Beurteilung von ausländischem Bienenhonig auf Grund des Hydroxymethylfurfurol- und Diastasegehaltes. *Mitt. Geb. Lebensmittelunters. u. Hyg.* 51(5): 373–390 [B, 312/62

Hadorn, H. & Zürcher, K. (1962a) Zur Bestimmung der Saccharase-Aktivität in Honig. *Mitt. Geb. Lebensmittelunters. u. Hyg.* 53(1): 6–28 [B, 192/64

Hadorn, H. & Zürcher, K. (1962b) Über Veränderungen im Bienenhonig bei der

grosstechnischen Abfüllung. *Mitt. Geb. Lebensmittelunters. u. Hyg.* 53(1): 28–34
[B, 597/68
Hadorn, H. & Zürcher, K. (1963a) Formolzahl von Honig. Gleichzeitige Bestim-
mung von Formolzahl, pH, freier Säure und Lactongehalt in Honig. *Mitt. Geb.
Lebensmittelunters. u. Hyg.* 54(1): 304–321 [B, 590/68
Hadorn, H. & Zürcher, K. (1963b) Über Zuckerfütterungshonig. *Mitt. Geb. Lebens-
mittelunters. u. Hyg.* 54(4): 322–330 [B, 930/64
Hadorn, H. & Zürcher, K. (1973) Erfahrungen mit einer neuen kinetischen
Methode zur Bestimmung der Diastasezahl in Honig und über die Eigenschaften
der Honigdiastase. *Apidologie* 4(1): 65–80 [B, 675/74
Hadorn, H., Zürcher, K. & Doevelaar, F. H. (1962) Über Wärme- und Lager-
schädigungen von Bienenhonig. *Mitt. Geb. Lebensmittelunters. u. Hyg.* 53(3):
191–229 [B, 582/67
Hadorn, H., Zürcher, K. & Strack, C. (1974) Gaschromatographische Bestimmung
der Zuckerarten in Honig. *Mitt. Geb. Lebensmittelunters. u. Hyg.* 65: 198–208 [B
Hahn, H. (1970) *Zum Gehalt und zur Herkunft der freien Aminosäuren in Honig.*
(Universität Stuttgart: Dissertation) [B, 754/71
Hajny, G. J., Hendershot, W. F. & Peterson, W. H. (1960) Factors affecting
glycerol production by a newly isolated osmophilic yeast. *Appl. Microbiol.* 8:
5–11
Hallermayer, R. (1969) Beitrag zur Beurteilung von Bienenhonig. *Gordian* 69(5):
230, 232–234 [B, 218/74
Hambleton, J. I. (1925) The effect of weather upon the change in weight of a
colony of bees during the honeyflow. *Bull. U.S. Dep. Agric.* No. 1339: 52 pp. [B
Hammer, O., Jørgensen, E. G. & Mikkelsen, V. M. (1948) Studier over danske
honningprøvers indhold af blomsterstøv [Studies on the pollen content of
Danish honey samples]. *Tidsskr. PlAvl* 52: 293–350 [B
Hamp, E. P. (1966) The position of Albanian. Pp. 97–121 from *Ancient Indo-
European dialects* ed. H. Birnbaum & J. Puhvel (Berkeley & Los Angeles:
California University Press)
Hanssen, E. & Hahn, F. (1963) Über Model. *Dt. LebensmittRdsch.* 59(12): 343–350 [B
Hansson, A. (1942) Honungens absorption av luftfuktighet [Absorption by honey
of moisture from the air]. *K. fysiogr. Sällsk. Lund. Förh.* 11: 18–41 [B
Hansson, A. (1949) Innehaller honungen C-vitamin? *Nord. Bitidskr.* 1(4): 99–101
[B, 71/51
Hansson, A. (1955) Fins honungsbiet kvar i vilt tillstånd i Norden? [Can wild
honeybees exist in the north?] *Nord. Bitidskr.* 7(4): 123–126 [B, 15/58
Hansson, A. (1961) Honungskonsumtionen i siffror [Honey consumption in figures].
Bitidningen 60(11): 367–371 [B
Hansson, A. (1966) Ein Messgerät für die Konsistenzbestimmung des Honigs.
Z. Bienenforsch. 8(6): 187–190 [B, 159/67
Haragsim, O. (1963) Medovice a její včelařské využití [Honeydew and its utiliza-
tion for beekeeping]. *Věd. Pr. výzk. Úst. včelař. v Dole* 3: 277–321 [B, 120/66
Haragsim, O. (1966) *Medovice a vcely* (Prague: Statni zemedelske nakladatelstvi)
Haragsim, O. (1974a) Nektarodarnost hrusni a cinnost vcel pri opylovani jejich
kvetu. *Věd. Pr. výzk. Úst. včelař. v Dole* 6: 9–23
Haragsim, O. (1974b) Nektarodarnost trnovniku akatu (*Robinia pseudoacacia* L.) v
dolnim Povltavi *Věd. Pr. výzk. Úst. včelař. v Dole* 6: 25–38
Haragsim, O. & Mácha, J. (1969) Cukry obsažené v nektaru jerlínu japonského
[Sugar contained in the Japanese sophora (*Sophora japonica* L.)]. *Sb. čsl. Akad.
zeměd. Věd. Rostlinná Výroba* 15: 659–664 [B, 709/72
Haragsim, O. & Slavikova, Z. (1968) Der Schnurbaum (*Sophora japonica* L.) und

seine Bedeutung für die Bienenzucht. *Z. Bienenforsch.* 9: 237-252   [B, 140/70

Harland, E. P. N. (1932) Old recipes for mead. *Br. Bee J.* 40: 108–109   [B

Harrer, H. (1953) *Seven years in Tibet* (London: Hart-Davis)   [B

Harris, W. F. & Filmer, D. W. (1947) A recent outbreak of honey poisoning. Part 6 *N.Z. Jl Sci. Technol.* 29: 134–143   [B

Harris, W. N. *et al.*, ed. (1865–1901) *Ancient laws of Ireland* (London: H.M.S.O.) Vol. II, p. 151

Harrison, W. (1587) *Description of England* (Ithaca, N.Y.: Cornell University Press) ed. G. Edelen (1968), p. 336

Hartley, D. (1964) *Water in England* (London: Macdonald), pp. 359–360   [B

Hasler, A. & Maurizio, A. (1949) Die Wirkung von Bor auf Samenasatz und Nektarsekretion bei Raps (*Brassica napus* L.). *Phytopath. Z.* 15: 193–207   [B, 149/51

Hasler, A. & Maurizio, A. (1950) Über den Einfluss verschiedener Nährstoffe auf Blütenansatz, Nektarsekretion und Samenertrag von honigenden Pflanzen, speziell von Sommerraps (*Brassica napus* L.). *Schweiz. landw. Mh.* 28: 201–211   [B, T, 161/51

Hawk, P. B., Smith, C. A. & Bergeim, O. (1921) The vitamine content of honey and honeycomb. *Am. J. Physiol.* 55: 339–348

Haydak, M. H. (1936) A prolonged test of milk and honey diet. *Minn. Med.* 19: 774–776

Haydak, M. H. (1955) The nutritional value of honey. *Am. Bee J.* 95(5): 185–191   [B, 248/55

Haydak, M. H. (1963) Activities of honey bees. Pp. 71–133 from *The hive and the honey bee* ed. R. A. Grout (Hamilton, Ill.: Dadant & Sons)   [B, see 541/64

Haydak, M. H., Palmer, L. S., Tanquary, M. C. & Vivino, A. E. (1942) Vitamin content of honeys. *J. Nutr.* 23: 581–588   [B

Haydak, M. H., Palmer, L. S., Tanquary, M. C. & Vivino, A. E. (1943) The effect of commercial clarification on the vitamin content of honey. *J. Nutr.* 26(3): 319–321   [B

Haydak, M. H., Vivino, A. E., Boehrer, J. J., Bjorndahl, O. & Palmer, L. S. (1944) A clinical and biochemical study of cow's milk and honey as an essentially exclusive diet for adult humans. *Am. Jour. Med. Sci.* 207: 209–219   [B

Haynie, J. D. (1953) Honey jelly: further studies on the effects of different grades of pectin and acids in making honey into a jelly. *Rep. Ia St. Apiar. for 1952*: 34–38   [B

Hazslinsky, B. (1938) Adatok a mez pollenanalitikai vizsgalatahoz [Contributions to pollen analysis of honey]. *Mezőgazd. Kutat.* 11: 143–159   [B

Hazslinsky, B. (1943) A nemes gesztenye, mint mezelö növeni [The sweet chestnut as a honey plant]. *A.m. Kir. Kertesz. Szölész. Föiskola* 9: 15–26   [B

Hazslinsky, B. (1952) Magyar akácmézek kvalitatív és pollenanalitikai vizsgálata [Qualitative and quantitative pollen analysis of Hungarian acacia honey]. *Mag. Tud. Akad. biol. Oszt. Közl.* 1(3): 317–417   [B

Hazslinsky, B. (1956) Toxische Wirkung eines Honigs der Tollkirsche (*Atropa belladonna* L.). *Z. Bienenforsch.* 3(5): 93–96   [B, 202/56

Heath, Ambrose (1956) *Honey cookery* (London: Neville Spearman)   [B, 423/57

Heiduschka, A. & Kaufmann, G. (1911) Über die flüchtigen Säuren im Honig. *Z. Unters. Nahr.- u. Genussmittel* 21(6): 375–378   [B

Heiduschka, A. & Kaufmann, G. (1913) Über die Säuren im Honig. *Süddt. ApothZtg* 53: 118–119

Helvey, T. C. (1953) Colloidal constitutents of dark buckwheat honey. *Fd Res.* 18(2): 197–205   [B, 35/84

Helvey, T. C. (1954) Study on some physical properties of honey. *Fd Res.* 19(3): 282–292 [B, 249/55

Hernández-Pacheco, E. (1924) Las pinturas prehistóricas de las cuevas de la araña (Valencia). *Mem. Com. Invest. paleont., Madr.* No. 34 [B

Herold, E. (1970) *Heilwerte aus dem Bienenvolk* (Munich: Ehrenwirth Verlag) [2nd ed. (1972) is reprint] [B, 1042/71

Hérouville, P. d' (1926/27) Virgile apiculteur. *Musée Belge* 30: 161; 31: 37, 777

Hertkorn, J. (1909) Beitrag zur Prüfung von Honig. *Chemikerzeitung* 33: 481

Hilger, A. (1904) Zur Kenntnis der im rechtsdrehenden Koniferenhonig vorkommenden Dextrine. *Z. Unters. Nahr.- u. Genussmittel* 8: 110–126

Hill, John (1759) *The virtues of honey in preventing many of the worst disorders* . . . (London: J. Davis & others) 2nd ed. [B, (photocopy)

Hillyard, T. N. (1965) Facts about beekeeping in England. *Bee Wld* 46(3): 77–85 [B, 52 L/66

Hinton, C. L. & Macara, T. (1924) The application of the iodimetric method to the analysis of sugar products. *Analyst, Lond.* 49(1): 1–23 [B

Hocking, Brian (1958) Making mead. *Can. Bee J.* 70: 11–18 [B

Hodges, D. (1952) *The pollen loads of the honeybee* (London: Bee Research Association Ltd.) 120 pp.; facsimile reprint with addenda 1974 [B, 221/52

Hodges, D. (1958) A calendar of bee plants. *Bee Wld* 39(3): 63–70 [B, 337/59

Hodges, R. & White, E. P. (1966) Detection and isolation of tutin and hyenanchin in toxic honey. *N.Z. Jl Sci.* 9(1): 233–235 [B, 236/68

Hoffmann, I. (1966) Gibt es bei Drohnen von *Apis mellifica* ein echtes Füttern oder nur eine Futterabgabe? *Z. Bienenforsch.* 8: 249–255 [B, 518/67

Højrup, O. (1957) Mjød—gudernes drik [Mead: drink of the gods]. *National-museets Arb.*: 53–62 [B, 208/58

Holthöfer, H., Juckenack, A. & Nuse, K.-H. (1963) *Deutsches Lebensmittelrecht* (Berlin: Heymanns) [B

Hoopen, H. J. G. ten (1963) Flüchtige Carbonylverbindungen in Honig. *Z. Lebens-mittelunters. u. -Forsch.* 119(6): 478–482 [B, 195/64

Horwitz, W. (ed.) (1960) *Official methods of analysis of the Association of Official Agricultural Chemists* 9th ed. (Washington: Association of Official Agricultural Chemists)

Horwitz, W. (ed.) (1965) *Official methods of analysis of the Association of Official Agricultural Chemists* 10th ed. (Washington: Association of Official Agricultural Chemists)

Howes, F. N. (1949) Poisoning from honey. *Fd Mf.* 24: 459–462

Hoy, R. (1788) *Proper directions how to manage bees in Hoy's octagon box bee-hives* (London: Honey Warehouse) [B

Hoyle, E. (1929) The vitamin content of honey. *Biochem. J.* 23(1): 54–60 [B

Huber, H. (1956) Die Abhängigkeit der Nektarsekretion von Temperatur, Luft- und Bodenfeuchtigkeit. *Planta* 48: 47–98 [B, 266/59

Hübschmann, H. (1962) *Altarmenische Grammatik* Vol. I, p. 455 (Hildesheim: Georg Olms)

Hughes, A. (1964) *The diary of a farmer's wife 1796–1797* (London: Countryside Books)

Hughes, G. B. (1973) Honey for the Georgian breakfast. *Country Life* 153(3949): 528–529 [B

Hugony, E. (1953) L'alluminio nell' imballaggio per prodotti, alimentari della Sicilia [Aluminum packaging of Sicilian food products]. *Conserve Deriv. agrum* 2(7): 107–112 [B, 198/64

Hungary. Magyar Népköztársasági Országos Szabvány (1966a). Méz. Minőségi

követelmény és mintavétel [Honey: quality requirements and sampling]. *Hungarian Standard* MSZ 6950/1: 1966: 5 pp. *In Hungarian*   [B, 510 L/72
Hungary. Magyar Népköztársasági Országos Szabvány (1966b). Méz. Vizsgálat [Honey: Methods for testing]. *Hungarian Standard* MSZ 6950/2: 1966: 12 pp. *In Hungarian*   [B, 511 L/72
Hyde, H. A. & Adams, K. F. (1958) *An atlas of airborne pollen grains* (London: Macmillan) 112 pp.   [B, 316/59

Ikuse, M. (1956) *Pollen grains of Japan.* (Tokyo) 303 pp.   [B, 161/58
Inglesent, H. (1940) Zymotic function of the pharyngeal, thoracic and postcerebral glands of *Apis mellifica. Biochem. J.* 34: 1415–1418   [B
Inglesent, H. (1974) Personal communication
International Commission for Bee Botany (1962) Method of pollen analysis of honey. *Bee Wld* 43(4): 122–123; *Z. Bienenforsch.* 6(4): 115–116; *Annls Abeille* 6(1): 75–76 (1963)   [B, 410 L/63
International Commission for Bee Botany of the I.U.B.S. (1970) Methods of melissopalynology. *Bee Wld* 51(3): 125–138; *Repr. BRA No.* M58: 14 pp.; *Apidologie* 1(2): 193–209, 211–227   [B, 1025/71
International Symposium on the Clinical and Metabolic Aspects of Laevulose (1971) Papers presented. (Crewe: Calmic Ltd)   [B, 530/72
Iparraguire, F. (1966) Economía apícola argentina. *Gac. Colmen.* 28(317): 352–360, 362–364, 366–372   [B, 643 L/66
Iran Standards Institution (1966) [Honey.] *Iran Standard* ISIRI 92: 1966: 7 pp. *In Persian*   [B, 515 L/72
Irvine, F. R. (1957) Indigenous African methods of beekeeping. *Bee Wld* 38(5): 113–128   [B, 74/59
Israel, R. J. (1972) The promised land of milk and date jam. *Nat. Jewish Monthly* 87(3): 26–28, 30   [B
Israel Standards Institution (1970) Honey. *Israel Standard* SII 373: 1970: 11 pp. *In English:* translation without guarantee from Hebrew (authentic version)   [B, 516 L/72
Italy. Laws & Statutes (1962) Legge n. 283. [Articles 5a, 8, 11b, 13 apply. English translations in *Fd agric. Legisl.* 13(3); *Fd Cosmetics Toxicol.* 1(1): 70]
Iversen, J. & Troels-Smith, J. (1950) Pollenmorphologische Definitionen und Typen. *Danm. geol. Unders. IV R.* 3(8): 1–60

Jabłoński, B. (1968) Wydajność miodowa ważniejszych roślin miododajnych w warunkach Polski. Część IV [An investigation of nectar secretion and honey production by the most important honey plants in Poland. Part IV]. *Pszczel. Zesz. Nauk.* 12(3): 117–126   [B, 404/72
Jackson, R. F. & Silsbee, C. G. (1924) Saturation relations in mixtures of sucrose, dextrose, and levulose. *Technologic. Pap. Bur. Stand., Wash.* 259: 277–304   [B
Jacobs, M. B. (1955) Flavoring with honey. *Am. Perfumer ess. Oil Rev.* 66(1): 46–47
Jamieson, C. A. (1954) Some factors influencing the crystallization of honey. *Rep. Ia St. Apiar.* 6954: 64–73   [B, 64/58
Jamieson, C. A. (1958) Facts about beekeeping in Canada. *Bee Wld* 39(9): 232–236   [B, 324/59
Jamieson, C. A. & Austin, G. H. (1958) Preference of honey bees for sugar solutions. *X Int. Congr. Ent., Montreal 1956* 4: 1059–1062   [B
Janscha, A. (1771) Abhandlung vom Schwärmen der Bienen. (Vienna)   [B

Jarvis, D. C. (1958) *Folk medicine: a Vermont doctor's guide to good health* (New York: Henry Holt) [B, 311/60

Jarvis, D. C. (1960) *Arthritis and folk medicine* (New York: Holt, Rinehart & Winston) [B

Jewell, W. R. (1931) Mineral constituents of some Victorian honeys. *J. Dep. Agric. Vict.* 29: 435–436

Joachim, A. W. R. & Kandiah, S. (1940) The analysis of Ceylon foodstuffs. IX. The composition of some Ceylon honeys. *Trop. Agric. Mag. Ceylon agric. Soc.* 95: 339–340

Johansen, C. A. (1966) Digest on bee poisoning, its effects and prevention. *Bee Wld* 47(1): 9–25 [B, 544/66

Johansen, C. A. (1969) The bee poisoning hazard from pesticides. *Bull. Wash. agric. Exp. Stn* No. 709: 14 pp. [B, 172 L/70

Johansson, T. S. K. & Johansson, M. P. (1967) Lorenzo L. Langstroth and the bee space. *Bee Wld* 48(4): 133–143 [B, 754/69

Johansson, T. S. K. & Johansson, M. P. (1969) The development of comb foundation. *Bee Wld* 50(2): 61–65 [B, 418/70

Johansson, T. S. K. & Johansson, M. P. (1971) Substitutes for beeswax in comb and comb foundation. *Bee Wld* 52(4): 146–156 [B, 381/73

Johnsen, P. (1954) Facts about beekeeping in Denmark. *Bee Wld* 35(6): 105–110 [B, 201/55

Johnson, L. H. (1946) Nectar secretion in clover. *N.Z. Jl Agric.* 73: 111 [B

Johnson, L. H. (1956) A cool room for honey. *N.Z. Jl Agric.* 93(2): 151, 153–156 [B, 239/59

Jones, J. E., Graham, A. J. & Sackett, L. H., appendix by Geroulanos, M. I. (1973) *Annual of the British School of Archaeology at Athens* 68: 355–452 [B

Jones, W. R. (1947) Honey poisoning. *Glean. Bee Cult.* 75: 76–77 [B

Joshi, C. G. & Godbole, N. N. (1970) The composition and medical properties of natural honey as described in Ayurveda. *Indian Bee J.* 32(3/4): 77–78 [B, 810/72

Journal of the American Medical Association (1954) An older pharmacopoeia. *J. Am. med. Ass.* 155(1): 26 [B

Jürgen, T. (1962) *Science and secrets of early medicine.* (London: Thames and Hudson)

Juritz, C. F. (1925) The problem of Noors honey. *J. Dep. Agric. Un. S. Afr.* 10(4): 334–337 [B

Kaart, S. (1961) Elektriline talvitusmee kvaliteedi määramine [Electrical methods for determining honey quality]. Pp. 89–92 from *Horticulture and apiculture. Materials of the 5th scientific session of the Tartu branch of the Estonian Society for Horticulture and Apiculture* [B, 169/69

Kakao und Zucker (1969) Weichkaramel und Erdnuss-Krokant. *Kakao Zuck.* 21(7): 379–380 [B, 216 L/72

Kalimi, M. Y. & Sohonie, K. (1964a) Mahabaleshwar honey. I. Proximate and mineral analysis and paper chromatographic detection of amino acids and sugars. *J. Nutr. Dietet.* 1(4): 261–264 [753/70

Kalimi, M. Y. & Sohonie, K. (1964b) Mahabaleshwar honey. II. Effect of storage on carbohydrates, acidity, hydroxymethylfurfuraldehyde, colour, and diastase content of honey. *J. Nutr. Dietet.* 1(4): 265–268 [754/70

Kalimi, M. Y. & Sohonie, K. (1965) Mahabaleshwar honey. III. Vitamin contents (ascorbic acid, thiamine, riboflavin, and niacin) and effect of storage on these vitamins. *J. Nutr. Dietet.* 2(1): 9–11 [755/70

Kalinowska, R. (1957) Przech odzenie cynku z naczyń i przedmiotów użytku do żywności [Transfer of zinc into foods from cooking and storage utensils]. *Roczn. państ. Zakl. Hig.* 8(5): 421–430 [B, 315/62]

Kalman, C. (1962) Modernizing the management of honey production in Israel. *Bee Wld* 43(3): 83–86 [B, 377 L/63]

Kaloyereas, S. A. & Oertel, E. (1958) Crystallization of honey as affected by ultrasonic waves, freezing and inhibitors. *Am. Bee J.* 98(11): 442–443 [B, 351/60]

Kardara, C. (1961) Dyeing and weaving works at Isthmia. *Am. J. Archaeology* 65: 261–266

Kardos, R. F. (1941) Über die reduzierenden Substanzen des Honigs. *Z. Unters. Lebensmittel* 82: 33–37

Karger, M. I., Zhukova, K. N. & Radzivon, E. N. (1944) [Content of iodine in food products.] *Trudy biogeokhim. Lab.* 7: 42–50. *In Russian*

Kask, M. (1938) Vitamin C-gehalt der estonischen Honige. *Z. Unters. Lebensmittel* 76: 543–545

Keast, A. (1960) The unique plants and animals of south-western Australia. *Aust. Mus. Mag.* 13(5): 152–157 [B, 153 L/68]

Keast, A. (1968) Seasonal movements in the Australian honeyeaters (Meliphagidae) and their ecological significance. *Emu* 67(3): 159–210 [B]

Kebler, L. F. (1896) Poisonous honey. *Proc. Am. pharm. Ass.* 44: 167–174

Kellogg, C. R. (1968) *Entomological excerpts from south eastern China (Fukien Province). Aborigines: silkworms, honeybees and other insects* (Claremont, Calif.: published by the author) [B, 476/68]

Kellogg, C. R. & Tang, Chung-chang (1963) Honey and its uses in China. *Am. Bee J.* 103(5): 176–178 [B, 190 L/64]

Kelly, F. H. C. (1954) Phase equilibria in sugar solutions. IV. Ternary system of water-glucose-fructose. *J. appl. Chem., Lond.* 4: 409–411

Kelty, R. H. (1920) Profit in honey vinegar. *Glean. Bee Cult.* 48: 659 only. [B]

Kennedy, J. G. (1963) Plastic film for food packaging. *Food Manuf.* 38(8): 415–417, 420 [B, 208 L/65]

Kenya Ministry of Agriculture (1967) *Kenya bee keeping pilot project (Oxfam)* (Publ. Kenya Minist. Agric.) 26 pages [B, 334 L/70]

Kenya Ministry of Agriculture (1969) *Kenya bee keeping pilot project (Oxfam).* Proposals and application for phase II, January 1969 (Publ. Kenya Minist. Agric.) 21 pages [B, 606 L/70]

Kerkvliet, J. D. & Putten, A. P. J. van der (1973) The diastase number of honey: A comparative study. *Z. Lebensm. Unters. u. -Forsch.* 153(2): 87–93 [B]

Kerr, W. E. (1960) Espécies e racas de abelhas. *Coopercotia* 18(127): 29–34 [B, 168/61]

Kerr, W. E. & Maule, V. (1964) Geographic distribution of stingless bees and its implications (Hymenoptera: Apidae). *J. N.Y. ent. Soc.* 72(1): 2–18 [B, 741/64]

Keys, J. (1796) *The antient bee-master's farewell* (London: G. G. and J. Robinson) p. 273 [B]

Khristov, G. & Mladenov, S. (1961) [Honey in surgical practice: the antibacterial properties of honey.] *Khirurgiya* 14(10): 937–946. *In Bulgarian* [B, 887/63]

Kiermeier, F. & Köberlein, W. (1954) Über die Hitzeaktivierung von Enzymen im Honig. *Z. Lebensmittelunters. u. -Forsch.* 98(5): 329–347 [B, 41/55]

Kierulf, B. (1957) Kan lyngtrekket forutsies? [Is it possible to predict the heather flow?]. *Nord. Bitidskr.* 9(2): 42–44 [B, 193/58]

Kifer, H. B. & Munsell, H. E. (1929) Vitamin content of honey and honeycomb. *J. agric. Res.* 39(8): 355–366 [B]

Killion, C. E. (1951) *Honey in the comb* (Paris, Ill.: Killion & Sons Apiaries) 114 pp.
[B, 153/5²

Kirkwood, K. C., Mitchell, T. J. & Ross, I. C. (1961) An examination of the occurrence of honeydew in honey. *Analyst, Lond.* 86(1020): 164–165 [B, 390/65

Kirkwood, K. C., Mitchell, T. J. & Smith, D. (1960) An examination of the occurrence of honeydew in honey. *Analyst, Lond.* 85(1011): 412–416 [B, 316/62

Kitzes, G., Schuette, H. A. & Elvehjem, C. A. (1943) The B vitamins in honey. *J. Nutr.* 26(3): 241–250 [B

Klassert, M. (1909) Kritische Betrachtungen über die Fiehe'sche Reaktion. *Z. Unters. Nahr.- u. Genussmittel* 17: 126–128

Kleber, E. (1935) Hat das Zeitgedächtnis der Bienen biologische Bedeutung? *Z. vergl. Physiol.* 22: 221–262 [B

Klek, J. & Armbruster, L. (1921) Columella und Plinius. Die Bienenkunde der Römer. *Arch. Bienenk.* 3(8): 251–255 [B

Kloft, W. (1958) Arbeitstagung über Honigtaufragen in Freiburg i. Br. *Z. Bienenforsch.* 4: 126–130 [B

Kloft, W. (1960a) Die Trophobiose zwischen Waldameisen und Pflanzenläusen mit Untersuchungen über die Wechselwirkungen zwischen Pflanzenläusen und Pflanzengeweben. *Entomophaga* 5: 43–54

Kloft, W. (1960b) Wechselwirkungen zwischen pflanzensaugenden Insekten und den von ihnen besogenen Pflanzengeweben. *Z. angew. Ent.* 45: 337–381; 46: 42–70

Kloft, W. (1960c) Die Honigtau-Erzeuger. Pp. 105–114 from *Biene und Bienenzucht* ed. A. Büdel and E. Herold (Munich: Ehrenwirth) [B, 4/61

Kloft, W. (1962) Praktisch wichtige Probleme der Honigtauforschung. *Dt. Bienenwirt.* 13: 240–244 [B, 369 L/63

Kloft, W. (1963) Problems of practical importance in honeydew research. *Bee Wld* 44(1): 13–18, 24–29 [B, 822/63

Kloft, W. (1965) Die Honigtauerzeuger des Waldes. Pages 35–155 from *Das Waldhonigbuch* by W. Kloft et al. (Munich: Ehrenwirth Verlag) [B, see 114/67

Kloft, W. (1969) Radioaktive Isotope und ionisierende Strahlung bei der Erforschung und Bekämpfung von Insekten. *Arb. Forsch. Land. Nordrhein-Westfalen* 196: 47–76 (Cologne & Opladen: Westdeutscher Verlag)

Kloft, W. & Ehrhardt, P. (1962) Studies on the assimilation and excretion of labelled phosphate in aphids. Radioisotopes and radiation in entomology. *Int. atom. Energy. Ag. Bull., Vienna:* 181–190

Kloft, W., Maurizio, A. & Kaeser, W. (1965) *Das Waldhonigbuch* (Munich: Ehrenwirth Verlag) [B, 114/67

Kluge, F. (1960) *Etymologisches Wörterbuch der deutschen Sprache* (Berlin: Walther de Gruyter)

Kluge, M. (1964) *Untersuchungen über die Zusammensetzung von Siebröhrensäften.* Universität Darmstadt: Dissertation

Knapp, F. W. (1967) Methyl anthranilate content of citrus and non-citrus honeys. *XXI Int. Beekeep. Congr. prelim. sci. Meet. Summ.* Paper 34: 67 only
[B, 788/67

Knee, W. J. & Medler, J. T. (1965) Sugar concentration of bumble bee honey. *Am. Bee J.* 105(5): 174–175 [B, 626/65

Koch, H. G. (1957) Zur Phänologie der Bienenvölker. *Z. angew. Met.* 3(2): 33–43
[B, 78/60

Koch, H. G. (1959) Witterung und Nektartracht der Bienenvölker. *Z. angew. Met.* 3(9): 278–292 [B, 336/61

Koch, H. G. (1961) Der Baubeginn der Bienenvölker als phänologisches Ereignis. *Z. angew. Met.* 4(3): 69–82 [B, 359/63

Koch, H. G. (1967) Der Jahresgang der Nektartracht von Bienenvölkern als Ausdruck der Witterungssingularitäten und Trachtverhältnisse. *Z. angew. Met.* 5(7/8): 206–216 [B, 720/69]

Kodoyne, M. I. [Kodyne, M. I., Codounis, M. I.] (1962) [*The crystallization of honey*]. (Athens: Ministry of Agriculture) 88 pp. *In Greek; long French summary* [B, 909/64]

Kodyne, M. I. [Kodoyne, M. I., Codounis, M. I.] (1967) The problem of honey crystallization and its refinement in fruit juice and preserves co-operatives in Greece. *XXI Int. Beekeep. Congr. Summ.* Paper 142: 94–95 [B, 794 L/67]

Kolb, H. (1972) World's oldest honey . . . ? *Am. Bee J.* 112(2): 53 [B

Komamine, A. (1960) Amino acids in honey. *Acta chem. fenn.* B 33: 185–187 [B, 153/62]

Kopřivová, M. (1961) *Zužitkování medu v domácnosti* [Domestic uses of honey] (Prague: Ceskoslovenský Svaz Včelařů) [B, 416 L/64]

Kornfeld, L. (1967) [*Honey in food and drink recipes*] (Tel Aviv: Israeli Bee-keepers' Association). *In Hebrew* [B

Kosmin, N. P. & Komarov, P. M. (1932) Über das Invertierungsvermögen der Speicheldrüsen und des Mitteldarmes von Bienen verschiedenen Alters. *Z. vergl. Physiol.* 17: 267–279 [B

Kottász, J. (1958a) A méz színének meghatározása. [Grading the colour of honey]. *Méhészet, Budapest* 6(6): 100 only [B, 94/59]

Kottász, J. (1958b) A méz szárazanyag-, illetve víztartalmának meghatározása [Testing honey for dry matter and water content]. *Méhészet, Budapest* 6(8): 137 only [B, 198/59]

Kovalev, A. M. & Burmistrov, A. N. (1969) [Characteristics of the honey-crop regions in R.S.F.S.R.] *Trudy nauchno-issled. Inst. Pchel.*: 165–176. *In Russian* [B, 415/72]

Kramer, S. N. (1958) *History begins at Sumer* (London: Thames & Hudson)

Kranz, B. (1967) Method for treating honey and the like. *U.S. Pat.* No. 3,297,453 [B, 383/67]

Kratky, E. (1931) Morphologie und Physiologie der Drüsen in Kopf und Thorax der Honigbiene. *Z. wiss. Zool.* 139: 120–200 [B, T

Kreis, H. (1915) Beitrag zur Honiguntersuchung nach der Präcipitinmethode. *Mitt. Geb. Lebensmittelunters. u. Hyg.* 6: 53–62

Kremp, G. O. W. (1968) *Morphologic encyclopedia of palynology* (Tucson: University of Arizona Press) [B, 802/70]

Kropáčová, S. (1963) Vztahy mezi klimatickými faktory, nektarem vojtěšky a náletem včel (*Apis mellifera* L.) na vojtěšku [Relations between climatic factors, lucerne nectar, and the number of honeybees working on lucerne]. *Sb. vys. Šk. zeměd. v Brně* A 4: 603–611 [B, 857/64]

Kropáčová, S. (1969) Příspěvek k pylovým analýzám medů jihovýchodní moravy [Contribution to the pollen analysis of honey produced in south-east Moravia]. *Sb. vys. Šk. zeměd. v Brně* A 17(4): 793–797 [B, 453/71]

Kropáčová, S. & Haslbachová, H. (1970) A study of the honeybee work (*Apis mellifera* L.) on sainfoin plants (*Onobrychis viciaefolia* s. *sativa* Thell.). *Sb. vys. Šk. zeměd. v Brně* A 18(1): 71–82 [B, 343/73]

Kropáčová, S. & Kropáč, A. (1968) [Importance of the chive (*Allium schoenoprasum* L.) as the source of nectar.] *Sb. vys. Šk. zeměd. v Brně* 16(2): 263–269 [B, 507/69]

Kropáčová, S. & Laitová, L. (1965) Rozbor obsahu cukrů v nektaru vojtěšky [Analysis of the sugars in lucerne nectar]. *Sb. vys. Šk. zeměd. v Brně* 12(3): 425–431 [B, 493/66]

Kropáčová, S. & Nedbalová, V. (1970) Nektárnost některých druhů vrb (*Salix* sp.).

[The nectar-bearing capacity of some willow species]. *Sb. čsl. Akad. zeměd. Věd, Lesnictvi* 16(43) (12): 1095–1100 [B, 700/72

Kropáčová, S. & Nedbalová, V. (1971) Pylové analýzy medů západní Moravy. [Pollen analysis of honey from western Moravia] *In Czech Sb. vys. Šk. zeměd. v Brně* 19(4): 703–710 [B, 425/73

Krumbholz, G. (1936) Neuere Forschungen über die Gärungsvorgänge in zucker-reichen Lebensmitteln. *Obst- u. Gemüse-Verw.-Ind.* 23: 70–72, 85–87, 96–99, 113–114

Krupička, P. (1959) Činnost včel při krmeni suchým cukrem [The activity of bees when fed dry sugar]. *Včelařstvi* 12(3): 37–38 [B, 16/61

Kuény, G. (1950) Scènes apicoles dans l'ancienne Egypte. *J. Near E. Stud.* 9: 84–93 [B, T

Kuliev, A. M. (1952) [*The study of plants yielding nectar and pollen*] (Moscow: Akademia Nauk S.S.S.R.). *In Russian* [B, 94/54

Kulinčević, J. (1959) Facts about beekeeping in Yugoslavia. *Bee Wld* 40: 241–250, 314–315 [B, 258/60

Kunkel, H. (1969) Über das Verhalten der Aphiden und verwandter Honigtauer-zeuger bei der Abgabe von Honigtau. *XXII Int. Beekeep. Congr. 1969, Munich:* 477–481 (1970) [B, see 853 L/70

Kunkel, H. (1972) Die Kotabgabe bei Aphiden (Aphidina, Hemiptera). *Bonn. zool. Beitr.* 23(2): 161–178

Kushida, T., Nakayama, O. & Koike, H. (1961) [Studies on the application of honey to alcoholic beverages. Part 1. The preliminary experiment on the application of honey to port-wine.] *Yamanashi Daigaku Hakko Kenkyu Hokoka* No. 8: 19–25. *In Japanese; English summary* [B, 190/66

Kushida, T., Nakayama, O. & Koike, H. (1963) [Studies on the application of honey to alcoholic beverages. Part 5. Experimental production of wines, applying honey and sugar to must.] *Yamanashi Daigaku Hakko Kenkyu Hokoka* No. 10: 43–47. *In Japanese; English summary* [B, 194/66

Küstenmacher, M. (1911) Zur Chemie der Honigbildung. *Biochem. Z.* 30:237–254

La Fay, H. (1956) New light on the Bible. *Town J.* (March): 34–35 [B

Lamb, K. P. (1959) Composition of the honeydew of the aphid *Brevicoryne brassicae* (L.) feeding on swedes (*Brassica napobrassica* DC). *J. Insect Physiol.* 3: 1–13 [B, 466/62

Lamond, E., ed. (1890) *Walter of Henley's 'Husbandry'* (London: Longmans, Green & Co.) pp. 80–81 [B (photocopy)

Lampitt, L. H. & Bilham, P. (1936) Notes on the absorption spectra of honey. *Chemy Ind.* 14: 71–72

Lampitt, L. H., Hughes, E. B. & Rooke, H. S. (1930) The diastatic activity of honey. *Analyst, Lond.* 55: 666–672 [B

Lane, J. H. & Eynon, L. (1923) Determination of reducing sugars by means of Fehlings solution with methylene blue as internal indicator. *J. Soc. chem. Ind., Lond.* 42: 32T, 143T, 463T

Langer, J. (1903) Fermente im Bienenhonig. *Schweiz. Wschr. Chem. Pharm.* 41: 17–18

Langer, J. (1909) Beurteilung des Bienenhonigs und seiner Verfälschungen mittels biologischer Eiweissdifferenzierung. *Arch. Hyg. Bakt.* 71: 308–330 [B

Langer, J. (1915) Das (serologisch fassbare) Eiweiss des Honigs stammt von der Biene (Langer) und nicht aus dem Blütenstaube (Küstenmacher). *Biochem. Z.* 69: 141–144

Langridge, D. F. (1961) Facts about beekeeping in Australia II. Victoria. *Bee Wld* 42(6): 153–156　　　　　　　　　　　　　　　　　　　　　　　[B, 758 L/63

Langridge, D. F. (1966) An investigation into some quality aspects of Victorian honey. *J. Agric. Vict. Dep. Agric.* 64: 81–90, 119–126, 139　　　　[B, 573/67

Langstroth, L. L. (1853) *Langstroth on the hive and the honey-bee, a bee keeper's manual* (Northampton, Mass.: Hopkins, Bridgman & Co.)　　[B (1914 reprint)

Lasceve, G. & Gonnet, M. (1974) Analyse par radioactivation du contenu minéral d'un miel. Possibilité de préciser son origine géographique. *Apidologie* 5(3): 201–223　　　　　　　　　　　　　　　　　　　　　　　　　　　　[B

Latif, A., Qayyum, A. & Manzoor-ul-Haq (1956) Researches on the composition of Pakistan honey. *Pakist. J. sci. Res.* 8(4): 163–166　　　　　[B, 29/59

Latif, A., Qayyum, A. & Manzoor-ul-Haq (1958) A contribution to bee-flora of Pakistan. *Pakist. J. sci. Res.* 10(2): 67–71　　　　　　　　　[B, 94/60

Latin America (1964) Código latinoamericano de alimentos. [honey, p. 160; artificial honey, p. 229; labelling, p. 61; containers, p. 53; trace metals, p. 33; health claims, p. 62]

Lavie, P. (1960) *Les substances antibactériennes dans la colonie d'abeilles (Apis mellifica* L.). Université de Paris: Thèses doctoraux; *Annls Abeille* 3: 103–183, 201–305　　　　　　　　　　　　　　　　　　　　　　　　[B, 761/63

Lavie, P. (1963) Sur l'identification des substances antibactériennes présentes dans le miel. *C.r. hebd. Séanc. Acad. Sci., Paris* 256: 1858–1860　　　[B, 880/63

Laxa, O. (1923) Méthode nouvelle et simple pour le dosage des albuminoides dans le miel. *Annls Falsif. Fraudes* 16: 286–289

Leclant, J. (1968) L'abeille et le miel dans l'Égypte pharaonique. Pp. 51–60 from *Traité de biologie de l'abeille*, Vol. V, ed. R. Chauvin (Paris: Masson et Cie)
[B, 325/68

Lecomte, J. (1961) Le comportement agressif des ouvrières d'*Apis mellifica* L., *Annls Abeille* 4: 165–270　　　　　　　　　　　　　　　　　[B, 428/62

Lederhouse, R. C., Caron, D. M. & Morse, R. A. (1968) Onion pollination in New York. *New York's Fd Life Sci.* 1(3): 8–9　　　　　　　　　[B, 17/69

Lee, R. B. & Devore, I. (ed.) (1968) *Man the hunter* (Chicago: Aldine Publishing Co.)

Lehrner, L. (1955) Többszörös atropinmérgezés méz fogyasztása után. *Népegészségügy* 36: 315–316

Leibovici, M. (1968a) L'abeille et le miel dans l'histoire des religions. Pp. 35–40 from *Traité de biologie de l'abeille*, Vol. V, ed. R. Chauvin (Paris: Masson et Cie)
[B, 325/68

Leibovici, M. (1968b) L'abeille et le miel selon la tradition babylonienne. Pp. 41–42 from *Traité de biologie de l'abeille*, Vol. V, ed. R. Chauvin (Paris: Masson et Cie)
[B, 325/68

le Maistre, W. G. (1936) The O.A.C. honey strainer. *Can. Bee J.* 44(11): 279–280 [B

Lennartz, T. (1947) Über antituberkulöse Stoffe: Antibiotische Eigenschaften von Derivaten des Honigs. *Z. Naturforsch.* 26: 7–9　　　　　　　　[B

Lensky, Y. (1961) Les échanges de nourriture liquide entre abeilles aux températures élevées. *Insectes soc.* 8: 361–368　　　　　　　　　　[B, 571/63

Lerner, F. (1963) *Aber die Biene nur findet die Süssigkeit* (Dusseldorf: Econ Verlag) 248 pp.　　　　　　　　　　　　　　　　　　　　　　　　　　[B

Le Sage, D. E. (1974) Bees in Indo-European languages. *Bee Wld* 55: 15–26, 46–52 [B

Leuenberger, F. & Morgenthaler, O. (1954) *Die Biene* (Aarau: Sauerländer)　[B

Lévi-Strauss, C. (1966) *Du miel aux cendres* (Paris: Plon)　　　　　[B, 184/73

Lévi-Strauss, C. (1973) *From honey to ashes* translated by J. & D. Weightman from *Du miel aux cendres* (1966) (London: Jonathan Cape Ltd.)　[B, 183/73

Lewis, C. Day, translator, *see* Virgil

Lewis, H. & Pedersen, H. (1937) *A concise comparative Celtic grammar.* (Göttingen: Vandenhoeck & Ruprecht)

Lieux, M. H. (1972) A melissopalynological study of 54 Louisiana (U.S.A.) honeys. *Rev. Palaeobot. Palynol.* 13(2): 95–124 [B, 474/74

Lindauer, M. (1953) Bienentänze in der Schwarmtraube (II). *Naturwissenschaften* 40(14): 379–385 [B, 29/55

Lindauer, M. (1956) Über die Verständigung bei indischen Bienen. *Z. vergl. Physiol.* 38: 521–557 [B, 325/58

Lindauer, M. (1957) Communication among the honeybees and stingless bees of India. *Bee Wld* 38(1): 3–14, 34–39 [B, 325/58

Lindner, K. E. (1962) Ein Beitrag zur Frage der antimikrobiellen Wirkung der Naturhonige. *Zentbl. Bakt. ParasitKde* 115(7): 720–736 [B, 381/65

Line, L. J. S. (1955) Substitute for a beekeeping hot room. *N.Z. Jl Agric.* 91(5): 461–462 [B, 128/59

Lochhead, A. G. (1933) Factors concerned with the fermentation of honey. *Zentbl. Bakt. ParasitKde* II Abt. 88: 296–302 [B

Lochhead, A. G. & Farrell, L. (1930a) Soil as a source of infection of honey by sugar-tolerant yeasts. *Can. J. Res.* 3(1): 51–64 [B

Lochhead, A. G. & Farrell, L. (1930b) Effect of preservatives on fermentation by sugar-tolerant yeasts from honey. *Can. J. Res.* 3: 95–103 [B

Lochhead, A. G. & Farrell, L. (1931a) Accessory food substances for osmophilic yeasts. I. A bioactivator in honey stimulating fermentation. *Can. J. Res.* 5: 529–538 [B

Lochhead, A. G. & Farrell, L. (1931b) The types of osmophilic yeasts found in normal honey and their relation to fermentation. *Can. J. Res.* 5: 665–672 [B

Lochhead, A. G. & Heron, D. A. (1929) Microbiological studies of honey. I. Honey fermentation and its cause. II. Infection of honey by sugar-tolerant yeasts. *Bull. Can. Dep. Agric.* No. 116: 47 pp. [B

Lochhead, A. G. & McMaster, N. R. (1931) Yeast infection of normal honey and its relation to fermentation. *Scient. Agric.* 11: 351–360 [B

Lockwood, W. B. (1969) *Indo-European philology* (London: Hutchinson University Library)

Loper, G. M. & Waller, G. D. (1969) Honeybee (*Apis mellifera* L.) selection among alfalfa flowers (*Medicago sativa*). *Agron. Abstr. A. Mtg Am. Soc. Agron. Detroit, Mich.*: 36

Loper, G. M. & Waller, G. D. (1970) Alfalfa flower aroma and flower selection by honey bees. *Crop Sci.* 10: 66–68 [B, 909/70

Lo Pinto, Maria (1957) *Eat honey and live longer* (New York: Twayne Publishers) [B, 5a28(2) L/60

Lothrop, R. E. (1932) Specific test for orange honey. *Ind. Engng Chem. analyt. Edn* 4: 395–396 [B

Lothrop, R. E. (1939) The composition of honey and its utilization-relation of composition and viscosity. *Am. Bee J.* 79: 130–133 [B

Lothrop, R. E. (1943) *Saturation relations in aqueous solutions of some sugar mixtures with special reference to higher concentrations.* George Washington Univ.: Thesis [B

Lothrop, R. E. & Gertler, S. I. (1933) Determination of amino acids and related compounds in honey. *Ind. Engng Chem. analyt., Edn* 5(2): 103–105 [B

Lothrop, R. E. & Holmes, R. L. (1931) Determination of dextrose and laevulose in honey by use of iodine oxidation method. *Ind. Engng Chem. analyt. Edn* 3: 334–339

Lothrop, R. E. & Paine, H. S. (1931a) Some properties of honey colloids and the

removal of colloids from honey with bentonite. *Ind. Engng Chem.* 23(3): 328–332
[B

Lothrop, R. E. & Paine, H. S. (1931*b*) Diastatic activity of some American honeys. *Ind. Engng Chem. analyt., Edn* 23: 71–74

Lothrop, R. E. & Paine, H. S. (1934) A new method of processing honey. *Am. Bee J.* 74(12): 542–543
[B

Lotmar, R. (1935) Abbau und Verwertung von Stärke und Dextrin durch die Honigbiene. *Arch. Bienenk.* 16(6): 195–204
[B

Louveaux, J. (1954) Application des méthodes de l'analyse pollinique à des produits au miel. *Annls Falsif. Fraudes* 47: 422–424

Louveaux, J. (1955) Introduction à l'étude de la récolte du pollen par les abeilles (*Apis mellifica* L.) *Phys. comp. Oecol.* 4(1): 1–54
[B, *see* 330/61

Louveaux, J. (1956) Remarques sur les facteurs conditionnant le choix par les abeilles (*Apis mellifica* L.) des plantes leur fournissant du pollen. *C.r. Acad. Sci., Paris* 242: 2994–2996
[B, 230/60

Louveaux, J. (1956*a*) Application des méthodes de l'analyse pollinique à des produits au miel. *Apiculteur* 100(5) *Sect. sci.:* 29–31
[B, 170/58

Louveaux, J. (1956*b*) Étude des miels français par l'analyse pollinique. *XVI Int. beekeep. Congr. prelim. sci. Meet.* Vienna Paper No. 637
[B, 260/56

Louveaux, J. (1958) Recherches sur l'origine dans le miel du pollen de plantes entomophiles dépourvues de nectaires. *Annls Abeille* 1(2): 89–92
[B, 346/59

Louveaux, J. (1958/1959) Recherches sur la récolte du pollen par les abeilles (*Apis mellifica* L.) *Annls Abeille* 1: 113–118, 197–221; 2: 13–111
[B, 330/61

Louveaux, J. (1961) Techniques améliorées pour l'analyse pollinique des miels. *Z. Bienenforsch.* 5(7): 199–204
[B, 209/63

Louveaux, J. (1966*a*) Essai de caractérisation des miels de callune (*Calluna vulgaris* Salisb.) *Annls Abeille* 9(4): 351–358
[B, 385/67

Louveaux, J. (1966*b*) Pollenanalyse einiger kanadischer Honige. *Z. Bienenforsch.* 8(6): 195–202
[B, 566/66

Louveaux, J. (1968*a*) L'analyse pollinique des miels. Pp. 325–362 from Vol. III *Traité de biologie de l'abeille* ed. R. Chauvin (Paris: Masson et Cie)
[B, 397/68

Louveaux, J. (1968*b*) Die französischen Heidehonige. *Z. Bienenforsch.* 9(5): 211–215
[B, 1016/70

Louveaux, J. (1970) *Atlas photographique d'analyse pollinique des miels.* Vol. III, Annexes microphotographiques aux méthodes officielles d'analyse (Paris: Service de la répression des fraudes et du contrôle de la qualité)
[B, 1026/71

Louveaux, J. & Trubert, E. (1958) A technical study on liquefying granulated honey. *Annls Abeille* 1(1): 19–30
[B, 385/59

Lund, R. (1909) Albuminate im Naturhonig und Kunsthonig. *Z. Unters. Nahr.- u. Genussmittel* 17: 128–130

Lund, R. (1910) Über die Untersuchung des Bienenhonigs unter spezieller Berücksichtigung der stickstoffhaltigen Bestandteile. *Mitt. Geb. Lebensmittelunters. u. Hyg.* 1: 38–58

Lunder, R. (1945) Pollenanalystiska undersökningar av Svensk honung. *Medd. Växtskyddsanst.* (45): 31 pp.
[B

Lunder, R. (1955) Der Einfluss von Honiglösapparaten auf das Pollenbild des Heidehonigs. *Z. Bienenforsch.* 3(3): 49–52
[B, 201/56

Lüttge, U. (1961, 1962*a*, 1962*b*) Über die Zusammensetzung des Nektars und den Mechanismus seiner Sekretion. *Planta* 56: 189–212; 59: 108–114, 175–194
[B, 130, 131/64

Lüttge, U. (1962*c*) Über den Vitamin C-Gehalt von Bienenhonig. *Z. Lebensmittelunters. u. -Forsch.* 117(4): 289–296
[B, 197/64

Lüttge, U. (1964) Die Nektarsekretion. *Dt. Bienenwirt.* 15: 238–243 [B, 306 L/55

Lüttge, U. (1969) Aktiver Transport. *Protoplasmatologia* 8(7b): 1–146 (Wien: Springer Verlag) [B, 661/71

Luxembourg. Laws & Statutes (1966) Règlement grand-ducal du 12 mars 1966 concernant le miel et les produits similaires. *Méml No. A* 18: 377 (8.4.66)

Maa, T. (1953) An enquiry into the systematics of the tribus Apidini or honey bees (Hym.). *Treubia* 21(3): 525–640 [B, 146/59

MacDonald, J. E. (1963) Honey pumps. *Glean. Bee Cult.* 91(2): 85–87 [B, 222/68

McDonald, J. L. (1964a) *Further studies on honey crystallization*. Cornell University: M.Sc. thesis [B, 579/67

McDonald, J. L. (1964b) Fine crystallized honey. *Glean. Bee Cult.* 92: 614–615, 689–700 [B

Maclachlan, R. G. (1935/36) Mystery of nectar. *Aust. Bee J.* 16: 113–116, 124–126, 138–140, 192–194; 17: 6–8, 24–26, 41–44, 54–56, 74–76 [B

Maclachlan, R. G. (1940) The starch content of plants in relation to honey flows. *Bee Wld* 21(9): 85–88 [B

Maeda, S., Mukai, A., Kosugi, N. & Okada, Y. (1962) [The flavour components of honey.] *J. Fd Sci. Tech.* 9(7): 270–274. *In Japanese; English summary* [B, 921/64

Maher, L. J., Jr. (1972) Nomograms for computing 0.95 confidence limits of pollen data. *Rev. Palaeobot. Palynol.* 13(2): 85–93

Majchrzak, R. (1962) [Stabilization of liquid foods such as musts, wines and meads]. *Pol. Pat.* 46,075 *In Polish* [B, 392/65

Makarochkin, B. A. & Yudenich, D. M. (1960) [Minerals in honey]. *Pchelovodstvo* 37(11): 34. *In Russian* [B, 903 L/63

Maksymiuk, I. (1960) Nektarowanie lipy drobnostnej *Tilia cordata* Mill. w rezerwacie Obrożyska kolo Muszyny [The nectar secretion of linden *Tilia cordata* Mill. at reserve Obrozyska near Muszyna (Carpathians)]. *Pszczel. Zesz. Nauk.* 4(2): 105–125 [B, 847/64

Mallik, A. K. (1958) Analysis of Indian honey. *J. Inst. Chem. (India)* 30(3): 171–173 [B, 168/60

Maltais, J. B. & Auclair, J. L. (1962) Free amino acid and amide composition of pea leaf juice, pea aphid haemolymph and honeydew following the rearing of aphids on single pea leaves treated with amino compounds. *J. Insect. Physiol.* 8: 391–399 [B, 46/65

Malyoth, E. (1951) Zur papierchromatographischen Darstellung des Honigs. *Naturwissenschaften* 38: 478 [B, 248/52

Manjo, G. (1970) personal communication

Markuze, Z. (1935) Zawartość witamin w miodzie [Vitamin content of honey]. *Archwm Chem. Farm.* 2: 175–182

Marpmann, G. (1903) Beiträge zur Prüfung und Beurteilung des Bienenhonigs. *Pharm. Ztg. Berl.* 48: 1010

Marquardt, P., Aring, E. & Vogg, G. (1953) Untersuchungen über das gemeinsame Vorkommen von Acetylcholin und Diastase im Honig. *Arzneimittel-Forsch.* 3: 446–448 [B, 40/55

Marquardt, P. & Spitznagel, G. (1956) Über den Gehalt des Honigs an Acetylcholin. *Fette Seif. AnstrMittel* 58(10): 863–865 [B, 702/62

Marquardt, P. & Vogg, G. (1952) Vorkommen, Eigenschaften, und chemische Konstitution des cholinergischen Faktors im Honig. *Arzneimittel-Forsch.* 2: 152–155 [B, 39/55

Marshall, C. R. & Norman, A. G. (1938) The analysis of mixtures of glucose and fructose with special reference to honey. *Analyst, Lond.*, 63: 315–323

Martens, N., Laere, O. van & Pelerents, C. (1964) Studie van de bijenflora in België door pollenanalyse. *Biol. Jaarb.* 32: 292–325 + 1 plate   [B, 489/66

Martensen-Larsen, O. (1954) Detoxication of drunkenness. *Brit. med. J. (ii)*: 464 only   [B, 65/55

Martimo, E. (1945) Suomalaisen hunajan ominaisuuksista ja alkuperästä [Composition and source of origin of Finnish honey]. *Maataloust. Aikakausk.* 17: 157–169

Martin, E. C. (1939) The hygroscopic properties of honey. *J. econ. Ent.* 32(5): 660–663   [B

Martin, E. C. (1958) Some aspects of hygroscopic properties and fermentation of honey. *Bee Wld* 39(7): 165–178   [B, 350/59

Marvin, G. E. (1928) The occurrence and characteristics of certain yeasts found in fermented honey. *J. econ. Ent.* 21(2): 363–370   [B

Marvin, G. E. (1930) Further observations on the deterioration and spoilage of honey in storage. *J. econ. Ent.* 23(2): 431–438   [B

Marvin, G. E. (1933) Methods for determining the weight per gallon of honey. *Am. Bee J.* 73(11): 426–428   [B

Marvin, G. E. (1934) Water content and weight per gallon of honey. *Am. Bee J.* 74(5): 212 only   [B

Marvin, G. E., Peterson, W. H., Fred, E. B. & Wilson, H. F. (1931) Some of the characteristics of yeasts found in fermenting honey. *J. agric. Res.* 43(2): 121–131 [B

Marvin, G. E. & Wilson, H. F. (1931) Some comparative data on moisture determination in honey by means of the refractometer and the vacuum drying oven. *J. econ. Ent.* 24: 603–604

Marvin, G. E. & Wilson, H. F. (1932) Some comparative data on moisture in top and bottom layers of honey after a year of storage, as indicated by the vacuum drying oven and refractometer. *J. econ. Ent.* 25(2): 514–520   [B

Maslowski, P. & Mostowska, I. (1963) Elektrochromatograficzne badania skladu wolnych aminokwasów w niektórych odmianach miodów [Electrochromatographic estimation of free amino acids in different honeys]. *Pszczel. Zesz. nauk.* 7(1): 1–6   [B, 919/64

Masson, L. & Schmidt-Hebbel, H. (1963) Contribución al estudio de los glucidos en mieles Chilenas. *Nutrición Bromatol., Toxicol.* 2: 85–88

Masterman, J. (1961) Facts about beekeeping in Australia 1. South Australia. *Bee Wld* 42(2): 29–36   [B, 756 L/63

Matile, P. (1956) Über den Stoffwechsel und die Auxinabhängigkeit der Nektarsekretion. *Ber. schweiz. bot. Ges.* 66: 237–266   [B, 92/62

Maugenet, J. (1964) L'hydromel. *Annls Abeille* 7: 165–179   [B, 189/66

Maurizio, A. (1936a) Gibt es Lindenhonig in der Schweiz? *Schweiz. Bienenztg* 59(3): 148–156   [B

Maurizio, A. (1936b) Schweizerische Honigtypen 2. Weidenhonig. *Schweiz. Bienenztg* 59(10): 547–556   [B

Maurizio, A. (1939) Untersuchungen zur quantitativen Pollenanalyse des Honigs. *Mitt. Geb. Lebensmittelunters. u. Hyg.* 30(1/2): 27–69   [B

Maurizio, A. (1940) Schweizerische Honigtypen 3. Vergissmeinnichthonig. *Schweiz. Bienenztg* 63 (1, 2, 3): 29–34; 87–93; 147–151   [B

Maurizio, A. (1941) Schweizerische Honigtypen. 4. Honig der Edelkastanie. *Schweiz. Bienenztg* 64(7/8): 351–362; 409–417   [B

Maurizio, A. (1946) Schweizerische Honigstatistik III. *Beih. schweiz. Bienenztg* 1(12): 571–873   [B

Maurizio, A. (1947) Schweizerische Honigtypen 5. Walliser Honigtypen. *Murithienne* 64: 38–50 [B

Maurizio, A. (1949*a*) Beiträge zur quantitativen Pollenanalyse des Honigs. *Beih. schweiz. Bienenztg* 2(18): 320–421 [B, T, 176/50

Maurizio, A. (1949*b*) Wird das Pollenbild des Honigs durch die Vorgänge in der Honigblase beeinflusst? *Beih. schweiz. Bienenztg* 2(18): 422–441 [B, T, 176/50

Maurizio, A. (1951) Pollen analysis of honey. *Bee Wld* 32(1): 1–5 [B, 47/53

Maurizio, A. (1952) Woher stammen die im Honig enthaltenen pflanzlichen Bestandteile? *Arch. Bienenk.* 29: 1–11 [B, 225/53

Maurizio, A. (1954*a*) Untersuchungen über die Nektarsekretion einiger polyploider Kulturpflanzen. *Arch. Julius Klaus-Stift. VererbForsch.* 29(3/4): 340–346 [B, 295/55

Maurizio, A. (1954*b*) Pollenernährung und Lebensvorgänge bei der Honigbiene (*Apis mellifica* L.). *Landw. Jb. Schweiz* 68: 115–182 [B, 7/56

Maurizio, A. (1955) Beiträge zur quantitativen Pollenanalyse des Honigs. 2. Absoluter Gehalt pflanzlicher Bestandteile in Tilia- und Labiaten-Honigen. *Z. Bienenforsch.* 3(2): 32–39 [B, 123/58

Maurizio, A. (1957) Zuckerabbau unter der Einwirkung der invertierenden Fermente in Pharynxdrüsen und Mitteldarm der Honigbiene (*Apis mellifica* L.) 1. Sommerbienen der Krainer- und Nigra-Rasse. *Insectes soc.* 4(3): 225–243 [B, 226/59

Maurizio, A. (1958*a*) Tipi di mieli della Svizzera italiana. *Riv. svizz. Apic.* 41(1): 20–26 [B, 279/59

Maurizio, A. (1958*b*) Beiträge zur quantitativen Pollenanalyse des Honigs. 3. Absoluter Gehalt pflanzlicher Bestandteile in Esparsette-, Luzerne-, Orangen- und Rapshonigen. *Annls Abeille* 1(2): 93–106 [B, 423/61

Maurizio, A. (1958*c*) Nouvelles recherches sur la sécrétion nectarifère de plantes cultivées polyploides: *Nicotiana*. *X Int. Congr. Ent. 1956, Montreal* 4: 1025 [B

Maurizio, A. (1959*a*) Zur Frage der Mikroskopie von Honigtau-Honig. *Annls Abeille* 2(2): 145–157 [B, 465/61

Maurizio, A. (1959*b*) Papierchromatographische Untersuchungen an Blütenhonigen und Nektar. *Annls Abeille* 2: 291–341 [B, 451/61

Maurizio, A. (1959*c*) Breakdown of sugars by inverting enzymes in the pharyngeal glands and midgut of the honeybee 2. Winter bees (Carniolan and Nigra). *Bee Wld* 40(11): 275–283 [B, 331/60

Maurizio, A. (1960*a*) *Blüte, Nektar, Pollen, Honig* (Nürnberg: Deutscher Imkerbund) [B, 217/61

Maurizio, A. (1960*b*) Das mikroskopische Bild jugoslawischer Importhonige. *Z. Bienenforsch.* 5(1): 8–22 [B, 464/61

Maurizio, A. (1960*c*) Bienenbotanik. Pp. 68-104 from *Biene und Bienenzucht* ed. A. Büdel & E. Herold (Munich: Ehrenwirth) [B, see 4/61

Maurizio, A. (1961) Zuckerabbau unter der Einwirkung der invertierenden Fermente in Pharynxdrüsen und Mitteldarm der Honigbiene (*Apis mellifica* L.). 3. Fermentwirkung während der Überwinterung bei Bienen der Ligustica-Rasse. *Insectes soc.* 8(2): 125–175 [B, 330/63

Maurizio, A. (1962) Das Pollenbild des Honigs einzelner Völker eines Standes. *Dtsch. Bienenw.* 13(8): 235–239 [B, 410/64

Maurizio, A. (1962*a*) Zuckerabbau unter der Einwirkung der invertierenden Fermente in Pharynxdrüsen und Mitteldarm der Honigbiene (*Apis mellifica* L.). 4. Sommerbienen der Italienischen, Kaukasischen und Griechischen Rasse. *Insectes soc.* 9(1): 39–72 [B, 541/63

Maurizio, A. (1962*b*) Zuckerabbau unter der Einwirkung der invertierenden Fer-

mente in Pharynxdrüsen und Mitteldarm der Honigbiene (*Apis mellifica* L.).
5. Einfluss von Alter und Ernährung der Bienen auf die Fermentaktivität der
Pharynxdrüsen. *Annls Abeille* 5(3): 215–232   [B, 797/64

Maurizio, A. (1962c) From the raw material to the finished product: honey. *Bee Wld*
43: 66–81   [B, 800/63

Maurizio, A. (1963) Pollenanalytische Beobachtungen 17–19. Honige aus Sumpf-
und Moorgebieten. *Grana palynol.* 4(2): 231–244   [B, 386/65

Maurizio, A. (1964) Mikroskopische und papierchromatographische Untersuchungen an Honig von Hummeln, Meliponinen und anderen zuckerhaltige
Säfte sammelnden Insekten. *Z. Bienenforsch.* 7(4): 98–110   [B, T, 387/65

Maurizio, A. (1965a) Honigtau—Honigtauhonig. Pp. 159–185 from *Waldhonigbuch*
by W. Kloft *et al.* (Munich: Ehrenwirth Verlag)   [B, T, 114/67

Maurizio, A. (1965b) Untersuchungen über das Zuckerbild der Hämolymphe der
Honigbiene (*Apis mellifica* L.). 1. Das Zuckerbild des Blutes erwachsener
Bienen. *J. Insect. Physiol.* 11: 745–763   [B, 72/66

Maurizio, A. (1966) Das Pollenbild europäischer Heidehonige. *Annls Abeille* 9(4):
375–387   [B, 386/67

Maurizio, A. (1968a) La formation du miel. Pp. 264–276 from *Traité de biologie de
l'abeille* Vol. III, ed. R. Chauvin (Paris: Masson et Cie)   [B, see 397/68

Maurizio, A. (1968b) Les diastases des glandes nourricières. Pp. 291–301 from
Traité de biologie de l'abeille Vol. I, ed. R. Chauvin (Paris: Masson et Cie)
                                                                [B, see 323/68

Maurizio, A. (1968c) Les plantes toxiques. Pp. 279–287 from *Traité de biologie de
l'abeille* Vol. IV, ed. R. Chauvin (Paris: Masson et Cie)   [B, 367/68

Maurizio, A. (1971) Le spectre pollinique des miels Luxembourgeois. *Apidologie*
2(3): 221–237   [B, 496/72

Maurizio, A. (1974) Die Bienenweide. Pp. 211–236 from *Der Schweizerische
Bienenvater* (Aarau: Sauerländer)   [B

Maurizio, A. & Grafl, I. (1969) *Trachtpflanzenbuch* (Munich: Ehrenwirth Verlag)
                                                                    [B, 396/70

Maurizio, A. & Louveaux, J. (1965) *Pollen de plantes mellifères d'Europe* (Paris:
Union des Groupements Apicoles Français) 148 pp. including 60 plates
                                                                    [B, 652/65

Maurizio, A. & Louveaux, J. (1967) Les méthodes et la terminologie en mélis-
sopalynologie. *Rev. Palaeobot. Palynol.* 3(1/4): 291–295   [B, 486 L/70

Meeks, R. W. (1968) A new method for removing comb from frames. *Glean. Bee
Cult.* 96(7): 399 only   [B, 745 L/68

Meineke, E. A. (1967) Honey candies for the retail trade. *Am. Bee J.* 107(12): 454–
457   [B, 776 L/68

Mellaart, J. (1963) Excavations at Catal Hüyük, 1962. *Anatolian Studies* 13: 43–103
                                                                    [B, 503/65

Mendelsohn, O. (1965) *A dictionary of drink and drinking* (London: Macmillan)

Merfield, F. G. & Miller, H. (1956) *Gorilla hunter* (New York: Farrar, Straus)

Merl, T. (1914) Über die Verwendbarkeit der 'Vakuum-Destillation' beim Ameisen-
säure-Nachweis. *Z. Unters. Nahr.- u. Genussmittel* 27: 733–743

Merz, J. H. (1963) The objective assessment of honey aroma by gas chromatography. *J. apic. Res.* 2(1): 55–61   [B, 896/63

Mexico. Dirección General de Normas (1953) Miel de abeja. *Mexico Norma Oficial
DGN F36: 1953:* 8 pp. *In Spanish*   [B

Mexico (1958) Norma oficial para miel de abejas, F36–1958 (Mexico, D.F.: Secreta-
taria de Economía)

Mexico. Laws & Statutes (1965) Codificación Sanitaria Mexicana. *Mexico D.F.:*

Ediciones Andrades [with amendments to 1967; labelling, p. 40; registration, p. 865]

Michel, E. (1942) Beiträge zur Kenntnis von *Lachnus* (*Pterolachnus*) *roboris* L. eines wichtigen Honigtau-Erzeugers an der Eiche. *Z. angew. Ent.* 29: 243–281

Michelotti, P. & Margheri, G. (1969) Studio sul contenuto in aminoacidi del miele. *Scienza Aliment.* 15(7): 179–180

Michener, C. D. (1974) *The social behavior of the bees* (Cambridge, Mass.: Belknap Press)   [B

Michener, C. D. & Michener, M. H. (1951) *American social insects: a book about bees, ants, wasps and termites* (New York, Toronto & London: D. Van Nostrand Co., Inc.)   [B, 2/53

Miller, D., White, J. W., Jr. & Johnson, J. A. (1960) Honey improves baked products. *Bull. Kans. agric. Exp. Stn.* No. 411: 23 pp.   [B, 5c 116(4)61

Milum, V. G. (1939) Why does honey discolor during processing and storage? *Am. Bee J.* 79(9): 445–447   [B

Milum, V. G. (1948) Some factors affecting the colour of honey. *J. econ. Ent.* 41(3): 495–505   [B, 52/52

Milum, V. G. (1956) Honey granulation and its control. *Glean. Bee Cult.* 84(2): 91–93   [B, 222/57

Minh, H. Van, Mendoza, B. G. & Laigo, F. M. (1971) The chemical composition of honey produced by *Apis dorsata*. *J. apic. Res.* 10(2): 91–97   [B, 197/72

Ministry of Agriculture and Fisheries (1934) Honey: grading and marking. *Mktg Leafl. Minist. Agric., Lond.* No. 3: 16 pp.   [B

Ministry of Agriculture, Fisheries & Food (1955–1971) *Bee health and beekeeping in England and Wales*, published annually   [B

Miśkiewicz, W. & Krauze, S. (1969) Badania nad miodami polskimi ze szczególnym unzglednieniem podstawowych skladnikow mineralnych [Investigations of Polish honeys, particularly the main mineral constituents]. *Roczn. państ. Zakl. Hig.* 20(1): 73–84   [B, 752/70

Mitchell, T. J., Donald, E. M. & Kelso, J. R. M. (1954) An examination of Scottish heather honey. *Analyst, Lond.* 79(940): 435–442   [B, 132/55

Mitchell, T. J., Irvine, L. & Scoular, R. H. M. (1955) An examination of Scottish heather honey. Part II. *Analyst, Lond.* 80(953): 620–622   [B, 281/56

Mitchener, A. V. (1955) Manitoba nectar flows 1924–1954, with particular reference to 1947–1954. *J. econ. Ent.* 48(5): 514–518   [B, 191/56

Mitra, S. N. & Mathew, T. V. (1968) Studies on the characteristics of honey. *J. Proc. Instn Chem. India* 40(1): 26–30   [B, 569/69

Mittler, T. E. (1953) Amino-acids in phloem sap and their excretion by aphids. *Nature, Lond.* 172: 207   [B, 249/54

Mittler, T. E. (1957, 1958a, 1958b) Studies on the feeding and nutrition of *Tuberolachnus salignus* (Gmelin, Homoptera, Aphididae). I. The uptake of phloem sap. II. The nitrogen and sugar composition of ingested phloem sap and excreted honeydew. III. The nitrogen economy. *J. exp. Biol.* 34: 334–341, 35: 74–84, 35: 626–638   [B

Mittler, T. E. (1962) What affects the amount of honeydew excreted by aphids? *XI Int. Congr. Ent., Vienna* 2: 540–541   [B, see AA 245/63

Mizusawa, H. & Matsumuro, H. (1968) [Studies on chemical composition of honey. IV. Composition of free amino acids.] *Bull. Fac. Agric. Tamagawa Univ.* No. 7–8: 194–196

Mladenov, S. (1968) [Major and trace elements in honey and their significance for humans]. *Hrana Ishrana* 9(7): 461–467. *In Croatian*   [B, 423/73

Moffett, J. O. & Parker, R. L. (1953) Relation of weather factors to nectarflow in honey production. *Tech. Bull. Kansas agric. Exp. Stn* No. 74: 27 pp. [B, 224/54

Mohl, H. (1834) *Beiträge zur Anatomie und Physiologie der Gewächse.* (1): Über den Bau und die Formen der Pollenkörner (Bern)

Molnár, V., Balogh, É., Szöcs, J. & Péter, É. (1966) Modificările timpului de reacţie sub efectul alcoolului şi influenţarea lui prin unele substanţe alimentare [Modifications of reaction time under the influence of alcohol, and the effect of certain alimentary substances [in man]]. *Revta med., Tirgu Mures* 12(3): 295–297. *In Rumanian* [B, 192/70

Mommers, J. (1966) Die Nektarabsonderung bei verschiedenen Rassen von Obstbäumen. *Z. Bienenforsch.* 8: 203–204 [B, 509/67

Moreau, E. (1911a) Identification et dosage des substances protéiques dans les miels. *Annls Falsif. Fraudes* 4: 36–41

Moreau, E. (1911b) Étude biologique des miels. *Annls Falsif. Fraudes* 4: 65–66

Moreau, E. (1911c) Étude biologique des miels. *Annls Falsif. Fraudes* 4: 145–148

Morgan, F. L. (1956–58) History of Australian beekeeping. *Aust. Bee J.* (36–39): long serial in many issues [B, 324/61

Morgenthaler, O. (1953) Was ist der Segelhalter? *Z. Bienenforsch.* 2: 141–146 [B, 134/56

Morris, D. (1967) *The naked ape* (London: Jonathan Cape)

Morse, R. A. (1953a) *The fermentation of diluted honey.* Cornell University: Thesis 89 pp. [B, 146/54

Morse, R. A. (1953b) Honey wine and vinegar. *Glean. Bee Cult.* 81: 461, 491, 508, 536–538 1912 ed. is *B.R.A. Bibliogr.* No. 13, 199 L/73 [B

Morse, R. A. (1961) *Annotated bibliography on honey wine (mead)* (Ithaca: Cornell University) 16 pp. [B

Morse, R. A. (1965) The effect of light on comb construction by honeybees. *J. apic. Res.* 4(1): 23–29 [B, 112/66

Morse, R. A. (1969) The varied fare of the honeybee. *Nat. Hist. N.Y.* 78(6): 58–65 [B

Morse, R. A. & Laigo, F. M. (1969) The potential and problems of beekeeping in the Philippines. *Bee Wld* 50(1): 9–14 [B, 345/70

Morse, R. A. & Steinkraus, K. H. (1971) Method of making wine from honey. *U.S. Pat. No.* 3,598,607 Aug. 10 [B, 955/73

Morton, I. D. & Sharples, E. (1959) Process of preparing a honey flavor. *U.S. Pat.* No. 2,916,382 [B, 140/61

Morus (Lewinsohn, R.) (1954) *Animals, men and myths* (London: Victor Gollancz Ltd)

Mosimann, J. E. (1965) Statistical methods for the pollen analyst: Multinomial and negative multinomial techniques. Pp. 636–673 from *Handbook of palaentological techniques*, B. Kummel & D. Raup (San Francisco: Freeman)

Mostowska, I. (1965) Aminokwasy w nektarach i jednogatunkowych miodach [Amino acids in nectar and honey]. *Zesz. nauk. wyźsz. Szk. roln. Olsztyn.* 20(3): 417–432 [B, 679/66

Moulé, L. (1910) Études zoologiques . . . la faune d'Homère. *Mém. Soc. zool. France* 23: 29–106 [First part is 22: 183–233 (1909)]

Mraz, C. (1955) The packing of cut comb honey. *Glean. Bee Cult.* 83(5): 271–274 [B, 309/55

Müller, H. (1961) Vorausschau und Erkundung von Waldtrachten. *Symp. Genet. Biol. Ital.* 11: 85–103 [B, 309/65

Munk, R. (1969) Function of filtercell of small cicadas in producing honeydew. *XXII Int. Beekeep. Congr., Munich*: 521 only [B, see 305/70

Munro, J. A. (1943) The viscosity and thixotropy of honey. *J. econ. Ent.* 36(5): 769–777 [B

Muttoo, R. N. (1944) Beekeeping in India: its past, present and future. *Indian Bee J.* 6(3/4): 54–77 [B

Muttoo, R. N. (1956) Facts about beekeeping in India. *Bee Wld* 37: 125–133, 154–157 [B, 341/57

Muzzati, G. (1953) Prugne e miele. *Apicoltore d'Ital.* 20(8): 190, 192 [B, 278/54

Mwaniki, H. S. K. (1970) Bee-keeping: the dead industry among the Embu. *Mila* 1(2): 34–41 [B, 353/72

Nafe, R. (1966) Retention of the flavour of roasted coffee. *Fr. Pat.* No. 1,432,831. *In French* [528 L/72

Naghski, J., Willits, C. O., Porter, W. L. & White, J. W., Jr. (1956) Maple-honey spread and process of making the same. *U.S. Pat.* No. 2,760, 870 [B, 55/59

Nair, P. K. K. (1964) A pollen analytical study of Indian honeys. *J. Indian bot. Soc.* 43(2): 179–191

Nair, P. V., Pillai, M. S. & Nair, C. S. B. (1950) Studies in the chemistry and utilisation of Travancore minor forest products. Part II. Clarification, storage, and composition of forest honey. *Bull. Res. Inst. Univ. Travancore Ser. A* 1(1): 69–78 [B, 165/54

Nakayama, O. & Koike, H. (1965) [Studies on the application of honey to alcoholic beverages. Part 6. Experimental production of mead from buck-wheat honey]. *Yamanashi Daigaku Hakko Kenkyu Hokoka No.* 12: 23–28. *In Japanese; English summary* [B, 583/67

Nakayama, O., Koike, H. & Kushida, T. (1961) [Studies on the application of honey to alcoholic beverages. Part 2. Preliminary experiments on mead production]. *Yamanashi Daigaku Hakko Kenkyu Hokoka No.* 8: 27–40. *In Japanese; English summary* [B, 191/66

Nakayama, O., Koike, H. & Kushida, T. (1962a) [Studies on the application of honey to alcoholic beverages. Part 3. Experimental production of sweet wines applying honey to must]. *Yamanashi Daigaku Hakko Kenkyu Hokoka No.* 9: 57–63. *In Japanese; English summary* [B, 192/66

Nakayama, O., Koike, H. & Kushida, T. (1962b) [Studies on the application of honey to alcoholic beverages. Part 4. Experimental production of sweet wines, applying honey to ordinary wine]. *Yamanashi Daigaku Hakko Kenkyu Hokoka No.* 9: 65–69. *In Japanese; English summary* [B, 193/66

Nakayama, O., Yanoshi, M. & Nagata, T. (1966) [Studies on the application of honey to alcoholic beverages. Part 7. On the browning reaction and the application of the 'solid antioxidants']. *Bull. Res. Inst. Ferment. Yamanashi Univ. No.* 13: 19–24. *In Japanese; English summary* [B, 493/70

National Academy of Sciences (1972) Final Report. Committee on the African Bee. *Washington, D.C.: National Academy of Sciences* [B

Nature, Lond. (1970) Uses for fructose. *Nature, Lond.* 226(5250): 1006–1007 [B, 211 L/72

Neger, F. W. (1918) Experimentelle Untersuchungen über Russtaupilze. *Flora, Jena* 10: 67–139

Nelson, E. K. (1930) The flavor of orange honey. *Ind. Engng Chem. Ind.* 22: 448 only [B

Nelson, E. K. & Mottern, H. H. (1931) Some organic acids in honey. *Ind. Engng Chem. Ind.* 23(3): 335 only [B

Nelson, E. V. (1967) History of beekeeping in the United States. *Agric. Handb. U.S. Dep. Agric. No.* 335: 2–4 [B, 643 L/67

Nelson, J. M. & Cohn, D. J. (1924) Invertase in honey. *J. biol. Chem.* 61: 193–224

Nelson, J. M. & Sottery, C. T. (1924) Influence of glucose and fructose on the rate of hydrolysis of sucrose by invertase from honey. *J. biol. Chem.* 62: 139–147

Nemoto, S. (1961) [Considerations on honey recorded in ancient literature of the Orient and the Far East]. *Pract. Pharm.* 12(9): 109–114. *In Japanese*
[B, 504/62]

Neprašova, L. & Svoboda, J. (1956) Kremeni včel suchým krystalovým cukrem [Feeding bees with dry granulated sugar]. *Včelařství* 9(3): 35 only   [B, 192/59]

Netherlands. Laws & Statutes (1965) Honigbesluit 14.9.65. *Staatsblad Text No.* 431: 1217. [*See* next reference for methods of analysis.]

Netherlands. Laws & Statutes (1971) Besluit van 15 juni 1971 tot wijziging van het honigbesluit. *Staatsblad* 488: 1119

Neuberg, C. & Roberts, I. S. (1946) Invertase. *Technol. Rep. Ser. Sug. Res. Fdn No.* 4: 62 pp.

Neumann, W. & Habermann, E. (1950–51) Über parasympathicomimetische Wirkungen des Bienenhonigs. *Arch. exp. Path. Pharmak.* 212: 163 only
[B, 46/53]

Newkirk, W. B. (1920) A picnometer for the determination of density in molasses. *Tech. Pap. Bur. Stand.* BS 13.T161

New South Wales. Laws & Statutes (1908) *Pure Food Act, and Regulations thereunder* [revised issue, 1960, and amendments to 1970]

New Zealand. Laws & Statutes (1973) *The food and drug regulations 1973.* Honey standards, Reg. 163; pesticides and trace metals, Reg. 24; labelling, Reg. 5, 6, 164

Nicol, H. (1937) A test of gas-tightness of honey jars. *Bee Wld* 18(9): 103–105 [B

Nicol, W. M. (1971) Sweeteners in foods. *Process Biochem.* 6(12): 17–19
[B, 432L/73]

Nicolaidis, N. J. (1955) Facts about beekeeping in Greece. *Bee Wld* 36(8): 141–149
[B, 263/56]

Nixon, H. L. & Ribbands, C. R. (1952) Food transmission in the honeybee community. *Proc. R. ent. Soc. Lond. B* 140: 43–50   [B, 147/53]

Nogueira-Neto, P. (1951) Stingless bees and their study. *Bee Wld* 32(10): 73–76
[B, 155/53]

Nogueira-Neto, P. (1953) *A criação de abelhas indígenas sem ferrão* (Meliponinae). (São Paulo: Chácaras e Quintais)   [B, 86/54]

Nogueira-Neto, P. (1962) O início da apicultura no Brasil [The introduction of beekeeping to Brazil]. *Biol. Agric., S. Paulo* 49: 5–14   [B, 260/65]

Nogueira-Neto, P. (1970) *A criação de abelhas indígenas sem ferrão* (Meliponinae). (São Paulo: Chácaras e Quintais) 2nd ed.   [B, 263/72]

Nordin, P. & Johnson, J. A. (1957) Browning reaction products of cake crumb. *Cereal Chem.* 34(3): 170–178   [B, 655/64]

Nordsiek, F. W. (1972) The sweet tooth. *Am. Scient.* 60: 41–45   [B, 807/72]

Nottbohm, F. E. (1928) Die Aschenbestandteile des Bienenhonigs. *Arch. Bienenk.* 8(5/6): 207–228   [B

Nussbaumer, T. (1910) Beitrag zur Kenntnis der Honiggärung nebst Notizen über die chemische Zusammensetzung des Honigs. *Z. Unters. Nahr.- u. Genusmittel* 20: 272–277

Nyárády, A. (1958) A méhlegelő és növényei [Plants providing bee forage]. *Bukarest: Mezőgazdasági és Erdészeti Allami Könyvkiadó. In Hungarian*
[B, 56/61]

Oertel, E. (1956) Nectar production by white clover. *Glean. Bee Cult.* 84: 461–463, 552   [B, 172/57]

Oertel, E., Emerson, R. B. & Wheeler, H. E. (1953) Transfer of radioactivity from worker to drone honeybees after ingestion of radioactive sucrose. *Ann. ent. Soc. Am.* 46: 596–598 [B, 28/55]

Oertel, E., Fieger, E. A., Williams, V. R. & Andrews, E. A. (1951) Inversion of cane sugar in the honey stomach of the bee. *J. econ. Ent.* 44: 487–492 [B, 79/52]

Okada, I., Oka, A. & Sugiyama, A. (1968) Honeybees and honey. I. Sugar inversion by honeybees and the number of pollen grains in honey. *Bull. Fac. Agric. Tamagawa Univ. No.* 7/8: 175–180 [B, 210/72]

O'Keefe, J. A. (1968) Bell and O'Keefe's *Sale of food and drugs* (London: Butterworth). Pp. 22, 27, 64, 70

Opie, I. & Opie, P. (1969) *Children's games in street and playground* (Oxford: Clarendon Press)

Oppen, F. C. & Schuette, H. A. (1939) Viscometric determination of moisture in honey. *Ind. Engng Chem. analyt. Edn* 11: 130–133

Orbán, G. & Stitz, J. (1928) Die Fluorescenz der Honige im ultravioletten Licht. *Z. Unters. Lebensmittel* 56(5/6): 467–471 [B

Örösi-Pál, Z. (1956) A mérgező méz titka nyomában [On the track of the poisonous honey]. *Méhészet* 4(2): 25–27 [B, 142/57]

Örösi-Pál, Z. (1968) Physiologie des glandes nourricières. Pp. 263–290 from Vol. I, *Traité de biologie de l'abeille*, ed. R. Chauvin (Paris: Masson et Cie) [B, 323/68]

Osaulko, G. K. (1953) [Use of honey in ophthalmology]. *Vest. Oftal., Mosk.* 32: 25 only. *In Russian;* summary in *Glean. Bee Cult.* 82(12): 739 only (1954) [B, 171/58]

Osborn, H. (1952) *A brief history of entomology* (Columbus, Ohio: Spahr & Glenn Co.) [B, 231/53]

Oschmann, H. (1961) Eine bienenkundliche Reise in die Volksrepublik China. *Arch. Geflügelz. Kleintierk.* 10(4): 235–255 [B, T, 769/64]

Pager, H. (1971) Ndedema. *Graz, Austria: Akademische Druck- u. Verlagsanstalt* [B, 163/73]

Pager, H. (1973) Rock paintings in Southern Africa showing bees and honey hunting. *Bee Wld* 54(2): 61–68 [B, 191/74]

Paillon, P. (1960) *La fabrication des produits alimentaires au miel* (Paris: Girardot) [B, 300/61]

Paine, H. S., Gertler, S. I. & Lothrop, R. E. (1934) Colloidal constituents of honey. Influence on properties and commercial value. *Ind. Engng Chem. Ind.* 26: 73–81 [B

Paine, H. S. & Lothrop, R. E. (1935) A small plant for filtering honey. *Am. Bee J.* 75(7): 326–330 [B

Palleson, H. R. (1952) Flash heating of honey using plate equipment. *Am. Bee J.* 92(5): 202–204 [B, 276/53]

Palmer-Jones, T. (1947a) A recent outbreak of honey poisoning. Part 1. Historical and descriptive. *N.Z. Jl Sci. Technol.* 29A: 107–114 [B

Palmer-Jones, T. (1947b) A recent outbreak of honey poisoning. Part 3. The toxicology of the poisonous honey and the antagonism of tutin, mellitoxin, and picrotoxin by barbiturates. *N.Z. Jl Sci. Technol.* 29: 121–125 [B

Palmer-Jones, T. (1953) Production of honey mead from New Zealand honeys. *N.Z. Jl Sci. Technol. A* 35: 375–389 [B, 168/55]

Panama. Laws & Statutes (1962) Law of 13.6.62 [quoted in *Fd Drugs Cosmetics Law J.*: 609 (1964)]

Panelatti, G. (1961) Quelques résultats de palynologie analytique et descriptive pour le Maroc. *Trav. Inst. scient. chérif. Sér. bot. (23) Rabat*

Pankiw, P. & Bolton, J. L. (1965) Characteristics of alfalfa flowers and their effects on seed production. *Can. J. Pl. Sci.* 45: 333–342 [B, 772/65

Papadakis, P. E. (1929) Further findings on invertase from honey. *J. biol. Chem.* 83: 561–568

Park, O. W. (1923) Water stored by bees. *Am. Bee J.* 63: 348–349 [B

Park, O. W. (1925) The storing and ripening of honey by honeybees. *J. econ. Ent.* 18: 405–410 [B

Park, O. W. (1927) Studies on the evaporation of nectar. *J. econ. Ent.* 20: 510–516

Park, O. W. (1928) Further studies on the evaporation of nectar. *J. econ. Ent.* 21: 882–887

Park, O. W. (1929) The influence of humidity upon sugar concentration in the nectar of various plants. *J. econ. Ent.* 22: 534–544

Park, O. W. (1932) Studies on the changes in nectar concentration produced by the honeybee. I. Changes which occur between the flower and the hive. *Res. Bull. Iowa agric. Exp. Stn No.* 151: 210–244 [B

Park, O. W. (1933) Studies on the rate at which honeybees ripen honey. *J. econ. Ent.* 26: 188–193

Park, O. W. (1949) How bees make honey. Pp. 125–152 from *The hive and the honey bee* ed. R. A. Grout (Hamilton, Ill.: Dadant & Sons) [B, see 107/51

Pasedach-Pöverlein, K. (1941) Über das 'Spritzen' der Bienen und über die Konzentrationsänderung ihres Honigblaseninhalts. *Z. vergl. Physiol.* 28: 197–210 [B, T

Paterson, C. R. (1947) A recent outbreak of honey poisoning. Part 4. The source of the toxic honey—field observations. *N.Z. Jl Sci. Technol.* 29A: 125–129 [B

Paterson, C. R. (1955) Construction of a beekeeper's cool room. *N.Z. Jl Agric.* 90(4): 355–360 [B, 260/57

Paterson, C. R. (1958) Recent developments in processing honey for soft, smooth grain. *N.Z. Jl Agric.* 97(6): 513–518 [B, 348/59

Paterson, P. D. (1966) The present economic status of *A.m. adansonii*: summary and bibliography. *Bee Wld* 47(4): 123–131; *Repr. BRA* M45 [B, 329 L/67

Patterson, R. W. & Bach, A. (1968) Contrasting land use reflected in pollen grain content of honeys from two Tasmanian districts. *J. Aust. Inst. agric. Sci.*: 91–92 [B, 1013/70

Paxton, J. (ed.) (1971) *The statesman's yearbook* (London: Macmillan)

Pearce, J. A. & Jegard, S. (1949) Measuring the solids content of honey and of strawberry jam with a hand refractometer. *Can. J. Res.*, F27: 99–103 [B, 167/54

Pedersen, M. W. (1953) Environmental factors affecting nectar secretion and seed production in alfalfa. *Agron. J.* 45: 359–361 [B, 78/55

Pedersen, M. W. (1958) Nectar secretion in relation to seed production in alfalfa. *X. Int. Congr. Ent., Montreal* 4: 1019–1024 [B

Pedersen, M. W., Le Fevre, C. W. & Wiebe, H. H. (1958) Absorption of C^{14}-labelled sucrose by alfalfa nectaries. *Science, N.Y.* 127(3301): 758–759 [B, 811/63

Peel, A. J. & Weatherley, P. E. (1959) Composition of sieve-tube sap. *Nature, Lond.* 184: 1955–1956

Pelimon, C. (1960) L'analyse pollinique des miels de la République Populaire Roumaine. *Annls Abeille* 3(4): 339–347 [B, 333/61

Pelimon, C. & Baculinschi, H. (1955) Cercetări asupra compozitiei mierii din R.P.R. [Researches on honey composition in Rumania]. *Anal. Inst. Cerc. zooteh.* 13: 621–637 [B, 215/60

Pellett, F. C. (1938) *History of American beekeeping* (Ames, Iowa: Collegiate Press) [B

Pellett, F. C. (1947) *American honey plants* (New York: Orange Judd Publishing Co.) [B

Penfold, A. R. & Willis, J. L. (1961) *The eucalypts* (London: Leonard Hill)
[B, 539/62

Percival, M. (1946) Observations on the flowering and nectar secretion of *Rubus fruticosus*. *New Phytol.* 45: 111–123 [B

Percival, M. (1961) Types of nectar in Angiosperms. *New Phytol.* 60: 235–281
[B, 844/64

Pérez, B. S. & Rodríguez, A. T. (1970) Composición química y espectro polínico de mieles españolas. *An. Bromat.* 22(4): 377–406 [B, 494/72

Perry, R. (1970) *Bears* (London: Arthur Barker Ltd.)

Pershad, S. (1967) Analyse de différents facteurs conditionnant les échanges alimentaires dans une colonie d'abeilles *Apis mellifica* L., au moyen du radio-isotope P^{32}. *Annls Abeille* 10: 139–197 [B, 350/68

Perti, O. N. & Pandey, P. (1967a) Study of some physico-chemical constants of *Apis dorsata* honey. Part I. *J. appl. Chem.* [referred to in (1967b) but may not have been published]

Perti, O. N. & Pandey, P. (1967b) Study of some physico-chemical constants of *Apis dorsata* honey. Part II. Specific gravity. *Indian Bee J.* 29: 1–3 [B, 480/72

Peru. Laws & Statutes (1963) Code of 19.6.63 [quoted in *Fd Drugs Cosmetics Law J.*: 609 (1964)]

Perušic, A. (1959) *Pčelinji med* [Honey]. (Zagreb). *In Croatian* [B, 280/60

Peterson, W. H., Hendershot, W. F. & Hajny, G. J. (1958) Factors affecting production of glycerol and D-arabitol by representative yeasts of the genus *Zygosaccharomyces*. *Appl. Microbiol.* 6: 349–357

Petrov, V. (1970) Mineral constituents of some Australian honeys as determined by atomic absorption spectrophotometry. *J. apic. Res.* 9(2): 95–101 [B, 750/71

Pfister, R. (1895) Versuch einer Mikroskopie des Honigs. *ForschBer. Lebensmitt. u. ihre Bez. z. Hygiene, forens. Chem., Pharmakogn.* 2(1/2): 1–9; 29–35

Phadke, R. P. (1962) Physico-chemical composition of major unifloral honeys from Mahabaleshwar (Western Ghats). *Indian Bee J.* 24(7/9): 59–65
[B, 907/64

Phadke, R. P. (1967a) Studies on Indian honeys. 1. Proximate composition and physico-chemical characteristics of Indian multifloral apiary honeys from *Apis indica* bees. *Indian Bee J.* 29: 14–26 [B, 478/72

Phadke, R. P. (1967b) Studies on Indian honeys. 2. Proximate composition and physico-chemical characteristics of unifloral honeys of Mahabaleshwar. *Indian Bee J.* 29: 33–46 [B, 479/72

Phadke, R. P. (1968) Studies on Indian honeys. 3. Proximate composition and physico-chemical characterizations of honeys from the wild honey bees *Apis dorsata*, *Apis florea* and *Trigona*. *Indian Bee J.* 30(1): 3–8 (1968) [B, 481/72

Philipsborn, H. von (1952) Über Calciumoxalat in Pflanzenzellen. *Protoplasma* 41(4): 415–424 [B, 227/53

Phillips, C. E. (1933) Honey for burns. *Glean. Bee Cult.* 61(5): 284 [B

Phillips, E. F. (1927) The utilization of carbohydrates by honeybees. *J. agric. Res.* 35(5): 385–428 [B

Pícha, S. (1965) Metody pro stanoveni obsahu vody ve včelich medech a jejich statistické zhodnoceni [Methods for determining the water content of honey, and their statistical value]. *Věd. Pr. výzk. Úst. včelař. v Dole* 4: 151–155
[B, 562/66

Pigman, W. W. & Goepp, R. M., Jr. (1948) *Chemistry of the carbohydrates* (New York: Academic Press Inc.)

Pique, R. (1914) Le mellimustimètre. *Bull. Ass. Chim. Sucr. Distill. Fr.* 31: 652–655

Plachy, E. (1944) Studie über die bakterizide Wirkung des Naturhonigs (Blüten- und Blatthonig) aus verschiedenen Höhenlagen sowie einige Untersuchungen über die Eigenschaft der antibakteriellen Hemmungsstoffe (Inhibine) im Naturhonig. *Zbl. Bakt.* 100: 401–419 [B

Plath, O. E. (1934) *Bumblebees and their ways* (New York: Macmillan) [B

Platts, J. T. (1960) *Dictionary of Urdu, classical Hindi and English* (London: Oxford University Press)

Plugge, P. C. (1891) Giftiger Honig von *Rhododendron ponticum. Arch. exp. Path. Pharmak.* 229: 552–554

Plutarch. North's translation, ed. R. H. Carr (1906) *Lives of Coriolanus, Caesar, Brutus and Antonius* (Oxford: Clarendon Press)

Plutarch. *Life of Alexander.* Loeb translation (1919) (London: Heinemann)

Pokorny, J. (1959) *Indogermanisches etymologisches Wörterbuch* (Bern: Francke Verlag)

Poland. Polski Komitet Normalizacyjny (1967) Miód pszczeli [Honey]. *Polska Norma PN-67 A-77626:* 16 pp. [B, 509 L/72

Poltev, V. I. (1956) Food toxicosis of bees and their diagnosis. *XVI Int. Beekeep. Congr. Vienna:* 73–74 (1956) [B, see 261/56

Poma Treccani, C. (1950) Interazione dell'acido alfanaftalen-acetico e miele nel radicamento di talle di vite. *Riv. Fruttic.* 12(3): 133–162 [B, 115/54

Popova, V. M., Rozental, M. D., Sadyrin, M. M., Trop, I. E. & Chulovskii, I. K. (1960) [Group poisoning with spring honey and a method of determination of the toxicity by the method of biological testing and pollen analysis.] *Gig. Sanit.* 25: 92–94. *In Russian*

Portugal Araújo, V. de (1954) Angola anda a desperdiçar a requeza que as abelhas lhe oferecem. *A Provincia de Angola* 31(8629): 1–2 [B, 263/55

Portugal Araújo, V. de (1955) Colmeias para 'abelhas sem ferrão' (Meliponini). *Bol. Inst. Angola No.* 7: 49 pp. [B, 76/58

Portugal Araújo, V. de (1957) Colmeias e utensílios para a cultura de 'abelhas sem ferrão' . . . *Gaz. agric. Angola* 1(12): 469–473; 2(1): 513–517 [B, 41/65

Poszwiński, L. & Warakomska, Z. (1969) Analiza pyłkowa miodów rzepakowych i wrzosowych województwa warszawskiego. *Pszczel. Zesz. nauk.* 13(1/3): 147–158 [B, 454/71

Pothmann, F. J. (1950) Der Einfluss von Naturhonig auf das Wachstum der Tb.-Bakterien. *Z. Hyg. InfektKrankh.* 130: 468–484 [B, 166/54

Pourtallier, J. (1962) Mise au point sur l'analyse des miels. *Bull. apic. Doc. sci. tech. Inf.* 5: 138–154 [B, 563/66

Pourtallier, J. (1964) Sugars in honey analyses. *XIX Int. Beekeep. Congr. 1963, Prague* 2: 572–577 [B, see 58 L/64

Pourtallier, J. (1967) Détermination quantitative des sucres des miels par chromatographie en phase gazeuse. *Bull. apic. Doc. sci. tech. Inf.* 10(2): 209–212 [B, 751/71

Pourtallier, J. (1973) Caractérisation physico-chimique des miels. *Simposio Internazionale di Apicoltura, Torino 1972:* 91–95 [B, 502/74

Pouvreau, A. (1974) Le comportement alimentaire des bourdons (Hymenoptera, Apoidea, *Bombus* Latr.): la consommation de solutions de sucre. *Apidologie* 5(3): 247–270 [B

Povchenko, G. P. (1950) [Poisonous honey from azaleas]. *Pchelovodstvo* 27(3): 188 only. *In Russian* [B

Prescott, S. C. & Dunn, C. G. (1959) *Industrial microbiology* 3rd ed. (London, New York: McGraw-Hill Book Co., Inc.) pp. 150–153

Prica, M. (1938) Über die baktericide Wirkung des Naturhonigs. *Z. Hyg. Infekt-Krankh.* 120: 437–443 [B

Pritchett, W. K. (1956) The Attic stelai. *Hesperia* 25: 178–344 (especially 260–261)

Pritsch, G. (1957*a*) Über Lindenpollen und Pollengehalt des Lindenhonigs. *Leipzig. Bienenztg* 71(6): 188–189 [B

Pritsch, G. (1957*b*) Über Raps- und Robinienpollen und den Pollengehalt von Raps- und Robinienhonig. *Leipzig. Bienenztg* 71(10): 324–328 [B

Pritsch, G. (1958) Untersuchungen über die Bienenweideverhältnisse verschiedener Standorte auf leichteren Böden unter besonderer Berücksichtigung der Untersuchungen mittels Honig-Pollenanalysen. *Arch. Geflügelz. Kleintierk.* 7(3/4): 184–247; (5/6): 282–310 [B, 218/61

Pryce-Jones, J. (1944) Some problems associated with nectar, pollen and honey. *Proc. Linn. Soc. Lond.* (2): 129–174 [B

Pryce-Jones, J. (1952) 'Stringiness' in honey and in sugar syrup fed to bees. *Bee Wld* 33(9): 147–150, 154–155 [B, 226/53

Pryce-Jones, J. (1953) The rheology of honey. Pp. 148–176 from *Foodstuffs: their plasticity, fluidity and consistency* ed. G. W. Scott Blair (Amsterdam: N.-Holland Publishing Co.) [B, 110/54

Puecher-Passavalli, P. (1964) *La legislazione sugli alimenti* (Rome: Puecher-Passavalli) Vol. I, p. 336

Pulewka, P. (1949) Andromedotoxin enthaltender Honig und eine biologische Methode zur Bestimmung seiner Giftigkeit. *Bull. Fac. med. Istanbul* 12: 275–286 [B, 96/56

Pyke, M. (1972) *Technological eating.* (London: John Murray) p. 10

Queensland. Laws & Statutes (1964) Minister for Health, The Food and Drug regulations [and amendments to 1971]

Rachmilevitz, T. & Fahn, A. (1973) Ultrastructure of nectaries of *Vinca rosea* L., *Vinca major* L. and *Citrus sinensis* Osbeck cv. Valencia and its relation to the mechanism of nectar secretion. *Ann. Bot.* 37(149): 1–9 [B

Rahmanian, M., Kouhestani, A., Ghavifekr, H., Ter-sarkissian, N., Olszyna-marzys, A. & Donoso, G. (1970) High ascorbic acid content in some Iranian honeys. Chemical and biological assays. *Nutr. Metab.* 12(3): 131–135 [B, 491/72

Ransome, H. M. (1937) *The sacred bee in ancient times and folklore* (London: George Allen & Unwin Ltd.) [B

Raumer, E. von (1908) Über die Fiehe'sche Reaktion zur Erkennung und Unterscheidung von Kunsthonigen und Naturhonigen. *Z. Unters. Nahr.- u. Genussmittel* 16: 517

Raumer, E. von (1909) Zur Beurteilung der Fiehe'schen Reaktion. *Z. Unters. Nahr.- u. Genussmittel* 17: 115–125

Rea, J. (1974) Some beekeeping observations in Ethiopia. *Bee Wld* 55(2): 61–63 [B

Renner, E. & Duisberg, H. (1968) Über den Zusammenhang zwischen einigen Qualitätsmerkmalen des Honigs und dessen Naturbelassenheit. *Z. Lebensmittelunters. u. -Forsch.* 136(3): 137–146 [B, 455/71

Reynolds, D. R. (1971) On the use of hyphal morphology in the taxonomy of sooty mould ascomycetes. *Taxon* 20: 759–768

Reynolds, V. & Reynolds, F. (1965) Chimpanzees of the Budongo forest. Chapter 11 of *Primate behavior,* ed. I de Vore (New York: Holt, Rinehart & Winston)

Rheinhardt, J. R. (1939) Ventilating the bee colony to facilitate the honey ripening process. *J. econ. Ent.* 32: 654–660   [B

Ribbands, C. R. (1953) *The behaviour and social life of honeybees* (London: Bee Research Association)   [B, 8/54

Rice, E. W. & Boleracki, P. (1933) Determination of moisture in sirups and viscous materials. *Ind. Engng Chem. analyt. Edn* 5: 11–12

Richards, O. W. (1953) *The social insects* (London: Macdonald)   [B, 257/53

Richter, A. A. v. (1912) Über einen osmophilen Organismus, den Hefepilz *Zygosaccharomyces mellis acidi* sp. n. *Mykol. Zentbl.* 1(3/4): 67–76   [B

Riggs, J. K. & Weaver, N. (1955) Unmarketable honey as an ingredient in steer fattening rations. *Progr. Rep. Texas agric. exp. Stn* No. 1826: 3 pp.   [B, 243/59

Rihar, J. (1960) Cebelarske postaje [Beekeeping stations]. *Socialist. Kmetystvo* 11(3/4): 185–187. *In Slovenian*   [B, 861/64

Rihar, J. (1963) Opazovalna, poročevavska in prognostična služba o medenju iglavcev v letu 1962 [Observations, records and prediction service on the honeydew flow from conifers in Slovenia in 1962.] *Slov. Čeb.* 65(5): 110–115. *In Slovenian*
                                [B, 862/64

Riley, D. & Young, A. (1966) *World vegetation* (London: Cambridge University Press)   [B

Rinaudo, M. T., Ponzetto, C., Vidano, C. & Marletto, F. (1973a) The origin of honey saccharase. *Comp. Biochem. Physiol.* 46B(2): 245–251   [B

Rinaudo, M. T., Ponzetto, C., Vidano, C. & Marletto, F. (1973b) The origin of honey amylase. *Comp. Biochem. Physiol.* 46B(2): 253–256   [B

Roberts, D. (1958) Improved method of fitting wax foundation to sections in comb honey production. *N.Z. Jl Agric.* 96(6): 593, 595, 597–598   [B, 299/59

Roberts, D. & Smaellie, E. (1958) Control of wax moth on comb honey and stored bee combs. *N.Z. Jl Agric.* 97(5): 464–468   [B, 370/59

Robinson, F. A. (1964) The effects of the December 1962 freeze on citrus honey production in Florida. *Florida Ent.* 47(1): 55–56   [B, 565/65

Roff, C. (1962) Facts about beekeeping in Australia: IV. Queensland. *Bee Wld* 43(2): 44–49   [B, 757 L/63

Romann, E. & Staub, M. (1961) Hydroxymethylfurfurol in Honig. I. Mitteilung: Quantitative Bestimmung und Umwandlung des Hydroxymethylfurfurols. *Mitt. Geb. Lebensmittelunters. u. Hyg.* 52(1): 44–58   [B, 207/63

Root, A. I. (1959) *The ABC and XYZ of bee culture* (Medina, Ohio: Root) [B, 145/60

Rope, C. G. (1962) Production of section comb honey. *N.Z. Jl Agric.* 105(4): 363, 365   [B, 379/63

Rope, C. G. (1963) Cut-comb honey. *N.Z. Jl Agric.* 106(3): 205, 207   [B, 199/64

Rosenberg, H. R. (1942) *Chemistry and physiology of the vitamins* (New York: Interscience Publishers Inc.)

Roski, D. (1968) Wie Honig so süss . . . *Garten u. Kleintierz. Imker* 7(26): 8–10
                                [B, 494 L/70

Rossi, J. (1959) Indagini microbiologiche sul tecc etiopico. *Ann. Microbiol.* 9(1/4): 150–160   [B, 271/61

Röttinger, A. C. (1926) Über die Bestimmung des Extraktes. Eine neue Mikromethode. *Öst. ChemZtg* 29: 1–4

Roussy, L. (1969) Le miel et le sucre de canne dans la confection de la pâtisserie et des médicaments pendant le haut Moyen-Âge. *Gaz. apic.* 70(746): 172–173
                                [B, 771 L/71

Rumania. Republica Socialista Romaniă. Standard de Stat (Rumanian Standard) (1969). Miere de albine. [Honey.] *R.S.R. Standard de Stat* STAS 784/1–69: 10 pp. *In Rumanian*   [B, 512 L/72

Ruttner, F. (1952) Alter und Herkunft der Bienenrassen Europas. *Öst. Imker* 2(1): 8–12 [B, T, 186/53
Ruttner, F. (1956) Oberösterreichische Honige. *Bienenvater* 77(3): 82–90 [B, 220/57
Ruttner, F. (1961) Der Pollen der Eichenmistel (*Loranthus europaeus* Jacq.) als Charakterform in österreichischen Honigen. *Z. Bienenforsch.* 5(7): 220–226
[B, 704/62
Ruttner, F. (1964) Zur pollenanalytischen Diagnose südosteuropäischer Honige. *Annls Abeille* 7(4): 321–327 [B, 169/67
Ruttner, F. (1969) The cause of the hybridization barrier between *Apis mellifera* L. and *Apis cerana* Fabr. (=syn. *A. indica* Fabr.). 1. Experiments with natural and artificial insemination. *XXII Int. Beekeep. Congr. prelim. sci. Meet. Summ.*: 170 only [B, 430/70
Rychlik, M. & Fedorowska, Z. (1960) Badania nad inwertaza miodowa: Czesc I. Modifikacja metody oznaczania inwertazy miodowej [Honey invertase. I. Modification of the method for determining honey invertase]. *Roczn. państ. Zakl. Hig.* 11(3): 413–422 [B, 205/63
Rychlik, M. & Fedorowska, Z. (1962a) O biochemicznej metodzie oznaczania dekstryn miodowych [Biochemical method for dextrin estimation in honey]. *Pszczel. Zesz. Nauk.* 6(3): 135–140 [B, 385/65
Rychlik, M. & Fedorowska, Z. (1962b) Badania nad inwertaza miodowa. Czesc II. Aktywnosc inwertazowa polskich miodow [Honey invertase. II. Invertase activity of Polish honeys]. *Roczn. państ. Zakl. Hig.* 13: 53–59 [B, 206/63
Rychlik, M. & Zborowski, J. (1966) Badania nad występowaniem związków flawonoidowych w odmianowych miodach pszczelich. I. Rutyna i kwercetyna [Occurrence of flavone compounds in honeys from different plants. I. Rutin and quercetin]. *Pszczel. Zesz. nauk.* 10(1/4): 123–130 [B, 217/68
Ryle, M. (1954) The influence of nitrogen, phosphate and potash on the secretion of nectar. *J. agric. Sci. Camb.* 44(4): 400–419 [B, 115/56

Sabatini, G. (1973) Spettro pollinico di alcuni mieli italiani. *Simposio Internazionale di Apicoltura, Torino, 1972*: 79–86 [B, 470 L/74
Sacchi, R. (1955) Mieli umbri e analisi rifrattometrica dell'umidità. *Ann. Sper. agr.* 10: 1029–1042 [B, 259/57
Sacchi, R. & Bosi, G. (1964) The heterogeneity of certain components of Italian kinds of honey and their specific rotation. *XIX Int. Beekeep. Congr. 1963, Prague* 2: 615–620 [B, see 58 L/64
Sackett, W. G. (1919) Honey as a carrier of intestinal diseases. *Bull. Colo. St. Univ. agric. Exp. Stn* No. 252: 18 pages [B
Sakagami, S. F. (1959) Some interspecific relations between Japanese and European honeybees. *J. Anim. Ecol.* 28: 51–68 [B, 576/62
Sanna, A. (1931) Su una qualità di miele della Gallura di sapore amoro. *Annali Chim. appl.* 21(8): 397–402
Santolaya, A. A. & Gentile, J. M. (1953) Análisis de la miel de *Polybia scutellaris*. *Rev. Fac. Cienc. Méd. Univ. Córdoba* 11: 227–232 [200/54
Santos, C. F. de O. (1961) Principais tipos de pólen encontrados em algumas amostras de mel. Nota prévia. *Rev. Agric., Piracicaba* 36(2): 93–96 [B, 908/64
Santos Ruiz, A., Lopez de Azcona, J. M. & Sampedro Piñeiro, A. (1949) Oligoelementos en alimentos españoles de origen animal. IV. Aves y varios. *Revta esp. Fisiol.* 4(4): 237–244 [B, 258/57
Savage, J. H. (1961) The production of section comb honey. *Scot. Beekpr* 38(11): 197–198 [B, 280 L/62

Schade, J. E., Marsh, G. L. & Eckert, J. E. (1958) Diastase activity and hydroxy-methyl-furfural in honey and their usefulness in detecting heat alteration. *Fd Res.* 23(5): 446–463 [B, 383/59]

Schafer, M. (1958) *Honey's nifty fifty recipes* (San Marino: California Honey Advisory Board) [B, 5d28 (1) 60]

Schafer, M. (1961) *The best from the West with honey* (South Laguna, Calif.: California Honey Advisory Board) [B

Schafer, M. (1966) *Honey; it's good everyway and every day* (Whittier, Calif.: California Honey Advisory Board) [B

Schaller, G. B. (1965) *The year of the gorilla* (London: Collins)

Schatzmayr, O. M. (*see* under Barth, O. M.)

Schenk, D. (1934) Refraktometrische Bestimmung des Zucker-(Extrakt-) Gehaltes in Marmeladen, Obstgelees, Malzextrakten, Honigen u. dergl. *Z. Unters. Lebensmittel* 67: 187–191

Schepartz, A. I. (1965a) The glucose oxidase of honey. II. Stereochemical substrate specificity. *Biochim. biophys. Acta* 96: 334–336 [B, 729/66]

Schepartz, A. I. (1965b) The glucose oxidase of honey. III. Kinetics and stoichiometry of the reaction. *Biochim. biophys. Acta* 99: 161–164 [B, 164/67]

Schepartz, A. I. (1966a) The glucose oxidase of honey. IV. Some additional observations. *Biochim. biophys. Acta* 118: 637–640 [B, 165/67]

Schepartz, A. I. (1966b) Honey catalase: occurrence and some kinetic properties. *J. apic. Res.* 5(3): 167–176 [B, 789/67]

Schepartz, A. I. & Subers, M. H. (1964) The glucose oxidase of honey. I. Purification and some general properties of the enzyme. *Biochim. biophys. Acta* 85: 228–237 [B, 203/65]

Schepartz, A. I. & Subers, M. H. (1965) A simplified method for the estimation of diastase in honey. *Glean. Bee Cult.* 93(6): 358–359, 378 [B, 163/67]

Schepartz, A. I. & Subers, M. H. (1966a) Observations on honey diastase. *J. apic. Res.* 5(1): 45–48 [B, 733/66]

Schepartz, A. I. & Subers, M. H. (1966b) Catalase in honey. *J. apic. Res.* 5(1): 37–43 [B, 734/66]

Scheunert, A., Schieblich, M. & Schwanebeck, E. (1923) Zur Kenntnis der Vitamine. I Mitteilung. Über den Vitamingehalt des Honigs. *Biochem. Z.* 139: 47–56

Scheurer, S. (1967a) Populationsdynamische Beobachtungen an auf *Pinus* lebenden Lachniden während des Jahres 1965. *Waldhygiene* 7(1): 7–22 [B, 309/69]

Scheurer, S. (1967b) Wir starten den ersten Versuch der Waldtrachtprognose. *Garten u. Kleintierz. Imker* 6(7): 11 only [B, 739 L/69]

Schmalfuss, H. & Barthmeyer, H. (1929) Diacetyl als Aromabestandteil von Lebens- und Genussmitteln. *Biochem. Z.* 216: 330–335

Schmidt-Nielsen, S. & Årtun, T. (1938) Reaktionskinetische Fragen bezüglich der Thermodestruktion der Amylasen im Honig. *Forh. Norske VidenskSelsk. Trondheim* 11: 101–104

Schmidt-Nielsen, S. & Engesland, S. (1938) Om termodestruksjon av amylase i honning. *Hyllningsskr. Bertil. Almgren Sextioarsdagen* 1938: 407–415

Schnepf, E. (1969) Sekretion und Exkretion bei Pflanzen. *Protoplasmatologia* 8(8): 1–181 (Vienna: Springer Verlag) [B, 497/71]

Schönfeld, A. (1927) Jak a jak rychle vytvářejí včely med [How, and within what time, do bees produce honey?]. *Věst. čsl. Akad. zeměd. Věd.* 3(3): 228–231. *In Czech & French*

Schöntag, A. (1953) Über den Einfluss von Mineralsalzen auf den Zuckergehalt des Nektars. *Z. vergl. Physiol.* 35: 519–526 [B, 182/54]

Schou, S. A. & Abildgaard, J. (1931) Om differentiering mellem honning og kunsthonning [A method of differentiating bee honey from artificial honey]. *Dansk Tidsskr. Farm.* 5: 89–105

Schremmer, F. (1969) Extranuptiale Nektarien, Beobachtungen an *Salix eleagnos* Scop. und *Pteridium aquillinum* (L.) Kuhn. *Öst. bot. Z.* 117: 205–222   [B, 128/72

Schremmer, F. (1972) Beobachtungen zur Biologie von *Apoica pallida* (Olivier 1791) einer neotropischen sozialen Faltenwespe (Hymenoptera, Vespidae). *Insectes Soc.* 19(4): 343–357   [B, 507/73

Schubring, W. (1940) World production and trade of honey and beeswax. *Int. Rev. agric.* 31(1): 578S–592S   [B

Schuette, H. A. (1935) Report on [the analysis of] honey. *J. Ass. off. agric. Chem.* 18: 162–164

Schuette, H. A. & Baldwin, C. L., Jr. (1944) Amino acids and related compounds in honey. *Fd Res.* 9: 244–249

Schuette, H. A. & Bott, P. A. (1928) Carotin: A pigment of honey. *J. Am. chem. Soc.* 50(7): 1998–2000   [B

Schuette, H. A. & Huenink, D. J. (1937) Mineral constituents of honey. II. Phosphorus, calcium, magnesium. *Fd Res.* 2: 529–538   [B

Schuette, H. A. & Pauly, R. J. (1933) Determination of the diastatic activity of honey. *Ind. Engng Chem. analyt. Edn* 5: 53–54

Schuette, H. A. & Remy, K. (1932) Degree of pigmentation and its probable relationship to the mineral constituents of honey. *J. Am. chem. Soc.* 54: 2909–2913   [B

Schuette, H. A. & Templin, V. (1930) Some observations on the application of the formol titration to honey. *J. Ass. off. agric. Chem.* 13: 136–142

Schuette, H. A. & Triller, R. E. (1938) Mineral constituents of honey. III. Sulfur and chlorine. *Fd Res.* 3(5): 543–547   [B

Schuette, H. A. & Woessner, W. W. (1939) Mineral constituents of honey. IV. Sodium and potassium. *Fd Res.* 4(4): 349–353   [B

Schuler, R. (1957) Wirkstoffe des Bienenhonigs II. Untersuchungen über die cholinergische Wirksamkeit des Bienenhonigs. *Arzneimittel-Forsch.* 7: 330–331   [B, 508/62

Schutz, A. H. (ed.) (1945) The romance of Daude de Pradas called *Dels Auzels Cassadors* [Concerning birds of the chase]. (Columbus: Ohio State University Press)   [B, 39/54

Schwarz, F. (1908) Beitrag zum Mineralstoffgehalt des Honigs, zugleich Erwiderung auf den Artikel von Korpsstabsapotheker Utz über denselben Gegenstand. *Z. angew. Chem.* 21: 436–439

Schwarz, H. F. (1948) Stingless bees (Meliponidae) of the Western Hemisphere. *Bull. Am. Mus. nat. Hist.* 90: 1–546   [B

Schwarz, H. F. (1949) The stingless bees (Meliponidae) of Mexico. *An. Inst. Biol. Univ. Méx.* 20(1/2): 357–370   [B, 3/52

Schweigart, H. A. (1961) Verfahren zum Haltbarmachen von Gemüse und Früchten. *D.B.R. Pat.* Nr. 1,101,118   [B, 665 L/63

Schweizerische Lebensmittelbuchkommission (1967) *Schweizerisches Lebensmittelbuch.* Methoden für die Untersuchung und Beurteilung von Lebensmitteln und Gebrauchsgegenständen. Teil 2, Kapitel 23, Honig und Kunsthonig (Bern: Schweizerische Lebensmittelbuchkommission)   [1023 L/71

Scott, P. M., Coldwell, B. B. & Wiberg, G. S. (1971) Grayanotoxins. Occurrence and analysis in honey and a comparison of toxicities in mice. *Fd Cosmet. Toxicol.* 9: 179–184   [B

Scott-Blair, G. W. (1953) *Foodstuffs: their plasticity, fluidity and consistency* (Amsterdam: North-Holland Publishing Co.)   [B

Sechrist, E. L. (1925) The colour grading of honey. *Dep. Circ. U.S. Dep. Agric.* No. 364 [B

Selling, O. H. (1947) Studies in Hawaiian pollen statistics. II. The pollens of the Hawaiian phanerogams. *Spec. Publ. B.P. Bishop Mus. Honolulu* 38: 1–430

Serwatka, J. (1958) Wyniki analizy pyłkowej miodów wrzosowych z 1956 r. [Results of pollen analysis of heather honey (1956 harvest)]. *Pszczel. Zesz. Nauk.* 2(2): 55–56 [B, 928/64

Seyffert, C. (1930) *Biene und Honig im Volksleben der Afrikaner* (Leipzig: Voigt-länder) [B, T

Sharma, P. L. (1960) Experiments with *Apis mellifera* in India. *Bee Wld* 41(9): 230–232 [B, 44/62

Shaw, F. R., Bourne, A. J. & Migliorini, R. (1957) The effect of various levels of fertilizers on the growth and nectar secretion of snapdragons (*Antirrhinum majus*). *Glean. Bee Cult.* 85: 598–599 [B, 265/58

Shaw, F. R., Shaw, M. & Weidhaas, J. (1956) Observations on sugar concentration of cranberry nectar. *Glean. Bee Cult.* 84: 150–151 [B, 116/58

Sherman, M. B. (1968) Honey connoisseur reaches inevitable goal—makes candy commercially. *Candy Ind. Confect. J.* 131(13): 5–6 [B, 767 L/71

Shimakura, M. (1973) Palynomorphs of Japanese plants. *Spec. Publ. Osaka Mus. nat. Hist.* 5: 1–60 [B

Shookhoff, M. W. (1957) Process for preparing free-flowing sugar powders. *U.S. Pat. No. 2,818,356* [B, 517 L/62

Short, G. R. A. (1960) Flavours and colours in medicines. *Pharm. J.* 185(5068): 565–570 [B, 9h224/61

Shuel, R. W. (1952) Some factors affecting nectar yield in red clover. *Pl. Physiol., Lancaster* 27: 95–110 [B, 162/52

Shuel, R. W. (1955) Nectar secretion in relation to nitrogen supply, nutritional status and growth of the plant. *Can. J. agric. Sci.* 35: 124–138 [B, 83/56

Shuel, R. W. (1956) Studies of nectar secretion in excised flowers. *Can. J. Bot.* 34: 142–153 [B, 272/56

Shuel, R. W. (1957) Some aspects of the relation between nectar secretion and nitrogen, phosphorus and potassium nutrition. *Can. J. Pl. Sci.* 37: 220–236 [B, 233/59

Shuel, R. W. (1959) Studies of nectar secretion in excised flowers. II. The influence of certain growth regulators and enzyme inhibitors. *Can. J. Bot.* 37: 1167–1180 [B, 79/64

Shuel, R. W. (1961) Influence of reproductive organs on secretion of sugars in flowers of *Streptosolen jamesonii* Miers. *Pl. Physiol., Lancaster* 36: 265–271 [B, 464/62

Shuel, R. W. (1964) L'influence des facteurs externes sur la production du nectar. *Annls Abeille* 7: 5–12 [B, 564/65

Shuel, R. W. (1966) Einige Fragen über die Nektarsekretion. *Z. Bienenforsch.* 8: 205–209 [B, 104 L/67

Shuel, R. W. (1967) Nectar secretion in excised flowers. IV. Selective transport of sucrose in the presence of other solutes. *Can. J. Bot.* 45: 1953–1961 [B, 708/67

Shuel, R. W. (1970) Current research on nectar. *Bee Wld* 51(2): 63–69 [B, 926 L/71

Shuel, R. W. & Pedersen, M. W. (1953) The effect of environmental factors on nectar secretion as related to seed production. *Proc. 6. Int. Congr. Grasland:* 867–871

Shuel, R. W. & Shivas, J. A. (1953) The influence of soil physical condition during flowering period on nectar production in snapdragon. *Pl. Physiol., Lancaster* 28: 645–651

Shutt, F. T. & Charron, A. T. (1903) Determination of water content of honey. *Chem. News, Lond.* 87: 195–196, 210–212

Siddiqui, I. R. (1965) Constitution of an arabinogalactomannan from honey. *Can. J. Chem.* 43: 421–425 [B, 628/65]

Siddiqui, I. R. (1970) The sugars of honey. *Adv. Carbohyd. Chem. Biochem.* 25: 285–309 [B, 191/72]

Siddiqui, I. R. & Furgala, B. (1967) Isolation and characterization of oligosaccharides from honey. Part I. Disaccharides. *J. apic. Res.* 6(3): 139–145 [B, 402/68]

Siddiqui, I. R. & Furgala, B. (1968a) Isolation and characterization of oligosaccharides from honey. Part II. Trisaccharides. *J. apic. Res.* 7(1): 51–59 [B, 352/69]

Siddiqui, I. R. & Furgala, B. (1968b) Centose: a branched trisaccharide containing a–D–(1→2) and a–D–(1→4) glycoside bonds. *Carbohyd. Res.* 6: 250–252 [B, 200/71]

Silva Ferreira, E. L. da (1970) *O mel e seus acúcares—processos actuais de analise.* Univ. Técnica Lisboa, Inst. Superior de Agronomia: Tese

Simonis, W. C. (1965) *Milch und Honig: Eine ernährungsphysiologische Studie* (Stuttgart: Verlag Freies Geistesleben) [B, 558 L/66]

Simpson, J. (1960) The functions of the salivary glands of *Apis mellifera. J. Insect Physiol.* 4: 107–121 [B, 413/62]

Simpson, J. (1962) The salivary glands of *Apis mellifera* and their significance in caste determination. *IV. Congr. U.I.E.I.S. Pavia*: 173–188

Simpson, J. (1963) The source of the saliva honeybees use to moisten materials they chew with their mandibles. *J. apic. Res.* 2: 115–116 [B, 799/64]

Simpson, J. (1964) Dilution by honeybees of solid and liquid food containing sugar. *J. apic. Res.* 3(1): 37–40 [B, 829/64]

Simpson, J., Riedel, J. B. & Wilding, N. (1968) Invertase in the hypopharyngeal glands of the honeybee. *J. apic. Res.* 7: 29–36 [B, 270/69]

Singh, S. (1962) *Beekeeping in India* (New Delhi: Indian Council of Agricultural Research) 214 pp. [B, 291/64]

Sipos, E. (1964) The latest problems of honey evaluation. *XIX Int. Beekeep. Congr. 1963, Prague* 2: 643–646 [B, see 58L/64]

Skibbe, B. (ed.) (1958) *Agrarwirtschaftsatlas der Erde* [World agricultural atlas]. (Gotha: Hermann Haack) xxiv + 248 pp. + 1 map [see 142L/64]

Skou, J. P. (1972) Ascosphaerales. *Friesia* 10(1): 1–24 [B, 13/74]

Sladen, F. W. L. (1912) *The bumble bee: its life-history and how to domesticate it* (London: Macmillan & Co.) [B

Slater, L. G. (1969) *Hunting the wild honeybee* (Olympia, Wash.: Terry Publishing Co.) [B, 705 L/71]

Smaellie, E. (1957) Hive management for section comb honey production. Preparing section comb honey for market. Preparation of cut-comb honey for market. *N.Z. Jl Agric.* 94: 289–292, 385–388, 495–497, 602–604 [B, 367/59]

Smith, F. G. (1953) Beekeeping in the tropics. *Bee Wld* 34(12): 233–245 [B, 261/55]

Smith, F. G. (1958a) Beekeeping observations in Tanganyika 1949–57. *Bee Wld* 39(2): 29–36 [B, 42/62]

Smith, F. G. (1958b) Communication and foraging range of African bees compared with that of European and Asian bees. *Bee Wld* 39(10): 249–252 [B, 69/62]

Smith, F. G. (1960) *Beekeeping in the tropics* (London: Longmans) [B, 362/60]

Smith, F. G. (1961) The races of honeybees in Africa. *Bee Wld* 42(10): 255–260 [B, 634/62]

Smith, F. G. (1962) *Beekeeping as a forest industry* (Dar es Salaam: British Commonwealth Forestry Conference) 6 pp. [B, 399 L/62

Smith, F. G. (1964) Beekeeping in Western Australia. *Bee Wld* 45(1): 19–31 [B, 546 L/64

Smith, F. G. (1967) Deterioration of the colour of honey. *J. apic. Res.* 6(2): 95–98 [B, 399/68

Smith, F. & Srivastava, H. C. (1957) Isolation of maltose from honeydew on alsike (*Trifolium hybridum*) seeds. *J. org. Chem.* 22: 987–988 [B, 357/58

Smith, L. B. & Johnson, J. A. (1951) The use of honey in bread products. *Bakers' Dig.* 25(6): 103–106 [B, 94/53

Smith, L. B. & Johnson, J. A. (1952) The use of honey in cake and sweet doughs. *Bakers' Dig.* 26(6): 113–118 [B, 303/53

Smith, L. B. & Johnson, J. A. (1953a) Honey: its value and use in popular cookies. *Bakers' Dig.* 27(2): 28–31 [B, 68/58

Smith, L. B. & Johnson, J. A. (1953b) Honey improves fruit cake quality. *Bakers' Dig.* 27(3): 52–54 [B, 133/55

Smith, M. R. (1963) *Chromatographic investigation of trace lipids in honey.* University of Arizona: Ph.D. dissertation [B, 214/68

Smith, M. R. & McCaughey, W. F. (1966) Identification of some trace lipids in honey. *Fd Res.* 31(6): 902–905 [B, 381–67

Smith, M. R., McCaughey, W. F. & Kemmerer, A. R. (1969) Biological effects of honey. *J. apic. Res.* 8(2): 99–110 [B, 1008/70

Snodgrass, R. E. (1949) The anatomy of the honeybee. Pp. 473–518 from *The hive and the honey bee* ed. R. A. Grout (Hamilton Ill.: Dadant & Sons) [B, see 107/51

Snodgrass, R. E. (1956) Anatomy of the honeybee (Ithaca, N.Y.: Comstock Publ. Associates, Cornell Univ. Press) [B, 90/57

Snyder, C. F. (1933) Report on drying, densimetric, and refractometric methods. *J. Ass. Off. agric. Chem.* 16: 173–175

Sols, A., Cadenas, E. & Alvarado, F. (1960) Enzymatic basis of mannose toxicity in honey bees. *Science, N.Y.* 131(3396): 297–298 [B, 280/65

Somogyi, M. (1952) Notes on sugar determination. *J. biol. Chem.* 195: 19–23

Sooder, M. (1952) *Bienen und Bienenhalten in der Schweiz* (Basel: G. Krebs) 341 pp. [B, 46/54

Sorges, F. (1933) Alcuni caratteri merceologici dei mieli siciliani. *Chimica Ind. Agric. Biol.* 9: 398–399

Sotavalta, O. (1954) On the fuel consumption of the honeybee (*Apis mellifica* L.) in flight experiments. (Contributions to the problem of insect flight III). *Ann. zool. Soc. zool. bot. fenn. Vanamo* 16(5): 1–27 [B, 260/58

Soudek, S. (1927) Htlanové žlázy včely medonosné (*Apis mellifica* L.)[The pharyngeal glands of the honeybee]. *Sb. vys. Šk. zeměd. v Brně* 10: 1–63 [B

South Australia. Laws & Statutes (1964) Food and Drugs Regulations [and amendments to 1970]

Spain. Laws & Statutes (1968) Código Alimentario español. *Boln of. Estado* [honey, p. 202; labelling, p. 34]

Spencer, J. F. T. & Sallans, H. R. (1956) Production of polyhydric alcohols by osmophilic yeasts. *Can. J. Microbiol.* 2: 72–79

Spencer, J. F. T. & Shu, P. (1957) Polyhydric alcohol production by osmophilic yeasts: effect of oxygen tension and inorganic phosphate concentration. *Can. J. Microbiol.* 3: 559–567

Spencer, J. F. T., Gorin, P. A. J., Hobbs, G. A. & Cooke, D. A. (1970) Yeasts isolated from bumblebee honey from Western Canada: identification with the aid of proton magnetic resonance spectra of their mannose-containing polysac-charides. *Can. J. Microbiol.* 16(2): 117–119 [B, 198/72

Spöttel, W. (1950) *Honig und Trockenmilch* (Leipzig: J. A. Barth) 323 pp. [B, 63/54

Springer, B. V. (1954) Bee-keeping in the Talmud. *Brit. Bee J.* 82: 62, 79, 99, 107, 149, 151, 182, 216 [B, 227/57

Srivastava, P. N. & Auclair, J. L. (1962a) Amylase activity in the alimentary canal and honeydew of the pea aphid *Acyrthosiphon pisum* Harr. (Homoptera, Aphididae). *J. Insect. Physiol.* 8: 349–355 [B, 137 L/64

Srivastava, P. N. & Auclair, J. L. (1962b) Characteristics of invertase from the alimentary canal of the pea aphid *Acyrthosiphon pisum* Harr. (Homoptera, Aphididae). *J. Insect Physiol.* 8: 527–535 [B, 19 L/64

Srivastava, P. N. & Auclair, J. L. (1963) Characteristics and nature of proteases from the alimentary canal of the pea aphid *Acyrthosiphon pisum* Harr. (Homop-tera, Aphididae). *J. Insect Physiol.* 9: 469–474

Standards Association of Central Africa (1971) Unprocessed honey and processed honey. *Central African Standard* CAS S32: 1971: 11 pp. [B, 517 L/72

Staudenmayer, T. (1939) Die Giftigkeit von Mannose für Bienen und andere Insekten. *Z. vergl. Physiol.* 26: 644–668 [B

Steinkraus, K. H. & Morse, R. A. (1966) Factors influencing the fermentation of honey in mead production. *J. apic. Res.* 5: 17–26 [B, 744/66

Steinkraus, K. H. & Morse, R. A. (1973) Chemical analysis of honey wines. *J. apic. Res.* 12(3): 191–195 [B, 226/74

Stephen, J. A. (1934). Ontario Agricultural College: Thesis

Stephen, W. A. (1941) Removal of moisture from honey. *Scient. Agric.* 22(3): 157–169 [B

Stephen, W. A. (1946) The relationship of moisture content and yeast count in honey fermentation. *Sci. Agric.* 26(6): 258–264 [P

Stephen, W. A. (1957) Chunk comb honey. *Penn. Beekpr* 32(1): 1–2, 4–7, 9, 11–12 [B, 361/57

Stephen, W. A. (1959) Some sugars secreted by the woolly alder aphid (*Prociphilus tesselatus*). *J. econ. Ent.* 52(2): 353 only [B, 202/60

Steyn, D. G. (1970) Honey as a food and in the prevention and treatment of disease. *Publ. Univ. Pretoria No.* 52: 16 pp. [B, 212/72

Stinson, E. E., Subers, M. H., Petty, J. & White, J. W., Jr. (1960) The composition of honey. V. Separation and identification of the organic acids. *Archs Biochem. Biophys.* 89(1): 6–12 [B, 150/62

Stitz, J. (1930) A méz fehérjetartalma. *Mezőgazd. Kutat.* 3: 25–29. *German summary*

Stitz, J. & Szigvárt, B. (1931a) Die Gefrierpunkt serniedrigung des Honigs. *Z. Unters. Lebensmittel* 62(4): 506–509 [B

Stitz, J. & Szigvárt, B. (1931b) Die elektrische Leitfähigkeit des Honigs. *Z. Unters. Lebensmittel* 63(2): 211–214 [B

Stomfay-Stitz, J. & Kominos, D. D. (1960) Über bakteriostatische Wirkung des Honigs. *Z. Lebensmittelunters. u. -Forsch.* 113(4): 304–309 [B, 384/61

Stone, R. J. (1967) Handling honey on a large scale. *XXI Int. Beekeep. Congr. Maryland*: 318–321 (1967) [B, 793 L/67

Strabo (1923) *Strabonis Geographia.* Loeb Classical Library (London and Cam-bridge, Mass.: Heinemann)

Straka, H. (1970) *Pollenanalyse und Vegetationsgeschichte* (Wittenberg Lutherstadt: A. Ziemsen Verlag) 2nd ed. [B, 722/73

Struthers, J. (1951) The Stewarton hive. *Scot. Beekeep.* 27(12): 239–241 [B

Sturm, W. & Hanssen, E. (1961) Über die Entwicklung der Begriffe Leb- und

Honigkuchen, sowie den Nachweis von Honig in Lebensmitteln. *Dt. Lebensmitt Rdsch.* 57(10): 261–270 [B, 770/70

Sturtevant, A. P. & Revell, I. L. (1953) Reduction of *Bacillus larvae* spores in liquid food of honeybees by the action of the honeystopper and its relation to the development of American foul brood. *J. Econ. Ent.* 46: 855–860 [B, 47/55

Subers, M. H., Schepartz, A. I. & Koob, R. P. (1966). Separation and identification of some organic phosphates in honey by column and paper chromatography. *J. apic. Res.* 5(1): 49–57 [B, 730/66

Sundberg, T. & Lundgren, A. (1930) Untersuchung über die Aschenbestandteile in schwedischem Honig. *Arch. Bienenk.* 11: 324–328

Sutherland, M. D. & Palmer-Jones, T. (1947a) A recent outbreak of honey poisoning. Part 2. The toxic substances of the poisonous honey. *N.Z. Jl Sci. Technol.* 29A: 114–120 [B

Sutherland, M. D. & Palmer-Jones, T. (1947b) A recent outbreak of honey poisoning. Part 5. The source of the toxic honey—laboratory investigations. *N.Z. Jl Sci. Technol.* 29A: 129–133 [B

Švec, J. (1968) Príspevok k výskumu sekrécie nektáru d'ateliny lúčnej (*Trifolium pratense* L.) [Contribution to the research on nectar secretion of red clover]. *Věd. Pr. výzk. Úst. včelař. v Dole* 5: 101–139 [B, 346/73

Sviderskaya, Z. I. (1959) [A case of food poisoning from honey.] *Gig. Sanit.* 24(5): 57 only. *In Russian* [B, 386/60

Svoboda, J. (1933) [Honeys from sub-Carpathian Russia with particular reference to those from *Fagopyrum sagitatum*]. *Chemické Listy* 27: 441–443. *In Russian*

Svoboda, J. (1935) *Výroba medoviny medových vin a likéru* [The manufacture of mead, wines and liqueurs from honey] (Prague: Zemského Ustředi včelařských spolku pro Cechy v Praze) [B

Svoboda, J. (1958) Facts about beekeeping in Czechoslovakia. *Bee Wld* 39(6): 137–150 [B, 35/62

Sweden. Laws & Statutes (1971) Food Act of 18.6.1971. Livsmedelslag *SFS 1971* 511, para. 13 *In Swedish*

Sweden (1972) National Food Administration notice on sugar and honey. SLV 36 (1972) *In Swedish*

Sweden. Sveriges Standardiseringskommission (1964) Honung. *Norm* VDN 1047: 1: 3 pp. *In Swedish* [B

Switzerland (1967) Honig und Kunsthonig. Vol. 2, Part II, Chp. 23 from *Schweizerisches Lebensmittelbuch* [B, 1052 L/72

Switzerland (1974) Die Honigkontrolle des Vereins Deutschschweizerischer Bienenfreunde [V.D.S.B.] (quoted in *Der Schweizerische Bienenvater*, ed. G. Fischer *et al.* Pp. 518–523. Fachschriften des V.D.S.B. (Aarau & Frankfurt am Main: Sauerländer)) [B

Switzerland. Laws & Statutes (1936–1972) Verordnung über den Verkehr mit Lebensmitteln und Gebrauchsgegenständen. 26.5.36, amended to 1972 [labelling, Art. 13, 14, 16, 19; honey, Art. 217–224; artificial honey, Art. 225–231; this publication is also available in French, Bundeskanzlei, Bern (1963)]

Szklanowska, K. (1957) Wstepne badania nad wydajnościa nektarowa pszczelnika moldawskiego (*Dracocephalum moldavicum* L.) [Preliminary investigation on nectar secretion in *Dracocephalum moldavicum* L.] *Pszczel. Zesz. Nauk.* 1(3): 129–138 [B, 586/63

Szklanowska, K. (1965) Factors influencing the secretion of nectar by *Dracocephalum moldavicum* L. *XX Int. Beekeep. Congr., Bucharest*: 374 only
[B, *see* 274/67

Szklanowska, K. (1967) [Influence of potassium fertilization in the nectar yield of

Dracocephalum moldavicum L.] *Annls Univ. Mariae Curie-Skłodowska* 32E: 107–121. *In Polish* [B, 719/72

Szklanowska, K. (1973) Nektarowanie roślin miododajnych w wybranych zbiorowiskach borowych nadleśnictwa Zwierzyniec i Kraśnik w województwie Lubelskim. *Rozp. habil. Akademia Rolnicza w Lublinie*

Takhtajan, A. (1969) *Flowering plants: origin and dispersal* (Edinburgh: Oliver & Boyd) Translated from 2nd Russian ed. (1961) [B, 231 L/70

Tamaki, Y. (1964*a*) Amino acids in the honeydew excreted by *Ceroplastes pseudoceriferus* (Green). *Jap. J. appl. Ent. Zool.* 8: 159–164 [B, 515 L/69

Tamaki, Y. (1964*b*) Carbohydrates in the honeydew excreted by *Ceroplastes pseudoceriferus* (Green). *Jap. J. appl. Ent. Zool.* 8: 227–234 [B, 514 L/69

Tamaki, Y. (1968) Constituents of honeydew produced by aphids. *Biol. Sci., Tokyo* 20(1): 17–25 [B, 668 L/70

Tasmania. Laws & Statutes (1960) Food and Drugs Regulations 1941 [reprint 1960, and amendments up to 1970]

Tatsuno, T., Shirotori, T., Iwaida, M. & Kawashiro, I. (1968) [Determination of harmful metals in foods. VII. Lead and copper contents in honey]. *J. hyg. chem. Soc. Japan* 14: 327–329 [B, 490/70

Täufel, K. & Müller, K. (1953) Zur Herkunft der Saccharide im Honig. *Z. Lebensmittelunters. u. -Forsch.* 96(2): 81–83 [B, 212/55

Täufel, K. & Müller, K. (1957) Analytische und Papierchromatographische Untersuchungen über die Saccharide von Honig und Kunsthonig. *Ernährungsforschung* 2(1): 70–82 [B, 27/61

Taylor, M. W. & Nelson, V. E. (1929) Molasses, sorghum and honey as sources of vitamin E. *Proc. Soc. exp. Biol.* 26(6): 521 only

Taylor, P. V. (1956) Melaleuca trees annoy Florida beekeepers. *Am. Bee J.* 96(11): 449 only [B, 223/58

Temnov, V. A. (1944) Bactericidal properties of honey and utilization of honey and other beekeeping products for the healing of wounds. *Bee World* 25(11): 86–87 [B

Terrier, J. (1953) Le dosage indirect de l'eau dans les produits sucrés tels que le miel, la confiture, le glucose industriel, les extraits concentrés de fruits et la purée de tomate. *Mitt. Geb. Lebensmittelunters. u. Hyg.* 44(3): 302–307 [B, 140/57

Thakar, C. V. (1973) A preliminary note on hiving *Apis dorsata* colonies. *Bee Wld* 54(1): 24–27 [B, 45/74

Thompson, F. (1939) *Lark rise* (Oxford: University Press)

Thompson, V. C. (1960) Nectar flow and pollen yield in southwestern Arkansas. *Rep. Ser. Ark. agric. Exp. Stn* No. 94: 38 pp. [B, 371/61

Thomson, R. H. K. (1936) Chemical composition of New Zealand honey. *N.Z. Jl Sci. Technol.* 18(2): 124–131 [B

Thöni, J. (1911) Die Verwendung der quantitativen Präzipitinreaktion bei Honiguntersuchungen. *Mitt. Geb. Lebensmittelunters. u. Hyg.* 2: 80–123 [B

Thöni, J. (1912) Die Verwendung der quantitativen Präzipitinreaktion bei Honiguntersuchung. II. *Mitt. Geb. Lebensmittelunters. u. Hyg.* 3: 74–94

Tilly, B. (ed.) (1973) *Varro the farmer: a selection from 'Res rusticae'.* (London: University Tutorial Press)

Tillmans, J. & Kiesgen, J. (1927) Die Formoltitration als Mittel zur Unterscheidung von künstlichen und natürlichen Lebensmitteln. *Z. Unters. Lebensmittel* 53: 131–137

Todd, F. E. & Vansell, G. H. (1942) Pollen grains in nectar and honey. *J. econ. Ent.* 35(5): 728–731 [B

Tokuda, Y. & Sumita, E. (1925) Studies on poisonous honey in Japan. *Bee Wld* 7: 4–5 [B

Tokyo Yakuhin Kaihatsu Co. Ltd. (1969) [Powdered honey]. *Jap. Pat. No.* 1359/69. *In Japanese* [213 L/72

Tone, E. & Coteanu, O. (1969) Spectrul polenic la unele sorturi de miere monoflora. [The pollen spectrum of some monofloral honey sorts]. *Anale Stat. cent. Seri. Apic.* 9: 39–48

Tonsley, C. (1969) *Honey for health* (London: Universal-Tandem Publishing Co. Ltd.) [B

Torr, D. (1967) Procédé de production d'un produit séché à base de miel et de lait et produit obtenu. *Fr. Pat. No.* 1,470,543 [B, 570/69

Torrent, J. A. (1949) Algunas consideraciones sobre la determinación refracto-métrica de agua en la miel. *Rev. Asoc. bioquim. argent.* 14(66): 1–5 [B, 192/55

Townsend, G. F. (1939) Time and temperature in relation to the destruction of sugar-tolerant yeasts in honey. *J. econ. Ent.* 32(5): 650–654 [B

Townsend, G. F. (1943) Moisture absorption by honey stored in various types of containers. Unpublished [B

Townsend, G. F. (1954) Private communication

Townsend, G. F. (1956) Honey, the crop without a surplus. *Can. Bee J.* 64(2): 19–21 [B, 205/57

Townsend, G. F. (1961) Preparation of honey for market. *Publ. Ont. Dept. Agric.* No. 544 [B, 411/63

Townsend, G. F. (1965) Preparation of honey for market. *Publ. Ont. Dept. Agric.* No. 544 revised [B, 185 L/66

Townsend, G. F. (1965a) Unpublished data on source of odour due to plastic containers [B

Townsend, G. F. (1966) Brood combs darken honey. Unpublished

Townsend, G. F. (1969a) Optical density as a means of colour classification of honey. *J. apic. Res.* 8(1): 29–36 [B, 197/70

Townsend, G. F. (1969b) How the beekeeper can influence the quality of honey. *XXII Int. Beekeep. Congr., Munich*: 593–596 [B, see 851/70

Townsend, G. F. (1970) Preparation of honey for market. *Publ. Ont. Dept. Agric.* No. 544 2nd revised ed. [B, 202 L/72

Townsend, G. F. (1971) The blending of honey at moderate temperatures. *J. apic. Res.* 10(2): 73–77 [B, 207/72

Townsend, G. F. & Adie, A. (1952) Melting honey for repacking. *Circ. Ont. agric. Coll.*, No. 121 [B, see 145/59

Townsend, G. F. & Adie, A. (1956) Uniform granulation of honey by continuous flow. *Circ. Ont. Agric. Coll.* No. 285 [B, 167/58

Townsend, G. F. & Barrington, M. G. (1953) Continuous honey processing. *Can. chem. Process.* 37:13, 24–28

Trautmann, A. & Kirchhof, H. (1932) Über das Vorkommen von Vitaminen im Honig. *Arch. Bienenk.* 13: 49–65 [B

Travaux et Documents de Géographie Tropicale (1974) *Pollen et spores d'Afrique tropicale No.* 16 [B

Turkot, V. A., Eskew, R. K. & Claffey, J. B. (1960) A continuous process for dehydrating honey. *Fd Technol., Champaign* 14(8): 387–390 [B, 516/62

Turnbull, C. M. (1966) *Wayward servants. The two worlds of the African pygmies* (London: Eyre & Spottiswoode)

Turnbull, C. M. (1968) The importance of flux in two hunting societies. Chapter 15 in *Man the hunter* ed. R. B. Lee & I. Devore (Chicago: Aldine Publishing Co.)

Turner, J. C. & Clinch, P. G. (1968) Estimation of tutin and hyenanchin in honey. I. A comparison of the thin-layer chromatography and intracerebral injection methods. *N.Z. Jl Sci.* 11(2): 342–345 [B, 565/69]

Turnow, G. R. (1893) *De apium mellisve apud veteres significatione et symbolica et mythologica* (Berlin: Weidmann)

Tutin, T. G., *et al.* (1964) *Flora Europaea* Vol. 1. Lycopodiaceae to Platanaceae. (London: Cambridge University Press) [B

Tutin, T. G., *et al.* (1968) *Flora Europaea* Vol. 2. Rosaceae to Umbelliferae (London: Cambridge University Press) [B, 396/69]

Ugarte, L. & Karman, G. (1945) Valor de la relación glucosa-levulosa en mieles argentinas. *An. Soc. quím. argent.* 33: 181–187

Ukkelberg, B. (1960) Temperaturen og lyngtrekket [Temperature and the heather flow]. *Birøkteren* 76(3): 42–43 *In Norwegian* [B, 273/60]

Ungan, A. (1940) Über giftige Honige aus Nord-Anatolien. *Türk Hifz. tecr. Biyol. Mecm.* 2(2): 166–169

United Kingdom. Laws & Statutes (1963) *The Weights and Measures Act* [Schedule 4 requires the declaration of quantity; regulations published in 1964 prescribe the manner of marking]

United Kingdom. Laws & Statutes (1970) *The labelling of food regulations 1970* (London: H.M.S.O.)

United Kingdom. Laws & Statutes (1972) *Trade Descriptions Act* (indication of origin) (exemption number 1) 1972. *See also Trade and Industry* 7.12.1972, p. 462

United States Department of Agriculture (1933) *United States standards for grades of comb honey* [effective 8.33; reprinted 4.57; summary in Root (1959) p. 327 and Grout (1963) p. 371]

United States Department of Agriculture (1951) *United States standards for grades of extracted honey* [effective 16.4.51; *Fed. Reg.* 18(52): 8005; summary in Grout (1963) p. 370; *Bee Wld* 35(5): 104 (1954)] [B, 101/54]

United States Department of Agriculture. Agricultural Marketing Service (1957a) *United States standards for grades of comb honey* (Washington: U.S. Department of Agriculture) 16 pp.; second amended issue 1967 [B, 227 L/68]

United States Department of Agriculture. Agricultural Marketing Service (1957b) *U.S. Standards for grades of extracted honey. Fed. Reg.* 22: 3535–3540 [B, 228 L/68]

United States Department of Agriculture (1963–1971) *Foreign Agricultural Circular* [on world honey production and trade]. Various titles; reference numbers are: 1963–65 FS 4–63, 5–64, 5–65; 1966 and on FHON 1–66, 1–67 etc. [B

United States Department of Agriculture. Foreign Agricultural Service (1969) U.S. honey in the West European market. *Publ. U.S.D.A. Foreign agric. Serv.* FAS M–206: 30 pp. [B, 196/71]

United States Department of Agriculture. Statistical Reporting Service. Crop Reporting Board (1967). Honey and beeswax: number colonies, yield per colony; production; price per pound; value of production; by states, 1955–64 revised estimates. *Statist. Bull. U.S. Dep. Agric. Statist. rep. Serv. No.* 388: 14 pp. [B, 988L/72]

United States Food and Drug Administration (1973) Recommended dietary allowances. *Fed. Reg.* 38(13): 2146 [B

United States. Laws & Statutes (1962) *Federal Food, Drug and Cosmetic Act*

[General provisions, Section 402; labelling, Section 403; quoted and summarized e.g. by Blanck, 1955]

Utz, F. (1908a) Welchen Wert hat die Bestimmung des Aschengehaltes und die Ausführung der Ley'schen Reaktion bei der Honiguntersuchung? *Z. Unters. Nahr.- u. Genussmittel* 15: 607–609

Utz, F. (1908b) Über die Verwendung des Refraktometers zur Bestimmung der Trockensubstanz und des spezifischen Gewichts des Honigs. *Z. angew. Chem.* 21: 1319–1321

Valin, J. (1956) L'analyse saccharimetrique du miel. *Annls Falsif. Fraudes* 49(573/574): 388–401 [B, 703/62

Vansell, G. H. (1939) The sugar concentration of Western nectars. *J. econ. Ent.* 32: 666–668

Vansell, G. H. (1940) Nectar secretion in *Poinsettia* blossoms. *J. econ. Ent.* 33: 409–410

Vansell, G. H. & Freeborn, S. B. (1929) Preliminary report on the investigations of the source of diastase in honey. *J. econ. Ent.* 22: 922–926

Varju, M. (1970) Mineralstoff-Zusammensetzung der ungarischen Akazienhonigarten und deren Zusammenhang mit der Pflanze und dem Boden. *Z. Lebensmittelunters. u. -Forsch.* 144(5): 308–312 [B, 789/72

Vasmer, M. (1953) *Russisches etymologisches Wörterbuch* (Heidelberg: Carl Winter)

Vavruch, I. (1952) Chromatografická studie včelího medu [Investigation of honey by chromatography]. *Chemické Listy* 46(2): 116–117 [B, 34/54

Vellard, J. (1939) *Une civilisation du miel: les Indiens guayakis du Paraguay* (Paris: Librairie Galimard) [B

Vergé, J. (1951) L'activité antibactérienne de la propolis, du miel et de la gelée royale. *Apiculteur* 95(6) Sect. sci.: 13–20 [B, 130/60

Vergeron, P. (1964) Interprétation statistique des résultats en matière d'analyse pollinique des miels. *Annls Abeille* 7(4): 349–364 [B, 170/67

Vergeron, P. & Louveaux, J. (1964) Étude du spectre pollinique de quelques miels espagnols. *Annls Abeille* 7(4): 329–347 [B, 376/67

Verordnung *see under* Germany, and Switzerland

Victoria. Laws & Statutes (1966) Department of Health, Food and Drug Standards Regulations [with amendments to 1968]

Vieitez, E. (1948) Palinologia hispana. Grana palynologica 5. *Svensk bot. Tidskr.* 42: 480–499

Vignec, A. J. & Julia, J. F. (1954) Honey in infant feeding. *Amer. J. Dis. Child.* 88(4): 443–451 [B, 109/55

Villumstad, E. (1951) Honningens hygroskopisitet og honningemballasjens tettingsevne [Hygroscopicity of honey and air-tightness of honey containers]. *Årsmeld. St. SmåbrLaerarsk.* 12: 20 pp. *In Norwegian* [B

Villumstad, E. (1952) Undersøkelser av behandlingens innflytelse på honning-kvaliteten [Investigations on the effect of treatment on the quality of crystalline honey]. *Årsmeld. St. SmåbrLaerarsk.* 13: 25 pp. *In Norwegian* [B, 100/54

Villumstad, E. (1954) The formation of crystals in honey. *XV Int. Beekeep. Congr. Copenhagen*: 2 pp. [B, *see* 237/54

Villumstad, E. (1956) Slynging og behandling av lynghonning [Extraction and treatment of heather honey]. *Årsmeld. St. SmåbrLaerarsk.* 16: 7 pp. *In Norwegian* [B, 23/58

Villumstad, E. (1960a) Vanninhold og gjaering i honning samlet av bifolk av forskjellige raser [Water content and fermentation of honey from colonies of different races]. *Birøkteren* 76(6): 103–109. *In Norwegian* [B, 352/60

Villumstad, E. (1960b) Senking av vanninnholdet i honning før slynging [Lowering the water content of honey before extracting]. *Birøkteren* 76(13): 216–219 [B, 229/61

Vinson, A. E. (1907) Honey vinegar. *Frmrs' Bull. U.S. Dep. Agric. No.* 276: 28–29 [B

Vinuesa, [A. G. de] (1968) En estas Navidades comeremos ocho millones de kilos de turrón. *Apicultura, Madrid* 18(200): 7–8 [B, 769 L/71

Vinuesa, [A.] T. G. de (1965) *Miel para su cocina* (Madrid: Instituto de Investigaciones Apícolas Mendes de Torres) [B

Virgil *The Georgics of Virgil* translated by C. Day Lewis (1940) (London, Toronto: Jonathan Cape) [B

Virtanen, A. I. & Kari, S. (1955) Free amino acids in pollen. *Acta chem. scand.* 9(9): 1548–1551 [B, 24/57

Vishnu-Mittre (1958) Pollen content of some Indian honeys. *J. sci. industr. Res.* 17C(7): 123–124 [B, 129/60

Vitez, P. (1965) Facts about beekeeping in Argentina. *Bee Wld* 46(1): 19–22 [B, 502 L/65

Vivino, A. E., Haydak, M. H., Palmer, L. S. & Tanquary, M. C. (1943) Antihemorrhagic vitamin effect of honey. *Proc. Soc. exp. Biol.* 53: 9–11

Voerman, L. & Bakker, C. (1911) Untersuchung einiger Proben echten Honigs. *Z. öff. Chem.* 17: 461–467

Vogel, B. (1931) Über die Beziehungen zwischen Süssgeschmack und Nährwert von Zuckern bei der Honigbiene. *Z. vergl. Physiol.* 14: 273–347 [B, T

Voigtlander, H. (1937) Honey for burns and scalds. *Bee World* 18(11): 128 [B

Voorst, F. T. van (1941) Biochemische suikerbepalingen. XII. Honing. *Chem. Weekbl.* 38: 538–542

Vorwohl, G. (1964) Die Beziehungen zwischen der elektrischen Leitfähigkeit der Honige und ihrer trachtmässigen Herkunft. *Annls Abeille* 7(4): 301–309 [B, 158/67

Vorwohl, G. (1966) Das mikroskopische Bild der Pollenersatzmittel und des Sediments von Futterteigen. *Z. Bienenforsch.* 8(7): 222–228 [B, 726/67

Vorwohl, G. (1968a) Natürliche Diastaseschwäche der Honige von *Apis cerana* Fabr. *Z. Bienenforsch.* 9(5): 232–236 [B, 750/70

Vorwohl, G. (1968b) Das Pollenbild der italienischen Honige. *Ber. dt. bot. Ges.* 81(11): 512–527 [B, 204/71

Vorwohl, G. (1970) Das mikroskopische Bild einiger Honige aus Florida. *Apidologie* 1(3): 233–269 [B, 793/72

Vorwohl, G. (1972) Das Pollenspektrum von Honigen aus den Italianischen Alpen. *Apidologie* 3(4): 309–340 [B

Vorwohl, G. (1973) Die Repräsentierung des *Citrus*-Pollens in Italianischen Orangenhonigen. *Apidologie* 4(3): 275–281 [B

Vouloir, G. (1935) *Nectar. L'industrie des hydromels, des eaux-de-vie de miel, et des vinaigres de miel* (Paris: Les Presses Universitaires de France) [T

Wahl, O. (1963) Vergleichende Untersuchungen über den Nährwert von Pollen, Hefe, Sojamehl und Trockenmilch für die Honigbiene (*Apis mellifica*). *Z. Bienenforsch.* 6: 209–280 [B, 873/64

Walde, A. (1954) *Lateinisches etymologisches Wörterbuch* (Heidelberg: Carl Winter)

Walker, H. S. (1917) A simplified inversion process for the determination of sucrose by double polarisation. *J. ind. Engng Chem.* 9: 490

Walsh, R. S. (1960) Anti-granulation method for honey packers. *N.Z. Jl Agric.*
101(3): 229, 231   [B, 413/63

Wanic, D. & Mostowska, I. (1964) Cukrowce w nektarze i miodzie [Sugars of
nectar and honey]. *Zesz. nauk. wyższ. Szk. roln. Olsztyn.* 17(4): 543–551 [B, 678/66

Wanner, H. (1953) Die Zusammensetzung des Siebröhrensaftes: Kohlenhydrate.
Ber. schweiz. bot. Ges. 63: 162–168

Watanabe, T. (1955a) Minute constituents of crude drugs. VIII. Acetylcholine and
related substances in honey 2. *J. pharm. Soc. Japan* 75: 83–85   [B, 393/57

Watanabe, T. (1955b) Minute constituents of crude drugs. IX. Acetylcholine and
related substances in honey 3. *J. pharm. Soc. Japan* 75: 86–88   [B, 393/57

Watanabe, T. & Aso, K. (1960) Studies on honey. II. Isolation of kojibiose,
nigerose, maltose, and isomaltose from honey. *Tohoku. J. agric. Res.* 11: 109–115
                                                                      [B, 151/62

Watanabe, T., Motomura, Y. & Aso, K. (1961) Studies on honey and pollen V.
On the sugar composition of honey (2). *Tohoku J. agric. Res.* 12(2): 187–190
                                                                      [B, 836/65

Watson, C. D. (1968) A step-by-step guide to making fondant from pure honey.
Candy Ind. Confect. J. 131(13): 7–8   [B, 768 L/71]

Watt, B. K. & Merrill, A. L. (1963) Composition of foods. *Agric. Handb. U.S. Dep.
Agric. No. 8*

Webb, B. H. & Walton, G. P. (1952) Dried honey-milk product. *U.S. Pat. No.*
2,621,128   [B

Weber, H. (1966) *Grundriss der Insektenkunde* (Stuttgart: Fischer)
                                                                      [B, 1949 ed. 92/50

Wedmore, E. B. (1932, 2nd ed. 1945) *A manual of bee-keeping for English-speaking
bee-keepers* (London: Edward Arnold & Co.)   [B

Wedmore, E. B. (1955) The accurate determination of the water content of honeys.
I. Introduction and results. *Bee Wld* 36(11): 197–206   [B, 360/57

Weishaar, H. (1933) Untersuchungen über Bestimmung, Mindestwert, und Herkunft
der Honigdiastase. *Z. Unters. Lebensmittel* 65: 369–399

Wells, P. H. & Giacchino, J. (1968) Relationship between the volume and the sugar
concentration of loads carried by honeybees. *J. apic. Res.* 7: 77–82 [B, 915/70

Werder, J. & Antener, J. (1938) Zur Vitamin-C Bestimmung in Nahrungsmitteln.
Mitt. Geb. Lebensmittelunters. u. Hyg. 29: 339–349

Western Australia. Laws & Statutes (1961) Food and Drug Regulations [and
amendments to 1967]

Wheler, G. A. (1682) *A journey into Greece* (London, for T. Cademan)

Whitcomb, W., Jr. & Wilson, H. F. (1929) Mechanics of digestion of pollen by the
adult honeybee and the relation of undigested parts to dysentery of bees. *Res.
Bull. Wis. agric. Exp. Stn No.* 92: 27 pp.   [B

White, J. W., Jr. (1950) New crystallized fruit spread shows commercial promise.
Fd Inds 22(7): 1216, 1298   [B, 55/52

White, J. W., Jr. (1952) The action of invertase preparations. *Archs. Biochem.
Biophys.* 39(1): 238–240   [B, 239/54

White, J. W., Jr. (1958) Coarse granulation of honey. *Glean. Bee Cult.* 86(12):
730–732   [B

White, J. W., Jr. (1959a) Report on the analysis of honey. *J. Ass. off. agric. Chem.*
42(2): 341–348   [B, 148/62

White, J. W., Jr. (1959b) Packaged honey and method of packaging same. *U.S.
Pat. No.* 2,902,370   [B, 131/60

White, J. W., Jr. (1961) A survey of American honeys. *Glean. Bee Cult.* 89: 230–
233   [B, 505 L/62

White, J. W., Jr. (1964a) Diastase in honey: the Schade method. *J. Ass. off. analyt. Chem.* 47: 486–488 [B, 204/65

White, J. W., Jr. (1964b) Dextrose determination in honey: a rapid photometric determination. *J. Ass. off. agric. Chem.* 47(3): 488–491 [B, 380/65

White, J. W., Jr. (1966) Methyl anthranilate content of citrus honey. *J. Fd Sci.* 31(1): 102–104 [B, 728/66

White, J. W., Jr. (1967a) Measuring honey quality—a rational approach. *Am. Bee J.* 107(10): 374–375 [B, 791/67

White, J. W., Jr. (1967b) Moisture determination in honey with an Eichhorn-type hydrometer. *J. apic. Res.* 6(1): 11–16 [B, 401/68

White, J. W., Jr. (1973) Toxic honeys. Pp. 495–507 from *Toxicants occurring naturally in foods.* National Research Council, 2nd ed. (Washington: National Academy of Sciences) [B

White, J. W., Jr. & Hoban, N. (1959) Composition of honey. IV. Identification of the disaccharides. *Archs Biochem. Biophys.* 80(2): 386–392 [B, 152/62

White, J. W., Jr. & Kushnir, I. (1966) Enzyme resolution in starch gel electrophoresis. *Analyt. Biochem.* 16(2): 302–313 [B, 202/71

White, J. W., Jr. & Kushnir, I. (1967a) Composition of honey. VII. Proteins. *J. apic. Res.* 6(3): 163–178 [B, 403/68

White, J. W., Jr. & Kushnir, I. (1967b) The enzymes of honey: examination by ion-exchange chromatography, gel filtration, and starch-gel electrophoresis. *J. apic. Res.* 6(2): 69–89 [B, 218/68

White, J. W., Jr., Kushnir, I. & Subers, M. H. (1964c) Effect of storage and processing temperatures on honey quality. *Fd Technol.* 18(4): 153–156 [B, 922/64

White, J. W., Jr. & Maher, J. (1951) Detection of incipient granulation in honey. *Am. Bee J.* 91(9): 376–377 [B, 160/52

White, J. W., Jr. & Maher, J. (1953a) Transglucosidation by honey invertase. *Archs Biochem. Biophys.* 42(2): 360–367 [B, 36/54

White, J. W., Jr. & Maher, J. (1953b) a-maltosyl β-D-fructofuranoside, a trisaccharide enzymically synthesized from sucrose. *J. Am. chem. Soc.* 75: 1259–1260 [B, 58/55

White, J. W., Jr. & Maher, J. (1954) Selective adsorption method for determination of the sugars of honey. *J. Ass. off. agric. Chem. Wash.* 37(2): 466–478 [B, 306/55

White, J. W., Jr., Petty, J. & Hager, R. B. (1958) The composition of honey. II. Lactone content. *J. Ass. off. agric. Chem.* 41(1): 194–197 [B, 146/62

White, J. W., Jr., Ricciuti, C. & Maher, J. (1952) Determination of dextrose and levulose in honey. Comparison of methods. *J. Ass. off. agric. Chem. Wash.* 35(4): 859–872 [B, 111/54

White, J. W., Jr. & Riethof, M. L. (1959) The composition of honey. III. Detection of acetylandromedol in toxic honeys. *Arch. Biochem.* 79(1/2): 165–167 [B, 155/62

White, J. W., Jr., Riethof, M. L. & Kushnir, I. (1961) Composition of honey. VI. The effect of storage on carbohydrates, acidity, and diastase content. *J. Fd Sci.* 26(1): 63–71 [B, 313/62

White, J. W., Jr., Riethof, M. L., Subers, M. H. & Kushnir, I. (1962) Composition of American honeys. *Tech. Bull. U.S. Dep. Agric. No.* 1261: 124 pp. [B, 655/63

White, J. W., Jr. & Subers, M. H. (1963) Studies on honey inhibine. 2. A chemical assay. *J. apic. Res.* 2(2): 93–100 [B, 914/64

White, J. W., Jr. & Subers, M. H. (1964a) Studies on honey inhibine. 3. Effect of heat. *J. apic. Res.* 3(1): 45–50 [B, 915/64

White, J. W., Jr. & Subers, M. H. (1964b) Studies on honey inhibine. 4. Destruction of the peroxide accumulation system by light. *J. Fd Sci.* 29(6): 819–828
[B, 483/70

White, J. W., Jr., Subers, M. H. & Schepartz, A. I. (1962) The identification of inhibine. *Am. Bee J.* 102(11): 430–431 [B, 193/64

White, J. W., Jr., Subers, M. H. & Schepartz, A. I. (1963) The identification of inhibine, the antibacterial factor in honey, as hydrogen peroxide and its origin in a honey glucose-oxidase system. *Biochim. biophys. Acta* 73: 57–70
[B, 621/65

White, J. W., Jr. & Walton, G. P. (1950) Flavour modification of low grade honey. *U.S. Bur. agric. ind. Chem. Circ. No.* AIC 272 [B, 62/55

Whitfield, B. G. (1956) Virgil and the bees. *Greece & Rome* 3(2): 99–117
[B, 228/57

Wigglesworth, V. B. (1950) Fuel consumption in the flying insect. *Sci. News Lett.* 16: 26–32 [B, 135/51

Wiley, H. W. (1892) Foods and food adulterants. Part Sixth. Sugar, molasses and sirup, confections, honey and beeswax. *Bull. U.S. Dep. Agric. Div. Chem. No.* 13(6): 633–874

Wilhelm, Maxine (1965) *Honey cook book* (Erick, Oklahoma: Wilhelm Honey Farm) [B

Williams, A. H. (1941) *An introduction to the history of Wales.* I. Prehistoric times to 1063 AD (Cardiff)

Willis, I. (1972) Adapting to a way of life. *Birds* 4(1): 11–15

Willson, R. B. (1953) Beekeeping in Mexico. *Glean. Bee Cult.* 81: 79–82, 143–146 [B

Willson, R. B. (1955) Meet the champions: Miel Carlota! *Glean. Bee Cult.* 83: 329–332; 408–410, 447; 473–476, 509 [B, 69/56

Wilson, C. A. (1973) *Food and drink in Britain: from the Stone Age to recent times* (London: Constable)

Wilson, H. F. & Marvin, G. E. (1929) On the occurrence of the yeasts which may cause the spoilage of honey. *J. econ. Ent.* 22: 513–517

Wilson, H. F. & Marvin, G. E. (1931) The effect of temperature on honey in storage. *J. econ. Ent.* 24: 589–597

Wilson, H. F. & Marvin, G. E. (1932) Relation of temperature to the deterioration of honey in storage. A progress report. *J. econ. Ent.* 25: 525–528 [B

Wilson Honey Company (1963) Continuous processing of honey simplified. *Fd Process.* 185–186 [B, 353/69

Wines and Vines (1964) Directory of the Wine Industry 1964–1965 (San Francisco)

Wines and Vines (1971) Directory of the Wine Industry (San Francisco)

Winkler, O. (1955) Beitrag zum Nachweis und zur Bestimmung von Oxymethyl-furfural in Honig und Kunsthonig. *Z. Lebensmittelunters. u. -Forsch.* 102(3): 161–167 [B, T, 127/56

Wodehouse, R. P. (1935) *Pollen grains* (New York–London: McGraw-Hill) [B

Wolf, J. P. & Ewart, W. H. (1955a) Carbohydrate composition of the honeydew of *Coccus hesperidum* L. Evidence for the existence of two new oligosaccharides. *Arch. Biochem. Biophys.* 58(2): 365–372 [B, 55/58

Wolf, J. P. & Ewart, W. H. (1955b) Two carbohydrases occurring in insect pro-duced honeydew. *Science, N.Y.*: 122: 973 [B, 294/56

Wolthers, P. (1955) Studier over inholdet af blomsterstøv i honningprøver fra danske lyngtraeksegne [Studies on the pollen content of honey samples from Danish heather regions]. *Tidsskr. PlAvl* 58: 683–721 [B, 65/58

Wolthers, P. (1956) Pollenanalyse af danske lynghonninger. *Tidsskr. Biavl* 90(5/6): 77–78, 95–97 [B

Woodburn, J. (1966) *The Hadza: an ethnographic film about an East African hunting and gathering tribe* (London: Hogarth Films)

Woodburn, J. (1968) An introduction to Hadza ecology. Chapter 5 in *Man the hunter* ed. R. B. Lee & I. Devore (Chicago: Aldine Publishing Co.)

Woźna, J. (1966) Obraz pyłkowy i barwa niektórych odmianowych miodów handlowych [Pollen spectrum and colour of some commercial unifloral honeys]. *Pszczel. Zesz. nauk.* 10(1/4): 139–153 [B, 229/68

Woźnica, J. (1966) Badania nad niektórymi czynnikami wpływającymi na dojrzewanie pszczelich pokarmów cukrowych [Some factors influencing the ripening of stores from syrup of different concentrations.] *Roczn. Nauk roln.* 89b(2): 233–238 [B, 740/72

Wulfrath, A. & Speck, J. J. (1955, 1958) Enciclopedia apícola. 2 vols. *Mexico, D.F.: Editora agrícola Mexicana* [B, 402, 403/59

Wyatt, G. R. & Kalf, G. F. (1957) The chemistry of insect hemolymph. II. Trehalose and other carbohydrates. *J. gen. Physiol.* 40: 833–847

Wykes, G. R. [1947] [Nectar secretion, with special reference to *Eucalyptus*]. *Unpublished report:* 22 + 33 pp. [B

Wykes, G. R. (1951) The preferences shown by honeybees for certain nectars. *Ann. appl. Biol.* 38: 546 [B, 139/53

Wykes, G. R. (1952a) An investigation of the sugars present in the nectar of flowers of various species. *New Phytol.* 51: 210–215 [B, 149/53

Wykes, G. R. (1952b) The preferences of honeybees for solutions of various sugars which occur in nectar. *J. exp. Biol.* 29(4): 511–518 [B, 5/54

Wykes, G. R. (1952c) The influence of variations in the supply of carbohydrates on the process of nectar secretion. *New Phytol.* 51: 294–300 [B, 160/58

Young, W. J. (1908) A microscopical study of honey pollen. *Bull. U.S. Dep. Agric. Bur. Chem. No.* 110

Yudkin, J. (1972) *Pure white and deadly: problem of sugar* (London: Davis-Poynter)

Zaiss (1934) Der Honig in äusserlicher Anwendung. *Münch. med. Wschr.* 11: 1891 [B

Zalewski, W. (1962) Porownanie metod oznaczania w miodzie suchej masy, zawartości cukrowców oraz aktywnosci α-amylazy [Comparison of methods for determining dry substance, carbohydrates and α-amylase activity in honey]. *Pszczel. Zesz. nauk.* 6(3): 121–133 [B, 384/65

Zalewski, W. (1965) Fosfatazy w miodach [Phosphatases in honey]. *Pszczel. Zesz. nauk.* 9(1/2): 1–34 [B, 565/66

Zander, E. (1932) Untersuchungen über die geformten Bestandteile des Honigs. *Z. Unters. Lebensmittel* 63(3): 313–329 [B

Zander, E. (1935–1951) *Beiträge zur Herkunftsbestimmung bei Honig*

(1935) I. Pollengestaltung und Herkunftsbestimmung bei Blütenhonig (Berlin: Reichsfachgruppe Imker) [B

(1937a) II. Pollengestaltung und Herkunftsbestimmung bei Blütenhonig (Leipzig: Liedloff, Loth & Michaelis) [B

(1941) III. Pollengestaltung und Herkunftsbestimmung bei Blütenhonig (Leipzig: Liedloff, Loth & Michaelis) [B

(1949) IV. Studien zur Herkunftsbestimmung bei Waldhonigen (Munich: Ehrenwirth) [B

(1951) V. Letzte Nachträge zur Pollengestaltung und Herkunftsbestimmung bei Blütenhonig (Leipzig: Liedloff, Loth & Michaelis) [B, 134/52

Zander, E. (1937b) Das Mikroskop im Dienste der Honiguntersuchung. *Leipzig. Bienenztg* 52: 205–210, 223–227 [B

Zander, E. & Maurizio, A. (1975) *Handbuch der Bienenkunde 6. Der Honig.* (Stuttgart: Verlag Eugen Ulmer)

Zander, E. & Weiss, K. (1964) *Handbuch der Bienenkunde 4. Das Leben der Biene* (Stuttgart: Verlag Eugen Ulmer) [B, 59/67

Zappi-Recordati, A. (1956) Facts about beekeeping in Italy. *Bee Wld* 37(12): 229–237 [B

Zauralov, O. A. (1966) [A simple method of predicting nectar secretion]. Pp. 111–114 from *Achievements of science and advanced experiment in beekeeping* ed. A. E. Feferman (Moscow: Rossel'khozizdat). *In Russian* [B, 731/69

Zezzos, R. (1969) I dolci della tradizione natalizia. *Apicolt. Ital.* 36(6): 133–135 [B, 763 L/71

Ziegler, H. (1955) Phosphataseaktivität und Sauerstoffverbrauch des Nektariums von *Abutilon striatum* Dicks. *Naturwissenschaften* 42: 259–260 [B, 358/60

Ziegler, H. (1956) Untersuchungen über die Leitung und Sekretion der Assimilate. *Planta* 47: 447–500 [B, 72/60

Ziegler, H. (1962) Die chemische Zusammensetzung des Siebröhrensaftes. *XI Int. Congr. Ent. 1956, Vienna* 2: 538–540 [B, see 245/63

Ziegler, H. (1968a) La sève des tubes criblés. Pp. 205–217 from Vol. III *Traité de biologie de l'abeille* ed. R. Chauvin (Paris: Masson et Cie) [B, see 397/68

Ziegler, H. (1968b) La sécrétion du nectar. Pp. 218–248 from Vol. III *Traité de biologie de l'abeille* ed. R. Chauvin (Paris: Masson & Cie) [B, see 397/68

Ziegler, H. & Kluge, M. (1962) Die Nucleinsäuren und ihre Bausteine im Siebröhrensaft von *Robinia pseudoacacia* L. *Planta* 58: 144–153

Ziegler, H. & Lüttge, U. (1960) Ueber die Resorption von C^{14}-Glutaminsäure durch sezernierende Nektarien. *Naturwissenschaften* 47: 305

Ziegler, H. & Lüttge, U. (1964) Die wasserlöslichen Vitamine des Nektars. *Flora, Jena* 154: 215–229

Ziegler, H. & Mittler, T. E. (1959) Über den Zuckergehalt der Siebröhren bzw. Siebzellensäfte von *Heracleum mantegassianum* und *Picea abies. Z. Naturf.* 14(4): 278–281 [B, 348 L/62

Ziegler, H. & Schnabel, M. (1961) Über Harnstoffderivate im Siebröhrensaft. *Flora, Jena* 150: 306–317

Ziegler, H. & Ziegler, I. (1962) Die wasserlöslichen Vitamine in den Siebröhrensäften einiger Bäume. *Flora, Jena* 152: 257–287 [B, 1/64

Zimmermann, G. (1961) Der Nachweis von Aminosäuren im Siebröhren- und Blutungssaft einiger Bäume. *StExamen Arb. Tech. Hochsch. Darmstadt*

Zimmerman, J. G. (1932) Über die extrafloralen Nektarien der Angiospermen. *Beih. bot. Cbl.* I, 49: 99–196

Zimmermann, M. (1953) Papierchromatographische Untersuchungen über die pflanzliche Zuckersekretion. *Ber. schweiz. bot. Ges.* 63: 402–429 [B, 94/54

Zimmermann, M. (1954) Über die Sekretion saccharosespaltender Transglukosidasen im pflanzlichen Nektar. *Experientia* 10(3): 145–146 [B, 277/55

Zimmermann, M. (1957) Translocation of organic substances in trees. I. The nature of the sugars in the sieve tube exudate of trees. *Plant Physiol.* 32(4): 288–291 [B

Zimmermann, M. (1958a) III. The removal of sugars from the sieve tubes in the white ash (*Fraxinus americana* L.). *Plant Physiol.* 33: 213–217 [B

Zimmermann, M. (1958b) Translocation of organic substances in the phloem of trees. Pp. 381–400 from *Physiology of forest trees* ed. K. V. Thimann (New York: Ronald Press) [B

Zimmermann, M. (1960) Absorption and translocation: transport in the phloem. *A. Rev. Pl. Physiol.* 11: 167–190 [B, 9j144(5)61

Zimmermann, M. (1961a) Movement of organic substances in trees. *Science, N.Y.* 133(3446): 73–79 [B, 9n 304(11)61

Zimmermann, M. (1961b) The removal of substances from the phloem. *Recent Advanc. Bot.* 1227–1229

Zimmermann, M. (1964) Sap movements in trees. *Biorheology* 2: 15–27 [B, 214 L/65

Zimna, J. (1959) Facelia blekitna jako roślina miododajna [*Phacelia tanacetifolia* as a nectar plant]. *Pszczel. Zesz. Nauk.* 3(2): 77–102 [B, 848/64

Zoneff, J. (1927) Bulgarische Bienenhonige und -wachse. *Z. Unters. Lebensmittel* 53: 353–376

Zürcher, K. & Hadorn, H. (1972) Eine einfache kinetische Methode zur Bestimmung der Diastasezahl in Honig. *D. LebensmittRdsch.* 68(7): 209–216 [B, 678/74

SUBJECT INDEX

Entries with the initial word 'honey' are used as little as possible, and information on e.g. Greek honey will be found under Greece, on honey processing under processing.

Except where otherwise stated or made clear by the context, entries usually refer to honey, e.g. specific heat = specific heat of honey. Most entries that are exceptions to this rule lead to information of interest in relation to honey, although not about it.

Apart from broad entries under continents, place names are indexed by country, and for countries with many entries there are sub-divisions under county, province, region or state. A few places whose location by country may not be obvious are entered direct, e.g. Alps, Sahara, Hymettus.

PLANT INDEX

Page numbers are cited under the botanical (Latin) name of the plant, and the common name is cross-referred to this. Any one entry in the text may include both names or only one. The index provides the botanical names of the few plants referred to in the text only by their common names.

Algae, fungi and yeasts are included in the Subject Index, together with bacteria and other micro-organisms.

PERSONAL NAME INDEX

Page numbers in italics refer to the Bibliography

594

www.ingramcontent.com/pod-product-compliance
Lightning Source LLC
Chambersburg PA
CBHW042110220326
41598CB00071BA/7321